Advances in Industrial Control

More information about this series at http://www.springer.com/series/1412

Vicenç Puig · Carlos Ocampo-Martínez
Ramon Pérez · Gabriela Cembrano
Joseba Quevedo · Teresa Escobet
Editors

Real-Time Monitoring and Operational Control of Drinking-Water Systems

Springer

Editors
Vicenç Puig
Research Center "Supervision, Safety and Automatic Control" (CS2AC-UPC)
Terrassa
Spain

Gabriela Cembrano
Institut de Robòtica i Informàtica Industrial
CSIC-UPC
Barcelona
Spain

Carlos Ocampo-Martínez
Institut de Robòtica i Informàtica Industrial
CSIC-UPC
Barcelona
Spain

Joseba Quevedo
Research Center "Supervision, Safety and Automatic Control" (CS2AC-UPC)
Terrassa
Spain

Ramon Pérez
Research Center "Supervision, Safety and Automatic Control" (CS2AC-UPC)
Terrassa
Spain

Teresa Escobet
Research Center "Supervision, Safety and Automatic Control" (CS2AC-UPC)
Terrassa
Spain

ISSN 1430-9491 ISSN 2193-1577 (electronic)
Advances in Industrial Control
ISBN 978-3-319-50750-7 ISBN 978-3-319-50751-4 (eBook)
DOI 10.1007/978-3-319-50751-4

Library of Congress Control Number: 2016959249

Printed on acid-free paper

This Springer imprint is published by Springer Nature
The registered company is Springer International Publishing AG
The registered company address is: Gewerbestrasse 11, 6330 Cham, Switzerland

Series Editors' Foreword

The series *Advances in Industrial Control* aims to report and encourage technology transfer in control engineering. The rapid development of control technology has an impact on all areas of the control discipline: new theory, new controllers, actuators, sensors, new industrial processes, computer methods, new applications, new design philosophies,…, new challenges. Much of this development work resides in industrial reports, feasibility study papers and the reports of advanced collaborative projects. The series offers an opportunity for researchers to present an extended exposition of such new work in all aspects of industrial control for wider and rapid dissemination.

The key networks that frame urban society include the utility networks of electricity, gas, water and wastewater. Along with transport networks, these have undergone and are still undergoing significant technological development of the way they are monitored. These changes are creating new opportunities for network organization and control. The technologies driving this change are those of smart sensors and wireless communications. Smart sensors are providing real-time information at many more physical locations, and wireless technology is making access to this information both immediate and inexpensive since there is no need to cable up the devices to central control rooms. This gives much more accurate information about resource supply and consumer demand. The potential for the reassessment and use of advanced control is widespread across these industries.

A look at the structure of the drinking water–wastewater cycle provides a glimpse of the areas with the potential to exploit these technological innovations.

Drinking-Water Supply

- Resource Management: This is the management of sources of water: lakes, rivers, aquifers and reservoirs. Monitoring levels, flows and rainfall are important aspects of this task.
- Treatment Process Stage: This covers the filtering, chlorination and biotreatment of the water resource to create a supply of drinking water quality.

- Distribution Network: The drinking water network is a pressurized supply network that covers a geographical region involving both rural and urban areas. Smart metering can give immediate readings of consumer demand. The design and construction of the physical network itself needs pumps, intermediate supply tanks and a large network of supply pipes, all of which need monitoring and control.

Wastewater Disposal

- Collection Network: This is the large network of sewage collection pipes, channels and tunnels. Gravity plays a significant role in the transport processes of wastewater but pumps, and adjustable weirs also feature in the network. The monitoring of the physical network is an important input to maintaining good wastewater flows.
- Treatment Process Stage: The treatment of wastewater is a multi-stage process with activated sludge processes and clarifier sequences playing an important role. Rainfall, which was an essential input to the drinking water resource, is a disturbance in the wastewater treatment process that can cause treatment process washout and may give rise to the need for holding ponds.
- Discharge to the Environment: Once the wastewater treatment is complete, discharge to the receiving waters occurs. But the monitoring continues to ensure no upset to the environment.

In the past, the *Advances in Industrial Control* monograph series has published contributions to the treatment of wastewater:

- *Control and Instrumentation for Wastewater Treatment Plants* by M. Reza Katebi, Michael A. Johnson and Jacqueline Wilkie (ISBN 978-1-85233-054-5, 1999); and
- *Model Predictive Control of Wastewater Systems* by Carlos Ocampo-Martinez (ISBN 978-1-84996-352-7, 2010).

However, the series has not had a contribution on the drinking water supply part of the water cycle before, so the editors are more than pleased to commend to readers the monograph *Real-Time Monitoring and Operational Control of Drinking-Water Systems* edited by Vicenç Puig, Carlos Ocampo-Martínez, Ramon Pérez, Gabriela Cembrano, Joseba Quevedo and Teresa Escobet.

The contributors report research performed at the research centre "Supervision, Safety, and Automatic Control" at the Technical University of Catalonia, and at Consejo Superior de Investigaciones Cientificas at the Institut de Robòtica i Informàtica Industrial in collaboration with various Spanish water companies. The book focuses on the pressurized drinking water network within the complete drinking water supply operation. The complexity of the network is readily grasped by a look at Fig. 2.2 that shows the water transport network for Barcelona, Spain. The editors present a comprehensive and detailed up-to-date account of the technical aspects to achieve the modelling, monitoring, and control of such a complex network. Much of this work is based on the editors' experience of working with real-world drinking water systems. The monograph concludes with chapter-length assessments of future trends, looking at such topics as the implications of "Big Data" for networks and coordination of regional and urban supply networks. This is an exemplary and very welcome addition to the *Advances in Industrial Control* monograph series.

Michael J. Grimble
Michael A. Johnson
Industrial Control Centre
University of Strathclyde
Glasgow, Scotland, UK

Preface

Drinking water utilities in urban areas are facing new challenges in their real-time operation because of limited water resources, intensive energy requirements, a growing population, a costly and ageing infrastructure, increasingly stringent regulations and increased attention towards the environmental impact of water use. Such challenges force water managers to monitor and control not only water supply and distribution, but also consumer demand. This book presents a set of approaches for the real-time monitoring and control of drinking water networks based on advanced ICT technologies of automation and telecommunications for largely improving their efficiency in terms of water use, energy consumption, water loss minimization and water quality guarantees.

The proposed approaches and tools presented in this book cover:

- decision support for real-time optimal control of the water transport network, operating the main flow and pressure actuators (pumping stations and pressure regulation valves) and intermediate storage tanks to meet demand using the most sustainable sources and minimizing electricity costs, thanks to the use of stochastic model predictive control algorithms that explicitly take into account the uncertainty associated with energy prices and actual demand;
- decision support for monitoring water balance and quality of the distribution network in real time via fault detection and diagnosis techniques, using information from hundreds of flow, pressure and water quality sensors, and hydraulic and quality-parameter evolution models, to detect and locate leaks in the network, possible breach in water quality, and sensor/actuator failures; consumer demand prediction, based on smart metering techniques, producing a detailed analysis and forecasting of consumption patterns and providing a service of communication to consumers, together with economic measures to promote a more efficient use of water at the household level.

All methods' approaches presented in the book are applied and illustrated using a real-life pilot demonstration based on the Barcelona drinking water network and Catalonia regional network (Spain). The results presented in the book are the results of the long collaboration of the Research Center "Supervision, Safety and Automatic

Control" (CS^2AC-UPC) at Technical University of Catalonia (UPC) and Consejo Superior de Investigaciones Cientficas (CSIC) at the Institut de Robòtica i Informàtica Industrial (IRI) with the Barcelona water company (AGBAR) and its technical centre (CETAQUA) in several Spanish and European research projects and private contracts as well as with *ATLL Concessionària de la Generalitat de Catalunya* in several private contracts.

Barcelona, Spain
2016

Vicenç Puig
Carlos Ocampo-Martínez
Ramon Pérez
Gabriela Cembrano
Joseba Quevedo
Teresa Escobet

Acknowledgements

The authors also wish to thank the support received by the companies *Aigües de Barcelona, ATLL Concessionària de la Generalitat de Catalunya, Aguas de Alicante* and *Empresa Municipal Mixta d'Aigües de Tarragona* in the development of this work.

This work has been partially funded by the Spanish Government (MINECO) through the project CICYT ECOCIS (ref. DPI2013-48243-C2-1-R), by MINECO and FEDER through the project CICYT HARCRICS (ref. DPI2014-58104-R) and by AGAUR of the Generalitat de Catalunya through the grant Industrial Ph.D. 2013-DI-041.

Contents

Part III Real-Time Control

List of Contributors

Ramon Ariño Aguas de Barcelona (AGBAR), Barcelona, Spain

Julián Barreiro-Gomez Departamento de Ingeniería Eléctrica y Electrónica, Universidad de los Andes, Bogotá, Colombia; Institut de Robòtica i Informàtica Industrial, CSIC-UPC, Barcelona, Spain

Joaquim Blesa Research Center "Supervision, Safety and Automatic Control" (CS2AC), UPC, Terrasa, Spain; Institut de Robòtica i Informàtica Industrial (IRI), CSIC-UPC, Barcelona, Spain

Gabriela Cembrano Institut de Robòtica i Informàtica Industrial, CSIC-UPC, Barcelona, Spain

Ramon Comasolivas Research Center "Supervision, Safety and Automatic Control" (CS2AC), UPC, Terrasa, Spain

Ramon Creus Aguas de Barcelona (AGBAR), Barcelona, Spain

Miquel Angel Cugueró Research Center "Supervision, Safety and Automatic Control" (CS2AC), UPC, Terrasa, Spain

Josep Cugueró Research Center "Supervision, Safety and Automatic Control" (CS2AC-UPC), Terrassa, Spain

Santiago Espin ATLL Concesionaria de la Generalitat S.A., Barcelona, Spain

Teresa Escobet Research Center "Supervision, Safety and Automatic Control" (CS2AC-UPC), Terrassa, Spain

Diego García Centro Tecnólogico del Agua (CETAQUA), Barcelona, Spain

Luis Eduardo Garza-Castañón Tecnológico Monterrey, Campus Monterrey, Monterrey, Mexico

Juan Manuel Grosso Institut de Robòtica i Informàtica Industrial (IRI), CSIC-UPC, Barcelona, Spain

Jordi Meseguer Centro Tecnólogico del Agua (CETAQUA), Barcelona, Spain

Meritxell Minoves Aguas de Barcelona (AGBAR), Barcelona, Spain

Fatiha Nejjari Research Center "Supervision, Safety and Automatic Control" (CS2AC), UPC, Terrasa, Spain

Carlos Ocampo-Martínez Institut de Robòtica i Informàtica Industrial, CSIC-UPC, Barcelona, Spain

Ramon Pérez Research Center "Supervision, Safety and Automatic Control" (CS2AC-UPC), Terrassa, Spain

Vicenç Puig Research Center "Supervision, Safety and Automatic Control" (CS2AC-UPC), Terrassa, Spain

Nicanor Quijano Departamento de Ingeniería Eléctrica y Electrónica, Universidad de los Andes, Bogotá, Colombia

Joseba Quevedo Research Center "Supervision, Safety and Automatic Control" (CS2AC-UPC), Terrassa, Spain

Gerardo Andrés Riaño-Briceño Departamento de Ingeniería Eléctrica y Electrónica, Universidad de los Andes, Bogotá, Colombia

Deneb Robles Tecnológico Monterrey, Campus Monterrey, Monterrey, Mexico

Juli Romera Research Center "Supervision, Safety and Automatic Control" (CS2AC), UPC, Terrasa, Spain

Jaume Roquet ATLL Concesionaria de la Generalitat S.A., Barcelona, Spain

Jordi Saludes Research Center "Supervision, Safety and Automatic Control" (CS2AC), UPC, Terrasa, Spain

Gerard Sanz Research Center "Supervision, Safety and Automatic Control" (CS2AC), UPC, Terrasa, Spain

David Saporta Aguas de Barcelona (AGBAR), Barcelona, Spain

Ramon Sarrate Research Center "Supervision, Safety and Automatic Control" (CS2AC), UPC, Terrasa, Spain

Congcong Sun Institut de Robòtica i Informàtica Industrial (IRI), CSIC-UPC, Barcelona, Spain

Maite Urrea Institut de Robòtica i Informàtica Industrial (IRI), CSIC-UPC, Barcelona, Spain

Fernando Valero ATLL Concesionaria de la Generalitat S.A., Barcelona, Spain

José Gabriel Valero Aguas de Barcelona (AGBAR), Barcelona, Spain

Ye Wang Institut de Robòtica i Informàtica Industrial (IRI), CSIC-UPC, Barcelona, Spain

Abbreviations

ACF	Autocorrelation function
AFC	Actuator-fault configuration
AMR	Automatic meter reading
AR	Autoregressive
ARIMA	Autoregressive integral moving average
ARR	Analytical redundancy relations
BOP	Basic optimization problem
CC-MPC	Chance-constrained MPC
CI	Confidence intervals
CIS	Critical infrastructure systems
CMPC	Centralized MPC
CSP	Constraint satisfaction problem
DAE	Differential algebraic equations
DB	Data base
DMA	District metering area
DMPC	Decentralized MPC
EGT	Evolutionary game theory
EMPC	Economic model predictive control
ETL	Extraction, transformation and loading
EV	Explained variance
FD	Fault diagnosis
FDI	Fault detection and identification
FHOP	Finite horizon optimization problem
FOSM	First-order second-moment
FS	Fault signature
FSM	Fault signature matrix
FTC	Fault-tolerant control
GA	Genetic algorithms
GIS	Geographical information systems
GPB	Graph-partitioning-based

GUI	Graphical user interface
HPC	High-performance computing
HW	Holt-Winters
ICT	Information and communications technology
KPI	Key performance indicator
LQR	Linear quadratic regulator
LSS	Large-scale system
MAE	Mean absolute error
MAPE	Mean absolute percentage error
MBD	Model-based diagnosis
MDB	Minimum-degree-based
MDG	Millennium development goals
ML-DMPC	Multi-layer distributed economic MPC
MPC	Model predictive control
MPI	Message passing interface
MSE	Mean-square error
MSO	Minimal structural overdetermined
NCMPC	Non-centralized MPC
NEMPC	Nonlinear economic MPC
NMPC	Nonlinear model predictive control
OOP	Open-loop optimization problem
ORP	Oxidation reduction potential
PACF	Partial autocorrelation function
PDF	Probability density function
PRV	Pressure reducing valve
RDBMS	Relational database management systems
RTC	Real-time control
SCADA	Supervisory control and data acquisition
SD	Sampling design
SM	Spatial models
SMPC	Stochastic model predictive control
SVD	Singular value decomposition
TB-MPC	Tree-based MPC
TSM	Time series models
UFW	Unaccounted-for-water
WDS	Water distribution system
WTS	Water transport system
WU	Water utilities

Notation

Throughout this book, as a general rule, scalars are denoted by lower-case letters (e.g., a, b, ...), vectors are denoted by boldface and small characters (e.g., $\mathbf{x}$), matrices are denoted by upper-case boldface letters (e.g., $\mathbf{A}$, $\mathbf{B}$, ...), and sets are denoted by calligraphic upper-case letters (e.g., $\mathcal{F}$, $\mathcal{G}$, ...). If not otherwise noted, all vectors are column vectors.

$\mathbb{R}$	Set of real numbers
$\mathbb{R}_+$	Set of non-negative real numbers, defined as $\mathbb{R}_+ \triangleq \mathbb{R}\setminus(-\infty,0]$
$\mathbb{Z}$	Set of integer numbers
$\mathbb{Z}_+$	Set of non-negative integer numbers
$\mathbb{Z}_{\geq c}$	Set defined as $\mathbb{Z}_{\geq c} \triangleq \{k \in \mathbb{Z} \,\vert k \geq c\}$, for some $c \in \mathbb{Z}$
$\mathbb{Z}_{[a,b]}$	Set of integer numbers from a to b, both limits included, i.e., $\{a, a+1, \ldots, b\}$
$\mathbb{Y}^m$	$\underbrace{\mathbb{Y} \times \mathbb{Y} \times \cdots \times \mathbb{Y}}_{m\,\text{times}}$
$\Vert q \Vert_p$	Arbitrary Hölder vector p-norm with $1 \leq p \leq \infty$
$\vert\mathcal{A}\vert$	Cardinality of set $\mathcal{A}$
$\mathcal{A} \subseteq \mathcal{B}$	$\mathcal{A}$ is a subset of $\mathcal{B}$
$\mathcal{A} \subset \mathcal{B}$	$\mathcal{A}$ is a proper subset of $\mathcal{B}$
$\mathcal{A} \cap \mathcal{B}$	Intersection between set $\mathcal{A}$ and $\mathcal{B}$
$\mathcal{A} \oplus \mathcal{B}$	Minkowski sum
$\mathcal{A} \sim \mathcal{B}$	Minkowski (or Pontryagin) difference
$\mathbf{A}^T$	Transpose matrix of $\mathbf{A}$
$\mathbf{A}^+$	Pseudoinverse matrix of $\mathbf{A}$
$\rightarrow$	Mapping
$\mapsto$	Maps to
$\mathbb{P}$	Probability operator
$\leftrightarrow$	If and only if
H_p	Prediction horizon
H_u	Control horizon
$\mathbf{x}$	Sequence of states $(\mathbf{x}_k)$, denoted by $\mathbf{x}_k \triangleq (x_1, x_2, \ldots, x_m)$

$J(\cdot)$	MPC cost or objective function
λ_i	i-th cost function weight
u^*	Optimal value of u
v	Tank volume (state variable)
q_u	Manipulated flow (control input)
d	Measured disturbance
Δt	Sampling time
$\bar{x}$	Mean value of variable$x$
diag	Diagonal matrix
min	Minimum
max	Maximum

Chapter 1
Real-Time Monitoring and Control in Water Systems

Jordi Meseguer and Joseba Quevedo

1.1 The Water Need

Water is a critical resource for supporting human activities and ecosystem conservation. As reported by the Food, Energy, Water (FEW) organization, there are both supply-side and demand-side threats to water. One supply-side threat arises from withdrawing freshwater from water surface sources and groundwater aquifers at rates that are faster than replenishment or recharge. Another supply-side problem is that even if there is enough water, its quality is not *good* enough to meet human needs; much of the freshwater around the world is being degraded. One of the most frequently cited statistics in discussion of water availability, presented in Fig. 1.1, shows that only about 2.5% of the Earth's water is fresh.[1] From that 2.5% freshwater available for the support of human life, agriculture, and most forms of non-ocean life, 30.1% corresponds to groundwater that is stored deep beneath and is non-renewable.

The demand-side concern arises from the following facts:

- An increasing number of people on the planet are geographically concentrated in regions that cannot sustain demand levels.
- Technologies that waste more water than alternative technologies and demand is often insufficiently discouraged because of inadequate price mechanisms and outdated regulation that set few limits on excessive use.
- Negative impacts of climate change are likely to give rise to uncertainties in water availability and water demands.

J. Meseguer
CETaqua, AGBAR Technival Center, Barcelona, Spain

J. Quevedo (✉)
Research Center "Supervision, Safety and Automatic Control" (CS2AC-UPC), Terrassa, Spain
e-mail: joseba.quevedo@upc.edu

[1] http://www.un.org/waterforlifedecade/scarcity.shtml

V. Puig et al. (eds.), *Real-Time Monitoring and Operational Control of Drinking-Water Systems*, Advances in Industrial Control,
DOI 10.1007/978-3-319-50751-4_1

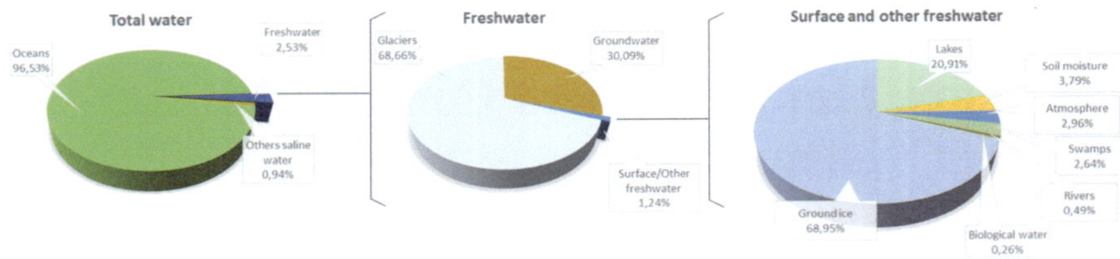

Fig. 1.1 Distribution of Earth's water

Water scarcity problems could happen in 2025, which means that, since then, in nearly half of the world a large number of people may not have access to safe and affordablewater for their basic needs [1]. The popular concept "Water is the new oil" stresses the importance and criticality of water.

Limited water supplies, conservation and sustainability policies, as well as the infrastructure complexity for meeting consumer demands with appropriate flow, pressure and quality make water management a challenging problem. This situation indicates the need for advanced monitoring and control in water supply, transport and distribution networks. Decision-support systems provide useful guidance for operators in complex networks, where actions for best resource management are not intuitive [2, 3].

Management of water systems involves different operational objectives, such as

- minimizing operational costs of pumping (which represents a significant fraction of the total expenditure as discussed in [4]),
- ensuring pressure regulation for providing suitable service, but avoiding overpressures, which can cause leaks or bursts,
- minimizing risks of either water shortage or water quality issues [5].

Both optimization and optimal control techniques provide an important contribution to strategy computation in water system management for efficient use of resources. Similarly, the problems related to modelling and control of water supply and distribution have been the subject of important research efforts during the last years as discussed in [6, 7].

1.2 Water Cycle and Networks

The water cycle includes all the activities that, on the one hand, allow to supply water to the end-user and, on the other hand, collect and treat the wastewater before delivering it to the natural environment or reusing it. The drinking-water system includes the catchment of water, its treatment, its transport and distribution. The sewage system includes the urban draining collection, storage, treatment and release or reuse.

In a large number of cities, drinking-water networks are managed using telemetry and telecontrol systems that provide, in real time, pressure, flow, quality and other measurements at several key locations in the network. They are also used to operate the main flow, pressure and storage control elements from a central dispatch in a centralized or decentralized scheme.

In some cases, advanced urban drainage systems include water control infrastructure, such as detention tanks, pumps, valves and weirs, monitored and controlled using telecontrol systems. These latter systems involve rain-gage networks, water-level meters in the sewers and actuators at the valves, pumps and weirs, which may be controlled from a central dispatch in order to control the water flow in the network towards wastewater treatment plants and, afterwards, to the receiving environment.

Controlling water cycle systems requires the development of adequate dynamic models to represent:

- open-channel elements, such as rivers, canals and aqueducts,
- pressurized water networks,
- combinations of both,

which have nonlinear responses to control actions, such as changing modes at different operating points. These systems also contain storage and control elements, such as tanks and valves, with a predetermined operational range, which producing hard constraints in the model. Additionally, some ON–OFF elements may exist. The use of hybrid modelling [8] to represent the dynamics of water systems for control objectives appears naturally as an appropriate option, and it constitutes a key issue. In addition, water system dynamics have stochastic components; mainly, consumer demand in the case of drinking-water systems and rain intensity in the case of urban drainage systems. Stochastic components in the models must also be represented using appropriate modelling and forecasting techniques (e.g., time series analysis).

Water system management must be carried out predictively. Control actions must be computed ahead in time, with an appropriate time horizon, based on real measurements and state estimation, as well as predictions of the stochastic variables. For drinking-water distribution networks, this *prediction* horizon is usually of the order of 24-h. Longer horizons are chosen for water supply and treatment management. For real-time control of urban drainage systems, the horizons depend on the average water transport time between the discharge points and the final collection/treatment/discharge points. A common order of magnitude for this horizon is 30 minutes.

The use of telemetry and telecontrol in systems of the water cycle is increasing constantly, due to the growing availability of reasonably priced sensors, telecommunication systems and computers in a situation of increasing awareness of water cycle management needs:

- to take maximum advantage of scarce water resources,
- to provide access to water to more regions, to control drinking-water quality,

- to minimize the use of water and energy,
- to cope with extreme events such as draughts and flooding,
- to reduce the impact of used water on the receiving environment, among others.

The increasing availability of control hardware and information systems cannot, on its own, contribute to coping with these challenges. The key issue is the use of the information. Appropriate water system modelling and control strategies must be developed at a similar fashion. Nevertheless, these two systems have specific characteristics (open-channel networks and chemical reactions) that maintain them beyond the scope of this book. This book is focused on the potable water in pressurized networks.

1.3 Real-Time Monitoring and Control

The incorporation of recent advances in the information and communications technology (ICT), in sensor and actuator reliable and cheap components, and in advanced metering of consumer demand, has a significant potential to improve efficiency in monitoring and management of quantity and quality of water, to achieve best strategies for water and energy use, to assess in real-time the efficiency of water networks, to avoid water losses because of leakage, to minimize risk of inadequate water quality, to understand consumer demands by taking into account the behaviours and attitudes of the consumers and to improve the plans of asset predictive maintenance. ICT technologies currently used by water utilities enable the reduction of development and deployment costs of monitoring and management solutions, increase their integration with other company processes, and improve the efficiency level required in urban water supply. At present, the continuous advances of ICT technology make real-time management of drinking-water a sector with quite high potentials for improvement.

The real-time operational control of a water network by means of a Supervisory Control and Data Acquisition (SCADA) system involves planning strategic operations of pumping stations and flow or pressure regulation valves over a 24-h horizon, in order to meet consumer demand at all times, with the appropriate levels of pressure and quality. This is generally achieved using intermediate water tanks, booster pumping and pressure reduction valves. Model predictive control (MPC) techniques are the best candidate to perform operational control, as they compute, ahead of time, the best admissible control strategies for valves, pumps or other control elements in a network to achieve minimum energy consumption, production cost minimization, environmental protection and other operational goals, that ensure current and future satisfaction of water demand.

The control of the network is quite sensitive to faults affecting sensors (flow meters, pressure meters, etc.), actuators (pumps, valves, etc.) and components (pipes and reservoirs). Thus, a diagnosis of faulty situations must be ensured by a monitoring module, so that the control system does not result in anomalous performance of the network, therefore compromising the success of its strategy and possibly causing the occurrence of serious consequences in terms of environmental impacts and economical losses. One way of achieving fault tolerance is to employ an online fault detection and isolation (FDI) scheme at the monitoring stage that is integrated with the real-time control system. When a fault is detected and isolated, the FDI module will trigger the controller to activate in response to some accommodation actions. MPC algorithms have the nice feature of being easily and automatically reconfigurable, so as to immediately provide the best control action after a structural change due to a fault.

Real-Time monitoring of water networks involves all the activities related to observing the network via telemetry systems. This includes real-time water balance analysis, for water loss reduction, leak detection and localization, and water quality surveillance for detection and localization of quality breach incidents. Detection and isolation of such faulty situations require the use of real-time sensor readings and hydraulic/quality mathematical models. Similarly, monitoring also involves the validation of telemetry data, the detection and replacement of missing data, and the validation of actuator status. Advanced automatic meter reading (AMR) at consumption points allows utilities to have detailed information of consumer demand with varying degrees of detail (daily and hourly). These data are quite useful for understanding time patterns of different types of consumers and to develop more accurate demand forecasting models. Smart meters allow identifying and resolving inefficiencies and irregularities in water use, by analysing water flows in real-time to improve customer service and to help conserve water. In fact, introducing a more frequent and highly accurate measurement of water consumption enables and supports a more efficient use of such a precious resource. One of the most important advantages to be expected from AMR is not only having information, but also sending this information to customers in order to promote modifications in their water consumption behaviour.

In drinking-water management, the key management levels of strategic operational control and real-time network monitoring are still routinely tackled separately. Several different technologies, information systems and experts are usually involved in these two levels of management. However, the combination of these two fundamental aspects of water management in an integrated environment allows to take advantage of their interaction enhancing the overall performance from the operational, socioeconomic and environmental point of view (i.e., reduce water resources and associated water supply cost while increasing the service level offered to the customer). Such an integrated software solution is a key decision-support system for water management, and it handles real-time interactions between SCADA, telemetry and AMR systems, with real-time databases and the three specific water management modules for operational control, monitoring and demand management, with appropriate graphical user interface layers and connections to geographical information systems (GIS). In addition, as these processes have an important influence on other utilities' processes, this integrated software solution has to be flexible enough to accommodate links to other utility information systems (i.e., customer, maintenance and investment planning information systems).

1.4 State of the Art

Nowadays, real-time monitoring and management systems of water supply and distribution networks are quite active areas of research, development and/or innovation of a large number of companies (e.g., Schneider Electric-Telvent, ABB, and DHI TaKaDu, among others). There is an increasingly high interest of water utilities worldwide in bridging operational activities and enterprise processes to improve real-time monitoring and management, e.g., of

- Water loss and operational costs
- Energy consumption and efficiency
- Compliance with specifications water supply
- Water quality and regulatory compliance of water distribution
- Data accuracy, reliability and availability
- Plans to locate and to incorporate new sensors to improve the knowledge about the real-time state of the network
- Plans to improve the advance maintenance of the assets and
- Emergency response.

1.4.1 Real-Time Monitoring of Water Networks

Water network monitoring refers to continuously observing water quantity and quality. Real-Time quantity monitoring allows to detect and isolate water leaks and other faults affecting the sensors (flow meters, pressure meters, etc.) and the actuators (pumps, valves, etc.). Real-Time quality monitoring allows to trace several water quality parameters along the networks for safety and water quality preservation and to react promptly in case that abnormal values are detected. Water network monitoring requires three elements to be implemented: sensors, communication devices and mathematical models.

Regarding network quantitative monitoring, the main aim is related to the interest of water utilities in avoiding water losses. The use of flow and pressure sensors together with hydraulic models of the water network for leak detection and isolation is a suitable approach for the online monitoring of water balance. A direct approach based on simulation has been proposed by [9, 10]. An alternative approach based on an inverse approach that formulates the leak detection and isolation problem as a parameter estimation approach was presented by [11] and further inverse approaches were investigated by [12–14]. Hydraulic model experience is extended in all water companies and the use of an approach based on a familiar tool is welcome by technicians. These models include flows and pressure variables. Pressure may usually be more easily monitored due to the low cost of sensors. Therefore, the detection and location of leakages based on differences between predicted and measured pressures is being explored. The sensitivity of pressures to the leakages is evaluated in the sensitivity matrix, which is the basis for leakage detection using pressure measurements and hydraulic models. Recently, a leak detection and isolation approach that makes explicit use of the sensitivity matrix has been proposed by [15].

Water quality monitoring is related to tracing a number of physical, biological and chemical indicators that describe the adequacy for consumer use. Like the hydraulic model of the water distribution network, the usefulness of water quality models not only depends on the quality of its calibration, but also depends on the hydraulic model calibration [16]. A water quality model cannot provide reliable results if real flows are not accurately represented by a hydraulic model. Management of water quality requires regular measurements and monitoring. Mostly, measurements of water quality are performed manually. The process can be slow and painstaking. Multiple point measurements are needed to cover an area. The process needs to be automated and extended to provide rapid and effective monitoring. In this line, the FP7-ICT project MOBESENS [17] provided a modular and scalable integrated solution based on communication technologies for water quality monitoring. Although this solution is applied to a river basin scale, certain aspects are potentially applicable to quality monitoring of drinking-water networks.

Several types of models to determine chlorine concentrations throughout a water distribution system have been described by [18, 19]. The currently used approaches for monitoring chlorine and detect changes of chlorine concentration in water networks can be classified into those that make use of a direct approach based on simulation [18, 20, 21] of conceptual input/output models [22, 23], and those using an inverse approach based on parameter estimation of the chlorine decay model by [19, 24, 25]. As in the case of leak monitoring, a chlorine-monitoring approach that makes use of a sensitivity analysis was proposed by [26, 27]. The water quality sensor placement problem has also been investigated from the security perspective.

The Battle of the Water Sensor Networks (BWSN) research challenge in 2006 instigated significant research results and elevated the importance of considering security issues in water distribution systems. Most of the research groups that participated in BWSN formulated a multiobjective integer optimization programming problem in [28]. Following this work, a mathematical formulation of the problem from a system and control perspective was presented in [29]. In addition, a dynamic manual sampling approach was proposed [30] for isolating an area in the network where the contamination source could have originated, using manual expanded sampling techniques.

Another problem related to water quality monitoring is the detection of contamination events. The underlying assumption behind contamination detection using surrogate water quality measurements is that contaminants injected in drinking-water will affect certain parameters which are monitored [31]. For example, a bacterial toxin may decrease the concentration of free chlorine, decrease the oxidation–reduction potential (ORP) and increase the conductivity of water. However, contamination-event detection in real water systems is challenging due to the large uncertainties in the dynamics and different factors that influence water quality and may cause a large number of false alarms (false positives).

Network monitoring requires the collection of data from the physical water network. Due to the scale of most urban water systems, it is practically impossible to install sensors or to take measurements at every candidate location. In order to obtain useful information for network monitoring, data must be collected from a subset of carefully chosen locations designed to maximize the performance with respect to specific design criteria. Several authors have started investigating the impact of sensor location on inferences regarding the system state such as [32–35] demonstrating that it is significant. The general sensor placement problem is combinatorial with a potentially enormous corresponding search space. Efficient algorithms that avoid complete search were proposed by [36]. Regarding quantitative monitoring facing leak detection and isolation, optimal sensor placement was addressed by [37]. In case of qualitative monitoring, optimal sensor placement methods were mostly developed for detecting contaminations events by [38–40].

In water networks, the telecontrol system must acquire, store and validate data from many flow meters and other sensors every few minutes to achieve an accurate monitoring of the whole network in real-time. Frequent operation problems in the communication system between sensors and data logger, or in the telecontrol itself, generate missing data during certain periods of time. The stored data are sometimes uncorrelated and of no use for historic records. Missing data must be therefore replaced by a set of estimated data. A second common problem is the lack of reliability of flow meters (offsets, drifts and breakdowns) producing incorrect flow data readings. These false data must also be detected and replaced by estimated data, since flow data is used for several network water management tasks, namely planning, investment plans, operations, maintenance and billing/consumer services, and operational control. The same type of problem can be found in gas or electricity networks in [41].

According to the nature of available knowledge, different approaches for data validation can be conceived, with varying degrees of sophistication. In general, one may distinguish between elementary signal-based (low-level) methods and model-based (higher level) methods in [42, 43]. Elementary signal-based methods use simple heuristics and limited statistical information of a given sensor in [44–46]. Typically these methods are based on validating either signal values or signal variations data validation. In the signal-based approach, data are assessed as valid or invalid according to two thresholds (a high one and a low one); outside these thresholds data are assumed invalid. On the other hand, methods based on signal variations look for strong variations (peaks in the curve) as well as lacks of variation (flat curves).

Model-based methods rely on the use of models to check the consistency of sensor data in [47] and more recently in [48]. The result of data validation may be either a binary variable indicating whether the data are considered valid or not, or a continuous validity index interpreted as a degree of confidence of the data. When the degree of confidence is too low, data can be either discarded or replaced by an estimate computed using a statistical or physical model [49]. Moreover, a subproduct of model-based approaches for sensor data validation is that the prediction provided by the model can be used to reconstruct the faulty state of a sensor. For example, a general learning methodology for fault diagnosis of nonlinear systems was presented by [50]. Results of fault detectability were later extended to the case of incipient faults by [51]. Under certain conditions, an upper bound on the fault detection time was derived for both abrupt and incipient faults, and a relationship was established between the location of the estimator pole and the upper bound on the detection time. More recently, some promising results on distributed fault diagnosis of a special class of nonlinear dynamical systems by [52, 53] were also obtained.

Finally, since large-scale drinking-water networks include a huge number of sensors and actuators, they are prone to some malfunctions (i.e., they are in a faulty situation). This problem calls for the use of an online fault detection and isolation system that is able to detect such faults and correct them (if possible) by activating fault-tolerant mechanisms, as the use of soft sensors, or using the embedded tolerance of the optimal/predictive controller, that avoids that the global real-time control should be stopped every time a fault appears. According to [54], this is one of the main reasons why today there is a small number of global real-time controls operating in the world. Some examples of these methods applied to the water domain are time series analysis techniques such as in [55–58], Kalman filters in [59–61], parity equations in [62–65] and more recently by [66] pattern recognition methods in [67], and principal component analysis in [68–70] and more recently in [71].

1.4.2 Real-Time Optimal Control of Water Networks

An efficient management of urban drinking-water networks requires a supervisory control system that optimally decides about the current operational configuration of the whole network. Such decisions are implemented automatically or offered as a decision support to operators and managers at the control centre. The control system must take into account operating constraints, consumer demands and operation costs (in particular electricity costs). The decisions of the control systems are translated into set points to individual, localized, lower-level control systems, that optimize the pressure profile to minimize losses by leakage and provide sufficient pressure, e.g., for high-rise buildings. The whole control system responds to changes in network topology (ruptures), typical daily/weekly profiles, as well as major changes in demand.

The main issues associated with such an operational control are the complexity of the network (tens or even hundreds of tanks, valves, pumping stations, points of water consumption, water sources), and how to operate the network optimally, that is by using water sources efficiently and minimizing operation costs, while satisfying water demand and quality standards. In particular, electrical energy is the main source of operation costs, both for water production (e.g., obtaining drinking-water in treatment plants may require an inverse osmosis process, which requires a fairly high amount of electrical energy) and water elevation by pumping stations (the consumption due to valves and tanks is usually considerably smaller). Currently, depending on the pumping station, its own importance, and the area of demand it covers, different bilateral contracts are established with energy supply companies, with a variety of prices and different cost periods (from two periods up to six different price periods per day, depending on working days/weekends and on seasons). Accordingly, current practice is to preallocate the pumping periods of each station when the energy prices arranged by the contract are the lowest possible for that station, making sure that expected demand is satisfied, with the help of intermediate water storage capacity.

Reducing operational costs by taking into account varying energy prices on the power exchange market, possible deviations from expected water demand, and the state of the physical network, in a combined and optimized way can lead to significant economic savings. In fact, even small percentage reductions can lead to large savings, given the large volumes of electrical energy involved in the operations (e.g., in the water network of the city of Barcelona in 2011, energy consumption was on average 245 MWh/day for water production and 92.5 MWh/day for water transport, the latter mainly due to pumping water from sea level up to more than 500 m above).

To design a control system to operate a complex water network and match consumer demand optimally with respect to energy costs requires simple, yet representative, mathematical models of the network dynamics (flows, storage, water sources and pressure) and of related operating constraints (pumping, valves, water quality and pressure), stochastic models of the uncertainty associated with water demand, possibly perturbed by active demand management, and stochastic models of varying electricity prices for the amount of energy purchased in the day-ahead energy market.

Complex nonlinear models are quite useful for offline operations (calibration and simulation, for instance). In fact, fine and detailed mathematical representations such as pressure flow models for drinking-water networks allow to simulate those systems with an enough high accuracy to observe different particular phenomena which are useful for designing and/or task planning. However, for online optimization purposes, simpler control-oriented models must be conveniently selected that capture the main dynamic behaviour of the water network, only taking into account representative hydrological/hydraulic variables and their response to actuation commands and to external disturbances. Such models must be simple, scalable, flexible, computationally simple and easily recalibrated online using data from the telemetry system, so that they can be easily embedded in real-time optimization algorithms. Several modelling techniques for drinking-water networks appeared in the scientific literature, including control-oriented flow-based models in [72, 73] and their extension to include pressure-based models in [74, 75].

Such models are used to solve the management problem of water networks efficiently, by means of optimization-based control, such as model predictive control (MPC). MPC is the leading concept in advanced process control of highly complex multivariable processes [76]. Its success is mainly due to its unique ability to handle physical constraints on the system (e.g., bounds on selected process variables and/or their rates of change) while optimizing a performance index (e.g., minimizing costs, or maximizing profits). The rest of the MPC design is automatic: the given (or most recently updated) model, constraints, and performance index define an optimal control problem over a finite time horizon in the future (for this reason, the approach is said predictive). This is translated into an equivalent optimization problem and solved online to obtain an optimal sequence of future control moves. Only the first of these moves is applied to the process, as at the next time step a new optimal control problem is solved, to exploit the information coming from fresh new measurements. In this way, an open-loop design methodology (i.e., optimal control) is transformed into a feedback one in [77, 78].

In the context of drinking-water network management, optimization-based scheduling was considered by [2] as a two-level optimization approach: the upper level solves an optimization problem based on flow-based models to get references for the lower level, which is instead based on pressure-based models. Recent studies have proven the effectiveness of MPC for control of water networks in [79, 80]. In particular, the effectiveness of decentralized and distributed MPC tools was demonstrated in the past FP7-ICT project WIDE [81] for the control of water distribution networks, taking into account large-scale deterministic models of the entire network, and decomposing the resulting optimal control formulation for efficiency of computations, scalability and maintenance.

On the other hand, management systems for water networks must necessarily cope with uncertainty. The most important uncertain variable is water demand, so that demand forecasting becomes a crucial component of the control system. Depending on the time horizon, there are short-term, mid-term and long-term forecasts [82]. The short-term forecasting is mainly used for operational control, considering a demand prediction for either one or more days ahead in [48, 83, 84]. The second main source of uncertainty is the price for electricity, in case (part of) the aforementioned fairly large amount of electrical energy is purchased on the power exchange market to exploit maximum convenience. Stochastic MPC formulations that explicitly exploit models of uncertainty to optimize expected revenues and penalize risk have been developed over the last decade in the academic community such as in [85–87]. As stochastic models of electricity prices dynamics can be derived from market data in [88, 89], stochastic MPC techniques were recently investigated within the FP7-ICT project EFFINET[2] for the management of uncertainty in water systems and for smart distribution grids in [90, 91] and for optimal bidding on energy markets in [92].

1.5 Outline of the Book

This book is made up of this introductory chapter presenting the scope, motivation and the state of the art. The second chapter describes several case studies that have motivated the development of the methodologies presented in the following chapters. The remainder of the book is divided into four main parts.

Part I consists of four chapters that cover the water network modelling topic. In particular, Chap. 3 presents the basis of water distribution network modelling by introducing hydraulic equations in a matrix form and the hydraulic solver to simulate the network, computing heads and flows from a predefined set of demands and boundary conditions. Only steady-state equations are considered in an extended period of simulation. Transients are not considered due to their low importance in pressurized water networks. Chapter 4 focuses on the different techniques to estimate the unknown parameters of the model from the data available of the network. The models presented in Chap. 3 are used in several tasks such as simulation, operational optimization and data validation, to compare the real state of the network provided by the sensors with the estimated state obtained by using the models in order to generate alarms in the case of faulty events. For all these tasks, the estimation of the model parameters to represent adequately the behaviour of the network becomes a fundamental task which determines the performance of rest. Chapter 4 analyses the problem of parameter estimation, the measurements required for guaranteeing the identifiability and the well-posedness of the problem. Then, both the parameterisation and the sampling design are presented, proposing a methodology that has given promising results with real water distribution networks. Chapter 5 presents the techniques to calibrate the nodal demands of water networks. Demands are unknown inputs that must be defined to solve the network's hydraulic equations. They are not physical elements of the network, but the driving force behind the hydraulic dynamics. A good calibration of demands is mandatory to obtain accurate results when simulating the hydraulic model. Chapter 5 provides an overview of the existing calibration methods and a detailed description of an online calibration procedure. And finally, Chap. 6 presents a wide spectrum of short-term demand algorithms to predict 24-h ahead the water consumes of the nodes in a water network using the available real data. This short-time forecasting demand is a key information for successful results of in the real-time operational control of the water networks. The selected

[2] http://www.EFFINET-project.eu.

forecast method must be easy to use and should be automatically calibrated. Moreover, the proposed algorithms should be adaptable, because in the network there are many demand time series to model and each floor of pressure has their own demand characteristics.

Part II analyses different key aspects of the real-time monitoring of water systems. It is organized into five chapters, which allow to give responses to a set of important questions, such as which are the minimum number of sensors to know the state of the network at any time, where is the best location of these sensors, how the reliability of the information provided by the sensors can be guaranteed, how is possible to reconstruct missing or anomalous sensor data, and how is possible to detect and isolate leaks, faults and quality events in a water network? The existence of a reliable real-time monitoring tool of complex systems is a key question for the management and operation of any modern water network. Chapter 7 focuses on the state of the art of the leak management including the real-time monitoring that allows the leak detection and localization techniques. Special attention will be devoted to model-based approaches and their application to real case studies. Chapter 8 deals with a problem that is of fundamental importance to health of water consumers, the monitoring in real-time of the quality of the water in the network allowing to raise an alarm whether a quality event is detected. Chapter 8 proposes a method to calibrate a water quality model that characterizes the chlorine concentration in the network using the real values of the chlorine sensors located in several nodes. This method has been applied to a real water network. Using this calibrated model, a method for water quality event detection and location-based chlorine sensitivity analysis of the nodes is proposed. Chapter 9 focuses on the methodology for sensor placement in a water network to assure the leak monitoring of the whole network. Two approaches have been described and compared in this chapter in order to perform a leak location task. These two approaches are applied to a district metered area (DMA) of a real water network. Chapter 10 presents a methodology composed of several tests to validate or not all the raw data acquired by the monitoring system. When a data pass all the set of tests, it can be considered that data are validated and reliable to be used for all the remaining tasks of the monitoring or control system. But, in case if the data are not validated, a methodology is presented to estimate and to reconstruct these data. Chapter 10 presents an application of this methodology to a water transport network with interesting results. Finally, Chap. 11 starts with a review of start of the art in fault diagnosis applied to drinking-water networks, using physical and analytical redundancies, and a model-based fault diagnosis methodology is formulated and applied to a real case study of water network using analytical redundancy relations.

Part III contains six chapters treating different aspects for an advanced operational real-time control for drinking transport and distribution water networks. The challenges of this part consist in developing an effective operational control system taking into account the complexity and uncertainty of the networks, the complexity of the network dimension with a number a large number of sensors, actuators and reservoirs and the complexity of the nonlinear behaviour of some components. The uncertainty of the demands and the difficulty to manage every day the network with a the probability of unpredictable fault events in one or several components of the network is a serious problem. Chapter 12 describes a model-based predictive control (MPC) for the supervisory flow management using a real-time monitoring system, described in Part II. MPC is used to generate flow control set points of the active components (valves and pumps) allowing to transport the water from the sources to the consumer areas to meet the demands, minimizing the economic cost of the operation and maintaining safety storage volumes in the tanks of the network. The model used in this chapter is a linear model based on the mass balances of the flows in the nodes with or without storage. Chapter 13 addresses the same problem in Chap. 12 but including pressure constraints and equations in the model of the controller. These additional equations are static, nonlinear relationships between the flows and the associated head loss. The optimal control technique presented in Chap. 13 is also based on MPC, but in this case because of the nonlinear restrictions the solution of the associated optimization problem becomes more involved since it is a nonlinear optimization problem. MPC approaches presented in both chapters have been applied to real water networks and the results are analysed and compared with the current operation based on the experience of the network managers. Chapter 14 deals with the uncertainty associated with the short-term demand forecast and the behaviour of the networks using stochastic MPC. In particular, in this chapter two approaches of stochastic MPC are considered. The first approach assumes a probability distributions of the future demands and state constraints and the second solution relies on tree-based scenarios to approximate the original uncertain problem without probabilities. Chapter 15 studies the MPC problem associated with the real-time operational control of water networks in the case of faulty situations. After a fault, the controller is redesigned to cope with the fault effect. Chapter 15 also presents a degradation analysis of the faulty system in terms of performances and reliability. The nature of large-scale systems of the water networks of big cities, such as the case of Barcelona city, leads to consider the partitioning of the network into subnetworks (subsystems) as presented in Chap. 16, where two different graph partitioning approaches are presented and applied in the Barcelona water distribution network. And finally, Chap. 17 focuses on the non-centralized model predictive control using distributed MPC techniques, where a set of local MPC controllers are in charge of controlling each one a subsystem of the entire system. This chapter is logically related to the previous chapter where the network is partitioned into subsystems with the aim of reducing the computational burden and to increase the scalability and modularity with respect to the centralized MPC control problems presented in the previous chapters.

Finally, Part IV presents several future trends in real-time monitoring and operational control of water networks in the last three chapters. Chapter 18 proposes the design of control strategy for water networks based on the interesting evolutionary-game theory (EGT), which allows to model the evolution of a population composed by a large and finite number of rational agents with capacity to make decisions. Chapter 19 proposes a multilayer MPC for the coordination of regional and urban networks. Chapter 19 aims not only to achieve the typical goals in the management of water urban networks, which are the minimization of the energy consumption while guaranteeing the quality of service (demand satisfaction), but also with the sustainable usage of the water resources decided at regional level. Finally, Chap. 20 deals with the application of the promising area of Big Data and Data Analytics to water networks. In particular, this chapter presents how these new technologies can address the new challenges created by the need of manipulating and analysing the huge number stored in the data centres collected of by automatic meter reading (AMR) connected to households. AMR offers continuously information about the water consumption of the users allowing to characterize their water usage, to detect leaks and fraud and to estimate nodal demand, among other. These data should be processed, validated and stored in a metadata model. Chapter 20 focuses on the application of data analytics and knowledge discovery tools to AMR data combined with other streams of information (e.g., billing system and call centre service). These last chapters of Part IV close the book suggesting a certain number of quite promising techniques to improve the real-time monitoring and operational control of drinking-water systems that it is expected that will be developed in the next years altogether with some other techniques that just currently being developed as, e.g., machine and deep learning, cloud computing, circular economy or factory 4.0.

References

1. Gosling SN, Arnell, NW (2016). A global assessment of the impact of climate change on water scarcity. Climatic Change 134(3):371–385
2. Ulanicki B, Brdys MA (1994) Operational control of water systems: structures, algorithms, and applications. Prentice Hall, New York
3. Simonovic SP (2000) Tools for water management. Water Int 25:76–88
4. Lopez-Ibanez M, Prasad TD, Paechter B (2008) Ant colony optimization for optimal control of pumps in water distribution networks. J Water Resour Plan Manage 7:337–346
5. Kurek W, Ostfeld FASCEA (2014) Multiobjective water distribution systems control of pumping cost, water quality, and storage-reliability constraints. J Water Resour Plan Manage 140:184–193
6. Cembrano G, Quevedo J, Puig V, Perez R, Figueras J, Ramon G, Rodryguez P, Barnet G, Casas M, Verdejo JM, Gil A, Marti J, Konig H (2005) Predictive optimal control of water supply and distribution. Control Eng Pract
7. Saul Montes, de Oca Armeaga (2011) Hierarchica multilayer and decentralized mpc control of the Barcelona water network. Master's Thesis, Universitat Politècnica de Catalunya
8. Labinaz G, Bayoumi MM, Rudie K (1997) A survey of modelling and control of hybrid systems. Ann Rev Control 21:79–92
9. Mandel D (1998) Diagnostic a base de redondance analytique: application a un réseau urbain de distribution d'eau potable. PhD Thesis, CRAN, Nancy, France
10. Almandoz J, Arregui F, Cabrera E, Cobacho R (2005) Leakage assessment through water distribution network simulation. J Water Resour Plan Manage 131(6):458–466
11. Pudar RS, Liggett JA (1992) Leaks in pipe networks. J Hydraul Eng 118(7):1031–1046
12. Liggett JA, Chen LC (1994) Inverse transient analysis in pipe networks. J Hydraul Eng 120(8):934–955

13. Liggett JA, Chen LC (1995) Monitoring water distribution systems. the inverse method as a tool for calibration and leak detection. In: Cabrera, E, Vela AF (eds) Improving efficiency and reliability in water distribution systems, vol 14. Kluwer, Boston, USA, pp 107–134
14. Kapelan Z, Savic DA, Walters GA (2003) A hybrid inverse transient model for leakage detection and rough-ness calibration in pipe networks. J Hydraul Res 41(5):481–492
15. Pérez R, Puig V, Pascual J, Quevedo J, Landeros E, Peralta A (2011) Methodology for leakage isolation using pressure sensitivity analysis in water distribution networks. Control Eng Pract 19(10):1157–1167
16. Munavalli GR, Mohan Kumar MS (2003) Water quality parameter estimation in a steady state distribution system. J Water Res Plan Manage 129(2):124–134
17. MOBESENS (Mobility for long term water quality monitoring). http://www.mobesens.eu/
18. Rossman LA, Boulos PF, Altman T (1993) Discrete volume element method for network water—quality models. J Water Resour Plan Manage ASCE 119(5):505–517
19. Islam MR, Chandhry MH, Clark RM (1997) Inverse modeling of chlorine concentration in pipe networks under dynamic condition. J Environ Eng 123(10):1033–1040
20. Ozdemir ON, Ger AM (1999) Unsteady 2-D chlorine transport in water supply pipes. Water Res 33(17):3637–3645
21. Clark RM, Rossman LA, Wymer LJ (1995) Modeling distribution system water quality: regulatory implica-tions. J Water Resour Plan Manage ASCE 121(6):423–428
22. Shang F, Uber JG, Polycarpou MM (2002) Particle backtracking algorithm for water distribution system analysis. J Environ Eng 128(5):441–450
23. Zierolf ML, Polycarpou MM, Uber JG (1998) Development and autocalibration of an input-output model of chlorine transport in drinking-water distribution systems. IEEE Trans Control Syst Technol 6(4):543–553
24. Al-Omari AS, Chaudhry MH, (2001) Unsteady-state inverse chlorine modeling in pipe networks. J Hydraul Eng 127(8):669–677
25. Munavalli GR, Mohan Kumar MS (2005) Water quality parameter estimation in a distribution system under dynamic state. Water Res 39:4287–4298
26. Gancel G, Mortazavi I, Piller O (2006) Sensitivity assessment for quality modelling for water distribution sys-tems. Appl Math Lett 10:1313–1319
27. Fabrie P, Gancel G, Mortazavi I, Piller O (2010) Computational study and sensitivity analysis for quality modelling in water distribution systems. J Hydraul Eng 34(1):34–45
28. Ostfeld A, Uber JG, Salomons E, Berry JW, Hart WE, Phillips CA, Watson JP, Dorini G, Jonker Gouw P, Kapelan Z, di Pierro F, Khu ST, Savic D, Eliades D, Polycarpou M, Ghimire SR, Barkdoll, Gueli R, Huang JJ, McBean EA, James W, Krause A, Leskovec J, Isovitsch S, Xu J, Guestrin C, VanBriesen J, Small M, Fischbeck P, Preis A, Propato M, Piller O, Trachtman GB, Wu ZY, Wal Ski T (2008) The battle of the water sensor networks (bwsn): a design challenge for engineers and algorithms. J Water Resour Plan Manage ASCE 134(6):556–568
29. Eliades DG, Polycarpou MM (2010) A fault diagnosis and security framework for water systems. IEEE Trans Control Syst Technol 18(6):1254–1265
30. Eliades DG, Polycarpou MM (2012) Water contamination impact evaluation and source-area isolation using decision trees. J Water Resour Plan Manage ASCE 138(5):562–570
31. Hall J, Zaffiro AD, Marx RB, Kefauver PC, Krishnan ER, Haught RC, Herrmann JG (2007) On-line water quality parameters as indicators of distribution system contamination. Security 99(1):66–77
32. de Schaetzen W, Randall-Smith M, Savic DA, Walters GA (1999) Optimal logger density in water distribu-tion network calibration. In: Savic DA, Walters GA (eds) Water industry systems: modelling and optimiza-tion applications, vol 1, pp 301–307
33. Yu G, Powell RS (1994) Optimal design of meter placement in water distribution systems. Int J Syst Sci 25(12):2155–2166
34. Lee BH, Deininger RA (1992) Optimal locations of monitoring stations in water distribution systems. J Environ Eng 118(1):4–16
35. Vitkovsky JP, Liggett JA, Simpson AR, Lambert MF (2003) Optimal measurement site locations for inverse transient analysis in pipe networks. J Water Resour Plan Manage 129(6):480–492

36. Rosich A, Sarrate R, Puig V, Escobet T (2007) Efficient optimal sensor placement for model-based fdi using and incremental algorithm. In: Proceedings of 46th IEEE conference on decision and control. New Orleans, USA, pp 2590–2595
37. Pèrez R, Puig V, Pascual J, Peralta A, Landeros E, Ll Jordanas (2009) Pressure sensor distribution for leak detection in Barcelona water distribution network. Water Sci Technol Water Supply 9(6):715–721
38. Ostfeld A, Salomons E (2004) Optimal layout of early warning detection stations for water distribution systems security. J Water Resour Plan Manage ASCE 130(5):377–385
39. Berry WJ, Fleischery LH, William E, Phillips C (2004) Sensor placement in municipal water networks. J Water Resour Plan Manage ASCE 131(1):237–244
40. Shastri Y, Diwekar U (2005) Optimal sensor placement for water distribution network security. J Water Resour Plan Manage ASCE pp 6713–6736
41. Matheson D, Jing C, Monforte F (2004) Meter data management for the electricity market. In: 8th international conference on probabilistic methods applied to power systems, Iowa State University, Ames, Iowa
42. Mourad M, Bertrand-Krajeswski JL (2002) A method for automatic validation of long time series of data in urban hydrology. Water Sci Technol 45(4–5):263–270
43. Denoeux T, Boudaoud N, Canu S, Dang VM, Govaert G, Masson M, Petitrenaud S, Soltani S (1997) High level data fusion methods. Technical report CNRS/EM2S/330/11-97v1.0, Universitè de Technologie de Compiegne, Compiegne, France
44. Burnell D (2003) Auto-validation of district meter data advances in water Supply Management-Maksimovic. In: Butler D, Memon FA (eds) Swets, Netherlands
45. Jörgensen HK, Rosenörn S, Madsen H, Mikkelsen PS (1998) Quality control of rain data used for urban run-off systems. Water Sci Technol 37(11):113–120
46. Maul-Kötter B, Einfalt T (1998) Correction and preparation of continuously measured rain gauge data: a standard method in north rhine-westphalia. Water Sci Technol 37(11):155–162
47. Tsang KM (2003) Sensor data validation using gray models. ISA Trans 42:9–17
48. Quevedo J, Puig V, Cembrano G, Blanch J (2010) Validation and reconstruction of flow meter data in the Barcelona water distribution network. Control Eng Pract 11:640–651
49. Petit-Renaud S, Denoeux TA (1998) Neuro-fuzzy system for missing data reconstruction. In: IEEE workshop on emerging technologies, Intelligent measurement and virtual systems for instrumentation and measurement, Saint-Paul, USA
50. Polycarpou M, Helmicki A (1995) Automated fault detection and accommodation: a learning system approach. IEEE Trans Syst Man Cybern 25(11):1447–1458
51. Zhang X, Polycarpou MM, Parisini T (2002) A robust detection and isolation scheme for abrupt and incipi-ent faults in nonlinear systems. IEEE Trans Autom Control 47(4):576–593
52. Ferrari R, Parisini T, Polycarpou M (2009) Distributed fault diagnosis with overlapping decompositions: an adaptive approximation approach. IEEE Trans Autom Control 54(4):794–799
53. Zhang X, Polycarpou M, Parisini T (2009) Decentralized fault detection in a class of large-scale nonlinear uncertain systems. In: Proceedings of the 48th IEEE conference on decision and control
54. Schütze M, Campisano A, Colas H, Schilling W, Vanrolleghe P (2004) Real time control of urban wastewater systems: where do we stand today? J Hydrol 299:335–348
55. Prescott SL, Ulanicki B (2001) Time series analysis of leakage in water distribution networks in water soft-ware systems theory and applications. Research Studies Press, England
56. Lobanova HV, Lobanova GV (2003) Approach for restoration of missing data, long-term time series and generalization of results
57. Bennis S, Berrada F, Kang N (1997) Improving single variable and multivariable techniques for estimating miss-ing hydrological data. J Hydrol 191(1–4):87–105
58. Bennis S, Kang NA (2000) New methodology for validating historical hydrometric data with redundant measurements. In: Blain WR, Brebbia CA (eds) Hydraulic engineering software viii
59. Piatyszek E, Voignier P, Graillot D (2000) Fault detection on a sewer network by a combination of a kalman filter and a binary sequential probability ratio test. J Hydrol 230(3–4):258–268

60. Pastres R, Ciavatta S, Solidoro C (2003) The extended kalman filter (ekf) as a tool for the assimilation of high frequency water quality data. Ecol Model 170(2–3):227–235
61. Ciavatta S, Pastres R, Lin Z, Beck MB, Badetti C, Ferrari G (2004) Fault detection in a real-time monitor-ing network for water quality in the lagoon of Venice (Italy). Water Sci Technol 50(11):51–58
62. Boukhris A, Giuliani S, Mourot G (2001) Rainfall-runoff multi-modelling for sensor fault diagnosis. Control Eng Pract 9(6):659–671
63. Hamioud F, Joannis C, Ragot J (2005) Fault diagnosis for validation of hydrometric data collected from sewer networks. In: 10th International conference on urban drainage, J10 ICUD, Copenhagen, Denmark, pp 21–26
64. Hamioud F, Joannis C, Ragot J (2005) Localisation de dèfauts de capteur pour la validation des mesures hy-dromètriques issues de rèseaux d'assainissement. 20eme colloque sur le traitement du signal et de l'image GRETSI 2005—Louvain la Neuve Belgique
65. Ragot J, Maquin D (2006) Fault measurement detection in an urban water supply network. J Process Control 16(9):887–902
66. Puig V, Stancu A, Escobet T, Nejjari F, Quevedo J, Patton RJ (2006) Passive robust fault detection using interval observers: application to the damadics benchmark problem. Control Eng Pract 14(6):621–633
67. Valentin N, Denoeux T (2001) A neural network-based software sensor for coagulation control in a water treat-ment plant. Intell Data Anal 5:23–39
68. Arteaga F, Ferrer A (2002) Dealing with missing data in mscp: several methods, different interpretations, some examples. J Chemom 16:408–418
69. Harkat MF, Mourot G, Ragot J (2006) An improved pca scheme for sensor fdi: application to an air quality moni-toring network. J Process Control 16(6):625–634
70. Nelson P, Taylor P, MacGregor J (1996) Missing data methods in pca and pls: score calculations with in-complete observations. J Chemom Intell Lab Syst 35:45–65
71. Gertler J, Romera J, Puig V, Quevedo J (2010) Leak detection and isolation in water distribution networks using principal component analysis and structured residuals. In: Conference on control and fault-tolerant systems (Sys-Tol), vol 16(6), pp 191–1964
72. Cembrano G, Quevedo J, Salamero M, Puig V, Figueras J, Martı J (2004) Optimal control of urban drainage systems. A case study. Control Eng Pract 12(1):1–9
73. Cembrano G, Wells G, Quevedo J, Pérez R, Argelaguet R (2000) Optimal control of a water distribution network in a supervisory control system. Control Eng Pract 8(10):1177–1188
74. Cembrano G, Quevedo J, Puig V, Pérez R, Figueras J, Verdejo JM, Escaler I, Ramón G, Barnet G, Rodríguez P, Plio M et al (2011) A generic tool for real-time operational predictive optimal control of water networks. Water Sci Technol 64(2):448–459
75. Mays L (2004) Urban stormwater management tools. McGrawHill, USA
76. Qin SJ, Badgwell TA (2003) A survey of industrial model predictive control technology. Control Eng Pract 11:733–764
77. Bemporad A (2006) Model-based predictive control design: new trends and tools. In: Proceedings of 45th IEEE conference on decision and control. San Diego, CA, pp 6678–6683
78. Rawlings J (2000) Tutorial overview of model predictive control. IEEE Control Syst Mag 38–52
79. Pascual J, Romera J, Puig V, Creus R, Minoves (2011) Operational predictive optimal control of Barcelona water transport network. IFAC Proc Volumes 44(1):10571–10578
80. Toro R, Ocampo-Martinez C, Logist F, Van Impe J, Puig V (2011) Tuning of predictive controllers for drinking-water networked systems. In: Proceedings of 18th IFAC world congress. Milano, Italy,
81. FP7-ICT-2007-2 WIDE—decentralized and wireless control of large-scale systems (2007). http://ist-wide.dii.unisi.it/
82. Billings RB, Jones CV (2008) Forecasting urban water demand. American Water Works Association, USA
83. Zhou SL, McMahon TA, Walton A, Lewis J (2002) Forecasting operational demand for an urban water supply zone. J Hydrol 259:189–202

84. Alvisi S, Franchini M, Marinelli A (2007) A short-term, pattern-based model for water-demand forecasting. J Hydroinformatics 9(1):39–50
85. Schwarm A, Nikolaou M (1999) Chance-constrained model predictive control. AIChE J 45(8):1743–1752
86. de la Penad DM, Bemporad A, Alamo T (2005) Stochastic programming applied to model predictive control. In: Proceedings of 44th IEEE conference on decision and control. Sevilla, Spain, pp 1361–1366
87. van Hessem D, Scherer CW, Bosgra OH (2001) Lmi-based closedloop economic optimization of stochastic pro-cess operation under state and input constraints. In: Proceedings of 40th IEEE conference on decision and control. Orlando, FL, pp 4228–4233
88. Swider D, Weber C (2007) Extended arma models for estimating price developments on day-ahead electricity markets. Electr Power Syst Res 77(5–6):583–593
89. Aggarwal S, Saini L, Kumar A (2009) Electricity price forecasting in deregulated markets: a review and evaluation. Int J Electr Power Energy Syst 31(1):13–22
90. Warrington J, Mariethoz S, Jones CN, Morari M (2010) Predictive power dispatch through negotiated loca-tional pricing. In: Innovative smart grid technologies conference Europe (ISGT Europe), IEEE PES, pp 1–8
91. Patrinos P, Trimboli S, Bemporad A (2011) Stochastic mpc for real-time market-based optimal power dis-patch. In: Proceedings of 50th IEEE conference on decision and control and European control conference. Orlando, FL, pp 7111–7116
92. Puglia L, Bernardini D, Bemporad A (2011) A multi-stage stochastic optimization approach to optimal bidding on energy markets. In: Proceedings of 50th IEEE conference on decision and control and European control conference. Orlando, FL, pp 1509–1514

Chapter 2
Case Studies

Ramon Ariño, Jordi Meseguer, Ramon Pérez and Joseba Quevedo

2.1 Introduction

As discussed in Chap. 1, this book presents a wide scope of research that combines multiple disciplines (as hydraulic and water quality modelling, data science, control, supervision, fault diagnosis) applied to the drinking-water systems. All the research presented and the book itself is oriented to the application of these methodologies. Thus, each chapter includes a section of simulation and results on real network models and data. These case studies are available because during the last two decades the authors have maintained a close collaboration with practitioners. This collaboration has helped in the design of the projects, the supervision of the decisions and finally the validation of the results. The experience of working for and with those who carry out the daily management of the system in study inspired this book as well. Hopefully, those readers who are working with water systems will appreciate the applicability of the approaches proposed. On the other hand, the examples help the understanding of the often complex methodologies presented. The research has been presented in international forums through conferences and journals, as shown in the references. This international exposure has guaranteed novelty and improvement beyond the state of the art. Nevertheless, due to the proximity, most of the techniques have been developed using case studies provided by the Catalan water companies and authorities. This regional focus in the applications provides homogeneity to the book. Besides which methodology fits in each kind of network is highlighted.

R. Ariño
Aguas de Barcelona (AGBAR), Barcelona, Spain

J. Meseguer
CETaqua, AGBAR Technical Center, Barcelona, Spain

R. Pérez (✉) · J. Quevedo
Research Center "Supervision, Safety and Automatic Control" (CS2AC-UPC), Terrassa, Spain
e-mail: ramon.perez@upc.edu

V. Puig et al. (eds.), *Real-Time Monitoring and Operational Control of Drinking-Water Systems*, Advances in Industrial Control,
DOI 10.1007/978-3-319-50751-4_2

2.2 Case Studies

The water supply system in Catalonia takes most of the water from two main river basins. Ebro river basin is managed by the Confederación Hidrográfica del Ebro (CHE) [2] as this river flows through other regions of Spain. The authors joined CHE in projects related to the river supervision [9]. This research on open channel systems, as stated before, is beyond the scope of this book. The other river basins are managed by the regional state holder Agència Catalana de l'Aigua (ACA) [1]. These river basins are called internal basins and they provide water to the Barcelona area (3.2 million people). Most of the water (around 80 %) that supplies the Barcelona metropolitan area (217000 m^3 in 2015) is surface water coming from rivers Ter and Llobregat. The rest comes from underground water, except for periods of water shortage, when a desalination plant provides up to 5 % of the water consumption. The use of desalinated water shows how critical water supply becomes in a densely populated (717 Habitants per km^2) area in Mediterranean coast. The catchment, treatment and transport of water in this area (Barcelona metropolitan) is managed by the company ATLL Concessionària de la Generalitat de Catalonia. Once water has been treated, it is distributed to 36 municipalities in the metropolitan area, through a distribution network comprising 5500 km of pipelines and 150 header tanks, which can store up to 540000 m^3 of drinking water. This distribution is carried out by nine different companies (public and private). The public–private consortium Aigües de Barcelona (AB) manages a part of the distribution networks in this region.

In next sections, different management levels in the supply (transport) and distribution networks will be considered. The characteristics that define these levels are the functionalities, the physical elements involved (e.g., tanks, pumps, valves), the size of these elements, the meshing grade of the network and the area they cover.

First of all, there is the regional supply network that covers a wide area and it links the water sources in the catchment with water treatment, usually containing river reaches, free-surface channels, reservoirs, pumping stations, etc. It has a large storage capacity and geographical extension, but it usually has a tree-like structure (low meshing) (see Chaps. 10 and 19 for more details). The transport network (second level) draws water from the regional network. It is managed by the water supply utility and has a structure similar to the regional one but with a smaller dimension. More importantly, this type of network is usually pressurized. It has storage capacity and pumping stations. It is normally organized into pressure levels, according to service needs and topographic elevation of the demand sectors. The output of this network is monitored (both pressure and flow) and corresponds to the input of the distribution network. The lowest level, the distribution network, is a meshed pressurized pipe network, which delivers water to individual consumers within the pressure levels. The instantaneous individual demands are the most uncertain parameter of the network, because in general they are not measured online.

2.3 Water Transport Network

The Barcelona water transport network (WTN) supplies water to approximately 3 million consumers, distributed in 23 municipalities in a 424-km^2 area. Water can be taken from both surface and underground sources. The most important ones in terms of capacity and use are Ter, which is a surface source, and Llobregat, where water can be taken from one surface source and one underground source. Water is supplied from these sources to 218 demand sectors through around 4645 km of pipes. The complete transport network has 63 storage tanks, 3 surface sources and 7 underground sources, 79 pumps, 50 valves, 18 nodes and 88 demands. The network is controlled through a SCADA system (Fig. 2.1). As it will be discussed in Chap. 12, for the predictive control scheme, a prediction horizon of 24 h is used. This record is updated at each time interval with a sampling time of 1 hour.

In Fig. 2.2, the whole network representation using the conceptual model used by the model predictive controller that corresponds to a simplification of the real system:

- Each demand sector corresponds to a subnetwork serving a given pressure level.
- Each actuator can integrate several pumps or valves working in parallel.

A water network system will generally contain a number of flow or pressure control elements, located at the supplies, at the water treatment plant inlets or within the network, and controlled through the telecontrol system.

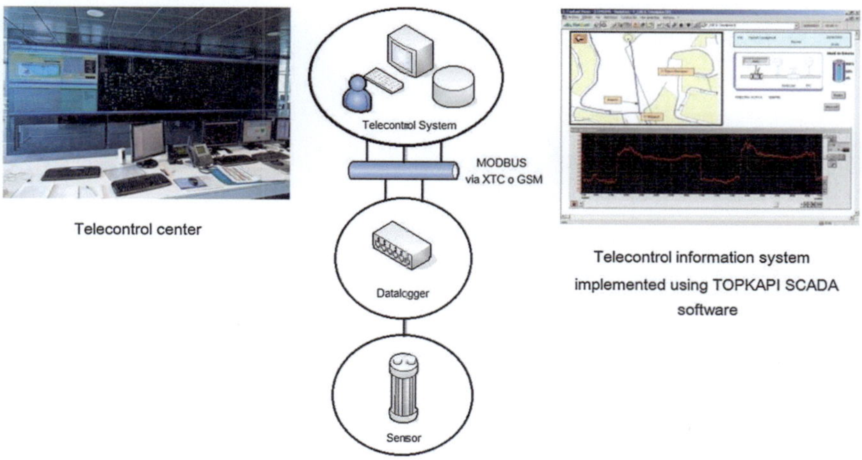

Fig. 2.1 Telecontrol of Barcelona water distribution system

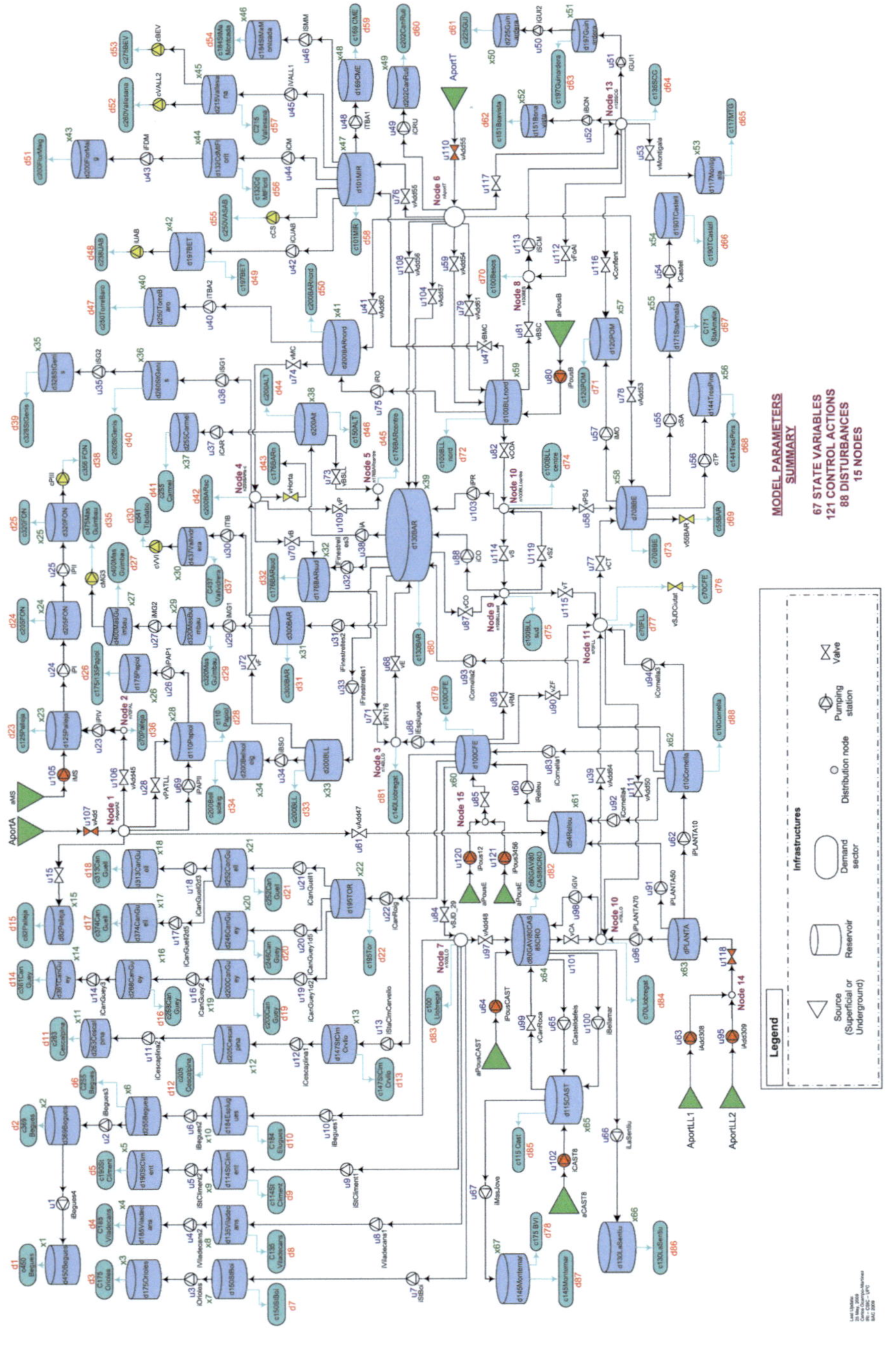

Fig. 2.2 Barcelona water transport network description

A section of the Barcelona water transport network has been selected in Chap. 13 to apply optimal control including pressure constraints (see Fig. 2.3). Specifically, 12 pressure zones that belong to the municipality of Barcelona (see Fig. 2.4), representing 18 % of the total network length and 23 % of the total annual water consumption. Within these pressure zones, there are 5 water tanks, 6 pumping stations and 7 regulation valves, with an annual energy consumption of 4.4 GWh. Energy consumption is highly dependent on the specific exploitation strategy decided by the utility (see Fig. 2.5).

This section of the network is representative of the complete network because it contains all the different kinds of state and control elements. The following figure shows a diagram of the study network, including the control elements, tanks, demands and connecting pipes. This network has five tanks (blue colour) and other five tanks that have the function of water sources (green colour) to substitute the rest of the network. It also has seven pumping stations, one flow valve, ten pressure valves and forty-seven demands.

Fig. 2.3 Simulation model of a portion of Barcelona WTN

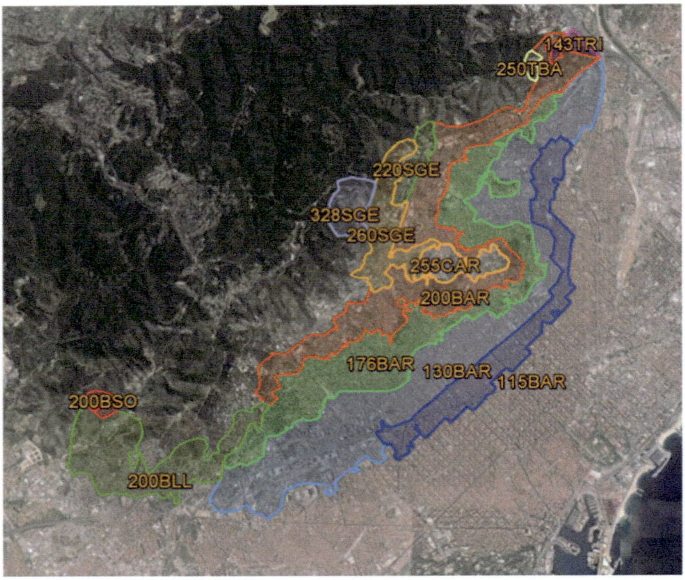

Fig. 2.4 Map of the portion of the Barcelona WTN

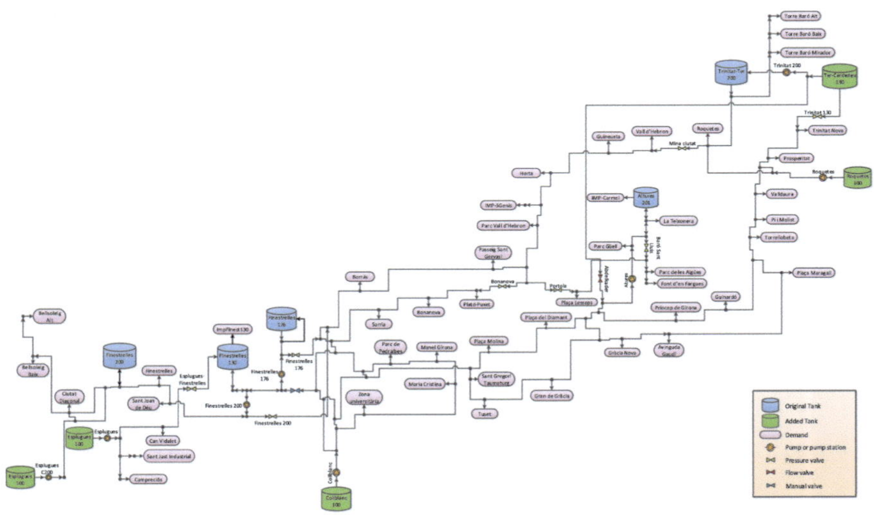

Fig. 2.5 Conceptual model of portion of Barcelona WTN

2.4 Water Distribution Network

The water distribution network of Barcelona (WDN) is organized in different pressure levels that supply water to the district metered areas (DMA). The elements of pressure levels and DMA are slightly different, and thus, it is worth to analyse them separately and highlight the chapters related to each of these systems. The organization of water distribution networks in DMA started in UK in 1980s and demonstrated to be a keystone for the performance improvement of networks. Both control and supervision benefit from the information provided by these units. Current research focus especially on the performance within these subsystems. In Barcelona, each sector (DMA) has the following configuration (see Fig. 2.6):

- One or two control points. In each of them, there is a continuous flow and pressure measurement (1 value/10 min) and, optionally, a pressure reducing valve.
- Optionally, one water quality control point, with the measurement of free residual chlorine, conductivity, pH and temperature (1 value/hour).
- Optionally, some internal pressure points (1 value/10 min).
- Some boundary closed valves.
- A data logger with a modem to get the signals from the field equipment and daily transmit data to the control centre.

Regarding customers consumption, there is a water meter for each customer. Some of them have automatic meter reading (AMR) and there is one sector with

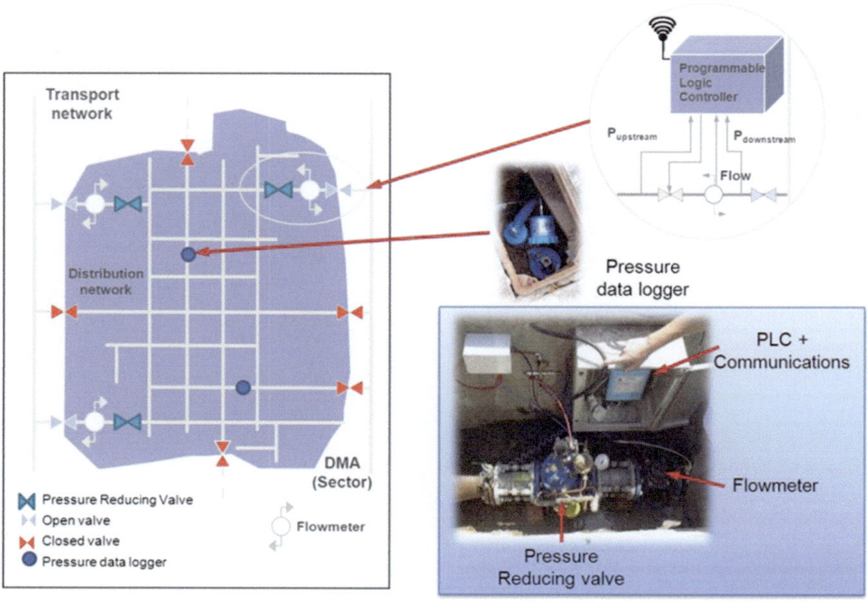

Fig. 2.6 Standard configuration for a DMA in Barcelona

a 100% of AMR that can provide detailed information about all the customer's consumption (hourly value). The control applied to the DMA is rather simple and its aim is to assure the supply with a guaranteed pressure. Nevertheless, the modelling of both hydraulic and chemical, presented in Chaps. 3–6, provides a wide research field in supervision. The efficiency of the DMA is assured in the simplest approach by searching leaks on field using acoustic devices. This leak monitoring is slow and costly and can be supported by more sophisticated methods. Chapter 8 presents how leakage management can be improved in the distribution network. Also the chlorine supervision based in models DMA includes some of the elements modelled and calibrated in Chaps. 3–6. These models lead to better performance through the supervision of the system. Two main issues in the water service that motivate this supervision:

- The efficiency in terms of balance between water delivered and produced. Leakage monitoring may be done on a routine basis or when major losses are suspected between night and day water demands [5]. Methods for locating leaks range from ground-penetrating radar to acoustic listening devices [4]. Some of these techniques require isolating and shutting down part of the system. Techniques based on locating leaks from pressure monitoring devices allow a more effective and less costly search in situ [8].

Fig. 2.7 Level 55 of the transport network

- The water quality in terms of assuring a good concentration of chlorine, the used chemical for disinfection. A good modelling approach of the chlorine decay allows to assure a correct concentration in all the service points in the DMA while reducing the high concentrations at the input and the use of chemicals. These models are used to detect abnormal situations related to this concentration [6].

Zone 55 is a part of the transport network presented in Sect. 2.4 and shown in Fig. 2.7. Its two inflows (Cantabria and Drassanes) and four outflows (Llull, Alaba, Joan de Borbó and Passeig Colon) have chlorine monitors. These chlorine measurements and the little meshing of the network make it quite suitable for the calibration of the chlorine decay model. This model is used within one of the DMAs that obtain

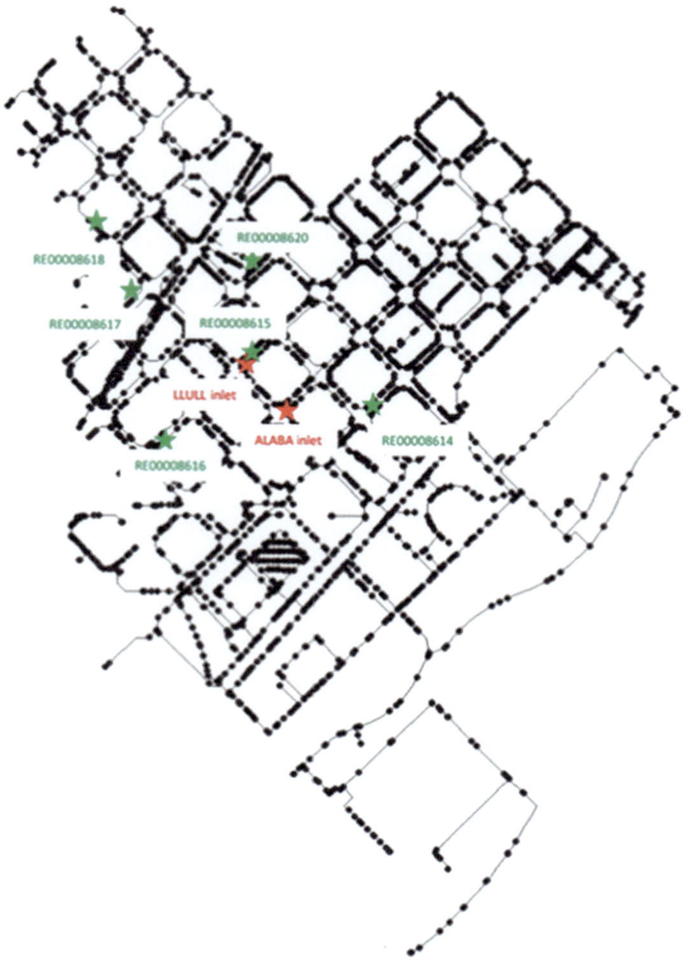

Fig. 2.8 Nova Icària DMA. In *red*, the two inflows, (flow and pressure sensors) and in *green*, the inner pressure sensors

the water from this zone, Nova Icària. This DMA has two inlets (Alaba and Llull), 1996 nodes and 3442 pipes. In Fig. 2.8, the water network of Nova Icària DMA can be seen from the EPANET file which contains the hydraulic model of this network. In this figure, the two DMA inlets have been highlighted using red star symbols. Nova Icària DMA is instrumented by flow and pressure sensors at every inlet. The sample time associated with all these sensors is set to 10 min. The results of the chlorine model calibration and its application for supervision are presented in Chap. 8. Six pressure sensors were installed for the leak detection and localization, they are highlighted in Fig. 2.8 using green star symbols. The methodology applied and the results in a pilot test with a real leak are described in Chap. 7.

A smaller DMA is used for illustrating the modelling, sample design and calibration procedures in Chaps. 4 and 5. Canyars DMA is situated in Castelldefels (Catalonia). The network model is composed of 721 pipes and 698 junctions. Water is supplied from the transport network through a pressure reduction valve, depicted in Fig. 2.9 with a blue triangle. Pressure and flow are monitored at the water inlet with a sample time of 10 min. The resolution is 0.3 l/s for the flow sensor, and 0.1 mwc for the pressure sensor. The minimum night flow is of about 3 l/s, and the peak hour flow is 27 l/s. Pressure control is applied to this network, fixing the pressure level at 38 metres during night-time and at 47 metres during daytime. The average daily maximum head loss in the network is 13.4 m. Three pressure sensors were installed, signalled in Fig. 2.9 with green stars.

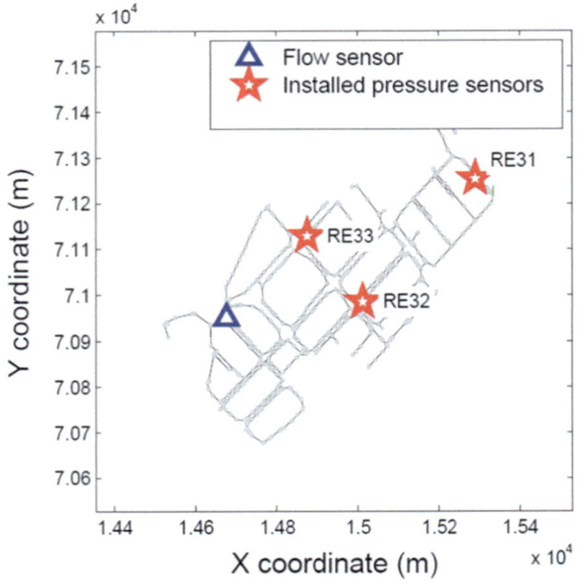

Fig. 2.9 Canyars DMA. In *blue triangle*, the inflow (flow sensor and pressure control), and in *green stars*, the inner pressure sensors

2.5 Software

The simulator used for the state estimation in the network throughout this book is EPANET.[1] EPANET is a widely used software in the academia to model water distribution piping systems. It is a public domain software that may be freely copied and distributed. It was developed by the US Environmental Protection Agency (EPA). EPANET performs extended period simulation of the water movement and quality behaviour within the pressurized pipe network as the transport and distribution networks. It tracks the flow of water in each pipe, the pressure at each node and the height of the water in each tank. The EPANET-MSX is the multi-species extension that allows prediction of chemical concentration throughout the network during a simulation period, water age, source and tracing. The use of the EPANET Programmer's Toolkit is a dynamic link library (DLL) of functions that allow developers to customize EPANET to their own needs. The functions can be incorporated into Windows applications written in C/C++, Delphi, Pascal, Visual Basic or any other language that can call functions within a Windows DLL. In the CS^2AC webpage ([3]), the adaptation for MATLAB (both 32 and 64-bit) is available. The combination of the power of simulation of EPANET together with the analytics of MATLAB allowed the application of the sophisticated algorithms described in following chapters.

Additionally, a simulator of the Barcelona water transport network has been built using MATLAB/Simulink and validated using real data coming from real scenarios (see Fig. 2.10). This model has been accepted by the company as a good representation of the actual water network behaviour and is used for its operational control [7] in Chap. 12, e.g., optimize water production and transport costs, guarantee a minimum volume in the tanks for eventual emergencies and smooth operations of the actuators to extend the life of the equipment. The *network* block is composed by different elements (blocks), such as tanks, nodes, pumps, valves and demands. Each demand

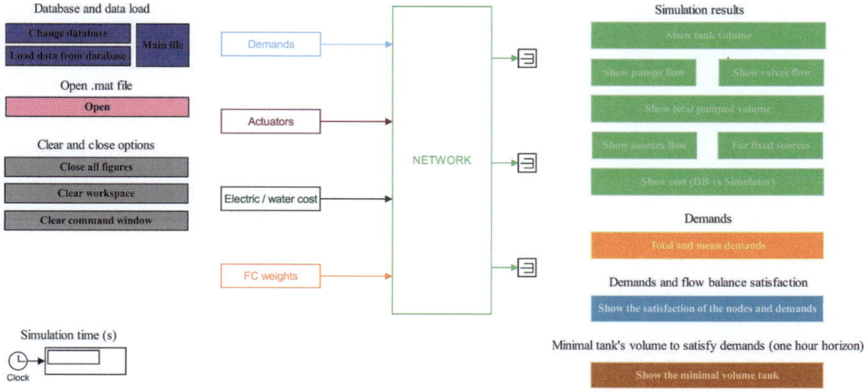

Fig. 2.10 Main Simulink screen of the Barcelona network simulator

[1] https://www.epa.gov/water-research/epanet.

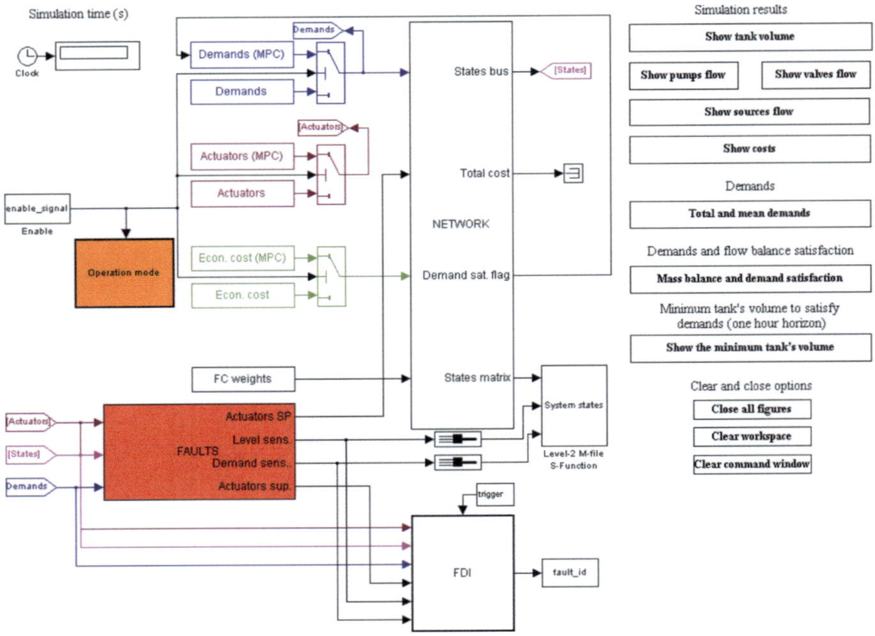

Fig. 2.11 MATLAB/Simulink Barcelona supply network simulator graphical user interface with a fault module

of this supply network is actually a district metered area of hundreds to thousands of users. Also, each actuator may integrate several pumps or valves working in parallel.

This simulator is equipped with a fault module (see Fig. 2.11) that allows to provide synthetic scenarios of the network under study and to design and test new control schemes and fault detection and identification (FDI) approaches as the one presented in Chap. 11 as well as the fault-tolerant control strategies presented in Chap. 14.

References

1. ACA—Agència Catalana de l'Aigua. http://aca-web.gencat.cat/aca/
2. CHEbro—Conferedación Hidrográfica del Ebro. http://www.chebro.es/
3. CS2AC—Research Center for Supervision, Safety and Automatic Control. https://cs2ac.upc.edu/en
4. Farley M, Trow S (2003) Losses in water distribution networks: a practitioner's guide to assessment, monitoring and control. IWA Publishing
5. Lambert M, Simpson A, Vtkovsk J, Wang X-J, Lee P (2003) A review of leading-edge leak detection techniques for water distribution systems. In: Proceedings of the 20th AWA convention, Perth, Australia, vol 610
6. Nejjari F, Pérez R, Puig V, Quevedo J, Sarrate R, Cugueró MA, Sanz G, Mirats JM (2012) Abnormal quality detection and isolation in water distribution networks using simulation models. Drinking Water. Eng Sci 5(1):67–72

7. Pascual J, Romera J, Puig V, Cembrano G, Creus R, Minoves M (2013) Operational predictive optimal control of Barcelona water transport network. Control Eng Pract 21(8):1020–1034
8. Pérez R, Puig V, Pascual J, Quevedo J, Landeros E, Peralta A (2011) Methodology for leakage isolation using pressure sensitivity analysis in water distribution networks. Control Eng Pract 19(10):1157–1167
9. Romera J, Ocampo-Martinez C, Puig V, Quevedo J (2013) Flooding management using hybrid model predictive control: application to the spanish ebro river. J Hydroinform 15(2):366–380

Part I
Modelling

Chapter 3
Modelling and Simulation of Drinking-Water Networks

Ramon Pérez and Gerard Sanz

3.1 Introduction

Models allow us to simulate the behaviour of WDN for many purposes: network optimization, background leakage modelling and control, smart demand metering, assessment, forecasting and management, asset management and performance modelling, real-time monitoring, modelling and control, network vulnerability, reliability, resilience and risk analysis, leakage and energy management, transient analysis, water quality, contaminant intrusion and water security, network operation and maintenance, etc. The type of model adopted depends on its intended use.

Water distribution network models can be classified depending on the dynamics involved: dynamic models and static models. Some WDN models characterize the transients in pipes, valves and pumps [1]. The analysis of transients is used to know the dynamic behaviour of the network. Dynamic models have been used for multiple objectives. In [2], these models are used for calibration of roughness, whereas in [3] the inverse transient models are used for leakage detection. The main drawbacks of dynamic models are the need of huge amounts of data, and the high computational power required.

Steady-state models are the most used in water companies for design, supervision and control. Steady-state simulations allow to determine the operating behaviour of a system under static conditions: fixed set of reservoir levels, tank levels and water demands. Steady-state simulations can be concatenated during the time to generate extended period simulations (EPS), where the only dynamics involved are the ones coming from tanks filling or emptying.

R. Pérez (✉) · G. Sanz
Research Center "Supervision, Safety and Automatic Control" (CS2AC-UPC), Terrassa, Spain
e-mail: ramon.perez@upc.edu

V. Puig et al. (eds.), *Real-Time Monitoring and Operational Control of Drinking-Water Systems*, Advances in Industrial Control,
DOI 10.1007/978-3-319-50751-4_3

Generally, water network models are automatically generated from geographic information system (GIS). This direct translation generates a model with a huge number of elements which do not have any impact on the network behaviour. The main aim of a reduced model is to preserve the nonlinear hydraulic behaviour of the original network and approximate its operation accurately under different conditions. There are different methods for reducing the complexity of the model, such as skeletonization, decomposition, usage of artificial neural networks (ANNs) metamodel or variables elimination.

Skeletonization is the process of selecting for inclusion in the model only the parts of the hydraulic network that have a significant impact on the behaviour of the system [4]. The level of skeletonization depends on the intended use of the model. The reduced models have been called "surrogate networks" or "grey boxes" [5]. In the first study of skeletonization [6], the authors systematically removed pipes from a model and the sensitivity of results. Brandon [7] suggested three heuristic rules that can be used to carry out the skeletonization process: (1) relatively small demands along any pipe were added to the node at the end of the pipe; (2) pipes with small diameters are eliminated, and the area that is fed by them is represented by a single node; and (3) a group of adjacent nodes with similar pressures is reduced to one node. Hamberg and Shamir [8] proposed an approach for reducing the size of the models for the preliminary design phase based in a step-wise combination of the system elements. Salarriaga et al. [9] skeletonized the network using the resilience concept. Walski et al. [4] proposed an automated skeletonization process.

Swamee and Sharma [10] proposed a simplification of the network by decomposing it in subsystems with one input in order to reduce the computational cost of the design of the WDN. The network reduction process can be treated as a decomposition of the network graph according to its connectivity properties [11].

A parameter fitting approach is presented in [12]. They reduced the network by calculating two parameters' vectors representing the nodal demands and the links conductances. An objective function was formulated for maximizing the accuracy of the simplified network. ANNs can capture the domain knowledge of hydraulic simulation model for predicting the consequences of different control settings on the performance of the WDN [13]. A systematic methodology using metamodels and ANNs is presented in [14].

Variable elimination is based on a mathematical formalism. Some of the system variables can be eliminated from the system of nonlinear differential equations that represent the mathematical model. An extended version of [15] is presented in [16], proposing an algorithm involving linearization, Gaussian elimination and a reconstruction of a reduced nonlinear model. The latter algorithm was implemented for integration of the model reduction module with an online optimization strategy [17].

3.2 Problem Statement

The set of methodologies presented in this book require the modelling and simulation of the WDN to achieve their objectives: parameter estimation, leak monitoring, sensor placement, predictive control, etc. Transient models provide a good representation of the dynamic behaviour of the network elements, but require a huge amount of data and are computationally expensive. When the number of pipes, pumps and valves increases, the network tends to become steadier and the transients lose importance. Furthermore, the used sensors' sampling times ($>10\,\text{min}$) tend to be much higher than the network elements' dynamics ($\approx 1\,\text{min}$), as WDN are pressurized systems.

This introduces the modelling of water distribution networks, focusing on the approach adopted by most of the authors on this book. Computer-based supervision and control applications in huge networks assume that the network behaviour is described by steady-state models concatenated in an extended period simulation (EPS) [4, 18]. This book considers this type of model for the benefits presented.

3.3 Proposed Approach

WDN models are composed of links connected to nodes. The behaviour of each of the elements is described by flow in links, and head in nodes. Head in a node is equivalent to the sum of the node pressure plus the node elevation, i.e., the height to which water would rise in an open-ended vertical pipe installed at the considered point.

Nodes represent junctions, tanks and reservoirs. Junctions are points where links join together and where water enters or leaves the network. Tanks are nodes with storage capacity, where the volume of stored water can vary along time during a simulation. Finally, reservoirs represent infinite external source or sink of water to the network. They are used to model lakes, rivers, transport networks, etc. As reservoirs are boundary points, their head cannot be affected by what happens within the network.

Links represent pipes, pump stations and control valves. Links are assumed to be always full of water, as pressurized networks are here considered. Pipes are links that convey water from one point in the network to another. Pumps provide energy to water, thereby raising its hydraulic head. Finally, valves are links that limit the pressure or flow at a specific point in the network. There are different types of valves: pressure reducing valve, flow control valve, pressure sustaining valve, etc.

3.3.1 Hydraulic Equations

The governing laws for flow in WDN under steady conditions are *conservation of mass* and *energy*. The *law of mass conservation* states that the rate of storage in a system is equal to the difference between the system's inflow and outflow. In pressurized WDNs, no storage can occur within the pipe network, although tank storage may vary over time. Therefore, in a pipe or a junction node, the inflow and outflow must be balanced. For a junction node,

$$\sum_{j \in \mathcal{J}_i} q_j = d_i, \tag{3.1}$$

where $\mathcal{J}_i$ denotes a set of pipes connected to node i, q_j is the flow of pipe j connecting with node i, and d_i is the consumption of node i. Equation (3.1) can be represented in matrix form as:

$$\mathbf{B} \cdot \mathbf{q} = \mathbf{d}, \tag{3.2}$$

where **B** is the incidence matrix that defines the connections between nodes and pipes. Coefficient b_{ij} indicates if pipe j is connected to node i:

- 1 if flow of pipe j enters node i.
- 0 if pipe j and node i are not connected.
- -1 if flow of pipe j leaves node i.

The *energy conservation law* states that the difference in energy between two nodes is equal to the energy added to the flow in the components between these points minus the frictional and minor losses. The relationship between pipe flow and energy loss caused by friction in individual pipes can be computed using one of three formulas: Hazen–Williams, Darcy–Weisbach and Chezy–Manning [4]. The general relationship is of the form:

$$\Delta h_{ij} = h_i - h_j = R_{ij} \cdot q_{ij}^r, \tag{3.3}$$

where Δh_{ij} is the head loss in pipe connecting nodes i and j; R_{ij} is the resistance coefficient that depends on the pipe's diameter, length and roughness; q_{ij} is the pipe flow rate; and r is the flow exponent. Expressions for the resistance coefficient and values for the flow exponent for each of the mentioned formulas are listed in [19]. The matrix representation of Eq. (3.3) is:

$$-\mathbf{B}^T \cdot \mathbf{h} = \mathbf{R}\,\mathbf{q}^r, \tag{3.4}$$

where **R** is a diagonal matrix with the resistances R_{ij} and **h** is a vector with the nodes head loss.

The energy balance for any path can be expressed as

$$\sum \Delta h_{ij} = \delta E_p, \tag{3.5}$$

where $\delta \mathbf{E}_p$ denotes the energy difference between the starting and final nodes of the path p. The summation is carried out over all links of the path. For loops, $\delta E_p = 0$, as the starting and ending node is the same.

The head-flow relationship (3.3), according to the Hazen–Williams formula

$$h_i - h_j = \frac{10.7 \cdot L}{C^{1.852} \cdot D^{4.87}} \cdot q_{ij}^{1.852}, \tag{3.6}$$

where L and D are the length and diameter (in m) of the pipe connecting both nodes, and C is the pipe roughness. The flow can be isolated, that is,

$$q_{ij} = G_{ij}^{0.54}(h_i - h_j)|h_i - h_j|^{-0.46}, \tag{3.7}$$

where G_{ij} is the pipe conductivity, calculated as

$$G_{ij} = \frac{1}{R_{ij}} = \frac{C^{1.852} \cdot D^{4.87}}{10.7 \cdot L}. \tag{3.8}$$

A diagonal matrix $\mathbf{C}$ containing the nonlinear parameter of each pipe can be defined as

$$c_{pp} = G_{ij}^{0.54}|h_i - h_j|^{-0.46}, \tag{3.9}$$

where c_{pp} is the nonlinear coefficient for pipe p, which connects node i with node j. By means of this matrix, the head-flow equation can be transcribed in matrix form as

$$\mathbf{q} = -\mathbf{C} \cdot \mathbf{B}^T \mathbf{h}. \tag{3.10}$$

The negative sign comes from $-\mathbf{B}^T$, as water flows through pipes always from the node with higher head to the node with lower head.

Finally, the matrix form of the equations governing the water distribution system can be obtained by joining (3.2) and (3.10), yielding

$$-\mathbf{B} \cdot \mathbf{C} \cdot \mathbf{B}^T \cdot \mathbf{h} = \mathbf{d}. \tag{3.11}$$

The system (3.11) is nonlinear due to the nonlinear elements in matrix $\mathbf{C}$. If all heads or flows are known, the system can be directly solved and demands can be computed. However, the inputs in the real system are demands (together with information from fixed head nodes, valves set points, among others).

3.3.2 Water Consumptions

Once the WDN model is available, a demand model has to be defined. Nodes in WDN models represent an aggregation of multiple users. Each of these users may be of different types, e.g., domestic and commercial. Users of the same type are usually assumed to consume water in the same way, following a predetermined diurnal demand pattern. The consumption of each user is computed by multiplying the pattern coefficients with the user's base demand, i.e., the user's average water consumption, computed from billing information. Once this is done, demands that are associated with a certain network node are aggregated, resulting in the total nodal consumption at a given point in time. To simplify, the demand at a network node, assuming a single user per node, is computed as

$$d_i(k) = \sum \frac{bd_i}{\sum_{j=1}^{n_d} bd_j} \cdot p_{a\to i}(k) \cdot q^{in}(k), \tag{3.12}$$

where bd_i is the base demand of node i, n_d is the number of nodes in the network, $p_{a\to i}$ is the value of diurnal pattern a associated with user i and $q^{in}(k)$ is the total network consumption metered at sample k.

However, the information on different types of users associated with a given network node and their diurnal patterns is not always available in practice. Quite often, the only information available is the consumption aggregated during a period of time (usually monthly or quarterly). This low temporal resolution information regarding the demands can still be used to compute the base demand of each user. The base demand of a node is computed from the sum of the base demands of consumers aggregated in this node. The basic model presented in (3.13) uses the nodal base demands, together with the total network consumption metered at the network inputs, to calculate the instantaneous demand of each node at each sample, that is,

$$d_i(k) = \frac{bd_i}{\sum_{j=1}^{n_d} bd_j} \cdot q^{in}(k). \tag{3.13}$$

The approach presented in (3.13) considers that all demands have the same behaviour, which is determined by q^{in}. Besides, the basic demand model cannot explain the daily variation of the relative pressure behaviour between two areas in the network. This book proposes a new approach to model nodal demands depending on their geographic location by means of the calibration of demand components. A detailed explanation is presented in Chap. 5.

Ideally, if individual meter readings are taken for every customer, they should be exactly equal to the amount of water that is measured leaving the treatment facility. In practice, however, this is not the case. Although inflow does indeed equal outflow, not all of the outflows are (or could be) metered. These “lost” flows are referred to as unaccounted-for-water (UFW) and can be assigned to leakage, theft, unmetered services or other causes. Leakage is frequently the largest component of UFW and includes distribution losses from supply and distribution pipes, trunk mains, services up to the meter and tanks. The amount of leakage varies depending on the system, but there is a general correlation between the age of a system and the amount of UFW. Newer systems may have as little as 5% leakage, while older systems may have 40% leakage or higher. Leakage tends to increase over time unless a leak detection and repair programme is in place (see Chap. 7). There are some methodologies to study the UFW by means of the minimal night flow [20] and the DMA performance. If better information is not available, UFW is usually spread uniformly around the system (in spatial and temporal terms). If UFW is reduced, then the utility will see higher peaking factors because UFW tends to flatten out the diurnal demand curve.

Leaks in this book are assumed to be located at the nodes of the network. This simplification implies a loss of accuracy of the order of the pipe length. Such simplification can be assumed if the maximum localization error required is greater than this length [21]. In order to simulate a leak, an emitter coefficient C_e is set in a node so that the leak size generated depends on the pressure of that node [19] as follows:

$$f_i = C_e \cdot p_i^{\gamma}, \tag{3.14}$$

where f_i is the leak water discharge at node i; C_e is the emitter coefficient; p_i is the pressure at the node i; and γ is an exponent of about 0.5 (Hazen–Williams, Darcy–Weisbach, Chezy–Manning formulas [19]).

3.3.3 Network Equations Solver

Simulating the water network model consists of determining the instantaneous values of nodes' heads and pipes' flows in the network under given boundary conditions (demands, control variables, reservoir heads, tanks levels, etc.). A key property of the nodal model is that it possesses a solution and the solution is unique [18]. Due to the nonlinear equations of the water distribution network, the solution requires the application of an iterative technique. These techniques estimate a solution and then improve it iteratively until the difference between the solutions falls within a specified tolerance. Examples of these techniques can be found in [22–26]. The most popular solver is obtained as a result of the application of the Newton–Raphson algorithm.

3.3.4 Chlorine Decay Modelling

Chlorine is the most popular water treatment disinfectant in municipal water distribution system. Chlorine is an oxidizing agent and it decays with time. Therefore, a minimum level of chlorine residual must be maintained in the distribution system to preserve both chemical quality and microbial quality of treated water [27]. In water distribution systems, chlorine decays over time as it reacts with organic materials in the water. A number of models have been developed to predict chlorine decay in drinking-water networks [28–30]. Generally, they can be divided into first-order and non-first-order reaction kinetic models. The first-order decay model has been mostly used because of its simplicity and its reasonable accuracy to represent chlorine decay in water systems. The first-order chlorine decay model includes expressions to describe reactions occurring in the bulk fluid and at the pipe wall. The differential form of the decay model is given by

$$\frac{dC(t)}{dt} = -k_S \cdot C(t), \tag{3.15}$$

where k_S is the decay rate and $C(t)$ is the chlorine concentration at a certain time t.

A second-order chlorine decay model based on the concept of competing reacting substances is developed in [30], while in [31], a semi-empirical combined first-order and second-order model is proposed, which provides a good description of the chlorine decay as follows:

$$\frac{dC(t)}{dt} = -k_R \cdot C(t)^2 - k_S \cdot C(t), \tag{3.16}$$

where the decay constants k_R and k_S are determined by deriving the best fitting of (3.16) with the experimental data. In Chap. 8, both models are calibrated, the results of calibration are compared with the real data and the resulting models are used for the quality monitoring of the network.

3.3.5 Network Skeletonization

Network skeletonization consists in selecting for inclusion in the model only the parts of the hydraulic network that have a significant impact on the behaviour of the system [4]. The level of skeletonization depends on the intended use of the model. Network skeletonization is proposed in this book for the following two main reasons:

- Reduction of the computational time,
- Elimination of mathematical inaccuracies.

For the first item, it is mandatory for many methodologies to be executed in real time. The number of elements in the WDN model is directly connected with the size of the matrices involved in the calibration problem, leak detection, etc., as well as the model inputs to be fixed before executing one simulation run. The reduction of the network model elements directly impacts on the computational time required for each of these processes.

On the other hand, network skeletonization helps to avoid mathematical problems in the computation of matrix **C**, whose parameters are calculated as shown in (3.9). When the flow through a pipe is null, the head loss on this pipe is zero, according to (3.3)

$$h_i - h_j = R_{ij} \cdot q_{ij} \cdot |q_{ij}|^{0.852} = R_{ij} \cdot 0 \cdot |0|^{0.852} = 0. \tag{3.17}$$

However, if (3.11) is used, matrix **C** is used to calculate the flow from the head loss, turning the previous equation into (3.9), where if the head loss is null, that is,

$$c_{pp} = G_{ij}^{0.54} |h_i - h_j|^{-0.46} = G_{ij}^{0.54} |0|^{-0.46} = \infty. \tag{3.18}$$

Consequently, an infinite value appears in matrix **C**, making the calculation of $\mathbf{C}^{-1}$ not possible. An intuitive solution is to replace this infinite value for a quite high one. However, the replacement of the infinite values may cause some inaccuracies in the results. Going in depth into the problem, the topology of the network seems to be the main cause, hence

- nodes connected with extremely short pipes that have a quite low resistivity factor may generate null head losses,
- series pipes which connect a null demand node.

The application of network skeletonization techniques eliminates these problems while keeping the same (or quite similar) hydraulic behaviour of the network model. As discussed in the introduction of this, there exist many network reduction techniques. In this book, skeletonization based on the steps listed in [4] is used:

- Removing simple pipes: Pipes are removed from the system based on the size or other criteria without any consideration of their effects on demand loading or hydraulic capacity.
- Removing branch pipes: Dead-end branches not containing tanks are trimmed back to a node that is part of a loop. This type of removal has no effect on the carrying capacity of the remainder of the system.
- Removing pipes in series: Pipes connected in series are replaced by an equivalent pipe which produces the same head loss. Removed nodes split their demands between the two nodes at the ends of the resulting pipe. A cut-off may be considered in order not to remove nodes with large demands.
- Removing parallel pipes: As in the previous case, an equivalent pipe replaces the parallel ones. New pipe's parameters have to be calculated. No effect on demands is produced in this process.
- Removing pipes to break loops: Pipes with the lowest carrying capacity are removed for breaking loops. This action produces a loss of the system capacity.

Non-pipe elements can also be removed but with some considerations [4]. Using these basic steps, automated skeletonization reduces the network until a stopping criterion defined by the user are achieved. This stopping criteria are chosen depending on the use of the model.

First, a basic skeletonization process which does not affect the hydraulic behaviour is performed:

1. Pairs of pipes connecting a null demand node are joined together.
2. Extrenal nodes connected by a unique pipe to the network are reduced.
3. Parallel pipes are replaced by a single equivalent pipe.

This reduction does not affect the hydraulic behaviour of the network. The second step consists in the reduction of short pipes with low resistivity factor. This modification may affect the hydraulic behaviour, depending on the intended use of the model. In this process, pipes with a resistivity factor lower than a specified threshold are reconstructed. The parameters of the pipe connected to the $n-1$ deleted pipes are recalculated, as follows:

$$C_r = (\frac{L_r}{D_r^{4.87}})^{0.54}(\sum_{i=1}^{n} \frac{L_i}{D_i^{4.87} C_i^{1.852}})^{-0.54}, \tag{3.19}$$

where C_r is the roughness of the new pipe; L_r is the length of the new pipe, equal to $\sum_i^n L_i$; D_r is the diameter of the new pipe; and L_i, D_i and C_i are the length, diameter and roughness of pipe i, respectively. The pipes i include the deleted and the updated pipe.

The process is performed iteratively, calculating the resistivity of each pipe of the network, assessing if the value is beyond a defined threshold and, if necessary, eliminating the pipe and reconnecting and re-parametrizing the appropriate elements of the network.

3.4 Simulation and Results

In this section, three examples are presented to illustrate the modelling and simulation of WDN. First, a simple network is presented to exemplify the construction of the matrix model containing the network equations. The DMA model of Nova Icària, presented in Chap. 2, is skeletonized and the hydraulic behaviour is analysed. Finally, the DMA model of Canyars, presented in Chap. 2, is simulated using the basic demand model of (3.13).

3.4.1 Matrix Model

The academic network is depicted in Fig. 3.1. It is a small network with six junctions and seven pipes. Water is supplied from a transport network through two pressure reduction valves. The transport network is represented by two reservoirs.

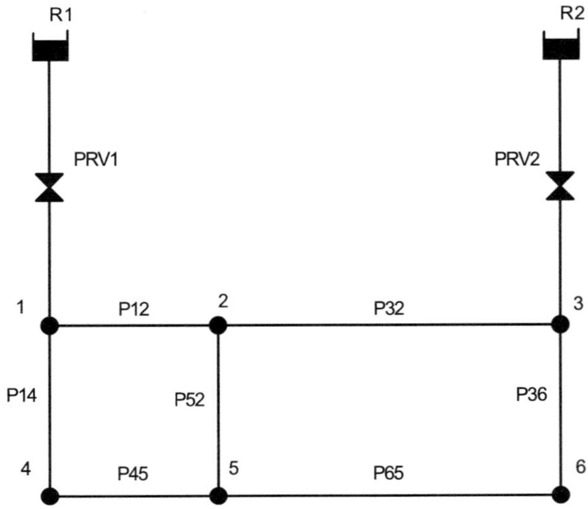

Fig. 3.1 Academic network

The flow continuity law defined in (3.1) states that the sum of inflows and outflows in every non-storage node is equal to zero. For the network in Fig. 3.1, the following equations can be written:

$$
\begin{aligned}
-q_{PRV1} &= d_{r_1},\\
-q_{PRV2} &= d_{r_2},\\
q_{PRV1} - q_{12} - q_{14} &= d_1,\\
q_{12} + q_{52} + q_{32} &= d_2,\\
q_{PRV2} - q_{32} - q_{36} &= d_3,\\
q_{14} - q_{45} &= d_4,\\
q_{45} + q_{65} - q_{52} &= d_5,\\
q_{36} - q_{65} &= d_6,
\end{aligned}
$$

where d_{r_1} and d_{r_2} are negative demands equal to the flows through PRV_1 and PRV_2. These eight equations can be easily presented in matrix form as follows:

$$
\begin{vmatrix}
-1 & 0 & 0 & 0 & 0 & 0 & 0 & 0 & 0\\
0 & -1 & 0 & 0 & 0 & 0 & 0 & 0 & 0\\
1 & 0 & -1 & -1 & 0 & 0 & 0 & 0 & 0\\
0 & 0 & 1 & 0 & 1 & 1 & 0 & 0 & 0\\
0 & 1 & 0 & 0 & -1 & 0 & -1 & 0 & 0\\
0 & 0 & 0 & 1 & 0 & 0 & 0 & -1 & 0\\
0 & 0 & 0 & 0 & 0 & -1 & 0 & 1 & 1\\
0 & 0 & 0 & 0 & 0 & 0 & 1 & 0 & -1
\end{vmatrix}
\cdot
\begin{vmatrix}
q_{PRV_1}\\ q_{PRV_2}\\ q_{12}\\ q_{14}\\ q_{32}\\ q_{52}\\ q_{36}\\ q_{45}\\ q_{65}
\end{vmatrix}
=
\begin{vmatrix}
d_{r_1}\\ d_{r_2}\\ d_1\\ d_2\\ d_3\\ d_4\\ d_5\\ d_6
\end{vmatrix}. \qquad (3.20)
$$

The matrix representation of the flow equations automatically generates the aforementioned incidence matrix **B** in (3.2).

The energy conservation equations for the seven pipes of the network in Fig. 3.1 can be written as

$$
\begin{aligned}
q_{12} &= G_{12}^{0.54}|h_1 - h_2|^{-0.46}(h_1 - h_2),\\
q_{14} &= G_{14}^{0.54}|h_1 - h_4|^{-0.46}(h_1 - h_4),\\
q_{32} &= G_{32}^{0.54}|h_3 - h_2|^{-0.46}(h_3 - h_2),\\
q_{52} &= G_{52}^{0.54}|h_5 - h_2|^{-0.46}(h_5 - h_2),\\
q_{36} &= G_{36}^{0.54}|h_3 - h_6|^{-0.46}(h_3 - h_6),\\
q_{45} &= G_{45}^{0.54}|h_4 - h_5|^{-0.46}(h_4 - h_5),\\
q_{65} &= G_{65}^{0.54}|h_6 - h_5|^{-0.46}(h_6 - h_5),
\end{aligned}
$$

where the nonlinear elements $G_{ij}^{0.54}|h_i - h_j|^{-0.46}$ will be referred as β_{ij}. The same equations can be written for the two pressure reduction valves, adding a control parameter V_{ij}. In this case, β_{PRV1} and β_{PRV2} include the control parameter. Moreover,

$$
\begin{aligned}
q_{PRV1} &= V_{PRV1} \cdot \beta_{PRV1}(h_{r_1} - h_1),\\
q_{PRV2} &= V_{PRV2} \cdot \beta_{PRV2}(h_{r_2} - h_3).
\end{aligned}
$$

The latter equations can be easily expressed in matrix form as follows:

$$
\begin{vmatrix} q_{PRV1} \\ q_{PRV2} \\ q_{12} \\ q_{14} \\ q_{32} \\ q_{52} \\ q_{36} \\ q_{45} \\ q_{65} \end{vmatrix}
=
\begin{vmatrix}
\beta_{PRV1} & 0 & 0 & 0 & 0 & 0 & 0 & 0 & 0 \\
0 & \beta_{PRV2} & 0 & 0 & 0 & 0 & 0 & 0 & 0 \\
0 & 0 & \beta_{12} & 0 & 0 & 0 & 0 & 0 & 0 \\
0 & 0 & 0 & \beta_{14} & 0 & 0 & 0 & 0 & 0 \\
0 & 0 & 0 & 0 & \beta_{32} & 0 & 0 & 0 & 0 \\
0 & 0 & 0 & 0 & 0 & \beta_{52} & 0 & 0 & 0 \\
0 & 0 & 0 & 0 & 0 & 0 & \beta_{36} & 0 & 0 \\
0 & 0 & 0 & 0 & 0 & 0 & 0 & \beta_{45} & 0 \\
0 & 0 & 0 & 0 & 0 & 0 & 0 & 0 & \beta_{65}
\end{vmatrix}
\begin{vmatrix}
1 & 0 & -1 & 0 & 0 & 0 & 0 & 0 \\
0 & 1 & 0 & 0 & -1 & 0 & 0 & 0 \\
0 & 0 & 1 & -1 & 0 & 0 & 0 & 0 \\
0 & 0 & 1 & 0 & 0 & -1 & 0 & 0 \\
0 & 0 & 0 & -1 & 1 & 0 & 0 & 0 \\
0 & 0 & 0 & -1 & 0 & 0 & 1 & 0 \\
0 & 0 & 0 & 0 & 1 & 0 & 0 & -1 \\
0 & 0 & 0 & 0 & 0 & 1 & -1 & 0 \\
0 & 0 & 0 & 0 & 0 & 0 & -1 & 1
\end{vmatrix}
\begin{vmatrix} h_{r_1} \\ h_{r_2} \\ h_1 \\ h_2 \\ h_3 \\ h_4 \\ h_5 \\ h_6 \end{vmatrix}.
\tag{3.21}
$$

3.4.2 *Skeletonization*

The high number of junctions and pipes in the network model increases considerably the computational effort required when applying any methodology using it. In the Nova Icària network case study, the skeletonization process presented in Sect. 3.3.5 is applied, considering only those reductions that do not affect the hydraulic behaviour of the network.

Figure 3.2 depicts the original model and the reduced model. The number of junctions has been reduced from 3377 to 1520 (45%), and the number of pipes from 3455 to 1644 (47.5%) (Table 3.1). The reduced model has been evaluated by comparing the pressures in the 1520 remaining nodes with the same nodes in the original network model. The highest pressure error during 144 samples (one day, with a sampling time of 10 min) is 3×10^{-6} mWC (meter water column), significantly lower than the sensors' resolution (0.1 mWC).

Table 3.1 Nova Icària network elements reduction through skeletonization

	Original	Reduced	Reduction (%)
Junctions	3377	1520	45
Links	3455	1644	47.5

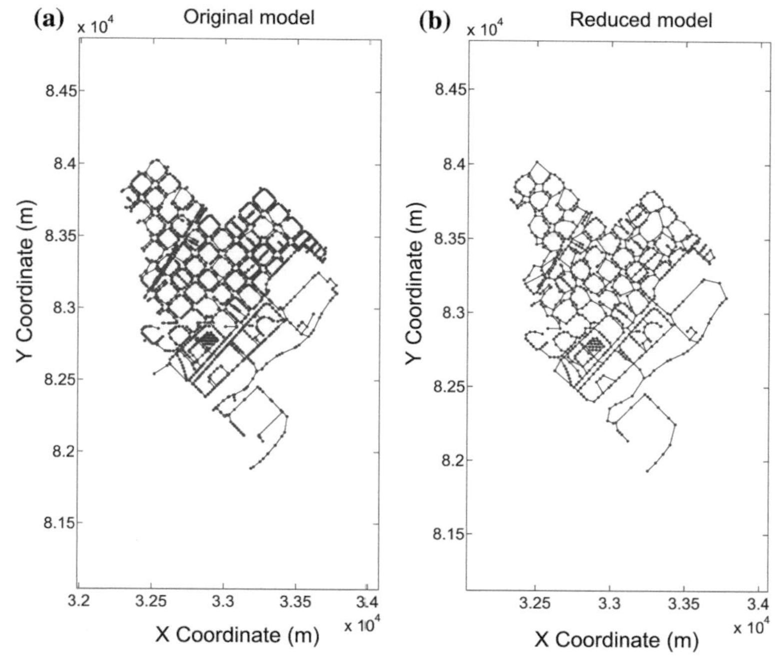

Fig. 3.2 Nova Icària network original and reduced EPANET models

3.4.3 Simulation

As it has been explained in previous sections, the solution of the sets of equations of the water distribution network allows to generate simulated predictions of the state of the network. Although the network solver can be manually implemented, there already exists many simulation-based software tools that ease this task. In this book, as stated in Chap. 2, EPANET is used.

As an illustrative example, the Canyars network, presented in Chap. 2 is simulated considering the basic demand model presented in (3.13). Figure 3.3 shows the pressure prediction error in the three available sensors ($RE31$, $RE32$ and $RE33$) when using the basic demand model. The blue thin line corresponds to the raw error using all data, and the red thick line represents the smoothed error, which has been computed by means of a smoothing spline. The green dashed line corresponds to the mean pressure prediction error. This error is treated as an offset that cannot be associated to the demand model. As suggested in [21], the offset has to be corrected to eliminate possible depths' errors, model nodes' elevations inaccuracies or badly calibrated sensors' offsets. Table 3.2 contains the specific correction for each sensor.

Figure 3.3 shows that the pressure prediction error when using the basic demand model follows the profile of the daily total consumption, as this demand model is not able to assign a different behaviour to each zone of the network. The prediction error

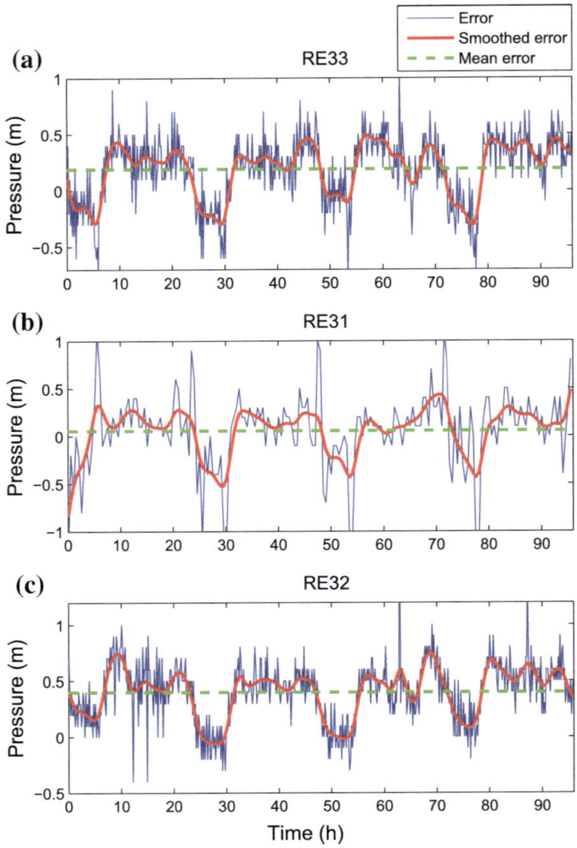

Fig. 3.3 Canyars network pressure prediction error in the three installed sensors during the precalibration week using the basic demand model. The *blue thin line* corresponds to the raw error, the *red thick line* corresponds to the smoothed error computed by means of a smoothing spline, and the dashed *green line* corresponds to the mean error

Table 3.2 Canyars sensors' sampling times (min), and offset corrections (metres)

Sensor ID	RE33	RE31	RE32 (%)
Sampling Time (min)	10	30	10
Offset correction (m)	0.19	0.05	0.4

can be improved by using of the demand components model (see Chaps. 4 and 5), which allows having multiple demand behaviours depending on the location of nodes in the network.

3.5 Conclusions

The equations presented in this chapter the steady-state model generally used by companies for the monitoring of the WDNs. These equations are a simplification since they do not take into account the transient. This approximation is justified both by the time constant of these transients and the complexity of including them in the computation. Even these steady-state models are often simplified following a skeletonization process that affects, when correctly applied, minimally to the hydraulic accuracy. The matrix formulation presented in this chapter is much convenient for the analysis of the network as it will be shown in Chap. 4. One of the main issues in all the networks is to calibrate the model if it has to be exploited. Demand calibration procedures are research topics of high interest as explained and illustrated in Chap. 5. In the following, the WDN models are intensively used for the monitoring and control of the networks with innovative methodologies.

References

1. Liggett J, Chen L Inverse transient analysis in pipe networks. J Hydraul Eng 8: 934–955
2. Vítkovský J, Simpson A, Lambert M. Leak Detection and calibration using transients and genetic algorithms. J Water Resour Plan Manage 4: 262–265
3. Kapelan Z, Savic D, Walters G A hybrid inverse transient model for leakage detection and roughness calibration in pipe networks. J Hydraul Res 5: 481–492
4. Walski T, Chase D, Savic D, Grayman W, Beckwith S, Koelle E (2003) Advanced water distribution modeling and management. Haestad Press
5. Shamir U, Howard C (1977) Engineering analysis of water-distribution systems. J Am Water Works Assoc 69(9): 510–514
6. Eggener C, Polwoski L (1976) Network models and the impact of modeling assumptions. J Am Water Works Assoc 68(4):189–196
7. Brandon T (1984) Water Distrib Syst. The insitute of water engineers and scientists, London
8. Hamberg D, Shamir U. Schematic models for distribution systems design. I: combination concept. J Water Resour Plan Manage 2: 129–140
9. Saldarriaga J, Ochoa S, Rodriguez D, Arbeláez J (2008) Water distribution network skeletonization using the resilience concept. In: Water distribution systems analysis. American society of civil engineers, Reston, VA, pp 1–13,
10. Swamee P, Sharma A. Decomposition of large water distribution systems. J Environ Eng 2: 269–283
11. Deuerlein J. decomposition model of a general water supply network graph. J Hydraul Eng 6: 822–832
12. Anderson EJ, Al-Jamal KH. Hydraulic-network simplification. J Water Resour Plan Manage 3: 235–240
13. Rao Z, Alvarruiz F. Use of an artificial neural network to capture the domain knowledge of a conventional hydraulic simulation model. J Hydroinformatics 1: 15
14. Broad D, Maier H, Dandy G. Optimal operation of complex water distribution systems using metamodels. J Water Resour Plan Manage 4: 433–443
15. Ulanicki B, Zehnpfund A, Martínez F (1996) Simplification of water network models. In: 2nd international conference on hydroinformatics, Zurich, pp 493–500
16. Martínez F, Ulanicki B, Salomons E (2012) A fast and practical method for model reduction of large scale water distribution networks. J Water Resour Plan Manage 2(1996):121119224314000
17. Paluszczyszyn D, Skworcow P, Ulanicki B (2011) Online simplification of water distribution network models. In: Savic DA, Kapelan ZS, Butler D (eds) 11th international conference on computing and control for the water industry, Centre for water systems, University of Exeter, Exeter, pp 749–754
18. Brdys MA, Ulanicki B (1994) Operational control of water systems: structures. Algorithms and applications, Prentice Hall

19. Rossman L (2000) EPANET 2 users manual. Water supply and water resources division, National Risk Management Research Laboratory
20. Lambert A. Accounting for losses: the bursts and background concept. Water Environ J 2: 205–214
21. Pérez R, Sanz G, Puig G, Quevedo J, Escofet MAC, Nejjari F, Meseguer J, Cembrano G, Tur JMM, Sarrate R. Leak localization in water networks: a model-based methodology using pressure sensors applied to a real network in Barcelona (applications of control). IEEE Control Syst 4: 24–36
22. Bhave P (1991) Analysis of flow in water distribution networks. Technomics, Lancaster, Pennsylvania
23. Lansey K, Mays Larry W (2000) Hydraulics of water distribution systems. McGraw-Hill, New York
24. Larock BE, Jeppson RW, Watters GZ (1999) Handbook of pipeline systems. CRC Press, Boca Raton, Florida
25. Ortega J, Rheinboldt W (1970) Iterative solution of nonlinear equations in several variables. Academic Press, New York
26. Todini E, Pilati S (1987) A gradient method for the analysis of pipe networks. In: Proceedings of the international conference and computer applications for water supply and distribution, Leicester, UK
27. Munavalli GR, Mohan Kumar MS (2003) Water quality parameter estimation in a steady state distribution system. J Water Resour Plan Manage ASCE 129:124–134
28. Boccelli DL, Tryby ME, Uber JG, Summers RS (2003) A reactive species model for chlorine de-cay and thm formation under rechlorination conditions. Water Res 37:2654–2666
29. Powell J, Hallam N, West J, Forster C, Simms J. Factors which control bulk chlorine decay rates. Water Res 1: 117–126
30. Clark RM (1998) Chlorine demand and TTHM formation kinetics: a second-order model. J Environ Eng ASCE 124:16–24
31. Hua F, West JR, Barker RA, Forster CF (1999) Modelling of chlorine decay in municipal water system. Water Res 33:2735–2746

Chapter 4
Parameter Estimation: Definition and Sampling Design

Gerard Sanz and Ramon Pérez

4.1 Introduction

According to [48], calibration "consists of determining the physical and operational characteristics of an existing system and determining the data that when input to the computer model will yield realistic results". In [2], the authors used the word *verified* in place of *calibrated* but described a process of calibration: "System simulation is considered verified during preliminary analysis for design when calculated pressures are satisfactorily close to observed field gauge readings for given field source send-out and storage conditions. If simulation is not satisfactory, the possibility of local aberrations, such as open boundary valves, is investigated. In the absence of other expected causative factors, the assumed local arterial network loads are adjusted until computed and observed field pressures are within reasonable agreement for various levels and extremes of demand, pumping, and storage". Walski [51] proposed a more precise definition: "Calibration of a water distribution model is a two step process consisting of: (1) Comparison of pressures and flows predicted with observed pressures and flows for known operating conditions (i.e., pump operation, tank levels, pressure reducing valve settings); and (2) adjustment of the input data for the model to improve agreement between observed and predicted valves. A model is considered calibrated for a set of operating conditions and water uses if it can predict flows and pressures with reasonable agreement".

A high degree of interest in this topic has been shown by researchers [47], but it has been considerably less covered by practitioners. A number of questions have to be answered, such as: (1) What parameters can be calibrated with confidence? (2) What is the acceptable level of discretization of calibration parameters and what is the acceptable level of agreement between measurements and model outputs? (3)

G. Sanz · R. Pérez (✉)
Research Center "Supervision, Safety and Automatic Control" (CS2AC-UPC), Terrassa, Spain
e-mail: ramon.perez@upc.edu

V. Puig et al. (eds.), *Real-Time Monitoring and Operational Control of Drinking-Water Systems*, Advances in Industrial Control,
DOI 10.1007/978-3-319-50751-4_4

How to parameterize the model when insufficient data are available? and (4) What objective function type to use?

Orsmbee [33] suggested a seven-step general calibration procedure as follows: (1) identification of the intended use of the model; (2) determination of initial estimates of the model parameters; (3) collection of calibration data; (4) evaluation of the model results; (5) macro-level calibration; (6) sensitivity analysis; and (7) micro-level calibration.

One of the most important issues in model calibration is the determination of the purpose of the model [53]. Seven possible purposes of a network model were identified as follows: pipe sizing for master planning, extended period simulations for planning studies, subdivision layout, rehabilitation studies, energy usage studies, water quality models and flushing programmes. In [50], a real system is modelled for daily pump scheduling and system expansion design to examine the impact of model purpose on the calibration process.

Battle of the water calibration networks is summarized in [34], the goal of this competition was to objectively compare the solutions of different approaches to the calibration of water distribution systems through application to real water distribution system. Interesting references have been extracted from this work and future work is well pointed:

- Due to the inherently *ill-posed or under-constrained* calibration problem in WDN, the solutions that provide a good match between measured and modelled data have to be validated with extra data.
- Uncertainty has to be included in the model parameters to explore the influence on the calibrated model outputs.
- Calibration size problem reduction is an important factor to be considered to avoid model overfitting, avoid unnecessary simulations or reducing the search space.
- Leakage data may be included in hydraulic calibration efforts because leakage directly affects nodal demand allocation and pump curve characterizations.
- The effect of different field data on model calibration should be investigated (use of flow and/or pressure measurements).

In [52], the author described the importance of good data collection. In [54], the same author classified data into three different degrees of usefulness:

- Good data are collected when there is sufficient head loss to draw valid conclusions about model calibration. It is necessary to have head loss in the system that is significantly greater than the error in measurement to avoid random adjustments [55].
- Bad data contain errors because of misread pressure gauge, incorrectly determined elevation of the pressure gauge or lack of information about which pumps were running when calibration data were collected. This type of data should be discarded.
- Useless data are collected when the head loss in the system is so low that head loss and velocity are of a similar order of magnitude as the errors in measurements. Such data can produce misleading models.

Ahmed et al. [1] developed a heuristic three-step procedure to assist in identifying the conditions under which useful data (good data) should be collected. The issue of data quality and quantity is closely related to that of sampling design, which will be addressed later in this chapter.

Formulas may assist the user in deciding whether to adjust roughness or water use and by how much [51]. They are based on fire flow test. To correct for inaccuracies in input data, it is necessary to first understand the sources of these inaccuracies. These can be grouped into several categories: (1) incorrect estimate of water use; (2) incorrect pipe carrying capacity; (3) incorrect head at constant head points (i.e., pumps, tanks, pressure reducing valves); or (4) poor representation of system in model (e.g., too many pipes removed in skeletonizing the system). The major source of error in simulation of contemporary performance will be in the assumed loadings distributions and their variations. On the other hand, [15] states: "the weakest piece of input information is not the assumed loadings condition, but the pipe friction factor". The certainties of a previous model must be stated so that the effort in calibration is in the good direction.

The most important uncertainty sources are demands and model simplifications [18], but uncertainty also originates from measurement errors, incorrect boundary conditions, inherent model structural errors or unknown status of valves [20, 55]. The calibration in this and next chapters focuses on demands due to their daily variability and continuous evolution depending generally on social and climate factors comparing to the more stable evolution of roughness.

The sensitivity matrix plays an important role in the solution of the direct/inverse problem [58], as well as in many of the methodologies developed in this book. Some of the existing general methods for the calculation of the sensitivity matrix are as follows [22]: (a) influence coefficient method (or perturbation method), (b) sensitivity equation method, (c) variational method (or adjoin method) and (d) automatic differentiation method.

The influence coefficient method uses the concept of parameter perturbation. At each simulation, one of the model parameters is perturbed [4], and the outputs measured. This method can be easily implemented, though computationally slow and relatively inaccurate when compared to other methods. $N + 1$ simulations are required, where N is the number of parameters in the model.

In the sensitivity equation method, a set of sensitivity equations are obtained by taking the partial derivatives with respect to each parameter in the governing equation and initial and boundary conditions. The same number of simulation runs as in the influence coefficient method is required. The method requires a solution of the forward problem (heads and flows) prior to the determination of unknown sensitivities. The calculated sensitivities are quite accurate [22].

The adjoin method computes relevant sensitivities once Lagrange multipliers are determined from a set of adjoin equations, which are derived from the basic WDN hydraulic model equations. This method also has a high accuracy and only requires N_s simulation runs, where N_s is the number of selected model's predicted variables.

The automatic differentiation method [19] is based on the differentiation of algorithms. Despite the good accuracy and computational performance, it produces a lengthy and complex computer code and requires a large number of changes to the source code of the appropriate hydraulic model [10].

Finally, a matrix analysis of the WDN linearized model where only one simulation is required at each iteration is proposed in [11].

The calculation of sensitivity matrices can be computationally demanding, as each element in the network generates an extra row or column in the matrix.

4.1.1 Identifiability

The calibration problem is often ill-posed. The ill-posedness is generally characterized by the non-uniqueness of the identified parameters. The uniqueness problem in parameter estimation is intimately related to identifiability [58].

Observability and identifiability terms are sometimes confused. System observability determines if the state of a system, i.e., the system variables (head, flow), can be estimated. On the other hand, system identifiability resolves if the parameters of the system (consumptions, roughness coefficients) can be calibrated. In conclusion, observability refers to system state (dynamic variables) while identifiability refers to system parameters (assumed constant in a certain time horizon).

An important contribution to the solution of the observability problem was made by [25], who formulated necessary and sufficient conditions for observability in power system state estimation in terms of meter location and network topology. According to their analysis, a network is observable if and only if it contains a spanning tree of full rank. The same problem for water systems is formulated in [5]. The identifiability can be classified as static and dynamic [36]. In [9], the study of identifiability is performed for the static problem using graph analysis based on [35]. The idea is that some operations in graphs are equivalent to operation on equations.

Conditions of identifiability for nonlinear dynamic systems can be found in the literature. The state-space formulation by means of the dynamic information of the system can be used [56]. For the linear case, the invertibility of the matrix of the equations set was studied by [49].

The complexity of the transient equations in dynamic identifiability makes their use difficult for real networks. The extended period identifiability is based on quasi-static equations, which allows to use simpler equations related from one time step to the next one by tank equations. The extended period identifiability is based on the sensitivity matrix rank in both linear and nonlinear cases [36]. The author stated that if many measurements are taken in the same conditions they will not add any information (without increasing the rank of the sensitivity matrix) but could be useful for filtering the noise in the measurements.

In [46], identifiability of the calibration problem is assured by defining a set of demand components to be calibrated that, considering the available measurements, generates a full-rank sensitivity matrix. This new parameterization is suitable for any element that abounds in a complex system. Both nodal demands and roughness of pipes are grouped [12, 26], and the hydraulic effects of roughness grouping are thoroughly studied in [30].

4.1.2 Sampling Design

Calibration accuracy should be judged both by the model's ability to reproduce data and by a quantitative measure of the uncertainty in calibrated parameter values. This uncertainty depends on the sampling design, including the measurement type, number, location, frequency and conditions existing at the time of sampling [8].

In the literature, the sampling design is defined as the procedure to determine the following [22]: (a) what WDN model predicted variables (pressures, flows, both, etc.) to observe; (b) where in the WDN to observe them; (c) when to observe (in terms of duration and frequency); and (d) under what conditions to observe.

In general, a sampling design may have one of the several purposes [29]: ambient monitoring, detection, compliance or research. Model calibration is considered research sampling, where the objective is to identify accurately the physical parameters of the system. A sampling design (SD) is a set of specified measurements, $\mathbf{y}$, at particular locations and times, along with the experimental conditions under which measurements are made [8].

One of the first sampling designs [51] suggested: (a) monitor pressure near the high demand locations; (b) conduct fire flow tests on the perimeter of the skeletal distribution system, away from water sources; (c) use as large as possible test flows at the fire hydrant; and (d) collect both head and flow measurements.

The importance of sensitivity in inverse problems comes from two primary reasons [28]. First, the need for the measurements to be made at a location where they are sensitive to the desired calibration parameters. Second, the degree of confidence that one has in the result depends on the sensitivity. Different approaches for solving the optimization problem have been developed. Usually, the main objective of finding the best locations for sensors is combined with other objectives (i.e., devices' cost). Genetic algorithms (GA), sensitivity matrix analysis or heuristic methods are some of the methodologies used.

The meter placement problem becomes a multi-objective optimization by seeking the best solution in terms of estimation accuracy and metering cost [59]. In this last reference, the authors developed a method employing dynamic analysis of the covariance matrix of state variables and the decision trees technique.

The potential location of the sensors may be ranked according to their overall relative sensitivity of nodal heads with respect to roughness coefficients [16]. Three general sensitivity-based methods are proposed in [8], and they are derived from the D-optimality criterion to rank the locations and types of measurements for estimating the roughness coefficients of a WDN model using pressure measurements, tracer concentration measurements and a combination of both. The authors outlined that the proposed methods, although suboptimal, may have some advantages over purely statistical methods that lack a physical basis. These three sensitivity-based methods are compared for selecting the worthwhile pressure and flow sensors' location in WDNs for calibrating roughness coefficients [14].

Pinzinger et al. [42] proposed three algorithms based on integer linear programming and greedy paradigm. The SD in [40] is formulated as an optimization problem which minimizes the influence of measurement errors in the state vector estimation subject to the constraint that the Jacobian matrix is of maximum rank. A greedy algorithm was used, which selected at each iteration the optimal location of the sensors. Some of the mentioned approaches used an iterative selection of the sensors, adding one sensor at each iteration to the set of already located ones. However, [22] demonstrated that the optimal set of locations for n monitoring points is not always a superset of the optimal set for $n-1$ monitoring points.

Sensitivity-based heuristic sampling design procedure for WDS model calibration to identify preferable conditions for data collection is developed in [27] accounting for uncertainty in measurements and its impact on both model parameters and predictions.

Three sampling design approaches are proposed in [13]. The first two were based on the shortest path algorithm, and set sensors' locations depending on the distance between the source and the set of potential sensors nodes. The third approach solved the optimization problem based on maximization of Shannon's entropy, locating sensors in the nodes with highest pressure sensitivity on roughness changes. The sampling design cost was also taken into account.

GA can find the combination of fire flow test locations that, when analysed collectively, stresses the greatest percentage of the hydraulic network, so the roughness parameters of grouped pipes can be calibrated [31]. Multi-objective sensitivity-based methods for sampling design minimize both uncertainty and SD cost objectives [22, 24]. Model accuracy was maximized and formulated as the D-optimal criterion, the A-optimal criterion and the V-optimal criterion. SOGA/MOGA (single/multi-objective GA) were used and compared, leading to the conclusion that the advantages in MOGA outweigh its disadvantages. The Jacobian matrix used was calculated prior to the optimization model run by assuming the model parameter values. Opposed to this deterministic approach, this latter assumption is handled by introducing parameter uncertainty using some predefined probability density function [6]. Results in studied cases [23, 24] assessed that the calibration accuracy based on prediction uncertainty (V-optimality) is preferred over parameter uncertainty (D-optimal and A-optimal criteria). Similarly, D-optimality is preferred over A-optimality.

The sampling design is posed in [21] as a multi-objective optimization problem, where the objective functions represented demand estimation uncertainty, pressure prediction uncertainty and demand estimation accuracy. The optimization problem was solved using MOGA based on Pareto-optimal solutions.

Not all sampling design approaches are addressed to parameter calibration. The sampling design is often based on the model application, for example a leakage detection methodology [37]. One sensor was located at each iteration of the procedure with the objective of minimizing the maximum number of nodes with the same binary signature (which cannot be isolated separately). The pressure sensitivity matrix analysis and an exhaustive search strategy produce an optimal sensor placement strategy [32]. Different sensor placement methodologies for demand calibration and leak detection are compared in [38]. Which performance criteria should be considered to place water quality and quantity sensors for both early detection and model calibration are investigated in [41].

4.2 Problem Statement

As explained above, the limited number of sensors together with the huge number of parameters requires a grouping of the parameters to make the calibration viable. In [44], the authors grouped demands depending on the type of user. Although good results were obtained with synthetic data, the analysis presented in [45] encourages the use of the demand components model.

The information extracted from the network depends on the type and location of the sensors. Each new sensor represents an additional equation in the system of equations to be solved. In order to have a determined system of equations, the number of measurements (sensors) has to be at least equal to the number of parameters, guaranteeing the system identifiability in the linear approximation.

In this chapter, a methodology both for the parameterization and the sampling design is pursuit. The questions to be answered are as follows:

- How can a huge number of parameters be grouped so that the system becomes identifiable with a reasonable number of field measurements?
- Where should these measurements be located so that a maximum of information is extracted for the calibration?
- Both questions use the information available in the sensitivity matrix.

4.3 Proposed Approach

The singular value decomposition (SVD) is a matrix decomposition method whereby a general $n_y \times n_x$ system matrix $\mathbf{A}$, relating model $\mathbf{x}$ and data $\mathbf{y}$:

$$\mathbf{A} \cdot \mathbf{x} = \mathbf{y}, \tag{4.1}$$

is factored into

$$\mathbf{A} = \mathbf{U} \cdot \boldsymbol{\Lambda} \cdot \mathbf{V}^T, \tag{4.2}$$

where $\mathbf{U}$ is a set of n_y orthonormal singular vectors that form a basis of the measured data vectorial space, $\mathbf{V}$ is a set of n_x orthonormal vectors that form a basis of the parameter vectorial space and $\boldsymbol{\Lambda}$ is an $n_y \times n_x$ diagonal matrix of singular values of $\mathbf{A}$, where the additional rows (more measurements than parameters) or columns (more parameters than measurements) are filled with zeros [3].

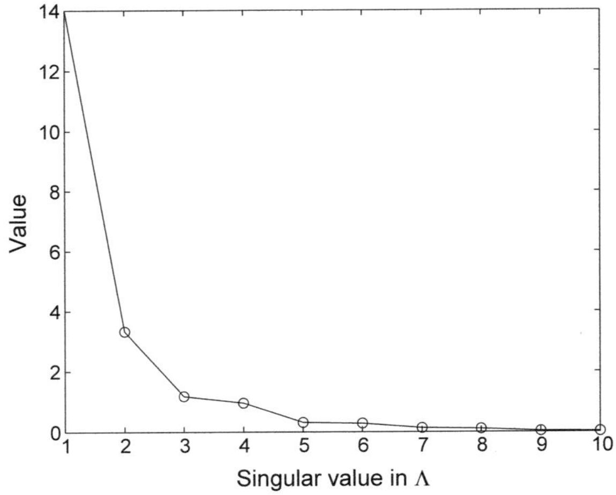

Fig. 4.1 λ singular values from the SVD of a 10 $\times$ 10 example sensitivity matrix

The SVD has many applications that can be useful for the parameter estimation. The key step to ensure the success of the calibration is the grouping of nodal demands into fewer parameters that, in the end, keep the network behaviour as close to the original behaviour as possible. This grouping ensures the identifiability of the system.

When calibrating parameters in nonlinear systems, the system matrix $\mathbf{A}$ in (4.1) is replaced by the system sensitivity matrix $\mathbf{S}$, which relates changes in data with changes in parameters. Explanations from now on focus on the sensitivity matrix $\mathbf{S}$.

The SVD allows to compute a reconstructed sensitivity matrix $\mathbf{S_r}$ from a subset of columns of $\mathbf{U}$ and $\mathbf{V}$, ignoring the information from these matrices that correspond to low relevant singular values. The singular values in matrix $\mathbf{\Lambda}$ from a 10 $\times$ 10 example sensitivity matrix are depicted in Fig. 4.1. We can observe quite low singular values

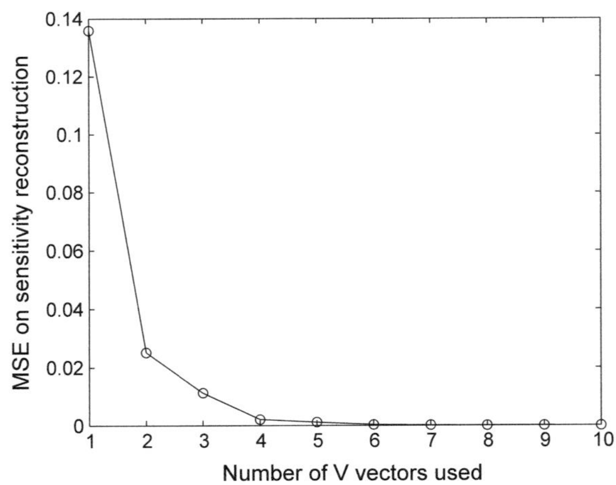

Fig. 4.2 MSE of the reconstructed matrix $\mathbf{S}$ from a different number of $\mathbf{U}$ and $\mathbf{V}$ vectors

from the fifth position, indicating that the corresponding columns in matrices **U** and **V** have low importance in the reconstruction of matrix **S**. Figure 4.2 presents the mean square error (MSE) in the reconstruction of the sensitivity matrix depending on the number of columns used from matrices **U** and **V**, corresponding to the same number of singular values in $\mathbf{\Lambda}$. It can be seen that when considering only the four first columns, the MSE falls to a quite low value.

The reduction of matrices **U** and **V** is used in the methodology presented to choose which parameters will be calibrated and which sensors will be used in the calibration process.

4.3.1 Parameter Definition

The grouping of parameters can be obtained from the analysis of the SVD of the system sensitivity matrix. "We can think of the eigenvectors $\mathbf{v}_i$, where $i = 1, \ldots, n$, as a new parameterization of the model. These vectors represent a set of n linear combinations of the old parameters that are fixed by the observations" [57]. Similarly, it is possible to reduce matrix **V** into $\mathbf{V_r}$, which is formed by the first n_c vectors $\mathbf{v}_i$, where n_c is the number of nonzero singular values of the sensitivity matrix. The new parameterization is obtained by defining a new parameter correction as follows:

$$\mathbf{x}^* = \mathbf{V_r}^T\mathbf{x}. \tag{4.3}$$

In WDNs, very low singular values appear (as seen in (4.1)); thus, n_c is defined in a way that all values below the n_c highest singular values are neglected. Furthermore, the consideration of quite low singular values leads to an increase of uncertainty [3]. The main drawback of this approach is the loss of the physical meaning of the calibrated parameters as they will be generated by a linear combination of the old parameters at each iteration. The sensors' data will be fitted, but the calibrated parameters will not have a direct relation with the WDN.

Consequently, the objective is to define the new parameterization as a static combination of the old parameters. The resolution matrix **R**, defined as

$$\mathbf{R} = \mathbf{V_r}\mathbf{V_r}^T, \tag{4.4}$$

describes how the generalized inverse solution smears out the original model **x** into a recovered model $\hat{\mathbf{x}}$. A perfect resolution is represented by the identity matrix, indicating that each parameter is perfectly resolved. When only n_c parameters corresponding to the highest n_c singular values are considered, the resolution matrix computed with $\mathbf{V_r}$ is not the identity matrix. Compact resolution appears, and parameters with similar sensitivities can be identified.

In the WDN particular case, compact resolution may appear but not being easily observable in the resolution matrix, as the parameter order in the sensitivity matrix **S** columns has no geographic order (in meshed networks, it is impossible to establish an order). The identification can be performed by means of the "*delta vector generation*" process by [57], which is adapted to define the matrix **M** with the membership of each individual demand to each demand component. The resulting parameterization is used to calibrate groups of demands.

Algorithm 4.1 presents the whole process to generate the matrix **M** from the reduced matrix $\mathbf{V_r}$. In lines 1–7, the delta vector generation process is performed, where the n_c vectors with the highest resolving power in the resolution matrix are obtained and normalized iteratively to generate the delta vectors.

Algorithm 4.1 Computation of nodal demands memberships to demand components

Require: $\mathbf{V_r}, n_c, n_d$
1: Compute $\mathbf{R} = \mathbf{V_r}\mathbf{V_r}^T$
2: **for** $z = 1 : n_c$ **do**
3: Find $j = \max(diag(\mathbf{R}))$
4: Compute $\mathbf{v}_z^* = \mathbf{R}_{(:,j)}/\sqrt{\mathbf{R}_{(j,j)}}$
5: Compute $\mathbf{R} = \mathbf{R} - \mathbf{v}_z^*\,\mathbf{v}_z^{*T}$
6: **end for**
7: Define $\mathbf{V}^* = [\mathbf{v}^*{}_1 \mid \mathbf{v}^*{}_2 \mid \cdots \mid \mathbf{v}^*{}_{n_c}]$
8: **for** $i = 1 : n_d$ **do**
9: Compute $\mathbf{M}_{(i,:)} = |\mathbf{V}^*{}_{(i,:)}| / \sum |\mathbf{V}^*{}_{(i,:)}|$
10: **end for**
11: **return** : **M**

In lines 8–11, matrix $\mathbf{V}^*$, which is formed by the $\mathbf{v}^*$ delta vectors, is used to generate the matrix **M**, associating each initial parameter to a new parameter (component) that produces the best resolution if n_c components are considered. The normalization of the rows in $\mathbf{V}^*$ is done so that the weights can be interpreted as memberships of each element parameter to each parameter component.

Three approaches were studied in [46] before reaching the final procedure: binary parameterization, positive hybrid parameterization and free hybrid parameterization

- The first approach assigns a single parameter component to each element parameter. After executing lines 1–7 in Algorithm 4.1, each demand is associated with the parameter component that has the highest value in the corresponding columns of the matrix $\mathbf{V}^*$.
- The second approach assigns a combination of demand components to each nodal demand with positive weights, exactly as presented in Algorithm 4.1.
- The free hybrid parameterization considers a combination of demand components that can include negative weights. For this approach, the absolute value in the numerator of line 9 of Algorithm 4.1 is ignored.

In all the proposed approaches, the solution tends to generate geographical patterns, as the topological information (incidence matrix **B**) is included in the sensitivity matrix. Results obtained in [46] concluded that the use of positive weights to perform the calibration of parameter components gave the best results in terms of error minimization.

4.3.2 *Sampling Design*

The sampling design is performed after the distribution of components, selecting the n_c best sensors. The process for locating the sensors uses matrix $\mathbf{U}$ in the same way as the parameterization process uses matrix $\mathbf{V}$. Initially, the sensitivity matrix $\mathbf{S}^*$ relating head and/or flow variations with demand components variations is computed and decomposed using the SVD. Matrix $\mathbf{U_r}$ is constructed with the first n_c columns of $\mathbf{U}$, as the information from the subsequent columns is negligible (they are multiplied by null rows of the $\mathbf{\Lambda}$ matrix). Then, the information density matrix $\mathbf{I_d}$ is computed as explained in [3], i.e.,

$$\mathbf{I_d} = \mathbf{U_r}\mathbf{U_r}^T, \tag{4.5}$$

describes how the generalized inverse solution smears out the original data $\mathbf{y}$ into a predicted data $\hat{\mathbf{y}}$. Since $\mathbf{I_d}$ has been constructed from n_c orthonormal vectors in $\mathbf{U_r}$, a set of n_c orthonormal vectors can be extracted from $\mathbf{I_d}$ in a way that they enhance the delta-like behaviour of the $\mathbf{I_d}$ matrix [57]. This "*delta-like*" vector generation process is presented in Algorithm 4.2 (lines 1–6). This process results in a set of delta-like vectors $\mathbf{u}^*$ that form matrix $\mathbf{U}^*$. Subsequently, the rows of matrix $\mathbf{U}^*$ are normalized (line 7), so that sensors with high sensitivity to multiple parameters are not selected. Finally, the sensor with the highest value in each of the n_c columns is selected as the sensor with highest information density to calibrate a particular parameter (lines 9–11). In the end, $n_s = n_c$ sensors are selected.

Algorithm 4.2 Sensor selection process

Require: $\mathbf{U_r}$, n_s
1: Compute $\mathbf{I_d} = \mathbf{U_r}\mathbf{U_r}^T$
2: **for** $z = 1 : n_s$ **do**
3: Find $j = \max(diag(\mathbf{I_d}))$
4: Compute $\mathbf{u}_z^* = \mathbf{I_d}_{(:,j)}/\sqrt{\mathbf{I_d}_{(j,j)}}$
5: Compute $\mathbf{I_d} = \mathbf{I_d} - \mathbf{u}_z^*\,\mathbf{u}_z^{*T}$
6: **end for**
7: Define $\mathbf{U}^* = [\mathbf{u}^*_1 \mid \mathbf{u}^*_2 \mid \cdots \mid \mathbf{u}^*_{n_s}]$
8: Normalize rows of $\mathbf{U}^*$
9: **for** $z = 1 : n_s$ **do**
10: Find $\mathbf{s}_{(z)} = \max(\mathbf{U}^*_{(:,z)})$
11: **end for**
12: **return** : $\mathbf{s}$

4.4 Simulations and Results

Following the idea of Chap. 3, first an academic example is used for illustrating the methodology, and afterwards, it is applied to a real network. Both examples will be used in the next chapter where the calibration problem is solved utilizing the results presented here.

4.4.1 Exemplification

The methodology presented above will be illustrated with the dummy networks presented in Figs. 4.3 and 4.4, which represent a meshed network and a tree-like network, respectively, where demands have to be calibrated. The simplicity of the networks will be useful to exemplify the methodology at each step.

Figures 4.5 and 4.6 show the output of the delta vector, $\boldsymbol{v}^*$, generation process (subfigure a)), and the memberships obtained after the normalization performed in lines 8–11 of Algorithm 4.1 (subfigure b)). Three sensors are considered, and therefore, three components are generated. The memberships represent the modulation of each nodal demand by each component and are produced from the delta vectors' directions.

Figures 4.7 and 4.8 depict in each of their subfigures, the memberships of each demand node to a particular demand component. The darker the colour in the map, the higher the membership to the depicted demand component.

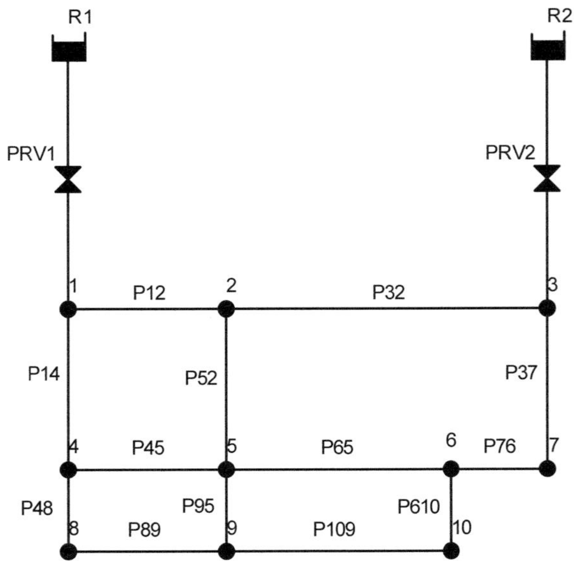

Fig. 4.3 Dummy meshed network

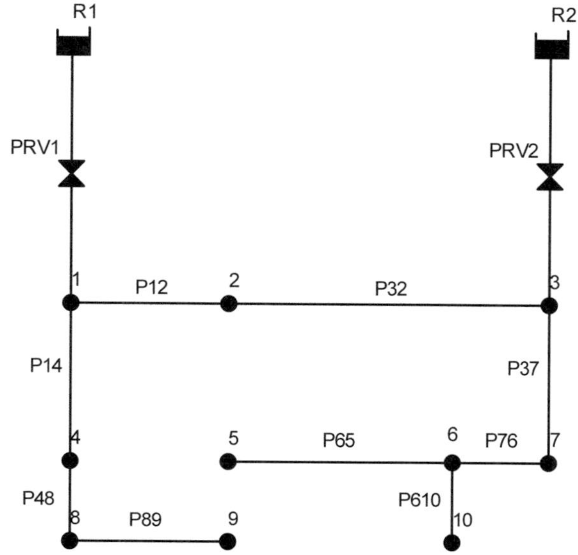

Fig. 4.4 Dummy tree network

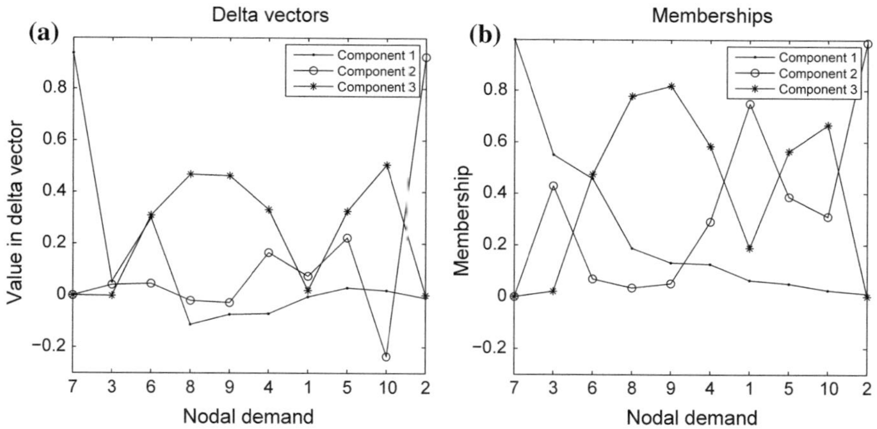

Fig. 4.5 Parameterization process applied to a meshed network: **a** Delta vectors and **b** memberships of each nodal demand to each demand component

Algorithm 4.1 uses the sensitivity matrix computed at a particular working point. The procedure can be applied considering multiple boundary conditions to make the membership definition process more robust. However, the static topology of the network is not expected to produce significant changes in the sensitivity matrix. The application of the same process using other working points for the dummy networks generates the same memberships with only $\pm 1\%$ variations in the memberships.

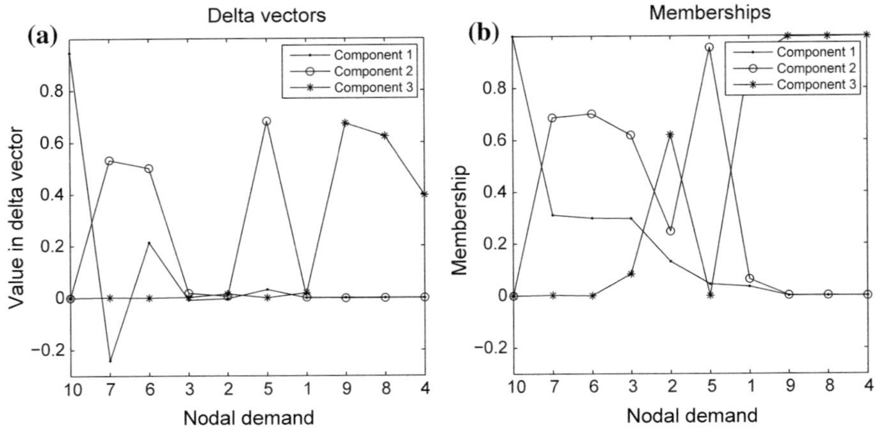

Fig. 4.6 Parameterization process applied to a tree-like network: **a** Delta vectors and **b** memberships of each nodal demand to each demand component

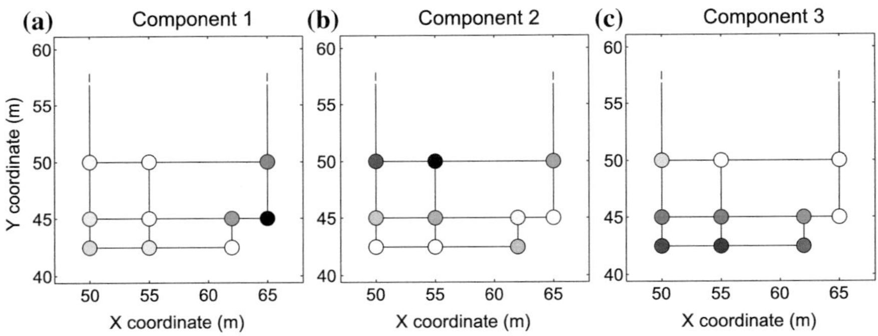

Fig. 4.7 Graphical representation of the nodal memberships to demand components in a meshed network

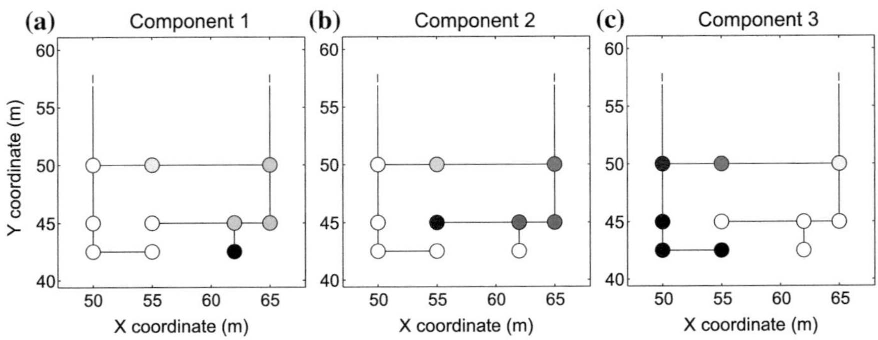

Fig. 4.8 Graphical representation of the nodal memberships to demand components in a tree-like network

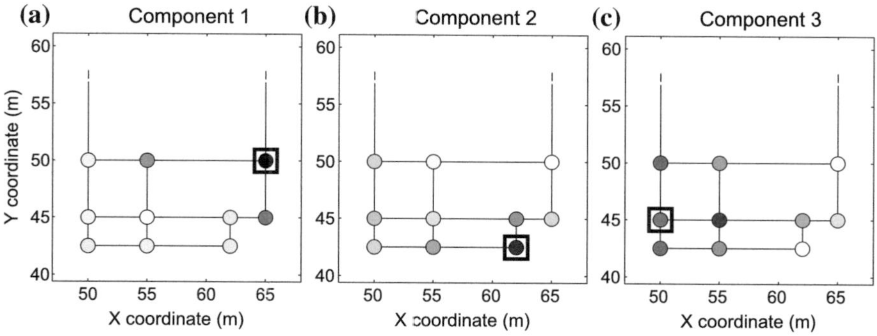

Fig. 4.9 Graphical representation of the nodal memberships to demand components in a meshed network considering three installed sensors

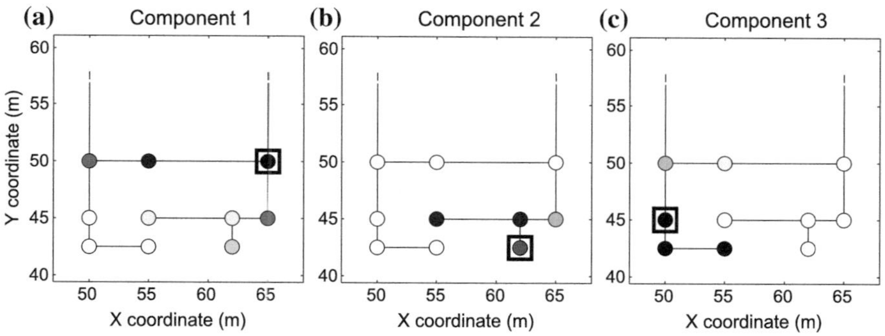

Fig. 4.10 Graphical representation of the nodal memberships to demand components in a tree-like network considering three installed sensors

The calibration methodology requires some inner sensors to be distributed through the sampling design. In case the network already has the sensors installed, the **S** matrix introduced in Algorithm 4.1 would be a reduced sensitivity matrix $\mathbf{S_r}$ where only the rows related to the available sensors would be considered. Figures 4.9 and 4.10 depict the parameterization of the two dummy networks considering that three sensors were already installed in the networks. These sensors are marked with a black square.

Figures 4.11 and 4.12 depict the final results of the parameterization and sampling design process. The sensor selection has been performed after the definition of the demand components (results from the previous section).

4.4.2 Demand Components' Model for a Real Network

In Chap. 3, it has been seen that the most widely used demand models are the basic demand model and the demand patterns' model. The basic demand model

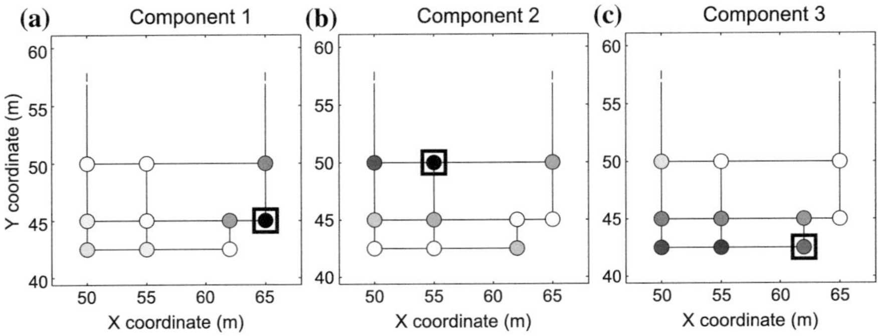

Fig. 4.11 Sensor selection results applied to a meshed network with three demand components

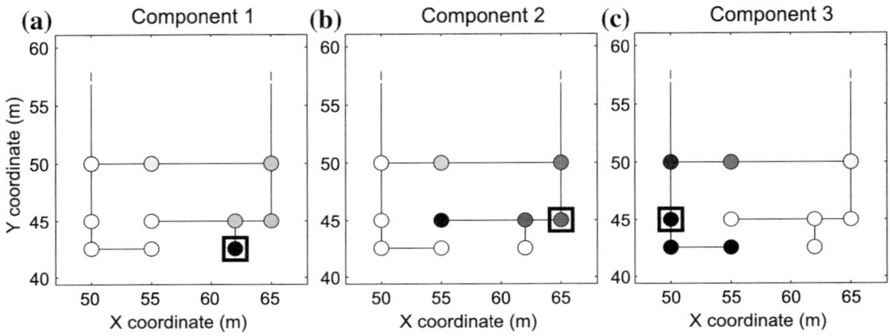

Fig. 4.12 Sensor selection results applied to a tree-like network with three demand components

cannot explain the daily variation of the relative pressure behaviour between two areas in the network, as it fixes the same behaviour to all demands. On the other hand, the demand patterns' model requires a lot of information that is not usually available (users associated with a given node, type of users) or does not fulfils the assumptions (incorrect predetermined diurnal demand patterns' values, users of the same type behaving differently). An example of the latter is presented in Fig. 4.13: automatic metre readings from two different segments (i.e., types of users) from a real network (Nova Icària) presented in Chap. 2 have been analysed. Each reading consists of the daily water consumption of a specific user, metered hourly. The correlation between every pair of readings within the same segment has been computed to assess the distance between their profiles, i.e., the similarity or dissimilarity of the users' behaviours. In each subfigure, the x-axis presents the users' telemetries, and each dot in the y-axis indicates the correlation between the user and all the other users in the same segment: the higher the correlation, the higher the similarity with its own segment's profiles.

Figure 4.13a presents a type of user with no similarity between its members, whereas Fig. 4.13b shows a type of user with more similarity between its members, but not enough to assume that all of them behave in the same way. In conclusion, the

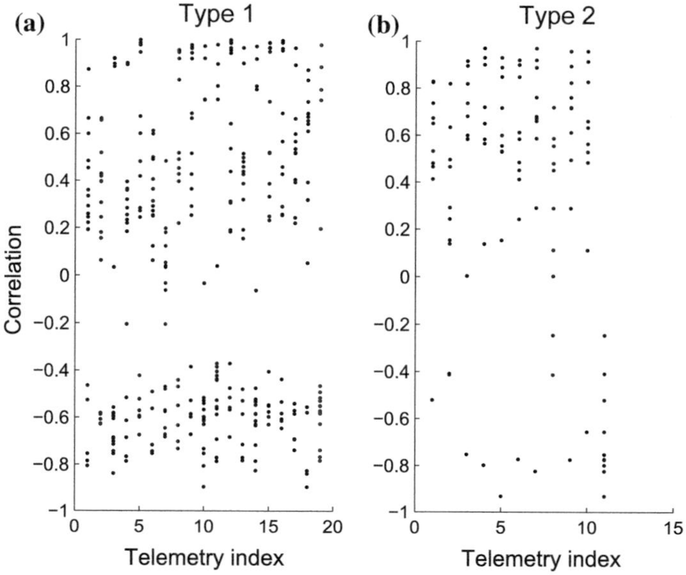

Fig. 4.13 Cross-correlations of Nova Icària DMA telemetries in users of segments **a** 1 and **b** 2

assumption of considering that all users of the same type behave in the same way can lead to incorrect results or high uncertainty in the calibrated parameters.

A new approach to model demands depending on their geographical location is presented, and their sensitivity to hydraulic variables. Initially, nodes in a specific zone of the network were assigned to a specific behaviour, which from now on will be called demand component. This produces a new model

$$d_i(t) = \frac{bd_i}{\sum_{k=1}^{n_d} bd_k} c_{j\to i}(t)\, q^{in}(t), \tag{4.6}$$

where $c_{j\to i}(t)$ is the value of the demand component j associated with node i depending on the node location. Demand components are calibrated demand multipliers that represent the behaviour of nodes in a determined geographical zone, avoiding the dependency on information of the user type and diurnal pattern behaviour. All nodes in the same area of node i have the same associated demand component. Consequently, all nodes in the same zone will have the same demand behaviour, weighted depending on their base demand. This demand model is capable of generating pressure variations in different zones of the network, as it happens in a real situation. Figure 4.14 presents a network where three demand components have been defined. Each subplot presents the set of nodes that are modulated by the same demand component according to (4.6).

However, the assumption that all nodes in the same area behave exactly in the same way is not realistic. For example, a node in the limit of the effect zone of two

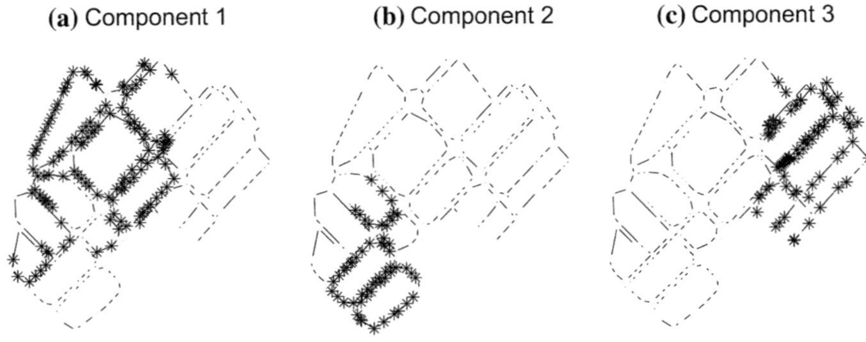

Fig. 4.14 Example of demand components with binary memberships

demand components should probably have a combination of the behaviour of the two demand components, instead of only one. To solve that, it is possible to redefine the demand model in (4.6), so that the degree to which each demand component is associated with each node is given as a membership, which depends on the nodes' geographical location. Thus, (4.7) represents the new demand model which can be written as follows:

$$d_i(t) = \frac{bd_i}{\sum_{k=1}^{n_d} bd_k} q^{in}(t) \left(\alpha_{i,1} c_1(t) + \alpha_{i,2} c_2(t) + \cdots + \alpha_{i,n_c} c_{n_c}(t)\right), \tag{4.7}$$

with

$$\alpha_{i,1} + \alpha_{i,2} + \cdots + \alpha_{i,n_c} = 1, \quad \forall i,$$

where $\alpha_{i,j}$ is the association of demand component j with node i, and n_c is the number of demand components. The membership $\alpha_{i,j}$ of each node to each demand component depends on the geographical location of the node and is computed by means of the sensitivity analysis presented in Sect. 4.3.1. The model in 4.7 can generate different behaviours in every demand, while only having to calibrate few (n_c) demand components.

This way of calibrating demands incorporates the usually ignored fact that demands depend in some ways of head status of the network [17]. For example, if the pressure in a specific zone of the DMA decreases, the calibration process will estimate demand component values that decrease the consumption of nodes in that zone. Demand components presented in this chapter should not be confused with the ones defined in [17], where demand components were generated with a previous knowledge of the use of water (human-based, volume-based, non-controlled orifice-based, leakage-based).

The calibrated demand components generate individual demands that may not be exactly as the real ones, but the aggregated demand in a zone at a specific sample and the cumulative demand of each individual node during a period of time (similar to the billing) should coincide with the real ones if other parameters (roughness, valve status, etc.) are well calibrated.

Figure 4.15 presents the nodes' memberships to three demand components defined in the network in Fig. 4.14. The first component is located on the north-west side of the DMA; the second component is located on the south-west of the DMA; and the third component is located on the east side of the network. The nodes' memberships are depicted in greyscale: the darker the colour of a node, the higher the membership of that node to the demand component. Table 4.1 contains the memberships of the two nodes highlighted in Fig. 4.15. Demand of node A is modulated (60%) by the value of demand component 1, while component 3 has a lower (35%) effect on it. On the other hand, demand of node B is completely (98%) modulated by demand component 3. Demand component 2 does not have any effect on both demands, as it is far (geographically and hydraulically) from the two example nodes. Note the similarity between binary demand components (Fig. 4.14) and hybrid demand components (Fig. 4.15).

A comparison of the calibration results between type of user-based demand patterns and pressure sensitivity-based demand components is presented in [43], with better results for the latter: the uncertainty in the calibrated parameters is reduced, while the geographical distribution is useful for applications requiring parameters to be related with zones of the network.

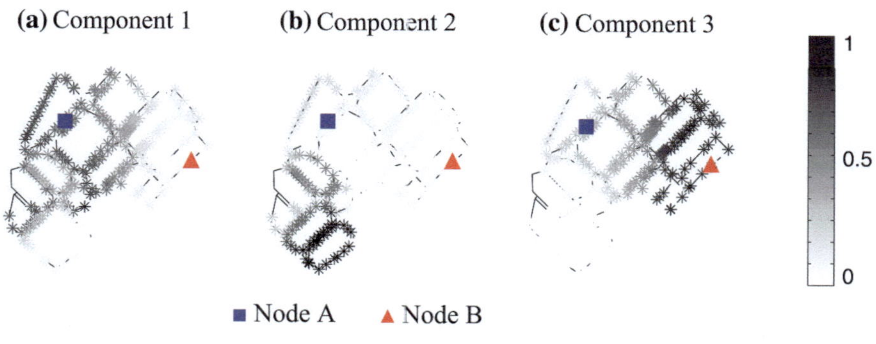

Fig. 4.15 Example of demand components and memberships in a network

Table 4.1 Memberships of nodes A and B of the example network

Node	A	B
Membership to $\mathbf{c}_1$	0.6	0.01
Membership to $\mathbf{c}_2$	0.05	0.01
Membership to $\mathbf{c}_3$	0.35	0.98

4.4.2.1 Sampling Design

The parameterization and sampling design processes were performed to propose the location of three pressure sensors and the parameter definition for demand calibration, as explained in Sects. 4.3.1 and 4.3.2. However, the proposed sensors' locations differ from the final ones, which have been obtained from the methodology in [7] (based on leak detection), developed by Cetaqua (Water Technology Centre of Aguas de Barcelona and the Suez group). The installed sensors can still be used to calibrate demands by defining the demand components depending on the available sensors' locations, thanks to the versatility of the proposed method (Sect. 4.3.1). Figure 4.16 depicts the proposed sensors' locations with circles and the final locations with stars.

The resolution of the sensors is 0.1 mwc (meters of water column), and the sampling times are defined in Table 3.2.

4.4.2.2 Data Analysis

Data from 9 March 2015 to 13 March 2015 (Monday–Friday) are used for the calibration process. Data from the following week, 16 March 2015 to 20 March 2015 (Monday–Friday), are used to validate and analyse the calibrated demand components. Previously, weekdays from 3 March 2015 to 6 March 2015 (Tuesday–Friday) are used to analyse and correct the data coming from the network, and to perform the parameterization process before the calibration starts. 2 March 2015 (Monday) is not used due to missing data. These three weeks will be referred, in current and next chapters, as precalibration week, calibration week and validation week. Weekends

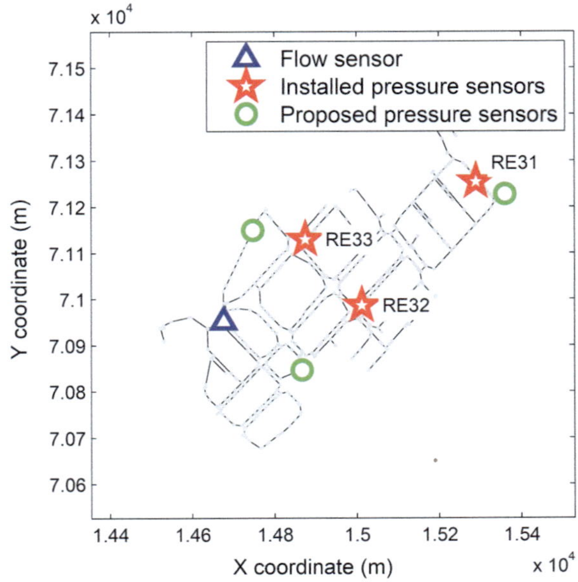

Fig. 4.16 EPANET network model of Canyars sector with highlighted sensor locations. The network water input is signalled with a *triangle*, the installed pressure sensors are signalled with *stars* and the proposed pressure sensors are signalled with *circles*. The flow sensor is installed at the input pressure reduction valve, so that the total flow consumed in the network is known

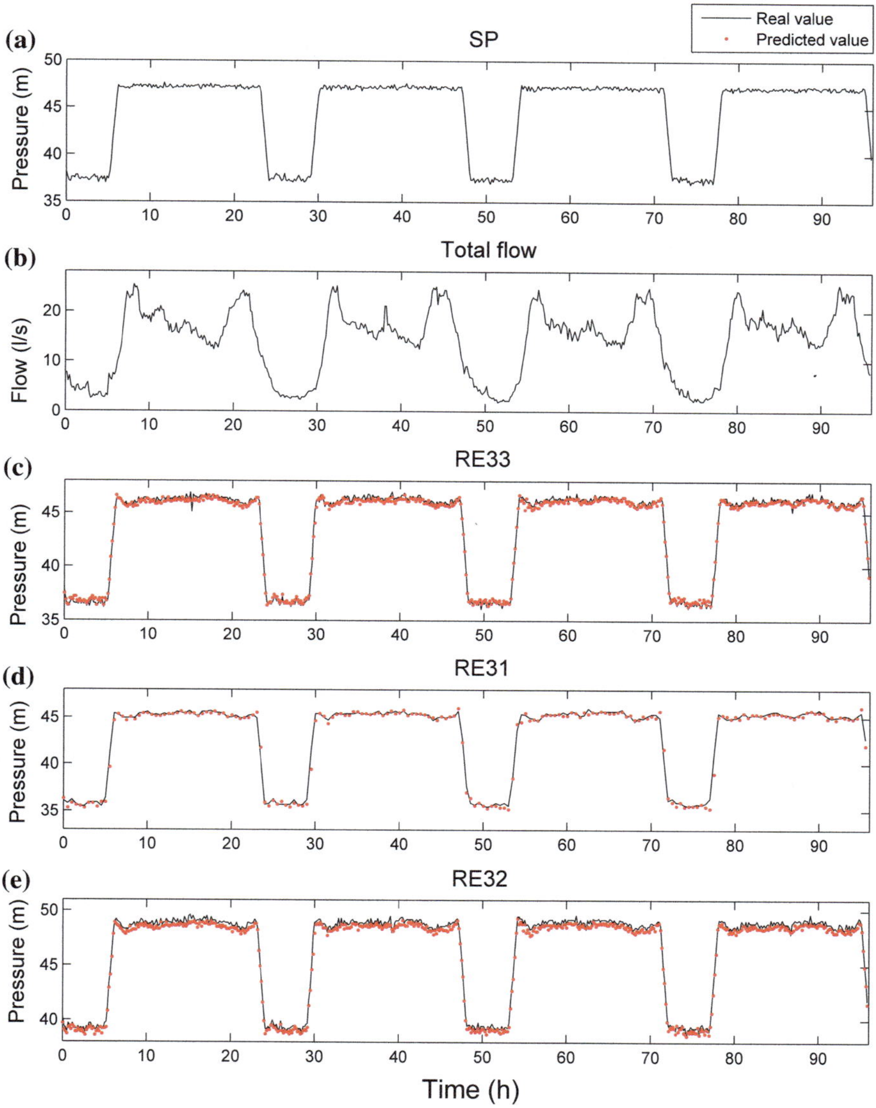

Fig. 4.17 Canyars network real and predicted data from 3 March 2015 to 6 March 2015 (precalibration week). *Black lines* and *red dots* refer to real and predicted data, respectively

are not considered in this case study but would follow the same calibration procedure as weekdays.

Figure 4.17 shows the complete set of data from the precalibration week, including boundary conditions (input valve's pressure set point (SP) and total flow) and the three pressure measurements. Black lines and red dots refer to real and predicted data, respectively. Predicted data have been obtained from simulating the network model with the given boundary conditions using the basic demand model presented in Chap. 3. Figure 4.17a shows the aforementioned pressure control at the DMA input.

Figure 3.3 shows the pressure prediction error in the three available sensors when using the basic demand model. The blue thin line corresponds to the raw error using all data, and the red thick line represents the smoothed error, which has been computed by means of a smoothing spline. The green dashed line corresponds to the mean pressure prediction error. This error is treated as an offset that cannot be associated with the demand model. As suggested in [39], the offset is corrected to eliminate possible depths errors, model nodes' elevations inaccuracies or badly calibrated sensors' offsets. The same correction in each sensor is also considered when using data from the calibration and validation weeks. Table 3.2 contains the specific correction for each sensor.

4.4.2.3 Parameterization

Data from the precalibration week are used to compute the sensitivity matrices to perform the parameterization process. The memberships of each nodal demand to three demand components are computed using Algorithm 4.1, considering the three installed sensors. Figure 4.18 depicts, in each of the network maps, the membership of each node to a particular demand component: the darker the node, the higher the membership to that component. Each map in Fig. 4.18 also includes the location of the sensor with the highest sensitivity to the component drawn.

The average percentage of consumption $\overline{d}_{c_j}$ of demand component j is computed from the billing information (nodal base demands **BDM**) and the recently computed memberships (**M**) as

$$\overline{d}_{c_j} = 100 \sum \mathbf{BDM}\,\mathbf{M}_{(:,j)}. \tag{4.8}$$

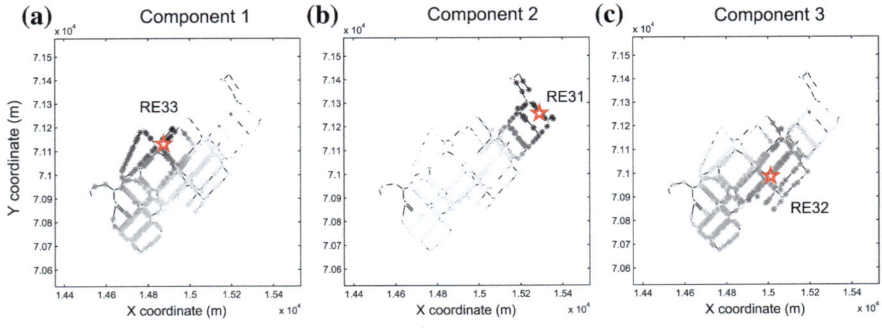

Fig. 4.18 Memberships of nodes to each demand component in Canyars network considering the three available sensors. Each representation of the network depicts a *greyscale* map with the membership of each node to a particular demand component: the *darker* the node in the map, the higher the membership of the node to the demand component. The sensor with the highest sensitivity to variations in each demand component is also depicted in each map

Table 4.2 Average percentage of demand components' water consumption in Canyars network computed from billing

Demand component	c_1	c_2	c_3
Average percentage of water consumption	40.2%	18.6%	41.2%

Table 4.2 sums up the average percentage of consumption of each demand component: demand component c_2 has the lowest percentage of consumption (18.6%), whereas c_1 and c_3 both have roughly a 40% of consumption. This information will be used to analyse the calibration results: errors in the average percentage of consumption of the calibrated demand components compared to the assumed consumption in Table 4.2 can be assigned to background leakage, burst, fraudulent consumptions, unknown status valves, non-metered users or wrong billing information.

4.5 Conclusions

One of the main issues in a calibration process of a complex system is to assure the identifiability. Here, a redefinition of the parameters to be calibrated in order to reduce its number is used. This new set of parameters is defined by the measurements that are available. If the sensor distribution is a part of the process, an optimal sensor distribution is provided in a straightforward way using the information matrix. Nonetheless, if the sensors are already installed, the parameter definition can adapt optimally to the information available.

The calibration problem is formulated as an optimization. The solution of this problem includes nonlinear equality constraints, and thus, it is not a convex one. Much research in this area is going on and results get seldom to the real applications because of the difficult trade-off between computational effort and reliability of the resulting models. Chapter 5 presents the state of the art and an original approach to this problem.

References

1. Ahmed I, Lansey K, Araujo J (1999) Data collection for water distribution network calibration. In: 2nd International conference on water pipeline systems, pp 271–278, Exeter
2. American Water Works Association Research Committee on Distribution Systems (1974) Water distribution research and applied development needs. J Am Water Works Assoc 66(6):385–390
3. Aster R, Borchers B, Thurber C (2005) Parameter estimation and inverse problems. Elsevier, New York
4. Bard Y (1974) Nonlinear parameter estimation. Academic Press, San Diego, California
5. Bargiela A (1985) An algorithm for observability determination in water-system state estimation. Control Theory Appl 132(6):245–250

6. Behzadian K, Kapelan Z, Savic D, Ardeshir A (2009) Stochastic sampling design using a multi-objective genetic algorithm and adaptive neural networks. Environ Modell Softw 24(4):530–541
7. Bonada E, Meseguer J, Mirats Tur JM (2014) Practical-oriented pressure sensor placement for model-based leakage location in water distribution networks. In: Piasecki M (ed) Informatics and the environment: data and model integration in a heterogeneous hydro world. New York
8. Bush C, Uber J (1998) Sampling design methods for water distribution model calibration. J Water Resour Plan Manage 124(6):334–344
9. Carpentier P, Cohen G (1991) State estimation and leak detection in water distribution networks. Civil Eng Syst 8(4):247–257
10. Chen LC (1995) Pipe network transient analysis—the forward and inverse problems. PhD thesis, Cornell University
11. Cheng W, He Z (2011) Calibration of nodal demand in water distribution systems. J Water Resour Plan Manage 137(1):31–40
12. Datta R, Sridharan K (1994) Parameter estimation in water distribution systems by least squares. J Water Resour Plan Manage 120(4):405–422
13. de Schaetzen WBF, Walters GA, Savic DA (2000) Optimal sampling design for model calibration using shortest path, genetic and entropy algorithms. Urban Water J 2(2):141–152
14. Del Giudice G, Di Cristo C (2003) Sampling design for water distribution networks. Trans Ecol Environ 61
15. Eggener C, Polwoski L (1976) Network models and the impact of modeling assumptions. J Am Water Works Assoc 68(4):189–196
16. Ferreri G, Napoli E, Tumbiolo A (1994) Calibration of roughness in water distribution systems. In: 2nd International conference on water pipeline systems, pp 379–396
17. Giustolisi O, Walski T (2012) Demand components in water distribution network analysis. J Water Resour Plan Manage 138(4):356–367
18. Goulet J-A, Coutu S, Smith IFC (2013) Model falsification diagnosis and sensor placement for leak detection in pressurized pipe networks. Adv Eng Inform 27(2):261–269
19. Griewank A, Juedes D, Mitev H, Utke J, Vogel O, Walther A (1998) ADOL-C: a package for the automatic differentiation of algorithms written in C/C++
20. Hutton CJ, Kapelan Z, Vamvakeridou-Lyroudia L, Savić DA (2014) Dealing with uncertainty in water distribution system models: a framework for real-time modeling and data assimilation. J Water Resour Plan Manage 140(2):169–183
21. Kang D, Lansey K (2010) Optimal meter placement for water distribution system state estimation. J Water Resour Plan Manage 136(3):337–347
22. Kapelan Z, Savic D, Walters G (2003) Multiobjective sampling design for water distribution model calibration. J Water Resour Plan Manage 129(6):466–479
23. Kapelan Z, Savic D, Walters G (2003) A hybrid inverse transient model for leakage detection and roughness calibration in pipe networks. J Hydraul Res 41(5):481–492
24. Kapelan Z, Savic D, Walters G (2005) Optimal sampling design methodologies for water distribution model calibration. J Hydraul Eng 131(3):190–200
25. Krumpholz G, Clements K, Davis P (1980) Power system observability: a practical algorithm using network topology. IEEE Trans Power Apparatus Syst PAS-99(4):1534–1542
26. Lansey K, Basnet C (1991) Parameter estimation for water distribution networks. J Water Resour Plan Manage 117(1):126
27. Lansey K, El-Shorbagy W, Ahmed I, Araujo J, Haan C (2001) Calibration assessment and data collection for water distribution networks. J Hydraul Eng 127(4):270–279
28. Liggett J, Chen L (1994) Inverse transient analysis in pipe networks. J Hydraul Eng 120(8):934–955
29. Loaiciga H, Charbeneau R, Everett L, Fogg G, Hobbs B, Rouhani S (1992) Review of ground water quality monitoring network design. J Hydraul Eng 118(1):11–37
30. Mallick K, Ahmed I, Tickle K, Lansey K (2002) Determining pipe groupings for water distribution networks. J Water Resour Plan Manage 128(2):130–139

31. Meier R, Barkdoll B (2000) Sampling design for network model calibration using genetic algorithms. J Water Resour Plan Manage 126(4):245–250
32. Nejjari F, Sarrate R, Blesa J (2015) Optimal pressure sensor placement in water distribution networks minimizing leak location uncertainty. Procedia Eng 119:953–962
33. Ormsbee L (1989) Implicit network calibration. J Water Resour Plan Manage 115(2):243–257
34. Ostfeld A, Salomons E, Ormsbee L, Uber J, Bros C, Kalungi P, Burd R, Zazula-Coetzee B, Belrain T, Kang D, Lansey K, Shen H, McBean E, Wu ZY, Walski T, Alvisi S, Franchini M, Johnson J, Ghimire S, Barkdoll B, Koppel T, Vassiljev A, Kim JH, Chung G, Yoo DG, Diao K, Zhou Y, Li J, Liu Z, Chang K, Gao J, Qu S, Yuan Y, Devi Prasad T, Laucelli D, Vamvakeridou Lyroudia L, Kapelan Z, Savic D, Berardi L, Barbaro G, Giustolisi O, Asadzadeh M, Tolson B, McKillop R (2012) Battle of the water calibration networks. J Water Resour Plan Manage 138(5):523–532
35. Ozawa T (1987) The principal partition of a pair of graphs and its applications. Discrete Appl Math 17(1–2):163–186
36. Pérez R (2003) Identifiability and calibration of water network models. PhD thesis, Universitat de Catalunya
37. Pérez R, Puig V, Pascual J, Peralta A, Landeros E, Ll Jordanas (2009) Pressure sensor distribution for leak detection in Barcelona water distribution network. Water Sci Technol: Water Supply 9(6):715
38. Pérez R, Sanz G (2014) Optimal placement of metering devices for multiple purposes. In: 11th International conference on hydroinformatics, New York
39. Pérez R, Sanz G, Puig V, Quevedo J, Escofet MAC, Nejjari F, Meseguer J, Cembrano G, Mirats Tur JM, Sarrate R (2014) Leak localization in water networks: a model-based methodology using pressure sensors applied to a real network in Barcelona [applications of control]. IEEE Control Syst 34(4):24–36
40. Piller O, Bremond B, Morel P (1999) A spatial sampling procedure for physical diagnosis in a drinking water supply network. In: Savic DA, Walters GA (eds) Water industry systems: modelling and optimization applications, pp 309–316. Exceter
41. Piller O, Deuerlein J, Gilbert D, Weber J-M (2015) Installing fixed sensors for double calibration and early-warning detection purposes. Procedia Eng 119:564–572
42. Pinzinger R, Deuerlein J, Wolters A, Simpson A (2011) Alternative approaches for solving the sensor placement problem in large networks. In: Water distribution systems analysis 2011
43. Sanz G, Pérez R (2014) Comparison of Demand Pattern Calibration in Water Distribution Network with Geographic and Non-Geographic Parameterization. In: 11th International conference on hydroinformatics, New York
44. Sanz G, Pérez R (2014) Demand pattern calibration in water distribution networks. Procedia Eng 70:1495–1504
45. Sanz G, Pérez R (2014) Parameterization and sampling design for water networks demand calibration using the singular value decomposition: application to a real network. In: 11th International conference on hydroinformatics, New York
46. Sanz G, Pérez R (2015) Sensitivity analysis for sampling design and demand calibration in water distribution networks using the singular value decomposition. J Water Resour Plan Manage 04015020
47. Savic D, Kapelan Z, Jonkergouw P (2009) Quo vadis water distribution model calibration? Urban Water J 6(1):3–22
48. Shamir U, Howard C (1977) Engineering analysis of water-distribution systems. J Am Water Works Assoc 69(9):510–514
49. Sorenson H (1980) Parameter estimation: principles and problems. Marcel Dekker, New York
50. Sumer D, Lansey K (2009) WDS calibration and assessment for alternative modelling objectives. Urban Water J 6(4):265–277
51. Walski T (1983) Technique for calibrating network models. J Water Resour Plan Manage 109(4):360
52. Walski T (1985) Assuring accurate model calibration. J Am Water Works Assoc 77(12):38–41

53. Walski T (1995) Standards for model calibration. In: American water works association computer conference, Norfolk
54. Walski T (2000) Model calibration data: the good, the bad, and the useless. J Am Water Works Assoc 92(1):94–99
55. Walski T, Sage P, Zheng W (2014) What does it take to make automated calibration find closed valves and leaks? World Environ Water Resour Congr 2014:555–565
56. Walter E, Pronzato L (1996) On the identifiability and distinguishability of nonlinear parametric models. Math Comput Simul 42(2–3):125–134
57. Wiggins R (1972) The general linear inverse problem: Implication of surface waves and free oscillations for Earth structure. Rev Geophys 10(1):251–285
58. William Y (1986) Review of parameter identification procedures in groundwater hydrology: the inverse problem. Water Resour Res 2:95–108
59. Yu G, Powell R (1994) Optimal design of meter placement in water distribution systems. Int J Syst Sci 25(12):2155–2166

Chapter 5
Parameter Estimation: Online Calibration

Gerard Sanz and Ramon Pérez

5.1 Introduction

In Chap. 4, the calibration problem was defined and posed in order to assure identifiability. The parameters to be estimated have been adapted to the measurements available. Furthermore, in case that more sensors could be installed, an optimal sensor placement algorithm has been presented based on the information provided by the sensitivity matrix. Nevertheless, even with a well-posed calibration problem, the optimization required in order to minimize the prediction error of the model requires a proper selection of the optimization algorithm. There are plenty of approaches in the literature that are presented in the following review. The SVD, used in previous chapter for the analysis of the system, is used in the proposed approach. This decomposition is useful for the uncertainty propagation evaluation needed in any parameter estimation process.

5.1.1 Calibration Methods

Global calibration problem is quite well presented by Savic et al. [38]. Methods are classified depending on their dynamics (static/transient) and depending on the optimization methods (prove/explicit/implicit).

Iterative calibration models [6, 35, 47, 48] are based on trial-and-error procedure. Unknown parameters are updated at each iteration using heads and/or flows obtained by solving the set of steady-state mass balance and energy equations. Iterative

G. Sanz · R. Pérez (✉)
Research Center "Supervision, Safety and Automatic Control" (CS2AC-UPC), Terrassa, Spain
e-mail: ramon.perez@upc.edu

V. Puig et al. (eds.), *Real-Time Monitoring and Operational Control of Drinking-Water Systems*, Advances in Industrial Control,
DOI 10.1007/978-3-319-50751-4_5

calibration models have been the base in the establishment of some fundamental principles and guidelines regarding WDN model calibration and have been utilized in the development of more sophisticated explicit and implicit methods.

Explicit calibration models [7, 12, 32] are based on solving an extended set of steady-state mass balance and energy equations. This extended set is solved explicitly, usually by the Newton–Raphson method. The main disadvantages and limitations are as follows: (1) The number of parameters to be calibrated must be equal to the number of measurements, (2) measurements errors are not taken into account, and (3) there is no way to quantify uncertainty in the estimated parameters. In conclusion, explicit methods only have historical significance and no apparent influence on the current practice of model calibration [38].

Implicit methods are formulated and solved using an optimization technique coupled with a hydraulic solver. The optimization tool sets/updates parameters and provides them onto the simulation model, which in turn returns the obtained predicted variables. The optimization tool employs an objective function to minimize the differences between measured and model-predicted variables. The type of optimization methods used vary from local search methods, through mathematical optimization to global search methods.

5.1.1.1 Non-evolutionary Optimization Methods

Multiple types of non-evolutionary methods exist, but gradient-based optimization seems to be dominant. General Reduced Gradient (GRG) is used in [23, 40]. The Gauss–Newton method and the improved version of Levenberg–Marquardt are the most used gradient type methods. These methods are used to solve the linear/nonlinear least squares formulation of the inverse problem [9–11, 18, 19, 24, 25, 33, 34, 36]. The influence of different choice of weights in the WLS case is investigated and a systematic procedure is given for the selection of suitable weights in [10, 36]. The nonlinear least squares problem is a non-convex problem with multiple optima for the objective function, and it is generally solved as an iterative procedure. To guarantee that the minimum found is the global minimum, the process should be started with widely differing initial values of the parameters. When the same minimum is found regardless of the starting point, it is likely to be the global minimum. Other non-evolutionary techniques used for calibration of hydraulic distribution models include the extended complex method of Box [31], linear and nonlinear programming [15], Kalman filtering [43] and simulated annealing [44].

5.1.1.2 Evolutionary Optimization Methods

Evolutionary optimization methods [13] were introduced in the area of calibration of WDN models by Savic and Walters [39]. Genetic Algorithms (GA), in particular, have been used later on in hydraulic models calibration works [20, 26, 42, 46, 49].

The main advantages of the evolutionary methods over non-evolutionary methods are the lack of complex mathematical apparatus to evaluate sensitivities or invert matrices, the ability to handle with large calibration problems, easy inclusion of additional calibration parameter types and constraints and the opportunity of being implemented in multiple computational machines due to their parallel nature [38]. However, [49] noted some disadvantages: not guaranteeing the achievement of the global optimum in large and complex systems, required for a careful setting up and parameter tuning to obtain a correct operation, not being suitable for small optimization problems, and the less computational efficiency when compared to the gradient-based methods.

The current status and future research directions of evolutionary algorithms and other meta-heuristics for better solving key water resources problems are presented in [27].

5.1.2 *Uncertainty*

In calibration, inaccuracy of the input data causes the results to be inaccurate too. Therefore, it is important to obtain not only the estimated values of the calibration, but also an indication of how reliable these estimations are.

A framework is presented in [16] where each stage of model development is considered, and the most promising methods available to quantify and reduce uncertainty at each of these stages are reviewed(calibration, data assimilation and model forecasting). A discussion linking the impact of calibration uncertainties to model decisions allows examining the implications of the uncertainties on future data collection efforts [41].

Three methods for confidence limit analysis are compared in [4]:

- Monte Carlo simulation: uncertainty in model predictions is calculated by a series of simulations where the input parameter's vector has random variations.
- Optimization-based approach: the confidence limits of the estimated values are calculated by means of an optimization problem with the linearized network equations as constraints.
- Sensitivity-based method: analysis of the sensitivity matrix generated from the linearized network equations.

The authors selected the latter approach as the better one due to the improvement on the computational requirements keeping similar results as the other methods. Furthermore, most of the reviewed bibliographies [1, 5, 8, 10, 18, 21, 22, 24, 28, 33, 36, 52] perform the quantification of the parameter and prediction uncertainties based on linear regression theory, using a method known in the literature as the FOSM model (First-Order Second-Moment) [3].

The FOSM model consists in the definition of the first-order approximation of the parameter covariance matrix $\mathbf{Cov}(\hat{\mathbf{x}})$ as

$$\mathbf{Cov}(\hat{\mathbf{x}}) = \mathbf{S}\,\sigma^2\,\mathbf{S}^T, \tag{5.1}$$

where σ^2 is the variance in measured parameters, and $\mathbf{S}$ is the matrix of the sensitivities of the measures relative to the estimated parameters $\hat{\mathbf{x}}$. Uncertainty in the parameter values is obtained from parameter variances in the ith diagonal element of the covariance matrix.

The prediction covariance matrix $\mathbf{Cov}(\hat{\mathbf{y}})$ can be also estimated to obtain the variance of the model prediction:

$$\mathbf{Cov}(\hat{\mathbf{y}}) = \mathbf{S_P}^T\,\mathbf{Cov}(\hat{\mathbf{x}})\,\mathbf{S_p}, \tag{5.2}$$

where $\mathbf{S_P}$ is the matrix of the sensitivities of the predicted values relative to the estimated parameters $\hat{\mathbf{x}}$.

We can compute 95% confidence intervals for the individual model parameters considering that each model parameter x_i has a normal distribution with mean $\hat{x}_i$ and variance $\mathbf{Cov}(\hat{\mathbf{x}})_{i,i}$. The 95% confidence intervals are given by

$$\hat{x}_i \pm 1.96\sqrt{\mathbf{Cov}(\hat{x}(i,i))}, \tag{5.3}$$

where the 1.96 factor is obtained from

$$\frac{1}{\sigma\sqrt{2\pi}}\int_{-1.96\sigma}^{1.96\sigma} e^{-\frac{x^2}{2\sigma^2}}\,dx \approx 0.95. \tag{5.4}$$

This information allows to represent the probability density function (PDF) of the calibrated parameters.

5.2 Problem Statement

The objective of the calibration problem, presented in Chap. 4, is to find the parameter vector $\mathbf{x}$ that minimizes the errors $\varepsilon = \mathbf{y_m} - \mathbf{y_p}(\mathbf{x})$, where $\mathbf{y_m}$ and $\mathbf{y_p}(\mathbf{x})$ are the vectors of measured and predicted values, respectively. Assuming the following linearized relationship between the parameters and the measurements:

$$\mathbf{A}\cdot\mathbf{x} = \mathbf{y}, \tag{5.5}$$

the corrections in parameters $\Delta\mathbf{x}$ that make $\varepsilon \to 0$ are obtained from:

$$\mathbf{S}\cdot\Delta\mathbf{x} = \mathbf{y_m} - \mathbf{y_p}(\mathbf{x}), \tag{5.6}$$

where $\mathbf{S}$ is the sensitivity matrix that relates errors in predictions to corrections in the models' parameters. In nonlinear problems, $\Delta\mathbf{x}$ is calculated iteratively and used to correct the parameter vector $\mathbf{x}$ as follows:

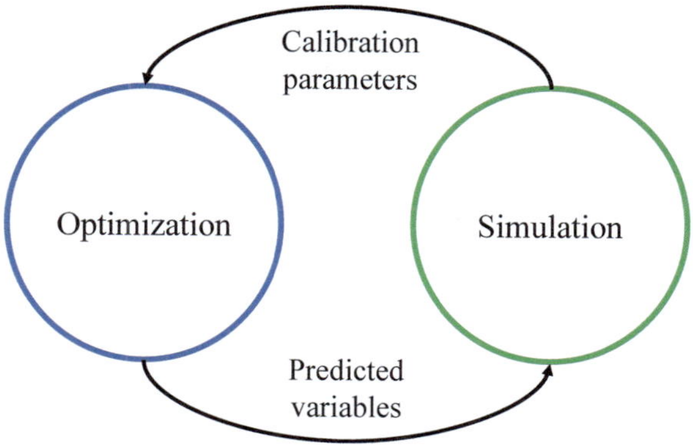

Fig. 5.1 Scheme of the implicit calibration procedure

$$\mathbf{x}(k+1) = \mathbf{x}(k) + \rho \Delta \mathbf{x}(k), \tag{5.7}$$

where k is the iteration number and ρ is a parameter to control the step size. The iterative scheme is continued until a termination criterion is achieved [3]. Figure 5.1 illustrates the calibration process where the optimization tool sets/updates parameters, passes them onto the simulation model, which returns back the model-predicted variables. The optimization tool computes the parameters that minimize a predefined objective function.

5.3 Proposed Approach

The inverse problem may be approached by operating on both sides of the original system of (5.5) with an $(n_x \times n_y)$ "inverse" matrix $\mathbf{H}$ such that the "solution" of model is obtained as follows:

$$\hat{\mathbf{x}} \equiv \mathbf{HAx} = \mathbf{Hy}. \tag{5.8}$$

The operator $\mathbf{H}$ will be a good inverse if it satisfies the following criteria [17]:

(a) $\mathbf{AH} \approx \mathbf{I_{n_y}}$. This is a measure of how well the model fits the data, since $\mathbf{Ax} = \mathbf{y}$ if $\mathbf{AH} = \mathbf{I_{n_y}}$. The information density matrix $\mathbf{I_d}$ presented in Chap. 4 is obtained from this reasoning.
(b) $\mathbf{HA} \approx \mathbf{I_{n_x}}$. This is a measure of the uniqueness of the solution, since there may exist only one solution if $\mathbf{HA} = \mathbf{I_{n_y}}$. The resolution matrix $\mathbf{R}$ presented in Chap. 4 is obtained from this reasoning.
(c) The uncertainties in $\mathbf{x}$ are not too large, i.e., $var(\mathbf{x})$ is small. For statistically independent data,

$$var(x_i) = \sum_{j=1}^{n_y} h_{ij}^2 var(y_j). \tag{5.9}$$

The system in (5.8) can be solved using methods such as Gaussian elimination (even-determined), least squares method (over-determined) or Penrose inverse solution (under-determined). However, none of these solution techniques can be used with rank-deficient or ill-conditioned matrices [2]. The SVD is capable of solving under-, over-, even- or mixed-determined problems with no rank conditions in **S**, as explained by Menke [29]. Equation (5.6) can be solved by using the SVD ($\mathbf{S} = \mathbf{U}\mathbf{\Lambda}\mathbf{V}^T$) as:

$$\Delta\mathbf{x} = \mathbf{V}\frac{1}{\mathbf{\Lambda}}\mathbf{U}^T\varepsilon. \tag{5.10}$$

The SVD determines the optimization direction $\Delta\mathbf{x}$ for a problem that minimizes $||\Delta\mathbf{x}||_2$ and $||\varepsilon||_2$. The SVD was used to solve the inverse problem by Wiggins [51] and Uhrhammer [45] for seismographic networks, by Wasantha Lal [50] for unsteady river flow networks and by [9] for WDN. The SVD provides a deep comprehension of the calibration problem, encouraging its adaptation in the current chapter to estimate demands in WDN.

The SVD matrices can also be used for the estimation of the parameter space covariance matrix to quantify the uncertainty of the calibrated model [45, 51]. The First-Order Second-Moment (FOSM) analysis for uncertainty quantification in (5.1) can be expressed in SVD terms as

$$\mathbf{Cov}(\mathbf{x}) = \mathbf{V}\frac{\sigma^2}{\mathbf{\Lambda}^2}\mathbf{V}^T, \tag{5.11}$$

which can be computed for a single parameter as

$$var(x_j) = \sum_{i=1}^{n_c}\{\sum_{l=1}^{n_c} v_{jl}\lambda_l^{-1}v_{il}\}^2 var(y_i), \tag{5.12}$$

where $1/\lambda_l$ is the lth diagonal element of $1/\mathbf{\Lambda}$, and λ_l is the lth diagonal element of $\mathbf{\Lambda}$ (for $\lambda_l = 0$, the corresponding element of $1/\mathbf{\Lambda}$ is set to 0). A cut-off level for small λ_l is set to avoid $1/\lambda_l$ becoming too large. In this work, the cut-off level is defined at a value of $\lambda = 10^{-3}$ as suggested by Wiggins [51] and Wasantha Lal [50]. If λ_l is quite small, var(x_j) will be quite large. Small λ_l values also have a direct effect on the resolution of the inverse problem. The generalized inverse solution of (5.10) can be written as

$$\Delta\mathbf{x} = \sum_{i=1}^{n_c}\frac{\mathbf{U}_{(:,l)}^T\varepsilon}{\lambda_l}\mathbf{V}_{(:,l)}. \tag{5.13}$$

In the presence of random noise, **y** will generally have a nonzero projection onto each of the directions specified by the columns of **U** (the rows of $\mathbf{U}^T$). The presence of a quite small λ_l in the denominator of (5.13) can thus give a quite large coefficient for the corresponding model space basis vector $\mathbf{V}_{(:,l)}$, and these basis vectors can thus dominate the solution. In the worst case, the generalized inverse solution is just a noise amplifier, and the answer is nonphysical and practically useless. The parametrization presented in Chap. 4 that reduces the parameters to be calibrated prevents the appearance of small λ values.

Online Application

The online calibration of the demand components process can be explained through the scheme in Fig. 5.2. Sensors in the network acquire periodic measurements that are stored in the database through the SCADA system. All data are analyzed to detect missing data, spurious measurements, trends, etc. At a particular day, d, and hour, h, the calibration process extracts from the database a set of measurements and boundary conditions, corresponding d and h. The number of measurements collected by each sensor within that hour depends on the sensor sampling time: the lower the sampling time, the higher the number of measurements taken in an hour. The multiple

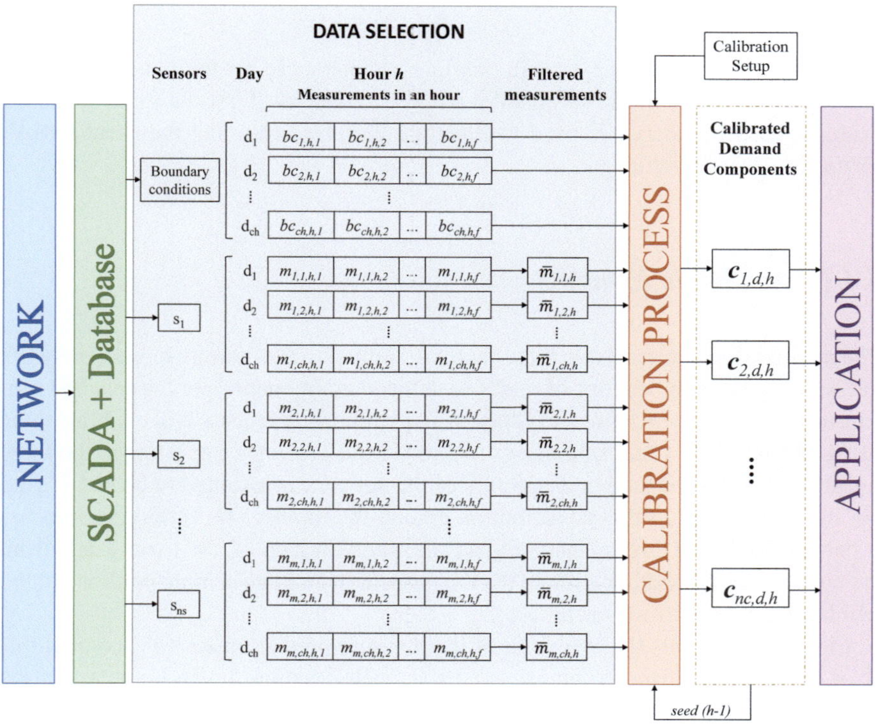

Fig. 5.2 Scheme of the whole calibration process that estimates a set of demand components values. Applications can make use of the calibrated components to improve the hydraulic model and the network performance

samples from the same hour at each sensor are filtered to reduce the sensors' noise and demand uncertainty effect.

Inside the calibration process, a simulation is run for each set of boundary conditions available. The predicted measurements obtained from the simulations are filtered, so that they can be compared with the real filtered measurements. The resulting prediction error is minimized by the correction of the demand components values.

Extra data from previous days at the same h can be used in the calibration process. The calibration horizon, cH, defines the number of days used to calibrate a single set of demand components values. The method assumes that the boundary conditions and demand behaviours from the cH days used are similar. When the calibration process finishes, the resulting demand components values can be used in any model-based application needing a calibrated hydraulic model such as leakage localization [14], quality modelling [30] or leak detection based on demand components analysis [37].

5.4 Simulations and Results

The examples presented next, as in previous chapters, are for both exemplification in a simple network and assessment in a real case studies. First the same academic example used in Chap. 4 is used to illustrate the procedure and then real DMA's demands are calibrated.

5.4.1 *Academic Example*

The methodology presented in this chapter is applied to the dummy meshed network in Chap. 4. The memberships of nodes to demand components and the selected sensors have already been defined in Chap. 4. The calibration process will calibrate three demand components that minimize the error in the predicted pressures at the three selected sensors. The total consumption of the network is assumed to be known, but not the distribution of this consumption among the inputs. The calibration process is performed during 24 h, considering a calibration horizon $cH = 1$ (only data from the current day is used to calibrate the components), and one sample per hour (there will be no filtering inside each hour).

Figure 5.3 presents the evolution of the pressure prediction RMSE and how this error makes the demand components vary at each iteration. It can be seen that the lower the RMSE, the lower the variation of the demand components values.

Once the termination criterion is achieved (maximum number of iterations, or number of iterations with no significant reduction in the RMSE), the calibration for that specific hour is finished. The calibrated demand components values are used as the starting point (seed) for the next hour. Figure 5.4 depicts the calibrated

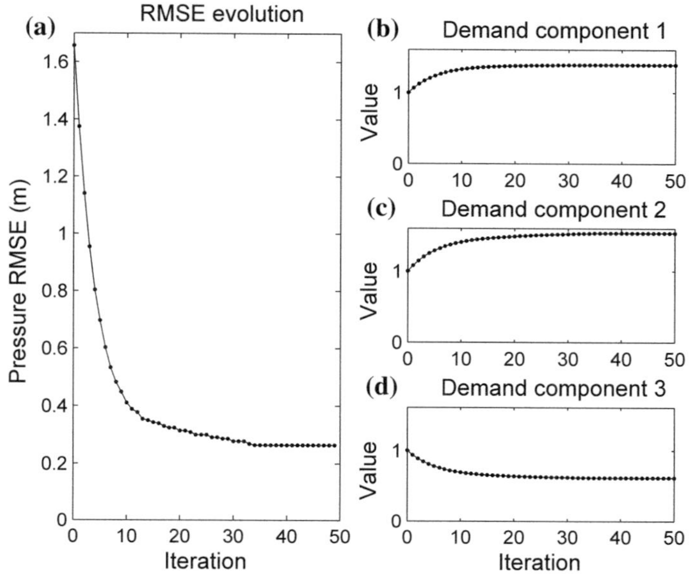

Fig. 5.3 Pressure prediction RMSE and demand components evolution during the iterative calibration applied to the dummy meshed network for $h = 15$

demand components during 24 h, and the 95% confidence intervals (CI) computed as explained in Sect. 5.1.2 and (5.11).

Table 5.1 collects the pressure prediction root mean squared error (RMSE) (first row), and the flow prediction RMSE (second row), for each demand model used (the basic demand model and the demand components model). The percentage of improvement with respect to the basic demand model is also presented. This improvement has been also computed for each type of measurement, which represents the error improvement when using the demand components model instead of the basic demand model.

The calibration process can be repeated, but this time adding the information from the flow distribution at the network inputs. Table 5.2 collects the same information as before, for the calibration that includes the flow measurements. It can be seen that the pressure RMSE improvement is lower compared to the previous results, but the flow RMSE improvement is much better.

Finally, if it is considered data of 5 days ($cH = 5$) with similar expected demand behaviours to calibrate simultaneously the demand components, the resulting uncertainty will be reduced. Figure 5.5 depicts the calibrated demand components during 24 h that minimize simultaneously the hourly pressure prediction error over five days. It can be seen that the confidence intervals when considering $cH = 5$ are narrower than the ones obtained when using $cH = 1$.

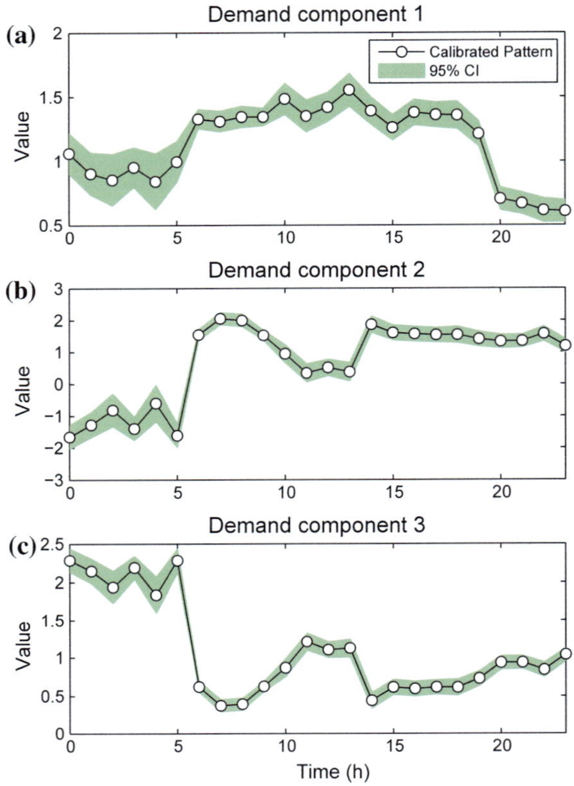

Fig. 5.4 Dummy meshed network calibrated demand components with 95% confidence intervals (*green* boundaries) using $cH = 1$

Table 5.1 Dummy meshed network pressure and flow prediction RMSE

	Basic model	Demand components model	
	RMSE	RMSE	Improvement
Pressure (m)	1.56	0.28	82%
Flow (l/s)	0.5	0.43	14%

Table 5.2 Dummy meshed network pressure and flow prediction RMSE obtained with the calibration that includes the flows measured at the network inputs

	Basic model	Demand components model	
	RMSE	RMSE	Improvement
Pressure (m)	1.56	0.55	64.7%
Flow (l/s)	0.5	0.11	78%

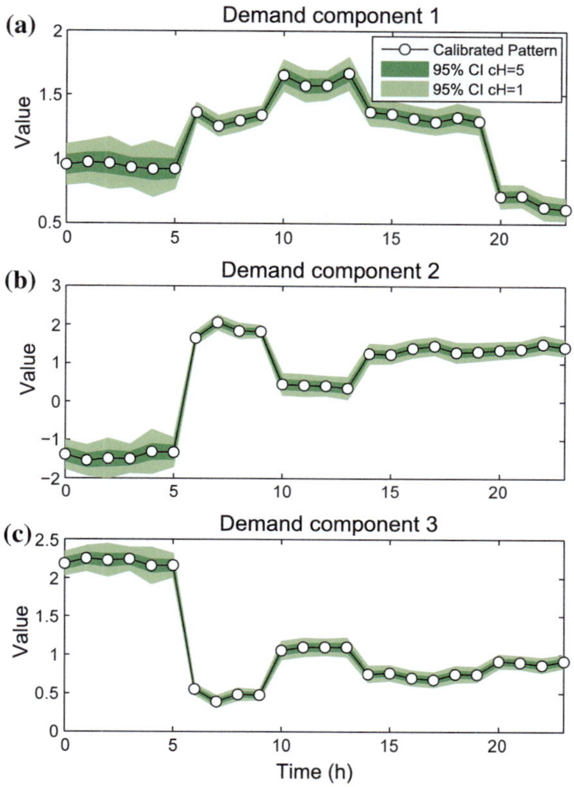

Fig. 5.5 Dummy meshed network calibrated demand components with 95% confidence intervals for $cH = 5$ and $cH = 1$

5.4.2 *Real DMA*

In the Canyars DMA, presented in Chap. 2, water is supplied from the transport network through a pressure reduction valve. Pressure and flow are monitored at the water inlet with a sample time of 10 min. The resolution is 0.3 l/s for the flow sensor, and 0.1 mwc for the pressure sensor. The minimum night flow is of about 3 l/s, and the peak-hour flow is 27 l/s. Pressure control is applied to this network, fixing the pressure level at 38 m during night-time and at 47 m during daytime. The average daily maximum head loss in the network is 13.4 m.

The organization of results in this section is as follows: initially, the sampling design process is performed to choose the sensors to be installed, comparing the solution with the water utility final decision. Second, available data are analyzed and classified for the multiple calibration stages. Then, the parametrization process is performed considering the existence of the installed sensors. Finally, gross errors in sensors are corrected and the online calibration process applied. Results using multiple calibration horizons are compared to analyze the advantages and disadvantages of their use.

Data used were presented in Fig. 4.17. Predicted data have been obtained from simulating the network model with the given boundary conditions using the basic demand model presented in Chaps. 3 and 4. Figure 4.17a shows the aforementioned pressure control at the DMA input. Figure 4.17a shows the aforementioned pressure control at the DMA input. Figure 3.3 shows the pressure prediction error in the three available sensors when using the basic demand model.

5.4.2.1 Calibration Results

Once the parameters to be calibrated are defined, and the information from the network sensors has been thoroughly analyzed and corrected, the online calibration starts. In the current study, the online part is not actually online, though the process and the execution are exactly the same.

Three different tests are considered. In each test, demand components values are calibrated with data from the calibration week and validated with data from the validation week. Furthermore, a new set of demand components values is calibrated using data from the validation week to compare the difference between updating the demand components values or keeping the same values once they are calibrated. Each test considers a different value of the calibration horizon parameter cH, which defines, for each sensor, the number of days from which data for a specific hour are going to be used to calibrate the demand components values for that particular hour:

1. Test 1: $cH = 5$ days. The online calibration starts on 13 March at 00:00 am (Friday). At each hour, data from the current day and four previous days (Monday–Thursday) are used to calibrate a unique value for each demand component. The online calibration continues until 13 March at 23:00 pm (same day), when a total of 24 values will have been calibrated for each demand component. The calibrated demand components values at a specific hour minimize the mean prediction error from that hour of the five days used to calibrate them.
2. Test 2: $cH = 1$ day. The online calibration starts on 9 March at 00:00 am (Monday). At each hour, only data from the current day are used to calibrate the demand components values. The online calibration continues until 13 March at 23:00 pm, when a total of 120 values will have been calibrated for each demand component. The calibrated demand components values at a specific hour minimize exclusively the prediction error for the hour of the day used to calibrate them.
3. Test 3: $cH = 3$ days. The online calibration starts on 11 March at 00:00 am (Wednesday). At each hour, data from the current day and two previous days are used to calibrate a single value for each demand component. The calibration continues until 13 March at 23:00 pm, when a total of 76 values will have been calibrated for each demand component. The calibrated demand components values at a specific hour minimize the mean prediction error from that hour of the three days used to calibrate them. The first 24 demand components values minimize the error in data from Monday to Wednesday, demand components values from 25 h to 48 h minimize the error in data from Tuesday to Thursday, and demand

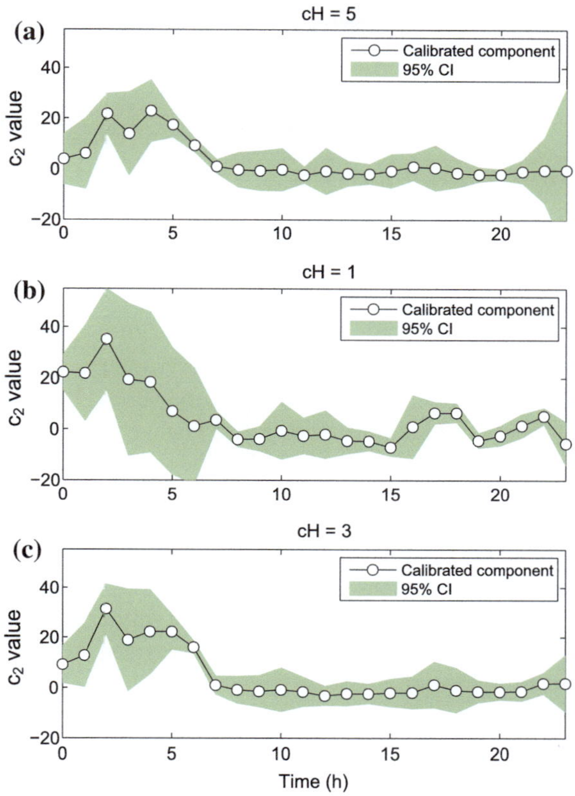

Fig. 5.6 Canyars calibrated demand component $\mathbf{c}_2$ with 95% confidence intervals: **a** $cH = 5$, **b** $cH = 1$ during 24 h, and **c** $cH = 3$ during 24 h

components values from 49 h to 72 h minimize the error in data from Wednesday to Friday.

Figure 5.6 presents an example of the calibrated daily values corresponding to the demand component $\mathbf{c}_2$ during the calibration week: (a) demand component values calibrated with $cH = 5$, that minimize the whole week pressure prediction error; (b) demand component values calibrated with $cH = 1$, that minimize the pressure prediction error during Friday; and (c) demand component values calibrated with $cH = 3$, that minimize the pressure prediction error for three consecutive days. Demand component values are presented as white circles surrounded by the 95% confidence intervals (green boundaries).

Table 5.3 collects the pressure prediction root mean squared error (RMSE) for each test. Each row corresponds to a different test. The three main columns represent the data used, and the use itself: column 1 contains the calibration results for the calibration week, column 2 contains the validation of these results using data from the validation week and column 3 contains the results considering a new calibration during the validation week. Each main column is composed of three subcolumns: the first subcolumn contains the pressure prediction RMSE when using the basic demand model, the second subcolumn contains the pressure prediction RMSE when using the calibrated demand components model and the third subcolumn contains the

Table 5.3 Canyars pressure prediction RMSE and percentage of improvement in the three calibration tests performed

Test	Calibration RMSE (m)			Validation RMSE (m)			Calibration RMSE (m)		
	March 9th–13th			March 16th–20th			March 16th–20th		
	Basic	Comp.	Imprv.	Basic	Comp.	Imprv.	Basic	Comp.	Imprv.
cH = 5	0.298	0.267	11%	0.304	0.282	7%	0.304	0.277	9%
cH = 1	0.298	0.250	16%	0.304	0.296	2%	0.304	0.252	17%
cH = 3	0.294	0.254	13%	0.296	0.294	1%	0.296	0.267	10%

error improvement when using the demand components model instead of the basic model, expressed as a relative percentage.

Figure 5.7 presents a comparison between the pressure prediction error in the calibration week using the basic demand model and the calibrated demand components model, both from Test 2 ($cH = 1$). The depths corrections have been used to correct the sensors' offsets. The columns of subfigures correspond to the basic demand model and demand components model, respectively. Each row of subfigures corresponds to each of the three sensors. The blue lines correspond to the raw errors using all data, and the red lines represent the smoothed errors, which have been computed by means of a smoothing spline. The green dashed lines correspond to the mean pressure prediction errors.

Finally, Fig. 5.8 presents the average percentage of consumption computed for each demand component. The black line corresponds to the assumed percentage previously shown Chap. 4, and the red lines correspond to the demand components' average percentage of consumption in the nine scenarios presented in Table 5.3.

5.4.2.2 Discussion

The basic demand model has been used to simulate the network during the precalibration week. Figure 3.3 shows that the pressure prediction error when using the basic demand model follows the profile of the daily total consumption in Fig. 4.17.b, as this demand model is not able to assign a different behaviour to each zone of the network. The prediction error can be improved by using the demand components model, which allows having multiple demand behaviours depending on the location of nodes in the network.

Results in Table 5.3 show that the use of calibrated demand components to model nodal demands minimizes the RMSE in the predicted pressures in the three tests performed, compared to the basic demand model. This is verified through the positive prediction error improvement in all tests. The comparison of the pressure prediction error between the basic demand model and the demand components model is shown in Fig. 5.7. The pressure prediction error is reduced in the three sensors. However, the prediction improvement in each sensor is different. The proximity of sensor

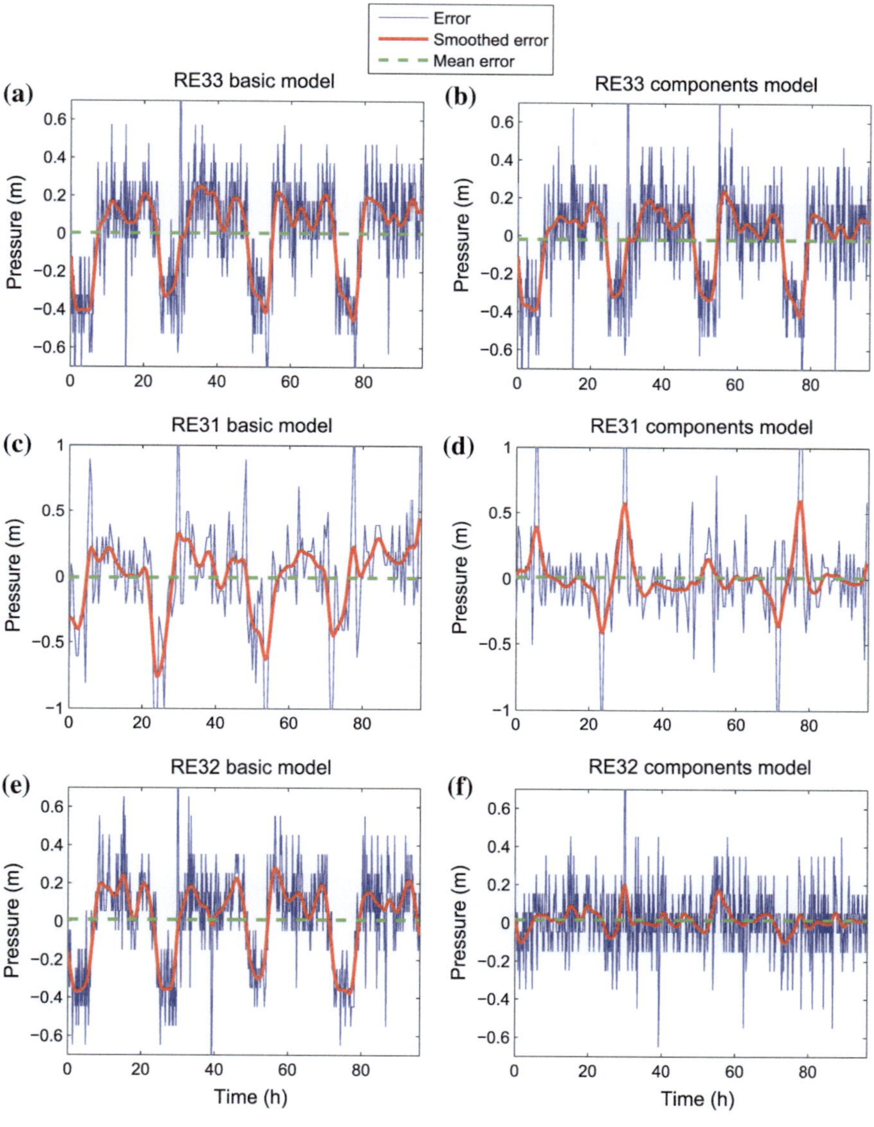

Fig. 5.7 Canyars network pressure prediction error during the calibration week using the basic demand model (column 1) and the demand components model (column 2) using $cH = 1$. The *blue thin line* corresponds to the raw error, the *red thick line* corresponds to the smoothed error computed by means of a smoothing spline and the *dashed green line* corresponds to the mean error. The depth correction has been applied to all sensors

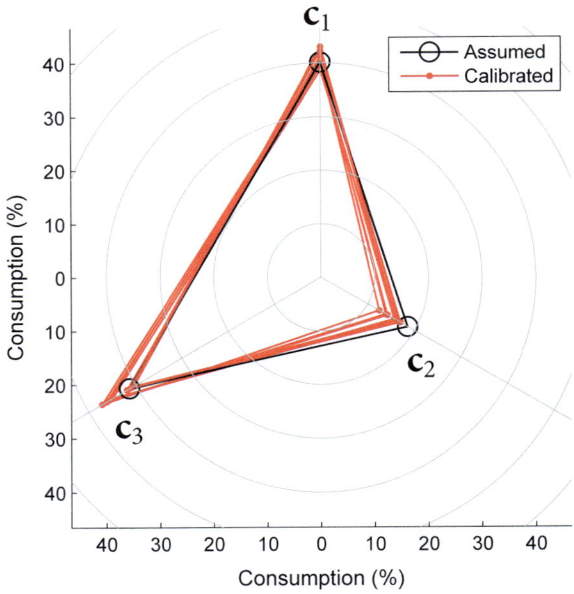

Fig. 5.8 Demand components' average percentage of consumption assumed from billing (*black line with circles*), and obtained with the calibrated demand components in the nine scenarios presented (*red lines with dots*)

"RE33" to the input, where the PRV sets the pressure level, can explain the little improvement. On the other hand, the pressure prediction error in sensors "RE31" and "RE32" is minimized considerably. Peaks in the smoothed prediction error in Fig. 5.7d are due to the single large errors in pressure predictions due to the PRV set point abrupt changes from night to day and vice versa. Summing up, the calibrated demand components model smooths the total water consumption-like profile of the pressure prediction error, as each area of the network has now a distinct behaviour.

The minimization of the prediction error depends on the calibration horizon used: the lower cH is, the higher the improvement on the predicted pressures. In both first and third main columns, Test 2 ($cH = 1$) has the greater improvement. Using data exclusively from one day allows the demand components to focus on minimizing a single error. If demand behaviours change in the real network, Test 2, that uses $cH = 1$, captures instantaneously these changes. On the other hand, Tests 1 and 3 accommodate the demand components values more slowly, since at each sample there are multiple prediction errors to be minimized at the same time. This is also observed in Fig. 5.6, where the profile from Test 2 ($cH = 1$) captures the daytime demand variations with more detail, opposite from Test 1 and 3, where the daytime profile is flatter.

Nevertheless, the validity of the demand components calibrated with higher cH values is longer, allowing the demand components to be applied in future data and still minimize the prediction error. That is because the use of multiple measurements to calibrate a unique set of demand components values leads to demand components that represent the usual behaviour in the network, ignoring special events in particular days that are not repetitive from one week to another. This fact is checked in the second main column in Table 5.3, where Test 1, that uses $cH = 5$, is the one with less error when using the demand components calibrated with data from the previous week.

The comparison between the second and third main columns in Table 5.3 shows that continuously updating the demand model with new online calibrated demand components is better than calibrating a set of demand components and using them with new data, in terms of pressure prediction RMSE. For example, in Test 2, the pressure prediction RMSE during the validation week (second main column) improved 2% compared to the basic model. On the other hand, if the online calibration is performed, the improvement on the pressure prediction RMSE is 17% (Test 2 in third main column). Hence, the continuous online calibration is required to capture the daily changes in the network demand behaviours, such as changes in users habits or new users coming online. Besides, the capture of changes in the network demand behaviours caused by leaks, unnoticed changes in valves status or other special events, can be analyzed and used to detect and locate them.

The uncertainty of the calibrated demand components can be analyzed through Fig. 5.6. During night-time, due to the nonlinear head loss/flow relation, pressure is less sensitive to demand variations. Consequently, the singular values of the sensitivity matrix are smaller, increasing the calibrated parameters variance as seen in (5.12). Using extra data to calibrate demand components (Fig. 5.6a, c) reduces the uncertainty in the calibrated demand components values compared to using data only from the current day (Fig. 5.6b).

Finally, Fig. 5.8 shows that the average percentage of consumption of each demand component in the nine scenarios tested is maintained. The variations observed (5–6%) respecting to the assumed percentage of consumption can be associated with the missing information from weekend demand components, which have not been calibrated in this work, small background leakage or fraudulent consumptions.

5.5 Conclusions

The basic demand model has been used to simulate the network during the pre-calibration week. The prediction error can be improved by using the demand components model, which allows to have multiple demand behaviours depending on the location of nodes in the network.

The minimization of the prediction error depends on the calibration horizon used: the lower cH is, the higher the improvement on the predicted pressures. Nevertheless, the validity of the demand components calibrated with higher cH values is longer, allowing the demand components to be applied in future data and still minimize the prediction error. In conclusion, the selection of a good calibration horizon depends on the final use of the model.

The demand modelling and calibration methodology has to be further tested, including weekends' datasets and multiple network conditions. Furthermore, alternative optimization methods, such as genetic algorithms, have to be tested and compared to the least squares process used.

References

1. Ahmed I, Lansey K, Araujo J (1999) Data collection for water distribution network calibration. In: 2nd International conference on water pipeline systems, pp 271–278, Exeter
2. Aster R, Borchers B, Thurber C (2005) Parameter estimation and inverse problems. Elsevier, New York
3. Bard Y (1974) Nonlinear parameter estimation. Academic Press, San Diego, California
4. Bargiela A, Hainsworth G (1989) Pressure and flow uncertainty in water systems. J Water Resour Plan Manage 115(2):212–229
5. Behzadian K, Kapelan Z, Savic D, Ardeshir A (2009) Stochastic sampling design using a multi-objective genetic algorithm and adaptive neural networks. Environ Modell Softw 24(4):530–541
6. Bhave P (1988) Calibrating water distribution network models. J Environ Eng 114(1):120–136
7. Boulos PF, Wood DJ (1990) Explicit calculation of pipe network parameters. J Hydraul Eng 116(11):1329–1344
8. Bush C, Uber J (1998) Sampling design methods for water distribution model calibration. J Water Resour Plan Manage 124(6):334–344
9. Cheng W, He Z (2011) Calibration of nodal demand in water distribution systems. J Water Resour Plan Manage 137(1):31–40
10. Datta R, Sridharan K (1994) Parameter estimation in water distribution systems by least squares. J Water Resour Plan Manage 120(4):405–422
11. Davidson J, Bouchart F (2006) Adjusting nodal demands in SCADA constrained real-time water distribution network models. J Hydraul Eng 132(1):102–110
12. Ferreri G, Napoli E, Tumbiolo A (1994) Calibration of roughness in water distribution systems. In: 2nd International conference on water pipeline systems, pp 379–396
13. Goldberg D (1989) Genetic algorithms in search, optimization and machine learning. Reading
14. Goulet J-A, Coutu S, Smith IFC (2013) Model falsification diagnosis and sensor placement for leak detection in pressurized pipe networks. Adv Eng Inform 27(2):261–269
15. Greco M, Del Giudice G (1999) New approach to water distribution network calibration. J Hydraul Eng 125(8):849–854
16. Hutton CJ, Kapelan Z, Vamvakeridou-Lyroudia L, Savić DA (2014) Dealing with uncertainty in water distribution system models: a framework for real-time modeling and data assimilation. J Water Resour Plan Managet 140(2):169–183
17. Jackson DD (1972) Interpretation of inaccurate, insufficient and inconsistent data. Geogphys J R Astron Soc 28(2):97–109
18. Kang D, Lansey K (2009) Real-Time demand estimation and confidence limit analysis for water distribution systems. J Hydraul Eng 135(10):825–837
19. Kang D, Lansey K (2011) Demand and roughness estimation in water distribution systems. J Water Resour Plan Manage 137(1):20–30
20. Kapelan Z (2002) Calibration of water distribution system hydraulic models. PhD thesis, University of Exeter
21. Kapelan Z, Savic D, Walters G (2003) Multiobjective sampling design for water distribution model calibration. J Water Resour Plan Manage 129(6):466–479
22. Kapelan Z, Savic D, Walters G (2005) Optimal sampling design methodologies for water distribution model calibration. J Hydraul Eng 131(3):190–200
23. Lansey K, Basnet C (1991) Parameter estimation for water distribution networks. J Water Resour Plan Manage 117(1):126

24. Lansey K, El-Shorbagy W, Ahmed I, Araujo J, Haan C (2001) Calibration assessment and data collection for water distribution networks. J Hydraul Eng 127(4):270–279
25. Liggett J, Chen L (1994) Inverse transient analysis in pipe networks. J Hydraul Eng 120(8):934–955
26. Lingireddy S, Ormsbee LE (1999) Optimal network calibration model based on genetic algorithms. In: WRPMD'99, Reston, VA. American Society of Civil Engineers, pp 1–8
27. Maier HR, Kapelan Z, Kasprzyk J, Kollat J, Matott LS, Cunha MC, Dandy GC, Gibbs MS, Keedwell E, Marchi A, Ostfeld A, Savic D, Solomatine DP, Vrugt JA, Zecchin AC, Minsker BS, Barbour EJ, Kuczera G, Pasha F, Castelletti A, Giuliani M, Reed PM (2014) Evolutionary algorithms and other metaheuristics in water resources: current status, research challenges and future directions. Environ Modell Softw 62:271–299
28. Mallick K, Ahmed I, Tickle K, Lansey K (2002) Determining pipe groupings for water distribution networks. J Water Resour Plan Manage 128(2):130–139
29. Menke W (1982) Geophysical data analysis: discrete inverse theory. Academic Press
30. Nejjari F, Pérez R, Puig V, Quevedo J, Cugueró M, Sanz G, Mirats J (2011) Abnormal quality detection and isolation in water distribution networks using simulation models. In: 11th International conference on computing and control for the water industry, pp 461–466, Exeter
31. Ormsbee L (1989) Implicit network calibration. J Water Resour Plan Manage 115(2):243–257
32. Ormsbee L, Wood D (1986) Explicit pipe network calibration. J Water Resour Plan Manage 112(2):166–182
33. Piller O, Bremond B, Morel P (1999) A spatial sampling procedure for physical diagnosis in a drinking water supply network. In: Savic DA, Walters GA (eds) Water industry systems: modelling and optimization applications, pp 309–316, Exceter
34. Pudar R, Liggett J (1992) Leaks in pipe networks. J Hydraul Eng 118(7):1031–1046
35. Rahal C, Sterling M, Coulbeck B (1980) Parameter tuning for simulation models of water distribution networks. ICE Proc 69(3):751–762
36. Reddy P, Sridharan K, Rao P (1996) WLS method for parameter estimation in water distribution networks. J Water Resour Plan Manage 122(3):157
37. Sanz G, Pérez R, Kapelan Z, Savic D (2015) Leak detection and localization through demand components calibration. J Water Resour Plan Manage 04015057
38. Savic D, Kapelan Z, Jonkergouw P (2009) Quo vadis water distribution model calibration? Urban Water J 6(1):3–22
39. Savic D, Walters G (1995) Genetic algorithm techniques for calibrating network models. Technical report, Centre for Systems and Control Engineering, Exeter
40. Shamir U (1974) Optimal design and operation of water distribution systems. Water Resour Res 10(1):27–36
41. Sumer D, Lansey K (2009) Effect of uncertainty on water distribution system model design decisions. J Water Resour Plan Manage 135(1):38–47
42. Tang K, Karney B, Pendlebury M, Zhang F (1999) Inverse transient calibration of water distribution systems using genetic algorithms. In: Savic D, Walters GA (eds) Water industry systems: modelling and optimization applications. Research Studies Press, Baldock, pp 317–326
43. Todini E (1999) Using a Kalman filter approach for looped water distribution network calibration. In: Savic D, Walters G (eds) Water Industry systems: modelling and optimization applications. Research Studies Press, Baldock, pp 327–336
44. Tucciarelli T, Criminisi A, Termini D (1999) Leak analysis in pipeline systems by means of optimal valve regulation. J Hydraul Eng 125(3):277–285
45. Uhrhammer R (1980) Analysis of small seismographic station networks. Bull Seismol Soc Am 70(4):1369–1379
46. Vítkovský J, Simpson A, Lambert M (2000) Leak detection and calibration using transients and genetic algorithms. J Water Resour Plan Manage 126(4):262–265
47. Walski T (1983) Technique for calibrating network models. J Water Resour Plan Manage 109(4):360
48. Walski T (1986) Case study: pipe network model calibration issues. J Water Resour Plan Manage 112(2):238–249

49. Walters G, Savic D, Morley M, Schaetzen W, Atkinson R (1998) Calibration of water distribution network models using genetic algorithms. Trans Ecol Environ 19
50. Wasantha Lal AM (1995) Calibration of riverbed roughness. J Hydraul Eng 121(9):664–671
51. Wiggins R (1972) The general linear inverse problem: implication of surface waves and free oscillations for Earth structure. Rev Geophys 10(1):251–285
52. Yeh W (1986) Review of parameter identification procedures in groundwater hydrology: the inverse problem. Water Resour Res 22(2):95–108

Chapter 6
Demand Forecasting for Real-Time Operational Control

Jordi Saludes, Joseba Quevedo and Vicenç Puig

6.1 Introduction

In the literature, the problem of water demand forecasting has been widely studied taking into account different timescales: long, medium and short terms (autoregressive integral moving average). The long-term forecasts allow building new infrastructure and foresee the possible future problems. Both short-term and medium-term forecasting are basically useful for the operational control [1] of the network. In large-scale infrastructures, such as the complex water transport network of Barcelona presented in Chap. 2, a telecontrol system acquires, stores and processes thousands of flowmeter and other sensor data every hour to achieve an accurate monitoring of the whole network in real-time. Additionally, the real-time network control needs an accurate prediction of the future consumption. These future values help the operational control system to decide the new actions in the future. The problem of short-term demand forecast coupled with the optimal operational control working at hourly timescale with a 24-h-ahead forecasting horizon has already addressed in the literature. In [2], a forecasting method based on patterns is proposed that combines the daily average estimation with an hourly demand pattern to provide hourly forecasts over the following 24 h. This approach is quite similar to the one proposed in [3] where the daily average estimation is provided by an ARIMA model, while the hourly estimation is also obtained using hourly patterns. Recently, the use of a bank of artificial neural networks (ANNs) has been proposed as an alternative to the approaches combining time series models and patterns [4].

The considered forecast methods are univariate. Although it is known that the water demands are strongly influenced by the meteorological variables, these are sometimes hard to predict. However, these variables tend to change in a smooth way, so that ARIMA models can capture their effect on the demand, taking pre-

J. Saludes (✉) · J. Quevedo · V. Puig
Research Center "Supervision, Safety and Automatic Control" (CS2AC-UPC), Terrassa, Spain
e-mail: jordi.saludes@upc.edu

V. Puig et al. (eds.), *Real-Time Monitoring and Operational Control of Drinking-Water Systems*, Advances in Industrial Control,
DOI 10.1007/978-3-319-50751-4_6

vious demand observations into account. So, the prediction has the meteorological effects included. Besides, online forecasting does not work well with multi-variate models. The selected forecast method should be easy to use and should be automatically calibrated. Moreover, this method should be adaptable, because in the network there are many demand time series to model and each sector has their own demand characteristics.

The main objective of this chapter is to study and compare several methods to predict the next 24-h-ahead demand in order to be integrated in the real-time operational control. In particular, the ARIMA and basic structural and exponential smoothing models are considered. However, in these three approaches, there are lot degrees of freedom to tune, in order to build a short-term demand prediction, that will be discussed along the chapter. Finally, two ARIMA, one structural, and one exponential smoothing models are implemented and compared with two naïve models in order to select the model that provides best results in Barcelona network case study.

6.2 Problem Statement

The main goal of the operational control of water networks is to meet the demands at consumer sites, but at the same time with minimum costs of operation and guaranteeing pre-established volumes in tanks (to preserve the satisfaction of future demands) and smooth operation of actuators (valves and pumps) and production plants. Model predictive control (MPC) [5] provides suitable techniques to implement the operational control of water systems to improve their performance as it will be presented later in this book, since it allows to compute optimal control strategies ahead of time for all the flow and pressure control elements. Moreover, MPC allows taking into account physical and operational constraints, the multi-variable input and output nature, the demand forecasting requirement and complex multi-objective operational goals of water networks. The optimal strategies are computed by optimizing a mathematical function describing the operational goals in a given time horizon and using a representative model of the network dynamics, as well as demand forecasts.

Water consumption forecasting for operational control of water networks in urban areas is usually managed on a daily basis, because common transport delay times between the supplies and the consumer sites allow operators to follow daily water request patterns. Therefore, this horizon is appropriate for evaluating the effects of different control strategies on the water network, with respect to operational goals.

6.3 Proposed Approach

In this section, several 24-h-ahead short-term demand forecasting methods will be presented and later compared when applied to the Barcelona water network case study.

6.3.1 Double-Seasonality ARIMA Models

ARIMA models (also known as integrated ARMA models) are quite suitable for modelling time series composed of a non-stationary trend component and a zero-mean component. This type of models is a broader class of ARIMA models that additionally to the autoregressive part (AR) and the moving average part (MA) includes differencing. These models will be denoted as `ARIMA (p,d,q)` where p, d and q are degrees of the models.

In the case of short-term demand forecasting, the time series presents additionally a daily and weekly seasonality when characterized at the hourly timescale. For this reason, to forecast the short-term water demand time series, a double-seasonal ARIMA model should be used [6]. This method was used over the years, but in recent years, this method has lost its popularity. The ARIMA prediction can be written as a polynomial of the past values and the past prediction errors. The characteristic of the seasonal ARIMA is that it needs one polynomial for the regular component and other for the seasonal one. Moreover, the double-seasonal ARIMA separates the seasonal polynomial in two polynomials. Each seasonal polynomial works only with one periodicity. This model is expressed as $\mathtt{ARIMA}\ (\mathtt{p,d,q}) \times (\mathtt{P1,D1,Q1})_{s_1} \times (\mathtt{P2,D2.Q2})_{s_2}$, where p, d, q, P_1, D_1, Q_1, P_2, D_2 and Q_2 are the degrees of the model and s_1 and s_2 are the number of periods in each seasonality.

The main problem of this method is that the difference operator orders and the distinct polynomials are not easy to determine. Another problem is the identification of the model. This process cannot be done in real-time. The large number of the seasons in the second seasonality, $s_2 = 168$ that corresponds to seven days, makes hard to obtain a good identification. So, an extension of this model $\mathtt{ARIMA}\ (\mathtt{0,1,1}) \times (\mathtt{0,1,1})_{24} \times\ (\mathtt{0,1,1})_{168}$ is proposed, and the three parameters are determined with the maximum-likelihood method.

6.3.2 Daily Seasonality ARIMA Model with Hourly Pattern

The second method relies on the works of [7]. The basic idea is to work in two timescales: daily and hourly. In each scale, a specific model is constructed. In the daily scale, the method works with the total day's consumption and the forecast is based on a seasonal ARIMA model. In the hourly scale, the method works with the hourly consumption and the forecast is based on the daily patterns.

The seasonal ARIMA models work with data that present a repeated stochastic pattern. This model only needs two polynomials and can be expressed as $\mathtt{ARIMA}\ (\mathtt{p,d,q}) \times (\mathtt{P,D,Q})_s$. The numbers p, d, q, P, D and Q are the degree of the distinct polynomials. The best model is selected with Bayesian information criterion [8] considering the set of models generated by $0 \le p \le 3$, $0 \le P \le 1$, $0 \le q \le 3$ and $0 \le Q \le 1$. Alternatively, the structure of the model can be derived using three main components:

- One-day-period oscillating signal with zero average value to cater for cyclic deterministic behaviour, implemented using a second-order (two parameters) model with two oscillating modes, in s-plane $s_{1-2} = +/-2\pi/24j$, or equivalently, in z-plane $z_{1-2} = cos(2\pi/24) + / - sin(2\pi/24)j$. The oscillating polynomial is

$$y(k) = 2cos(2\pi/24)y(k-1) - y(k-2). \tag{6.1}$$

- An integrator that takes into account possible trends and nonzero mean values of the flow data described by

$$y(k) = y(k-1). \tag{6.2}$$

- An autoregressive component of order 21 to consider the influence of previous values within the series

$$y(k) = -a_1 y(k-1) - a_2 y(k-2) - a_3 y(k-3) - ... - a_{21} y(k-21). \tag{6.3}$$

The three components considered lead to a final order of 24 (i.e., the number of samples within a day for sampling period of 1 h). The time series model has the following structure:

$$y_p(k) = -b_1 y(k-1) - b_2 y(k-2) - b_3 y(k-3) - ... - b_{24} y(k-24), \tag{6.4}$$

which is consistent with the typical daily demand pattern of 24 h.

Once a daily prediction for specific day is obtained, it is distributed along the day hours using a demand pattern. The demand pattern is generated such that the sum of hourly consumptions at the end of the day is one, because the sum of the hourly predictions is the daily forecast. Previous works [7] show that several types of demand patterns should be used: one for weekdays, one for Saturdays and one for Sundays/holidays. Then, the hourly prediction is obtained by distributing the daily prediction using the demand pattern. The hourly prediction based on the pattern demand has a big problem: it does not use the new available information since the hourly prediction is computed for the whole day without re-estimating it taking into account the registered demand of the hours of the day already elapsed. To solve this problem, a new procedure for hourly demand forecast based on the consumption of the previous hours and the daily consumption has been developed.

6.3.3 *Basic Structural Model*

Basic structural models (BSMs) can be used taking into account that the time series can be divided in additive independent components. These models gained popularity in the mid-1980s, because the modelization is simple and the Kalman filter helps to optimize few parameters [9]. In this approach, it is assumed that the time series can be divided into three additive independent parts: level, seasonal and irregular components. Thus, the model can be written as follows:

$$d(k) = L(k) + S(k) + e(k), \tag{6.5}$$

where L is the level component, S is the seasonal component and e is the irregular component. Two regular components are random walks, and their best prediction is the recent past value. This model can also be written as state-space model. This allows using the Kalman filter and the maximum-likelihood procedure to obtain the variances. In the case that the double seasonality is considered, the two seasonal components can be included. In this case, a huge number of dummy variables appear. To alleviate this problem, one seasonal can be alternatively considered using periodic splines to characterize the intraday demands. Harvey and Koopman [10] studied this method to forecast the hourly electrical demand.

6.3.4 Exponential Smoothing Method

The exponential smoothing methods were proposed by Brown [11] in the 1950s. Holt and Winters improved the method to work with tendency and seasonal components. After the Winters paper, their method is known as Holt–Winters (HW) method, although it is a kind of exponential smoothing method. Its main characteristic is their simplicity. So, this method does not need offline identification phase and only an optimization with least squares or other method is required. For this reason, it is used as an automatic forecasting method. Nowadays, it is a standard method in the electricity and water demand forecast.

The prominent problem of the HW method is that it works with just one periodicity (additive or multiplicative). Taylor [12] proposed a new extension with a multiplicative double seasonality. The ℓ-step-ahead Holt–Winters prediction with multiplicative one seasonal periodicity is

$$d(k + \ell|k) = (L(k) + \ell T(k))\, S\left(k + \ell - \left[\frac{\ell}{s} + 1\right] s\right), \tag{6.6}$$

where L is the level component, T is the trend component, S is the seasonal component, s is the period and $[\cdot]$ is the integer part of the dot content. These components can be modelled as follows:

$$L(k) = \alpha \frac{d(k)}{S(k-s)} + (1-\alpha)(L(k-1) + T(k-1)), \tag{6.7a}$$

$$T(k) = \gamma(L(k) - L(k-1)) + (1-\gamma)T(k-1), \tag{6.7b}$$

$$S(k) = \delta \frac{d(k)}{L(k)} + (1-\delta)S(k-s), \tag{6.7c}$$

where α, γ and δ are the parameters. The least square parameter estimation method provides these parameters. Often, the residuals of the model are correlated. In this case, a simple `AR (1)` is added to improve the prediction residuals. As a basic structural model, the classical HW method only has one seasonal component. As already discussed, to handle both seasonal periodicities present in the hourly water demand, the double-seasonal HW method introduced by Taylor [12] will be used. This extended model has the following form:

$$d(k+\ell|k) = (L(k) + \ell T(k))S_1\left(k+\ell-\left[\frac{\ell}{s_1}\right]s_1\right)S_2\left(k+\ell-\left[\frac{\ell}{s_1}\right]s_1\right), \tag{6.8}$$

where $S_1(k)$ is the first seasonal component, $S_2(k)$ is the second seasonal component and s_1 and s_2 are the number of seasons of each period. These components can be modelled as

$$L(k) = \alpha\frac{d(k)}{S_1(k-s_1)S_2(k-s_2)} + (1-\alpha)(L(k-1)+T(k-1)), \tag{6.9a}$$

$$T(k) = \gamma(L(k)-L(k-1)) + (1-\gamma)T(k-1), \tag{6.9b}$$

$$S_1(k) = \delta_1\frac{d(k)}{L(k)S_2(k-s_2)} + (1-\delta)S_1(k-s_1), \tag{6.9c}$$

$$S_2(k) = \delta_2\frac{d(k)}{L(k)S_1(k-s_1)} + (1-\delta)S_2(k-s_2), \tag{6.9d}$$

where α, γ, δ_1 and δ_2 are the parameters that should be estimated with least square method. As mentioned above, an error model is included to address the structure still present in the residuals. In this case, the residuals are modelled with $\mathrm{AR}_{s_1}(1) \times \mathrm{AR}_{s_2}(1)$. The new forecasting method performs better than the former without the correction residuals.

6.3.5 *Naïve Methods*

All the proposed forecasting methods will be compared against two naïve methods. These methods are used as benchmarking methods. If the considered method performs worse than these reference approaches, it cannot able to capture the variability of the data. The first naïve method is the random walk. The random walk is widely used as benchmark method. The random walk prediction is a constant forecast with the last real value. The second method is a seasonal version of the random walk. This method is used when the time series contains seasonality. Their prediction is the $t-s$ value where t is the predicted time and s is the number of periods: for the weekly period, $s=7$ or $s=168$, depending on the considered timescale (daily or hourly).

6.4 Simulations and Results

As described in Chap. 2, the Barcelona water transport network is composed of about 100 demand sectors associated with different pressure levels. Currently, the information system receives, in real time, data from 100 control points, mainly flowmeters and a few pressure sensors. It should be noticed that only few pressure levels concentrate almost 75% of the water consumption. So, the demand forecasting approaches will be tested on in these sectors, since improving prediction on them could lead to major economic benefits. Here, the results will be presented for those pressure sectors of the highest demand. The other major sectors have a similar behaviour than those that are presented in this chapter.

To compare the distinct forecasting methods, a set of comparative indicators are used. These indicators help to decide which forecast model is performing better. Since the best model will presumably depend on the timescale considered, it is expected that the best model for the daily prediction will not give the best hourly forecast. The repeatability of historical data also influences the prediction quality of each model. The set of indicators are as follows:

- *Explained variance* (EV) measures the not-modelled variance

$$EV = 1 - \frac{\mathrm{Var}(e_\ell(i))}{\mathrm{Var}(x_\ell)}, \tag{6.10}$$

  where $e_\ell(i)$ is the ℓ-step prediction error from time i and x_ℓ is the time series. If the $EV = 1$, the model captures the whole process variance.
- *Mean absolute error* (MAE) is evaluated as follows

$$MAE = \frac{1}{N}\sum_{i=1}^{N} |e_\ell(i)|. \tag{6.11}$$

- *Mean square error* (MSE) can be expressed as

$$MSE = \frac{1}{N}\sum_{i=1}^{N} (e_\ell(i))^2. \tag{6.12}$$

- *Mean absolute percentage error* (MAPE) that rescales the errors respect the process mean

$$MAPE = \frac{100}{N}\sum_{i=1}^{N} \frac{|e_\ell(i)|}{\mu}, \tag{6.13}$$

  where μ is the mean of the time series.

A good prediction should satisfy EV $\approx$ 1, MAPE $\approx$ 0% and present small MAE (or MSE). The MAPE is widely used in many comparisons, and their interpretations are similar to MAE and MSE. The advantage of MAPE is that it is adimensional.

The results presented for the considered pressure levels are representative of the results obtained in most of the pressure levels. The considered pressure levels are selected, because it represents the 25% of the total consumption in the city. The other major pressure levels have similar characteristics as the considered one. Figure 6.1 presents the daily consumption and the one-step forecast errors of the considered methods.

The first graph presents the real data. The daily forecast for the hourly model is the sum of the forecast from one to twenty-four steps. The second graph presents the double ARIMA forecast error for the daily consumption. The third graph presents the daily ARIMA error. In this example, the selected model structure is $\mathtt{ARIMA}$ $\mathtt{(0,1,1)} \times \mathtt{(0,1,1)}_7$. The fourth graph presents the HW forecast error. The fifth graph presents the basic structural model. From these graphs, it can be observed that the double ARIMA has the biggest error. The basic structural model presents a heteroscedasticity error that corresponds to an error with a not-constant variance. The daily ARIMA and double HW errors are similar, and the error peaks are related to the special events. In the future works, the selected procedure should be improved to take into account the influence of these events.

In Fig. 6.2, a piece of the hourly time series with their one-step forecast errors is presented. The first graph is the hourly data. The second graph corresponds to double ARIMA errors. The third graph is the error forecast using the adaptable demand patterns. The fourth graph presents the double HW forecast errors. The last graph presents the basic structural model errors. It can be observed that the basic structural model errors have the biggest variance. The pattern distribution of the daily consumption and double HW is similar. The double ARIMA has important error peaks that suggest that the model is not correct.

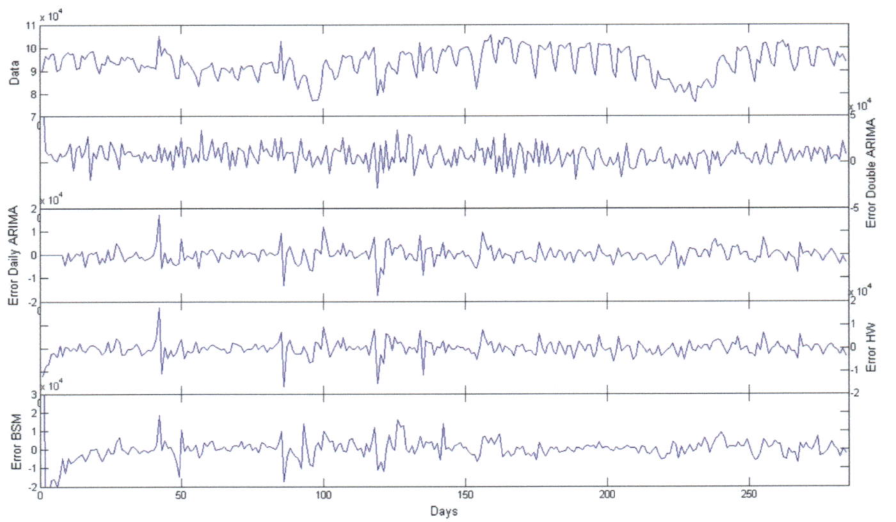

Fig. 6.1 Daily consumption of the 70BBE pressure level with their forecast errors

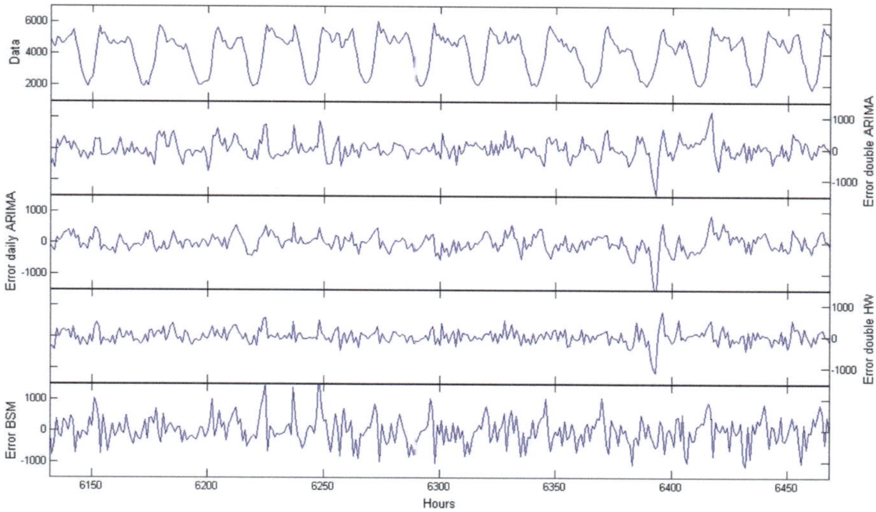

Fig. 6.2 A piece of hourly consumption of the 70BBE pressure level with forecasting errors

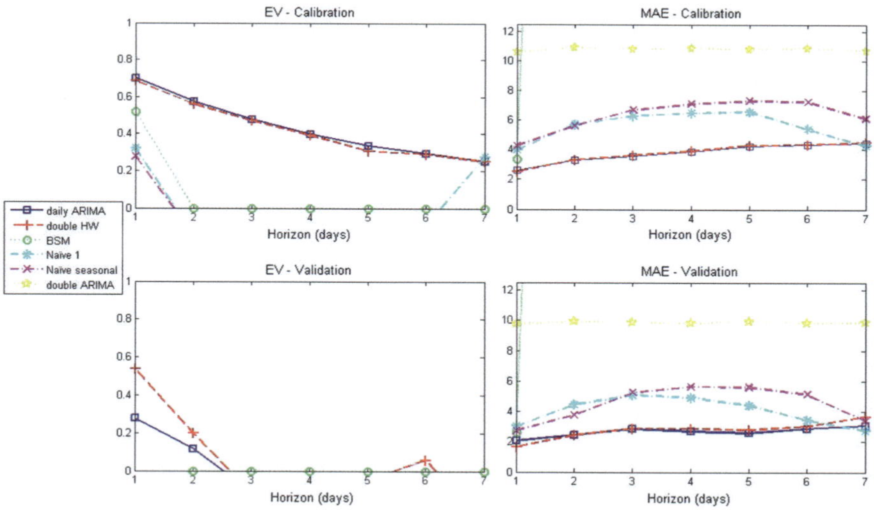

Fig. 6.3 The EV and MAPE indicators for each method in the daily scale for the considered pressure level

It can also be noticed that the double ARIMA and the basic structural models are worse than the other two. The daily ARIMA and the double HW are similar and visually cannot be decided which is the best. In Figs. 6.3 and 6.5, the EV and MAPE indicators for each model are presented in the daily and hourly scales. In Tables 6.1 and 6.2, there are MAPE indicators for the calibration phase corresponding to both scales.

Table 6.1 MAPE indicator for each method for the daily time series for the considered pressure level

Horizon	Double ARIMA	Daily ARIMA	HW	BSM	Naïve 1	Naïve 7
1	10.5612	2.6005	2.4900	3.3478	3.9869	4.2656
2	10.9078	3.2740	3.2855	100.0000	5.6993	5.6183
3	10.8099	3.5517	3.6025	100.0000	6.2783	6.6960
4	10.8586	3.8584	3.9056	100.0000	6.5002	7.1116
5	10.7913	4.2101	4.2880	100.0000	6.5699	7.3217
6	10.8718	4.3275	4.3482	100.0000	5.4119	7.2524
7	10.7100	4.4230	4.4744	100.0000	4.2656	6.1098

Table 6.2 MAPE indicator for each method for the daily time series of the 70BBE

Horizon	Double ARIMA	Pattern	HW	BSM	Naïve 1	Naïve 7
1	5.6846	5.7577	4.8709	8.8440	10.8495	8.2799
2	14.5969	5.8433	5.1908	12.0605	18.2944	12.3703
3	12.9981	5.9047	5.3459	12.7374	24.6421	18.7919
4	13.1956	5.9508	5.4046	12.6250	30.0094	24.9268
5	13.1033	5.9863	5.4201	12.4775	34.3851	30.2886
6	13.2732	6.0045	5.3957	12.3761	37.5536	34.6361
12	13.0239	6.0373	5.4401	12.0890	42.9569	43.8022
24	11.1971	6.4395	5.4437	12.0379	8.2938	13.9054

It can only be noticed that in the calibration phase, the daily ARIMA and the double HW have a good performance according to the EV indication. But, in the validation phase, the EV indicator shows that any model is unable to capture the variance. In the calibration phase, the daily ARIMA and the daily forecast of the double HW are able to explain the 70% of the variance with one-step forecast, but they only explain the 20% with the seven-step forecasts. This is a typical result because the seven-step forecast uses intermediate predictions, while the single-step one does not. Considering the MAPE indicator, the behaviour in the two phases is similar. The daily ARIMA and the daily forecast of the double HW have the best MAPE indicators. It can also be observed that the MAPE of the daily ARIMA and the daily prediction of the HW are smaller than 5%. For the other methods, their MAPE is larger than 5%. So, in the daily scale, the daily ARIMA and the HW have a similar behaviour.

Focusing on the hourly scale, the most surprising fact is the change in behaviour between the calibration and validation phases when using the daily ARIMA with demand pattern distribution (see Fig. 6.4). In the calibration phase, the indicators for this method are close to the double HW indicators. But, in the validation phase, the

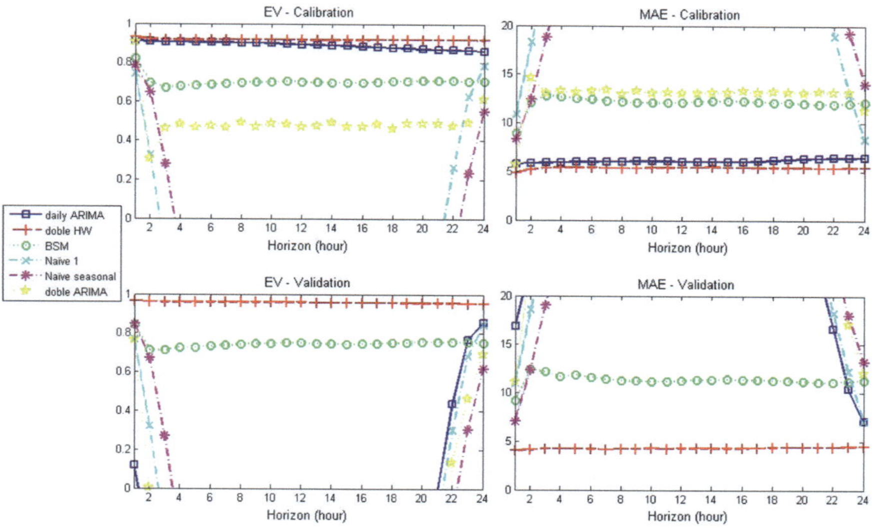

Fig. 6.4 The EV and MAPE indicators for each method in the hourly scale for the 70BBE pressure level

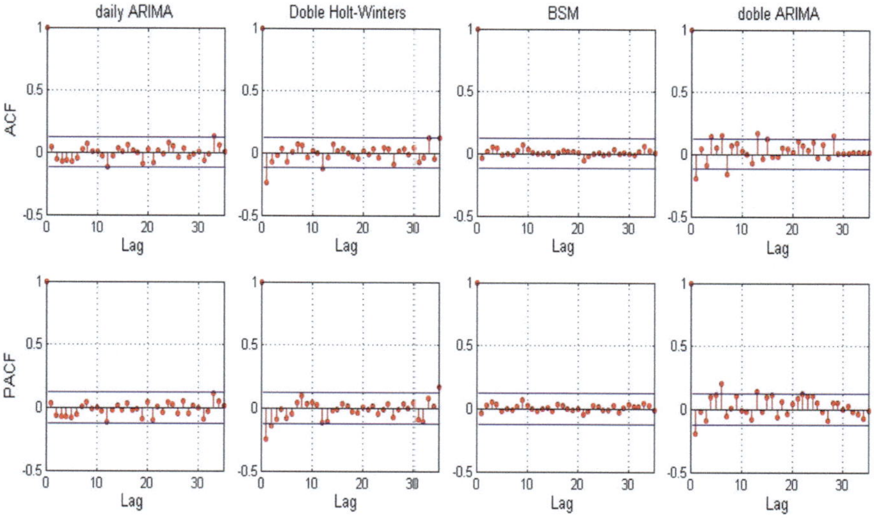

Fig. 6.5 ACF and PACF for the distinct methods

ARIMA with pattern distribution indicators is significantly worse. In the other cases, the double HW is similar in both phases. The worsening in the daily ARIMA forecast may be due to the change in the patterns. So, the behaviour is different during the two phases, and the pattern model is unable to forecast these changes. The rest of the methods have worse indicators than HW.

Finally, the last assessment involves verifying whether the daily errors were correlated or not. In case the residuals were correlated, the method will not capture all the time series information. The daily error is considered, because it provides information about future events. Figure 6.5 presents the autocorrelation (ACF) and partial autocorrelation (PACF) for each method. It can be noticed that the daily ARIMA and the basic structural model are not correlated. The double HW presents a correlation for few lags. Finally, the double ARIMA presents the lags autocorrelated because the residuals are not white noise.

Thus, it can be concluded that the double HW provides the best daily and hourly forecast. These methods are better than the naïve methods in every timescale and phase.

6.5 Conclusions

In this chapter, four methods to predict a future demand values in two timescales (hourly and daily) are studied and compared. These methods are tested in Barcelona water transport network in a representative pressure level. The first conclusion is that the basic structural method is the slowest, because the model contains a matrix with large dimension. This fact leads to slow the prediction. The second is that the double ARIMA produces bad predictions, because it can be easily influenced by particular events. The third method is based on the combination of daily ARIMA with pattern distribution. This method presents distinct behaviour depending on the timescale and phase. In the daily scale, the forecasting is good. In the hourly scale, the prediction goodness depends on the phase. In the calibration phase, forecasting is good, but in the validation phase, it is not. Finally, the double HW method seems to be the most robust forecasting method and it is easy to implement too. The main daily ARIMA problem is in the calibration phase since while computation time is high, the selected method has better prediction. Another problem is the sensitivity to outliers. To reduce the outlier effect, an outlier detection algorithm could be implemented for the daily scale. This method is only good for daily forecasting because the pattern does not provide an accurate forecast in the larger timescale.

The double HW is a deterministic method. So, in typical pressure floors, where seasonal periodicity changes smoothly, it provides good forecasting. If it is assumed that the residuals follow an ARIMA model, the forecast improves. The best residual ARIMA model is `AR(1)` × `AR24(1)`. The parameters are estimated using the least square method. Since the majority of pressure levels change their periodicities smoothly, it can be concluded that the best method is the double HW in both timescales.

References

1. Pascual J, Romera J, Puig V, Cembrano G, Creus R, Minoves M (2013) Operational predictive optimal control of Barcelona water transport network. Control Eng Pract 21(8):989–1156
2. Alvisi S, Franchini M, Marinelli A (2007) A short-term, pattern-based model for water-demand forecasting. J Hydroinformatics 9(1):39–50
3. Quevedo J, Puig V, Cembrano G, Blanch J, Aguilar J, Saporta D, Benito G, Hedo M, Molina A (2010) Validation and reconstruction of flow meter data in the Barcelona water distribution network. Control Eng Pract 18(6):640–651
4. Romano M, Kapelan Z (2014) Adaptive water demand forecasting for near real-time management of smart water distribution systems. Environ Model Softw 60:265–276
5. Ocampo-Martinez C, Puig V, Cembrano G, Quevedo J (2013) Application of mpc strategies to the management of complex networks of the urban water cycle. IEEE Control Syst Mag 33(1):15–41
6. Box G, Jenkins G, Reinsel C (1994) Time series analysis: forecasting and control, 3rd edn. Prentice Hall, Englewood Cliffs
7. Quevedo, J, Puig, V, Cembrano, G, Aguilar, J, Isaza, C, Saporta, D, Benito, G, Hedo, M, Molina, A (2006) Estimating missing and false data in flow meters of a water distribution network. In: 6th IFAC symposium on fault detection, supervision and safety of technical processes. Shangai, China
8. Schwarz G (1978) Estimating the dimension of a model. Ann Stat 6(2):461–464
9. Harvey A (1989) Forecasting, structural time series models and the Kalman filter. Cambridge Univesity Press, Cambridge (UK)
10. Harvey A, Koopman S (1993) Forecasting hourly electricity demand using time-varying splines. J Am Stat Assoc 88(424):1228–1236
11. Brown, RG (2004) Smoothing, forecasting and prediction of discrete time series. Courier Corporation
12. Taylor J (2003) Short-term electricity demand forecasting using double seasonal exponential smoothing. J Oper Res Soc 54:799–805

Part II
Real-Time Monitoring

Chapter 7
Leak Monitoring

Ramon Pérez, Josep Cugueró, Gerard Sanz, Miquel A. Cugueró and Joaquim Blesa

7.1 Introduction

Waste and loss of water have been sometimes disregarded due to the low water price and ease of exploitation in developed countries. However, both users and utilities are increasing their concern to avoid present and future water scarcity. Individual users can optimize their daily routines to reduce water waste, but burst and background leakage will be present independently of it [1].

Leakage in water distribution systems has attracted a lot of attention by both practitioners and researchers over the past years. Leak identification is divided into leakage awareness and leak localization, as suggested in the review of leakage management-related methods in distribution pipe systems from detection and assessment to efficient control [2]. Leakage awareness focuses on leak detection in the network, but does not give any information about its precise location [3–7]. On the other hand, leakage localization is an activity that identifies and prioritizes the areas of leakage to make pinpointing of leaks easier [8]. Leak localization techniques can be divided into two categories: external and internal [9]. The use of external methods such as acoustic logging [10], penetrating radar [11] or liquid detection methods [12] has some drawbacks like needing a large number of sensors, not being suitable for application in large urban areas, or being invasive. Internal methods use continuously monitored data to infer the position of leaks using models. Many techniques in the literature are based on transient analysis, which is mainly used on single, grounded pipelines due to the high effect of the system uncertainty on results [13–16]. Non-transient model-based leakage localization techniques have been also developed during the last years [17–20]. These techniques analyse the difference between measurements

R. Pérez (✉) · J. Cugueró · G. Sanz · M.A. Cugueró · J. Blesa
Research Center Supervision, Safety and Automatic Control (CS2AC-UPC), Terrassa, Spain
e-mail: ramon.perez@upc.edu

J. Blesa
Institut de Robòtica i Informàtica Industrial CSIC-UPC, Barcelona, Spain

V. Puig et al. (eds.), *Real-Time Monitoring and Operational Control of Drinking-Water Systems*, Advances in Industrial Control,
DOI 10.1007/978-3-319-50751-4_7

and estimated values from leaky scenarios to signal the probability of a zone to contain leakage.

The use of models for monitoring and supervising DWNs is a common practice in water companies. A proper calibration of these models is required to obtain reliable results when using them [21], as it has been analysed in Chaps. 3, 4 and 5. Once the model is calibrated, the model-based leak detection and localization methodologies reviewed can make use of it. However, these methodologies do not consider the evolution of demands in the real system. This evolution should be taken into account because demands are parameters that change continuously and leakages may be masked with their evolution [22, 23].

7.2 Problem Statement

Given a model for the non-transient behaviour of a network and a sequence of measurements from it, the problem is to locate a node in the network where there may be a leak. Two kind of measurements can be distinguished: boundary conditions, which are pressure $\mathbf{h}^S$ and total inflow q^{in} at the n_h inputs of the network, and additional n_y head measurements $\mathbf{y} = (y_1 \dots y_{n_y})$ from selected nodes in the network. Measurements are acquired at sampling time T_s, but as the effect of a leak over the measurements is usually small and has fluctuations due to sensor resolution, the localization methodology is applied at a larger time T_L. Thus, $N_L = T_L/T_s$ measurements are available between two consecutive iterations. Assuming that the boundary conditions have not changed significantly during T_L, the mean of the measurement y_i at instant kT_L ($\bar{y}_i(kT_L)$) is computed as

$$\bar{y}_i(kT_L) = \frac{\sum_{j=0}^{N_L-1} y_i(kT_L - jT_s)}{N_L}. \tag{7.1}$$

For simplicity and clearness, the time dependence of the variables is only made explicit when necessary.

7.2.1 *Model of the Network*

In the absence of leakage, the total inflow q^{in} is distributed among the network nodes according to a given demand pattern. The demands of the nodes are represented by a vector $\mathbf{d} = (d_1, \dots, d_{n_d})$ with n_d equal to the number of nodes in the network. In a non-leakage scenario, the total inflow is equal to the sum of demands, i.e.,

$$q^{in} = \sum_{i=1}^{n_d} d_j, \tag{7.2}$$

where each demand d_i can be related to q^{in} by a weight, i.e.,

$$d_i(q^{in}) = \alpha_i q^{in}. \tag{7.3}$$

Given the boundary conditions, the computation of a prediction $\hat{\mathbf{y}}_{\mathbf{nf}}$ for a non-leakage scenario is denoted by

$$\hat{\mathbf{y}}_{\mathbf{nf}} = \mathbf{g}_{nf}(q^{in}, \mathbf{h}_S, \mathbf{d}(q^{in})), \tag{7.4}$$

where $\hat{\mathbf{y}}_{\mathbf{nf}} \in \mathbb{R}^{n_y}$, $\mathbf{g}_{nf} : \mathbb{R} \times \mathbb{R}^{n_h} \times \mathbb{R}^{n_d} \rightarrow \mathbb{R}^{n_y}$, $\mathbf{h}_S \in \mathbb{R}^{n_h}$ and $\mathbf{d} \in \mathbb{R}^{n_d}$. Subscript nf indicates non-faulty, i.e., non-leakage scenario. The difference

$$\mathbf{r} = \mathbf{y} - \hat{\mathbf{y}}_{\mathbf{nf}}, \tag{7.5}$$

which quantifies the consistency of the measurement with the model prediction, is called a *residual*. There will be two concepts to be defined afterwards: *observed residual* and *predicted residual*. If there is no uncertainty in model (7.4), the absence of leakage implies $\mathbf{r} = 0$.

In a leakage scenario, only the possibility of one leak of nominal value f^0 in an unknown node of the network is considered. The nominal value of the leak affects (7.2) and (7.3), yielding

$$\begin{aligned} q^{in} &= \sum_{i=1}^{n_d} d_j + f^0, \\ d_i(q^{in} - f^0) &= \alpha_i(q^{in} - f^0). \end{aligned} \tag{7.6}$$

Consider the n_d predictions $\hat{\mathbf{y}}_{\mathbf{f_i}}$, each one corresponding to a leak of nominal value f^0 in node i

$$\hat{\mathbf{y}}_{\mathbf{f_i}} = \mathbf{g}_{f_i}(q^{in}, \mathbf{h}_S, \mathbf{d}(q^{in} - f^0), f^0), \qquad i = 1 \ldots n_d, \tag{7.7}$$

where $\hat{\mathbf{y}}_{\mathbf{f_i}} \in \mathbb{R}^{n_y}$, $\mathbf{g}_{f_i} : \mathbb{R} \times \mathbb{R}^{n_h} \times \mathbb{R}^{n_d} \times \mathbb{R} \rightarrow \mathbb{R}^{n_y}$. Subscript f_i indicates a faulty scenario consisting of a leak in node i. The differences $\hat{\mathbf{r}}_{\mathbf{f_i}} = \hat{\mathbf{y}}_{\mathbf{f_i}} - \hat{\mathbf{y}}_{\mathbf{nf}}$ are the predicted residuals for the nominal leak f^0.

7.3 Proposed Approach

Given a set of measurements and models that represent different leakage scenarios, the proposed methodology aims to select the model that is most consistent with these measurements. Each scenario considers only one leak in a different location (node) of the network. Therefore, selecting the most consistent scenario is equivalent to selecting the most consistent location for the leak. Different authors have studied the problem of leak localization from different perspectives. In [24], a complete mathematical view of the consistency problem as an inverse problem is given. In [19, 25], the point of view of model-based fault diagnosis is followed [26] In all mentioned works, the idea is to solve the consistency problem with algorithms that can use existing efficient network solvers for the forward problem. This work takes the approach described in [27], and it differs from the one in [25, 28] only in the correlation measure. In this approach, a leak in a node will be considered to be a fault that is to be localized. The algorithm gives the most consistent location for a leak given a set of measurements from the network. First, the leak localization problem will be tackled for one time step and the time argument will be omitted. Later, the use of information from more than one time steps will be described.

For clarity, it is firstly explained how the sensor location would be performed if only measurements from one time step were available. If there is no uncertainty in model (7.7) and the value of the unknown leak to be located is small enough, then the dependency of the observed residual $\mathbf{r}$ can be assumed to behave approximately linear with respect to the leak f, i.e.,

$$\mathbf{r} = \hat{\mathbf{r}}_{\mathbf{f_i}} \cdot \frac{f}{f^0}, \qquad i = 1 \ldots n_d. \tag{7.8}$$

Then, the residual leak sensitivities collected in the *sensitivity matrix* denoted by Ω is computed as follows:

$$\boldsymbol{\Omega} = \begin{bmatrix} \frac{\partial r_1}{\partial f_1} & \cdots & \frac{\partial r_1}{\partial f_{n_d}} \\ \vdots & \ddots & \vdots \\ \frac{\partial r_{n_y}}{\partial f_1} & \cdots & \frac{\partial r_{n_y}}{\partial f_{n_d}} \end{bmatrix}. \tag{7.9}$$

Following the ideas reported in [24], this matrix can be approximated by

$$\boldsymbol{\Omega} \simeq \frac{1}{f^0}[\hat{\mathbf{r}}_{\mathbf{f_1}}, \ldots, \hat{\mathbf{r}}_{\mathbf{f_{n_d}}}]. \tag{7.10}$$

Because of the linearity of $\mathbf{r}$ with respect to f, if vectors $\hat{\mathbf{r}}_{\mathbf{f_i}}$ are linearly independent, then each $\hat{\mathbf{r}}_{\mathbf{f_i}}$ characterizes a different leak. Therefore, a correlation measure to test the linear dependency between $\mathbf{r}$ and $\hat{\mathbf{r}}_{\mathbf{f_i}}$ can be used to select the most consistent leak with $\mathbf{r}$ with the measurements and the model. Thus, the selected leak is the one maximizing the correlation measure

$$\rho(\mathbf{r}, \hat{\mathbf{r}}_{\mathbf{f_i}}) = \frac{\mathbf{r}^T \hat{\mathbf{r}}_{\mathbf{f_i}}}{\parallel \mathbf{r} \parallel \parallel \hat{\mathbf{r}}_{\mathbf{f_i}} \parallel}, \tag{7.11}$$

where $\parallel . \parallel$ denotes the norm associated with the vector dot product. In this work, the 2-norm is used.

Algorithm 7.1: Leak localization using a single time step

Require: q^{in} , $\mathbf{h_S}$, $\mathbf{y}$, $(\alpha_1, \ldots, \alpha_{n_d})^T$, f^0
  $\mathbf{d} = q^{in} (\alpha_1, \ldots, \alpha_{n_d})^T$
  $\hat{\mathbf{y}}_{\mathbf{nf}} = g_{nf}(q^{in}, \mathbf{h_S}, \mathbf{d})$
  $\mathbf{r} = \mathbf{y} - \hat{\mathbf{y}}_{\mathbf{nf}}$ to discard a leakage scenario
  if $\mathbf{r} = 0$ **then**
    return 'No leakage scenario'
  else
    $\mathbf{d} = (q^{in} - f^0) (\alpha_1, \ldots, \alpha_{n_d})^T$
    $\hat{\mathbf{y}}_{\mathbf{f_i}} = g_{\mathbf{f_i}}(q^{in}, \mathbf{h_S}, \mathbf{d}, f^0), \quad i = 1 \ldots n_d$
    $\hat{\mathbf{r}}_{\mathbf{f_i}} = \hat{\mathbf{y}}_{\mathbf{f_i}} - \hat{\mathbf{y}}_{\mathbf{nf}}, \quad i = 1 \ldots n_d$
    Select the node index i^* that maximizes $\rho(\mathbf{r}, \hat{\mathbf{r}}_{\mathbf{f_i}})$
  end if
  return i^*

Algorithm 7.1 summarizes the leak localization procedure for one time step.

7.3.1 Including Temporal Information

As described in Sect. 7.2, measurements are acquired at sampling time T_s, but the location methodology is applied at a larger time T_L considering the average of the last N_L measurements. Then, every time T_L, an average residual $\bar{\mathbf{r}}(kT_L)$ is computed as the difference of the average measurements $\bar{\mathbf{y}}(kT_L)$ and estimations $\bar{\hat{\mathbf{y}}}_{nf}(kT_L)$. In the same way, average predicted residuals $\bar{\hat{\mathbf{r}}}_{\mathbf{f_i}}(kT_L)$ $i = 1, \ldots, n_d$ can be computed.

To improve the characterization of a persistent leak, instead of only $\bar{\mathbf{r}}(kT_L)$ and $\bar{\hat{\mathbf{r}}}_{\mathbf{f_i}}(kT_L)$, the concatenation of these two vectors over the last M samples $\bar{\mathbf{r}}((k - M + 1)T_L : kT_L)$ and $\bar{\hat{\mathbf{r}}}_{\mathbf{f_i}}((k - M + 1)T_L : kT_L)$ is considered, where

$$\bar{\mathbf{r}}((k - M + 1)T_L : kT_L) = (\bar{\mathbf{r}}^T((k - M + 1)T_L), \ldots, \bar{\mathbf{r}}^T(kT_L))^T,$$

$$\bar{\hat{\mathbf{r}}}_{\mathbf{f_i}}((k - M + 1)T_L : kT_L) = (\bar{\hat{\mathbf{r}}}_{\mathbf{f_i}}^T((k - M + 1)T_L), \ldots, \bar{\hat{\mathbf{r}}}_{\mathbf{f_i}}^T(kT_L))^T.$$

Therefore, the correlation to be maximized to select the most consistent leak at time kT_L considering a past time horizon $H_M = M \cdot T_L$ is

$$\rho(\bar{\mathbf{r}}((k - M + 1)T_L : kT_L), \bar{\hat{\mathbf{r}}}_{\mathbf{f_i}}((k - M + 1)T_L : kT_L)). \tag{7.12}$$

Algorithm 7.2 summarizes the leak localization procedure in steady state, i.e., after a time $M \cdot T_L$.

Algorithm 7.2: Leak localization in steady state

Require: $(\alpha_1, \ldots, \alpha_{n_d})^T, f^0$

At each jT_s :

Obtain the sensor values and save in $\mathbf{y}(jT_s), q^{in}(jT_s), \mathbf{h_S}(jT_s)$

Compute $\mathbf{d}(jT_s), \hat{\mathbf{y}}_{\mathbf{nf}}(jT_s), \hat{\mathbf{y}}_{\mathbf{f_i}}(jT_s), \quad i = 1 \ldots n_d$

Compute $\mathbf{r}(jT_s) = \mathbf{y}(jT_s) - \hat{\mathbf{y}}_{\mathbf{nf}}(jT_s)$

Compute $\hat{\mathbf{r}}_{\mathbf{f_i}}(jT_s) = \hat{\mathbf{y}}_{\mathbf{f_i}}(jT_s) - \hat{\mathbf{y}}_{\mathbf{nf}}(jT_s), \quad i = 1 \ldots n_d$

At each kT_L :

Compute the means $\bar{\mathbf{r}}(kT_L)$ and $\bar{\hat{\mathbf{r}}}_{\mathbf{f_i}}(kT_L)$

Construct vectors $\bar{\mathbf{r}}((k - M + 1)T_L : kT_L)$ and $\bar{\hat{\mathbf{r}}}_{\mathbf{f_i}}((k - M + 1)T_L : kT_L)$

Select the leak $f_I(kT_L)$ that maximizes

$\rho(\bar{\mathbf{r}}((k - M + 1)T_L : kT_L), \bar{\hat{\mathbf{r}}}_{\mathbf{f_i}}((k - M + 1)T_L : kT_L)), \quad i = 1 \ldots n_d$

Output $f_I(kT_L)$ to the user

7.4 Simulations and Results

The leak localization methodology presented here has been tested in a DMA of Barcelona, called Nova Icària, and described in Chap. 2, under a real leak scenario. The Nova Icària DMA has flow and pressure sensors at every inlet and six inner pressure sensors, whose placement is also marked in Fig. 7.2.

Generally, flow and pressure sensors existing in the DMA networks are integrated with a SCADA system used to supervise the DWN. The SCADA system monitors the pressure and flow at the inlets of every DMA. This monitoring process is carried out by multichannel data loggers linked to every inlet. On the one hand, these data loggers register these measurements with a sampling time T_s of 10 min, and on the other hand, they are integrated with the SCADA through a GSM (Global System for Mobile Communication) network. Thereby, every day at 7:00 a.m., the SCADA system retrieves the inlet measurements of all DMAs from 00:00 h to 23:50 h of the previous day. After these data are retrieved, a data validation process fills database with validated data. The inner pressure sensor measurements are required to carry out the leak localization process. These measurements are recorded in a similar way as DMA inlets. Pressure measurement resolution is 0.1 mWC. The oversampling described in Sect. 7.3 is done through the leak localization period T_L of 1 h.

To ease the access to the DMA measurements stored in the SCADA-validated database, once the DMA measurements of the day before are available, these data are packed in a Excel file in .xls format and are sent by e-mail to those workstations where the leak localization models are available. Figure 7.1 shows the conceptual integration between the Nova Icària instrumentation data and the model-based leak localization methodology.

To assess the leak localization methodology, a leak was forced in the Nova Icària DMA using a discharge component. The experiment took place on 20 December 2012 at 00:30 and lasted around 30 h. The exact location of the leak is indicated in Fig. 7.2. The leak effect can be observed in Fig. 7.3a, where the time evolution of the DMA total inflow on 20 December 2012, affected by the leakage event, and on

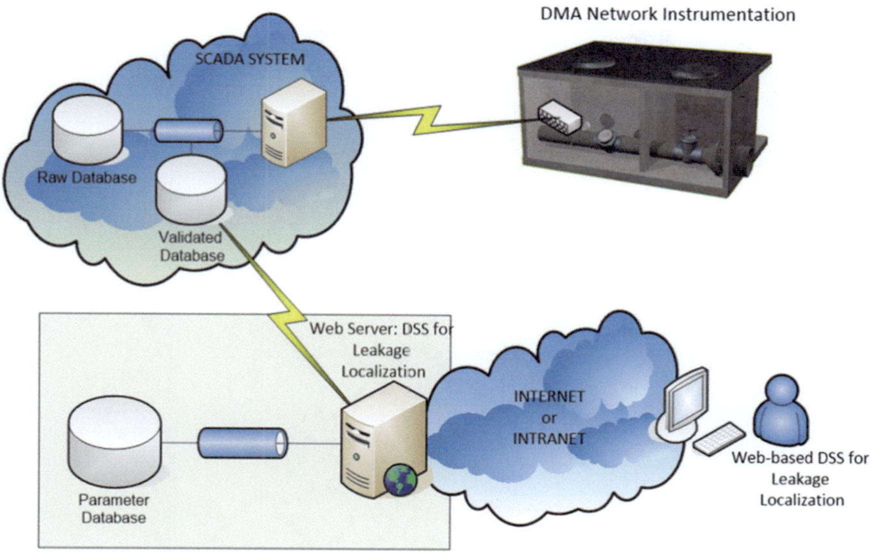

Fig. 7.1 A conceptual scheme of the decision support system for leak localization highlighting the main processes: online measurements gathered by the district metered area (DMA) instrumentation, data transmission from sensors' data loggers to the water company supervisory control and data acquisition system (SCADA), data validation process and validated database population and leak location analysis carried out in the leakage location tool using the DMA EPANET hydraulic model (parameter database)

19 December 2012, unaffected by the leakage event, has been plotted, showing the significant flow increase caused by the leakage.

The first stage in the leak localization process is to detect the occurrence of a new leakage scenario in the DMA. In general, a detection procedure followed by water utilities is based on the analysis of the difference between night flows. Although leakage is pressure dependent, night-time pressure is lower. The fact that at night the demand uncertainty is smaller makes the analysis of night flows more reliable than that of day flows. As shown in Fig. 7.3a, the total DMA inflow significantly increased on December 20 when the leak occurred compared to the previous day. The difference between these two flows (Fig. 7.3b) and the average difference is an estimate of the leak. In this case, the estimated size of the leak is about 5.6 l/s. The model-based leak location methodology requires the estimation of the emitter coefficient Ce, which according to Chap. 3 can be obtained using the estimated average size of the leakage (5.6 l/s, Fig. 7.3b) and an estimate of the average pressure at the leak location. This pressure value has been estimated to be about 50 mWC by averaging the measurements of the DMA inner pressure sensors (Fig. 7.4) corresponding to December 20. As a result, and using $\gamma = 0.5$ (the Darcy–Weisbach formula in Chap. 3), the estimated emitter coefficient is 0.8. The peaks in the leakage observed in Fig. 7.3 are modulated by the network pressure.

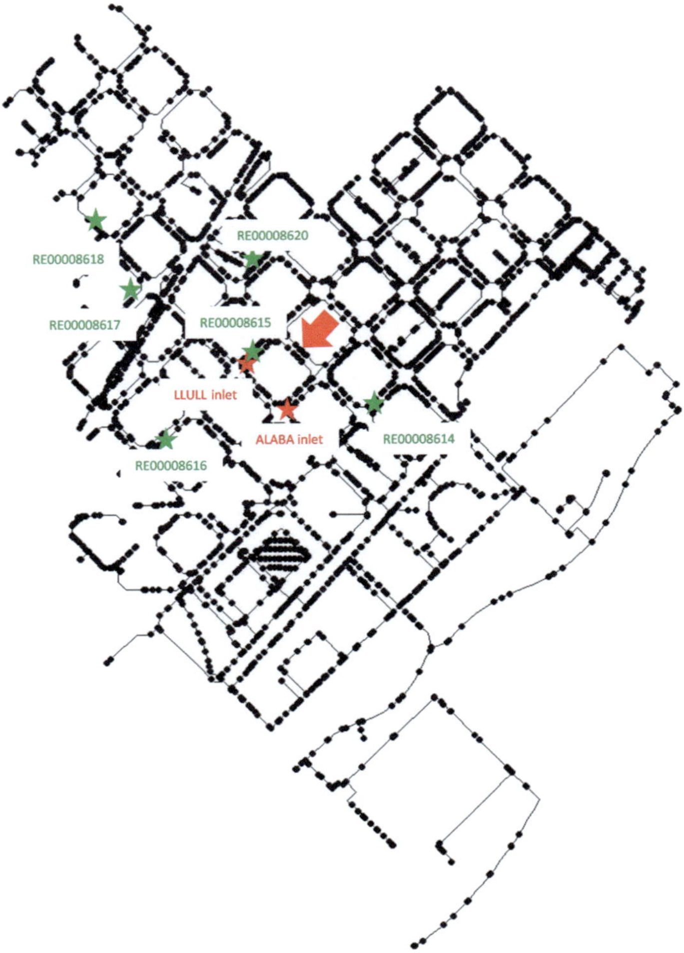

Fig. 7.2 The water network of Nova Icària district metered area (DMA) (EPANET model) highlighting inner pressure sensors (*green stars*), DMA inlets (*red stars*) and the exact location of the leak (*red arrow*) that was intentionally introduced to test the method

After the detection and size estimation of the leak, the calibration of the DMA hydraulic model and the inner pressure sensors is compared for verification, since existing model errors or poor calibrations may lead to low confidence in the performance of the leak localization methodology. To carry out the model verification, the data of December 19 were used since no major leaks were present that day. The general procedure to calibrate the DMA hydraulic model is derived from [25] where the pressure in the DMA inlets at time instant k is fixed, while the flow value in the inlets at this time instant is distributed among all the DMA inner nodes according to the values of their base demand and related demand patterns. The water demand

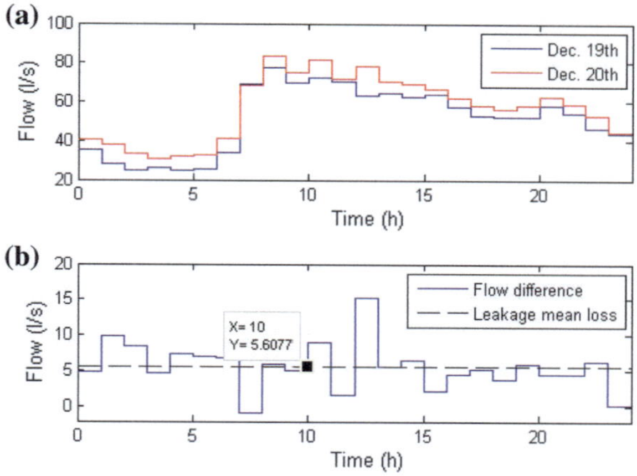

Fig. 7.3 Leak size estimation: **a** time evolution of the Nova Icària DMA total inflow on 20 December 2012 affected by the leak event and on 19 December 2012 unaffected by the leak and **b** time evolution of the difference between these two previous flows and its average value (leak size estimation)

model described in Chap. 3 is one of the main sources of uncertainty that may lead to inaccurate performance, and consequently, special attention should be paid to their calibration. In the application considered in this chapter, the base demand of the network nodes has been obtained from the billing information of this DMA by Aigües de Barcelona (see Chap. 5). Each base demand corresponds to the aggregation of consumers attached to a single node, assuring stability in demands.

As an output of the whole model calibration process, a calibrated model of the DMA hydraulic network at every time instant k is obtained, which can be used to predict flow values in the DMA inlets and the pressure in the monitored DMA inner nodes. In Fig. 7.5, the time evolution of the measured and predicted inflows at (a) Alaba and (b) Llull inlets for December 19 is observed, showing the degree of accuracy achieved after the calibration of the hydraulic model. According to the obtained results (Fig. 7.5), certain bounded modelling errors still exist, which may be due to the existing uncertainty in the hydraulic network parameters and the considered demand model [29]. Nonetheless, these inflows are acceptable to obtain a reliable performance of the leak localization methodology.

Regarding the DMA inner pressure sensors, divergence between real measurements and model-simulated values arises, as shown in Fig. 7.5, due to an inaccurate estimation of the depth of the measurement points. Indeed, before carrying out the leakage scenario considered here, the inner pressure sensor RE00008615 presented an abnormal behaviour and consequently was considered unreliable and excluded from the analysis. Thereby, assuming that no major leaks were present on December 19 and that the rest of the pressure sensors were reliable, the differences between measured and model-simulated pressures have been adjusted to correct topographic errors in the model. In Fig. 7.4, DMA inner pressure sensor measurements' time series, their resulting values after model correction and the corresponding model

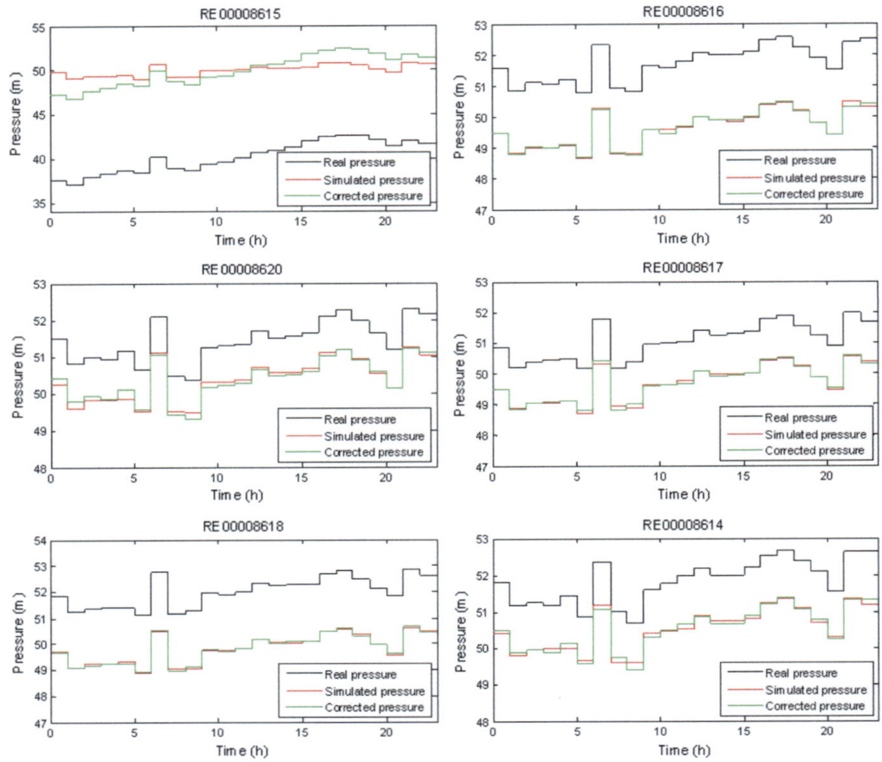

Fig. 7.4 The time evolution of the measurements of the inner pressure sensors (*black*), their model-simulated values (*red*) and the corrected measurement values (*green*). Only sensor RE00008615 shows a different trend and shape when simulated, in comparison with the measurement. Sensor RE00008615 is considered faulty because its mismatch cannot be attributed to bad topographic data, which was confirmed by the operators of the network. Thus, sensor RE00008615 should not be used in further leak localization analyses

predictions have been plotted for December 19. This figure shows that there is a mismatch between the sensor measurements and the predictions given by the model. After estimating the average value of this mismatch for every sensor and correcting the sensor measurements accordingly, the corrected measurements describe the model predictions. The correcting factor used to adjust the sensor measurements and the model predictions have been used to update the known, but inaccurate sensor depths when the leak localization methodology is applied on December 20 to perform leak localization.

Applying the data analysis described above, the occurrence of leakage on December 20 was detected and the quality of the calibrated DMA hydraulic model was evaluated using sensor measurements from December 19. The leak localization methodology was applied to analyse the sensor data of December 20 to obtain the most probable locations of the detected leak. This methodology has been packaged in

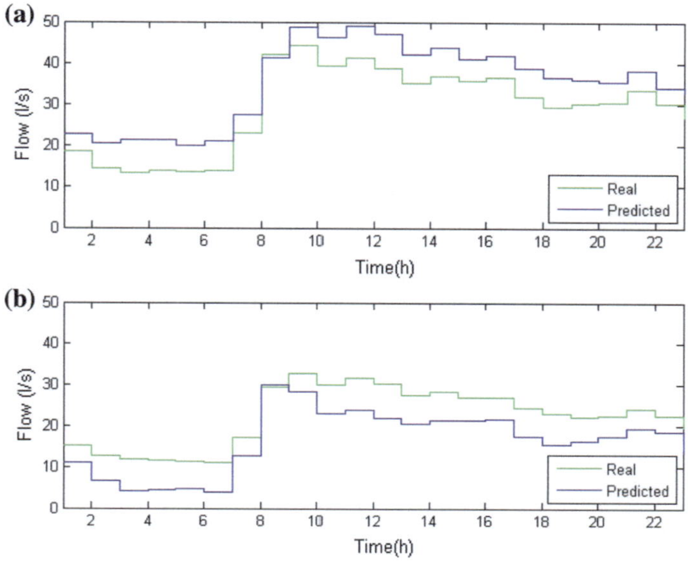

Fig. 7.5 The time evolution of the measured (*green*) and predicted (*blue*) inflows in **a** Alaba and **b** Llull inlets for December 19 showing the degree of accuracy achieved once the hydraulic model has been calibrated

a model-driven tool to make it easy to apply to different scenarios [30]. The tool provided an hourly result in the Nova Icària leakage scenario on December 20.

When applying this procedure to obtain the result for a certain hour, it must be taken into account that the used inner pressure sensors have a finite resolution, as discussed previously. This fixed resolution, together with the existing noise in the measurements, may be a source of inaccuracy in the computed results. To overcome the undesired effects derived from the sensor resolution, two main strategies have been considered. First, the sensor measurement considered at a certain hour is the result of applying an average filter to the measured values during the last $T_L = 1$ h (six measurements with a $T_s = 10$-min in (7.1)). Second, pressure measurements and model predictions from consecutive hours can be accumulated along a cumulative time window of a given length to obtain an accumulated observed residual and an accumulated sensitivity matrix. In the present case, a $H_M = 10$-h-length cumulative window has been used, so the observed residuals (r in (7.5)) and the sensitivity matrix (Ω in (7.9)) from the last 10 h are used to generate the resulting correlation vector (7.11) at each step, so that those nodes with the highest leak probability can be determined. To analyse the leakage scenario, data from December 20 to December 21 have been used, obtaining a leak correlation vector at each time instant (i.e., one per hour).

The value of the jth component determines the correlation between the observed residual and the theoretical fault signature (jth column of the Ω) predicted by the model for a leak placed in the jth node of the network (7.10). The correlation vector

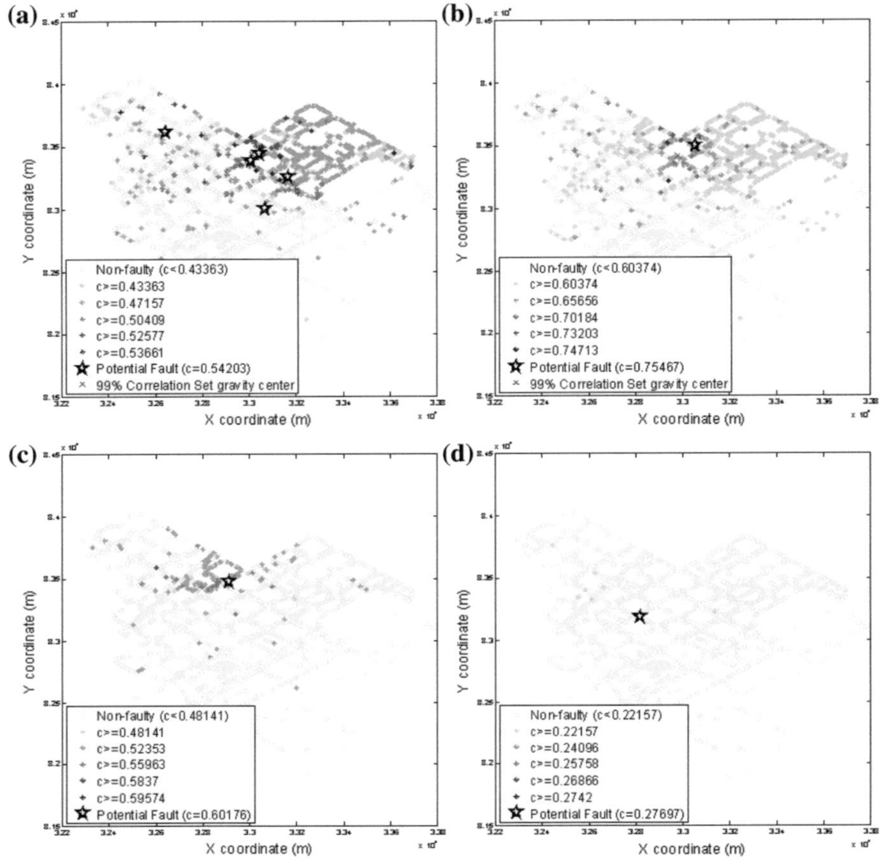

Fig. 7.6 Correlation vector obtained with the leak isolation methodology for December 20 at **a** $T_L = 14$ h, **b** $T_L = 15$ h and **c** $T_L = 22$ h and December 21 at **d** $T_L = 20$ h

can be represented graphically on the top of the DMA using a grey map where the highest correlations are darker than lower correlations. The level of grey depends on the degree of correlation obtained at every time instant, which means that the graphical representation associated with a certain time instant cannot be directly compared with the one of another time instant since the associated highest correlation value may be different. In this graphical representation, those nodes with the highest correlation value (c_{max}) are depicted with a black star. Additionally, a cross point to the centre of gravity of the set of nodes with high correlation (in this case, those whose correlation value c is greater than $0.99c_{max}$).

Figure 7.6 shows four graphical representations of the correlation vector obtained with the leak isolation methodology for December 20 at $T_L = 14$ h (Fig. 7.6a), $T_L = 15$ h (Fig. 7.6b) and $T_L = 22$ h (Fig. 7.6c), when the leak is still present, and December 21 at $T_L = 20$ h (Fig. 7.6d), when the leak is already fixed. Thus, Figs. 7.6a–c signals a

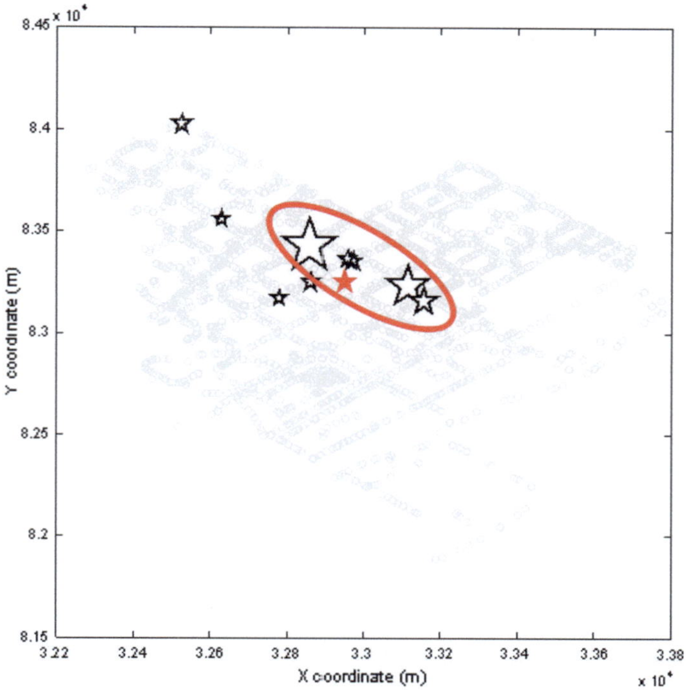

Fig. 7.7 A graphical representation (*black stars*) of the most probable localizations of the leak according to the accumulation of the resulting correlation vectors related to every node through a 48-h time window (December 20 and 21): the bigger the star size, the bigger are the corresponding accumulated correlations. Additionally, the real localization of the leak (*red star*) and the area containing those nodes with the highest correlation values (*red outline ellipsoid*) are also shown

potential leak moving around a small zone of the network with correlations oscillating between 0.6 and 0.75 (maximum correlation value is one). Figure 7.6d depicts the correlation vector after the leak is fixed, pointing out the meaningful decrease of the resulting highest correlation value regarding the cases when the leakage is still present.

In Fig. 7.7, the resulting correlation vectors obtained at every time instant during December 20 and 21 have been accumulated to determine the nodes with the highest correlation. Consequently, the most probable leak locations according to this 48-h time window are determined (only those correlation values higher than 0.5 are considered). The star size depends on the resulting value of the accumulated correlation. The bigger the star is, the bigger the corresponding accumulated correlations. Additionally, in Fig. 7.7, the real location of the leak has also been signalled using a red star and that area containing the nodes with the higher accumulated correlation values has been marked using an ellipsoid with a red outline. Comparing the leak localization indications given by the method to the real location of the leak, the resulting error is considered acceptable in the sense that the predicted area of the leak has an acceptable size containing the real location of the leak. It must be considered that the resulting error is mainly due to the inconsistency between the hydraulic and demand models and the sensor measurements. Note that a nodal leak localization using a small set of sensors determines potential network areas where the leak is located, rather than the exact node where the leak is. This situation occurs because, when using few sensors, there could be certain leaks causing the same pressure disturbance from the point of view of the used sensor network, and consequently, isolation among the potential leaks cannot be carried out.

7.5 Conclusions

This chapter presents a model-based methodology for leak localization in DWNs using pressure measurements. The method presented uses residuals obtained from the pressure measurements and their estimates from the network hydraulic model that characterizes the behaviour of the DMA without leakage. The residuals are compared with the leak sensitivity matrix that contains the predicted pressure disturbance caused by each potential leak in all of the monitored network's inner nodes (theoretical fault sensitivity). Leak isolation relies on correlating the observed residuals with the theoretical fault sensitivity contained in the leak sensitivity matrix. The leak localization methodology has been implemented in a software tool that interfaces with a geographic information system and allows the easy use by water network managers. Simulation results obtained by applying the method to a DMA of the Barcelona DWN highlight the effectiveness of the approach. Finally, a real application of this method on the Nova Icària DMA pilot test has been presented showing satisfactory results in a real fault scenario.

Regarding the future research related to this subject, several issues remain open. One research issue is to quantify the effect of uncertainty in demands, sensors and leak magnitude estimation on the methodology and accuracy of the leak localization procedure. Another related issue is to reduce the impact of uncertainty on the whole process. It is also of interest to extend the methodology to the detection and isolation of multiple leaks and to complement the methodology with a sensor's fault detection process, to guarantee that only valid sensor data are used for leak localization.

References

1. The World Bank (2008) Reducing water loss in developing countries using performance-based service contracting 44722. P-notes 2006(4):2006–2009
2. Puust R, Kapelan Z, Savic DA, Koppel T. A review of methods for leakage management in pipe networks. Urban Water J (1):25–45
3. Kapelan Z, Savic D, Walters G, Covas D, Graham I, Maksimovic C (2003) An assessment of the application of inverse transient analysis for leak detection: part I. In: International conference on advances in water supply management, London

4. Mounce SR, Boxall JB, Machell J. Development and verification of an online artificial intelligence system for detection of bursts and other abnormal flows. J Water Resour Plan Manage (3):309–318
5. Mounce SR, Mounce RB, Boxall JB. Novelty detection for time series data analysis in water distribution systems using support vector machines. Journal of Hydroinformatics (4):672
6. Palau CV, Arregui FJ, Carlos M. Burst detection in water networks using principal component analysis. J Water Resour Plan Manage (1):47–54
7. Romano M, Kapelan Z, Savić DA. Automated detection of pipe bursts and other events in water distribution systems. J Water Resour Plan Manage (4):457–467
8. Michele R, Zoran K, Savić DA (2013) Geostatistical techniques for approximate location of pipe burst events in water distribution systems. J Hydroinformatics 15:634–651
9. ADEC (2000) Technical review of leak detection technologies—vol. 1—Crude oil transmission pipelines. Technical report, Alaskan Department of Environmental Conservation, Alaska
10. Pilcher R (2007) Leak location and repair guidance notes and... the never ending war against leakage. In: Water loss 2
11. Hugenschmidt J, Kalogeropoulos A. The inspection of retaining walls using GPR. J Appl Geophys (4):335–344
12. Henault J-M, Moreau G, Blairon S, Salin J, Courivaud J-R, Taillade F, Merliot E, Dubois J-P, Bertrand J, Buschaert S, Mayer S, Delepine-Lesoille S. Truly distributed optical fiber sensors for structural health monitoring: from the telecommunication optical fiber drawling tower to water leakage detection in dikes and concrete structure strain monitoring. Adv Civ Eng 1–13
13. Colombo AF, Lee P, Karney BW. A selective literature review of transient-based leak detection methods. J Hydro-environ Res (4):212–227
14. Kim SH. Extensive development of leak detection algorithm by impulse response method. Journal of Hydraul Eng (3):201–208
15. Liggett J, Chen L. Inverse transient analysis in pipe networks. J Hydraul Eng (8):934–955
16. Vítkovský J, Simpson A, Lambert M. Leak detection and calibration using transients and genetic algorithms. J Water Resour Plan Manage (4):262–265
17. Farley B, Mounce SR, Boxall JB. Field validation of "Optimal" instrumentation methodology for burst/leak detection and location. In: Water distribution systems analysis (2010) American Society of Civil Engineers. Reston, VA, pp 1093–1102
18. Goulet J-A, Coutu S, Smith IFC. Model falsification diagnosis and sensor placement for leak detection in pressurized pipe networks. Adv Eng Inform (2):261–269
19. Pérez R, Puig V, Pascual J, Quevedo J, Landeros E, Peralta A. Methodology for leakage isolation using pressure sensitivity analysis in water distribution networks. Control Eng Pract (10):1157–1167
20. Pérez R, Sanz G, Puig V, Quevedo J, Escofet MAC, Nejjari F, Meseguer J, Cembrano G, Mirats Tur JM, Sarrate R. Leak localization in water networks: a model-based methodology using pressure sensors applied to a real network in Barcelona [Applications of control]. IEEE Control Syst (4):24–36
21. Sumer D, Lansey K. Effect of uncertainty on water distribution system model design decisions. J Water Resour Plan Manage (1):38–47
22. Sanz G, Perez R, Kapelan Z, Savic D (2015) Leak detection and localization through demand components calibration- ACCEPTED. J Water Resour Plan Manage
23. Wu Z, Sage P (2006) Water loss detection via genetic algorithm optimization-based model calibration. In: ASCE 8th international symposium on water distribution system analysis, Cincinnati
24. Pudar R, Liggett J. Leaks in pipe networks. J Hydraul Eng (7):1031–1046
25. Quevedo J, Cugueró M, Pérez R, Nejjari F, Puig V, Mirats J (2011) Leakage location in water distribution networks based on correlation measurement of pressure sensors. 8th IWA symposium on systems analysis and integrated assessment. IWA, San Sebastián, pp 290–297
26. Staroswiecki M, Blanke M, Kinnaert M, Lunze J (2006) Diagnosis and fault-tolerant control, 2nd edn. Springer, Berlin

27. Sanz G, Pérez R, Cugueró M-A, Cugueró J (2013) Accuracy assessment of leak localisation method depending on available measurements. In: CCWI2013, Perugia
28. Pérez R, Cugueró M-A, Cugueró J, Sanz G. Accuracy assessment of leak localisation method depending on available measurements. Procedia Eng 1304–1313
29. Pérez R, Nejjari F, Puig V, Quevedo J, Sanz G, Cugueró M, Peralta A (2011) Study of the isolability of leaks in a network depending on calibration of demands. In: 11th international conference on computing and control for the water industry, Exeter, pp 455–460
30. Meseguer J, Mirats-tur JM, Cembrano G, Puig V, Sanz G, Ibarra D, Quevedo J (2014) Environmental Modelling & Software A decision support system for on-line leakage localization n P e. 60:331–345

Chapter 8
Quality Monitoring

Fatiha Nejjari, Ramon Pérez and Vicenç Puig

8.1 Introduction

Model-based water quality monitoring is a reliable tool only if the water quality model is able to predict the real DWN behaviour [18]. Water quality modelling is not an easy task because of the complexity of the processes involved. Moreover, a hydraulic model analysis has to be performed previously in order to provide the resulting flow distribution to the water quality module to transport the chlorine through the network. The chlorine concentrations within a DWN are governed by bulk and wall reaction parameters. Bulk decay coefficients for chlorine depend on the nature of the water source and the treatment it has received, while wall decay coefficients depend on the pipe material and its condition. It is generally assumed that chlorine evolution in a DWN can be described by a first-order kinetic model. But, the bulk decay parameter can also be not first order, and some more reliable alternatives have been presented in some recent works as [5] or [3].

In the water quality model calibration, some of the parameters can be determined by laboratory tests and others have to be estimated by means of field measurements. Since manual trial-and-error method [6] is time-consuming, automated calibration procedure for those parameters is expected to provide better results. Zeirolf et al. [19] illustrated the use of input–output models for chlorine transport to estimate the first-order (global and zoned) wall reaction parameter. The model is applicable only for first-order reaction kinetics and does not incorporate storage tanks and multiple water quality sources. Al-Omari and Chaudhry [1] used finite difference procedures for the determination of the overall first-order chlorine decay coefficient. Munavalli and Kumar [12] developed an inverse model, which estimates the various reaction parameters in a multi-source steady-state distribution system.

F. Nejjari (✉) · R. Pérez · V. Puig
Research Center "Supervision, Safety and Automatic Control" (CS2AC-UPC), Terrassa, Spain
e-mail: fatiha.nejjari@upc.edu

V. Puig et al. (eds.), *Real-Time Monitoring and Operational Control of Drinking-Water Systems*, Advances in Industrial Control,
DOI 10.1007/978-3-319-50751-4_8

The changes in water quality that take place throughout the DWN are still quite difficult to reproduce by using mathematical models due to the complexities arising from varying hydraulic conditions and non-applicability of universal chlorine reaction kinetics. Several models that determine chlorine concentrations throughout a DWN have been described in the literature (see, e.g., [15] or [8] among others). Nowadays, three types of numerical models are used to calculate the changes of chlorine concentration in DWNs: the forward simulation model [6, 15], the inverse model [1, 8, 12] and the input–output model [17, 19].

The forward simulation of a water quality model, which tracks changes in chlorine concentration depending on time and localization, consists of the hydraulic model and elements describing chlorine transport along the pipe system. Chlorine propagation is described by a simplified transport equation and calculated numerically by Lagrangian time-driven method [10]. The predicted chlorine concentrations depend both on chlorine and on bulk and wall reaction parameters. All necessary input parameters needed for this model are either an overall first-order reaction parameter or bulk reaction parameter along with the wall reaction parameters depending upon the type of reaction rate expressions. The decay parameters are determined experimentally by the bottle test, and wall and overall reaction parameters are obtained by the calibration of data gathered in the field. The forward simulation model provides a great tool to calibrate the water quality model since it can estimate efficiently the parameters involved in overall first-order, first and zero-order wall reaction kinetics [14]. This model shows suitable correspondence of results and measured values in fixed hydraulic conditions and for turbulent flows ($Re > 10000$). The quality of correspondence increases with the distance between nodes and the water source. The source chlorine concentration is calculated by a trial-and-error procedure. But, the same trial-and-error procedure used for model calibration appears to be complicated and time-consuming.

Inverse modelling also denoted as automatic calibration consists in estimating the parameter values. An optimal parameter set is calculated automatically by minimizing an objective function, often defined as the summed squared deviation between the calibration targets (field data) and their simulated counterparts. Most inverse model techniques in addition to estimating optimal parameter values also produce calibration statistics in terms of parameter and observation sensitivities, parameter correlation and parameter uncertainties. The inverse chlorine decay model was developed by [8]. In this model, chlorine decay coefficients for each pipe in the network were known, chlorine concentrations at some of the nodes are known by specification, while chlorine concentration at the source was unknown. The objective was to determine the unknown source concentration, so that the specified and the calculated concentrations match. Islam et al. [8] presented an inverse model for directly calculating the chlorine concentrations needed at the system sources in unsteady flow conditions for meeting a specified concentration value at a particular node in the network. The model used a one-dimensional chlorine transport equation, which was discretized by using a four-point implicit finite difference scheme, and solved further simultaneously with mass balance equations and appropriate boundary conditions. The main drawback of [8] is that it is suitable only for an even-determined case in which the number of unknowns (nodal and source concentrations) equals the number of equations. Al-Omari and Chaudhry [1] extended Islam's model to the under-determined case where the number of unknowns is greater than the number of available equations and where there is more than one solution for which the prediction error is zero. Additional a priori information is added to the problem, and by minimizing the Euclidean length of the solution vector subject to the equations that describe chlorine transport in the network, a unique solution is generated. Laird et al. [9] presented an origin-tracking algorithm for solving the inverse problem of contamination source identification based on a nonlinear programming framework. The model considered the time delays of each pipe individually and scaled efficiently to large networks.

On the other hand, the input–output model developed by [19] is used to track chlorine propagation in a DWN and to calibrate the wall demand coefficient in a network without water storage tanks. This numerical model identifies all pipes covering flow paths between upstream and downstream sampling points and determines the wall reaction parameters for these pipes. Shang et al. [17] enhanced the basic input–output model with particle backtracking algorithm which made including water storage tanks and multiple water sources possible. In the input–output model using a particle backtracking algorithm, the output concentrations are modelled as a variable depending on input concentrations, network hydraulics and physical characteristics of the pipe network. The algorithm was mainly developed for feedback control providing information on all paths between pairs of nodes and less toward contamination source identification using monitoring stations' information.

Another important issue in water quality monitoring is the chlorine event that can be manifested by added variability and lower chlorine concentrations at sensor locations in the network. A DWN can be perceived as a complex chemical reactor in which various processes occur simultaneously. Some of these processes take place in the bulk phase and others at the pipe wall and can degrade water quality. Water quality events can be caused directly or indirectly by an internal corrosion due to an oxidation/reduction reaction, by detachment and leaching of pipe material or biofilm formation, by regrowth of micro-organisms on the internal surface and/or by a loss of disinfectant and formation of by-products (DBPs), or an intrusion of contaminants.

This chapter presents a methodology that enables to efficiently calibrate a water quality model such that the field-observed water quality values match with the model simulated ones. The calibration of the chlorine model consists in estimating the unknown parameters by comparing the measured and simulated chlorine concentrations at the monitoring nodes within the distribution system in a least square sense with a normalized quadratic cost function. The method is applied to a real DWN in Barcelona network and demonstrated that a water quality model can be optimized for managing adequate water supply to consumers and to perform further monitoring tasks (e.g., abnormal chlorine levels) in a more reliable way. The strategy consists in dividing the area of study in different zones and then estimating the chlorine bulk coefficients taking into account one-source or two-source DWN. The methodology represents accurately the process in the network improving the water quality prediction in the area of study and allowing the establishment of zones where the chlorine decay was significant. The simulations have been performed using the EPANET-MSX software package, MATLAB and the EPANET toolkit (see Chap. 2). Another important issue that is considered in this chapter is the water quality event detection and location. The proposed methodology is based on chlorine measurements and chlorine sensitivity analysis of the nodes of the network. Simulations of the water quality of the network done with realistic bulk decay and with an abnormal one provide an approximation of this sensitivity. An event location algorithm that correlates online the residuals, generated by comparing the available chlorine concentration measurements with their estimation values using a model with the sensitivity matrix is used. The correlation between the observed residual event signature and each column of the sensitivity matrix is a measure of the similarity of the residual effect concerning pipe bulk decay event due to pipe material detachment.

8.2 Problem Statement

To ensure the safe supply of drinking-water, the quality needs to be monitored online. The consequence of inadequate monitoring can result in substantial health and economic risks. Model-based monitoring requires modelling water quality in order to ensure the delivery of high-quality drinking-water. To design a model-based monitoring system, the first step consists in developing a water quality model that is able to reproduce the behaviour of the water quality when transported in the DWN. After selecting the type of model to be used from the set of existing models described in the introduction of this chapter, the next step is to calibrate the quality model. The calibration of the quality model for monitoring (e.g., chlorine) consists in estimating the unknown parameters by comparing the measured and estimated concentrations at the monitoring nodes within the DWN in a least square sense with a normalized quadratic cost function.

Another important issue is the detection and location of water quality events that can be manifested by added variability and lower chlorine concentrations at sensor locations in the network. Chemical reactions that take place both in the bulk phase and at the pipe walls can degrade water quality. Water quality events can be caused directly or indirectly by an internal corrosion due to an oxidation/ reduction reaction, by detachment and leaching of pipe material or biofilm formation, by regrowth of micro-organisms on the internal surface and/or by a loss of disinfectant and formation of by-products (DBPs), or an intrusion of contaminants.

8.3 Proposed Approach

This chapter presents a model-based approach for monitoring the chlorine concentration. As discussed in the previous section, the first step of this methodology is to calibrate a water quality model such that the field water quality measurements match with the estimations provided by the model. The calibration of the quality (chlorine) model consists in estimating the unknown parameters by comparing the measured and estimated chlorine concentrations at the monitoring nodes within the DWN in a least square sense with a normalized quadratic cost function. Since this function involves a non-explicit expression of the model, a GA is applied to optimize the model parameters by minimizing the difference between the model-predicted values and the field-observed ones.

The second phase of the proposed methodology is a water quality event detection and location approach based on sensor measurements and sensitivity analysis. Simulations of the water quality of the network carried out with normal and abnormal bulk decays provide an approximate way to perform such sensitivity analysis. An event location algorithm that correlates online the residuals, generated by comparing the available chlorine concentration measurements with their estimation values using the quality model, with the sensitivity matrix is used. The correlation between the observed residual quality event signature and each column of the sensitivity matrix is a measure of the similarity of the residual effect concerning pipe bulk decay event due to pipe material detachment. The validation and application of this methodology are illustrated with a district metering area (DMA) in the Barcelona network.

8.3.1 Chlorine Decay Model Calibration

The chlorine decay model has been calibrated by means of solving a least square problem that leads to an optimization problem. The least square problem with a normalized quadratic cost function given by

$$J(\theta) = \frac{1}{Nn_s} \sum_{k=1}^{N} \sum_{s=1}^{n_s} \left(C_s(k) - \hat{C}_s(k, \theta) \right)^2, \tag{8.1}$$

where θ are the quality parameters (bulk decay coefficients) to be tuned, N is the number of measurements, n_s is the number of sensors, $C_s(k)$ is the chlorine measurement for a sensor s at a certain instant k and $\hat{C}_s(k, \theta)$ is its estimation using a particular chlorine decay model. To solve the previous least square problem, the following optimization problem in (8.1) is solved:

$$\begin{aligned} &\min_{\theta} J(\theta) \\ &s.t. \\ &\qquad \hat{C}_s(k, \theta) = EPANET\_MSX\_simulation(k, \theta). \end{aligned} \tag{8.2}$$

Notice that since the estimation of the sensor measurement $\hat{C}_s(k, \theta)$ using a particular chlorine decay model requires a simulation using EPANET-MSX, the previous optimization problem does not have an analytical expression. Thus, it must be solved using a heuristic approach such as GAs [11].

The percentage error is obtained using the validation period of data as follows:

$$e_v = 100\left[1 - \left(\frac{\max\left(C_{s_v}(k)\right) - \sqrt{J_v}}{\max\left(C_{s_v}(k)\right)}\right)\right], \tag{8.3}$$

where $\max\left(C_{s_v}(k)\right)$ is the maximum measured value in the validation period among all the sensors used for calibration and J_v is the cost index in (8.1) obtained using validation data. The performance index (8.3) is useful in order to compare between different models and to evaluate the fit obtained between a particular calibrated model and the measured data.

8.3.2 *Quality Event Detection and Location*

This section proposes a detection and location method of abnormal water quality concentrations based on chlorine measurements and chlorine sensitivity analysis. Simulations of the water quality of the network done with realistic bulk decay and by introducing an abnormal one at a time at each pipe segment (link) provide an approximation of this sensitivity. An event isolation algorithm that correlates the residuals, generated by comparing the available chlorine concentration measurements with their estimation values using a model, with the event sensitivity matrix is used. The correlation between the observed residual event signature and each column of the sensitivity matrix is a measure of the similarity of the residual effect concerning pipe bulk decay event due to pipe material detachment. The same methodology has been applied to detect and isolate leakages in nodes of the network (see Chap. 6).

8.3.2.1 Event Detection Procedure

Model-based event detection and location rely on the use of analytical redundancy techniques to monitor the changes in the network water quality [7]. These techniques use the consistency check based on residuals $r(k)$ computation, obtained from measured input $u(k)$ and output $y(k)$ signals and the analytical relationships which are obtained by system modelling that provides an estimated output

$$r(k) = y(k) - \hat{y}(k). \tag{8.4}$$

At each time step k, the residual is compared with a threshold value (zero in the ideal case and almost zero in a real case). The threshold value is typically determined using statistical or set-based methods that take into account the effect of noise and model uncertainty [2]. When a residual is bigger than the threshold, it is determined that there is an event in the system; otherwise, it is considered that the system is working properly. In practice, because of the input–output noise, nuisance inputs and modelling errors affecting the model considered, robust residual generators must be used. Robustness can be achieved at the residual generation phase (active) or at the evaluation phase (passive) [7]. Robust residual evaluation allows obtaining a set of observed event signatures

$$\boldsymbol{\Phi}(k) = \left[\phi_1(k),\ \phi_2(k),\ \ldots \phi_{n\phi}(k)\right], \tag{8.5}$$

where each event indicator is obtained as follows:

$$\phi_i(k) = \begin{cases} 0 & if \quad |r_i(k)| \leq \tau_i(k), \\ 1 & if \quad |r_i(k)| > \tau_i(k), \end{cases} \tag{8.6}$$

and where $\tau_i(k)$ is the detection threshold associated with the residual $r_i(k)$.

8.3.2.2 Event Location Using the Correlation Method

Event location is carried out on the basis of observed event signatures, $\phi_i(k)$, generated by the detection module and its relation with all the considered events, $\mathbf{f}(k) = \left[f_1(k),\ f_2(k),\ \ldots,\ f_{nf}(k)\right]$, which are compared with the theoretical event signature matrix. The use of the information associated with the relationship between the residuals and events, by means of the residual event sensitivity, allows improving the isolation results. The sensitivity matrix contains the sensitivity of the chlorine sensor residuals for the different possible events f_j affecting the system, which is written as

$$\boldsymbol{\Omega} = \begin{matrix} & \begin{matrix} f_1 & f_2 & \cdots & f_m \end{matrix} \\ \begin{matrix} r_1 \\ r_2 \\ \vdots \\ r_n \end{matrix} & \begin{bmatrix} \omega_{11} & \omega_{12} & \cdots & \omega_{1m} \\ \omega_{21} & \omega_{22} & \cdots & \omega_{2m} \\ \vdots & \vdots & \ddots & \vdots \\ \omega_{n1} & \omega_{n2} & \cdots & \omega_{nm} \end{bmatrix} \end{matrix}, \tag{8.7}$$

where $\omega_{ij} = \frac{C_{ij}^f - C_{ij}^{nf}}{f_j}$ with C_{ij}^f and C_{ij}^{nf} being the mean values of the chlorine concentration measured by sensor i in normal situation and when an event occurs in node j at a certain time, when the bulk decay is abnormal or normal, respectively. The main idea of correlation-based event isolation consists in comparing the columns (events) of the sensitivity matrix (8.7) with the corresponding residual vector at time k by using the correlation function. The correlation coefficient $\rho_{r,\Omega_{(:,j)}}(k)$ between $\mathbf{r}(k)$ and each column j of $\boldsymbol{\Omega}$ (i.e., $\boldsymbol{\Omega}_{(:,j)}(k)$) is computed by means of the Pearson's correlation coefficient, which is defined as

$$\rho_{r,\Omega_{(:,j)}} = \frac{\mathrm{cov}(\mathbf{r}, \boldsymbol{\Omega}_{(:,j)})}{\sqrt{\mathrm{cov}(\mathbf{r}, \mathbf{r})\mathrm{cov}(\boldsymbol{\Omega}_{(:,j)}, \boldsymbol{\Omega}_{(:,j)})}}, \tag{8.8}$$

where $\mathrm{cov}(\mathbf{r}, \boldsymbol{\Omega}) = E\left[(\mathbf{r} - \bar{\mathbf{r}})\left(\boldsymbol{\Omega}_{(:,j)} - \bar{\boldsymbol{\Omega}}_{(:,j)}\right)\right]$ is the covariance function between $\mathbf{r}$ and $\boldsymbol{\Omega}_{(:,j)}$ being $\bar{\mathbf{r}} = E(\mathbf{r})$ and $\bar{\boldsymbol{\Omega}}_{:,j}(:,j) = E(\boldsymbol{\Omega}_{(:,j)})$, respectively. The columns of $\boldsymbol{\Omega}$ having higher correlation values with the residual vector $\mathbf{r}$ at time k are the most probable elements to have an event. The correlation between the observed residual fault signature (i.e., $\mathbf{r}(k)$) and each column of the matrix $\boldsymbol{\Omega}$ is a measure of the similarity between the real event residual effect (with unknown magnitude) and the events considered in matrix $\boldsymbol{\Omega}$ in (8.7) (with known magnitude) that allows discovering which is the column of this matrix (event) having the same behaviour. For events with similar size to those used to obtain the sensitivity matrix $\boldsymbol{\Omega}$, the correlation function obtains the maximum similarity (shape and form), i.e., $\rho = 1$ in the element having the event, for any magnitude of the real event. Because of the nonlinearity of the water network system, if the real magnitude of the event is far from the event size used to compute matrix $\boldsymbol{\Omega}$, the similarity of the correlation function decreases, but a high correlation between the residuals corresponding to a particular event and the corresponding column of the sensitivity matrix still exists. The vector obtained is the decision vector that will be used to figure out which is the event occurring in the system. More concretely, the maximum correlation value in this vector will point out the corresponding $\boldsymbol{\Omega}$ column (event) as the most probable node presenting the event, i.e.,

$$\max_{j} \left(\rho_{r,\Omega_{(:,j)}}(k)\right). \tag{8.9}$$

8.4 Simulation and Results

The calibration is applied to a part of the Barcelona DWN in order to estimate a water quality model that can be used for managing adequate water supply to consumers and to perform further monitoring tasks (e.g., abnormal chlorine levels) in a more reliable way [13]. The strategy consists in dividing the area under study into different zones and then estimating the chlorine bulk coefficients taking into account one- or two-source network and a first-order chlorine decay model (see Chap. 3). The methodology represents accurately the process in the network, improving the water quality prediction in the area of study and allowing the establishment of zones where the chlorine decay was significant. The simulations have been performed using the EPANET-MSX software package [16].

8.4.1 Calibration Case Study

8.4.1.1 Network Description

In order to estimate the chlorine decay coefficient, the origin of water and its path has to be known. Thus, the transport network is more suitable for determining the parameters of the quality model as it is less meshed than the distribution network (see Chap. 2). Nevertheless, the coefficients can be generalized to the whole network assuming the same water origin and pipe materials. The case study in this chapter is a part of the Barcelona DWN, a vast network of 4480 km of pipes which is divided into 113 different pressure levels. It corresponds to the pressure level 55th and has 438 nodes and 453 pipes (see Fig. 8.1). There are two inflows (i.e., Cantàbria and Drassanes) where flow, pressure and chlorine concentration are measured and four outputs (Llull, Àlaba, Passeig Colom and Joan de Borbó) where these variables are monitored as well. The flow and pressure data of these control points together with the data provided for some relevant demands allow the hydraulic adjustment of the model. Once the hydraulic model is available, the chlorine information is used for the chlorine model calibration.

8.4.1.2 Hydraulic Model Validation

To achieve a suitable water quality calibration, a properly calibrated hydraulic model is essential before starting water quality calibration. The accuracy of water quality simulation relies on the hydraulic simulation results.

From the hydraulic point of view, the network selected has two inputs, called Cantàbria and Drassanes, and 21 measured points. In order to simulate it isolated, the

Fig. 8.1 Pressure level 55th of Barcelona DWN

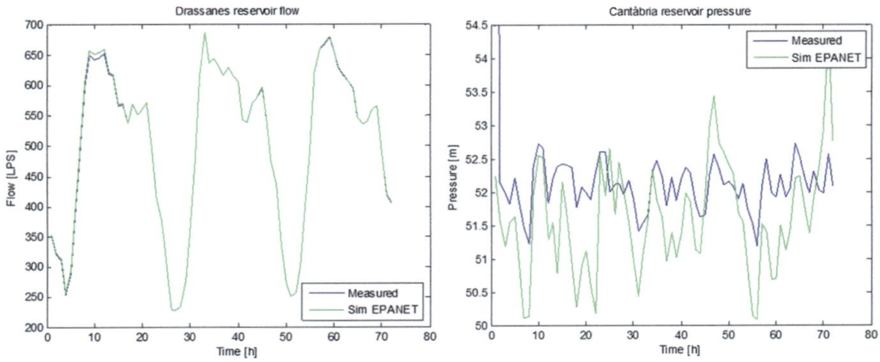

(a) Drassanes reservoir flow (b) Cantàbria reservoir pressure

Fig. 8.2 Comparison between the measured and simulated flows and pressures in Drassanes and Cantàbria reservoirs

boundary conditions have to be fixed. Boundary conditions are pressures and flows that are measured every hour at the network inputs. Pressures are fixed in the reservoir of Drassanes and flow in the reservoir of Cantàbria. After the hydraulic simulation, the resulting flows and pressures at Drassanes and Cantàbria are obtained and compared with the measured values, sampled every hour. In Fig. 8.2, a comparison between flows and pressures corresponding to the time range from 07/09/2010 (at 00:00 h) to 09/09/2010 (at 23:00 h) in Drassanes and Cantàbria reservoirs is depicted. The results obtained with the EPANET simulator show a proper validation and calibration of the hydraulic model when comparing with the measured data.

8.4.1.3 Chlorine Decay Calibration

The chlorine concentration data used are hourly sampled, measured from 03/09/2010 (at 00:00 h) to 14/09/2010 (at 23:00 h) in the sensors installed in the actual network (see Fig. 8.1). In order to perform the chlorine model calibration, the first 24 h of data are used to let the chlorine getting stabilized in the network. Hence, the data within this period are not used to calibrate the chlorine decay model (from 03/09/2010 (00:00) to 03/09/2010 (23:00)). The measurements comprised between 04/09/2010 (00:00) and 10/09/2010 (23:00) (7 days) have been used to calibrate the chlorine model, and the data comprised between 11/09/2010 (00:00) and 14/09/2010 (23:00) (4 days) have been used to validate the model. In this calibration process, real chlorine measurements in Cantàbria and Drassanes have been considered in the simulations as the injected chlorine in the network. These simulations have been used, in the model error function of the calibration procedure, to obtain the chlorine decay model parameters. The GA and Direct Search Toolbox included in the MATLAB R2009a release have been used in this work. Measurements in the four currently available sensors in Fig. 8.1 have been considered to compute the cost index (8.1). The figures in this section show the results obtained using the whole period of data (i.e., calibration and validation) by testing the model obtained in the calibration period (i.e., from 04/09/2010 (00:00) to 10/09/2010 (23:00)). In all these figures, the boundary between calibration and validation periods is depicted with a dashed line.

8.4.1.4 Case 1: Single Zone

Measurements in the four currently available sensors in Fig. 8.1 have been considered to minimize the cost index in (8.1). After the optimization process, the estimated chlorine decay constant is $K_{b_1} = 1.8024\,\mathrm{d}^{-1}$. The results obtained in the whole period of data (i.e., calibration and validation) are shown in Fig. 8.3. The cost value obtained for the calibrated single-zone first-order chlorine decay model is $J = 0.030669$. The corresponding average percentage error is $e_v = 19.41\%$. As it may be observed from Fig. 8.3, a reasonable proper fit is obtained for the measurement points in Joan de Borbó and Passeig Colom, but the fit is not as suitable as for measurement points in Àlaba and Llull.

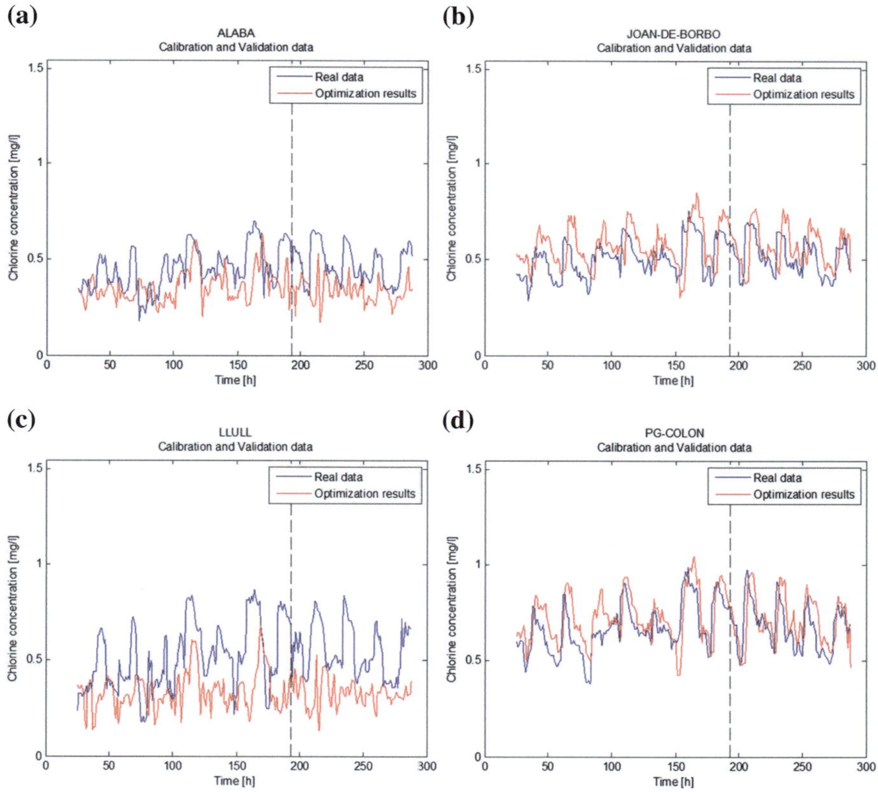

Fig. 8.3 Validation of a one-zone chlorine decay model in pressure level 55th of Barcelona DWN

8.4.1.5 Case 2: Two Zones

The results presented for the Case 1 show how, while in Joan de Borbó and Passeig Colon measurement points a reasonable proper fit between real data and the computed values is achieved, in Alaba and Llull measurement points the model fit is not that suitable. This could be explained because Joan de Borbó and Passeig Colon always receive chlorinated water from a single water source, while Alaba and Llull may receive chlorinated water from two different sources (e.g., Drassanes and Cantàbria) depending on the network demands in a particular time. A possible and more accurate approach to model this behaviour is to divide the network into two different zones, southern (Zone I) and northern (Zone II), and assign a different chlorine decay model to each one. The two selected zones are depicted in Fig. 8.4.

As in the previous calibrations, real chlorine measurements in Cantàbria and Drassanes have been considered as the injected chlorine points in the network. The obtained estimated chlorine decay constants after the calibration process for the two different zones are as follows: $K_{b_1} = 1.5469$ d^{-1} and $K_{b_2} = 1.9213$ d^{-1}. Results obtained in the whole period of data (i.e., calibration and validation) are shown in Fig. 8.5. The cost value obtained for the calibrated two zones using the first-order chlorine decay model is $J = 0.02913$. The corresponding average percentage error is $e_v = 18.64\%$.

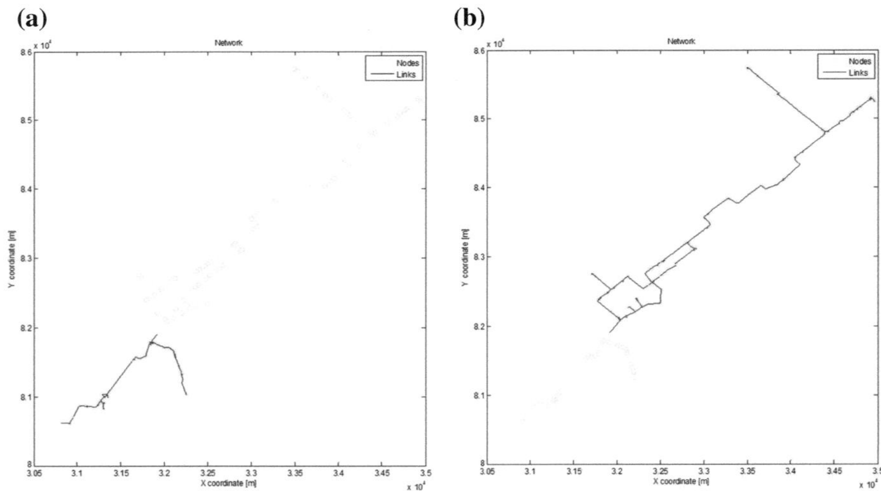

Fig. 8.4 Two zones selected in pressure level 55th of Barcelona DWN, with a different chlorine decay model each

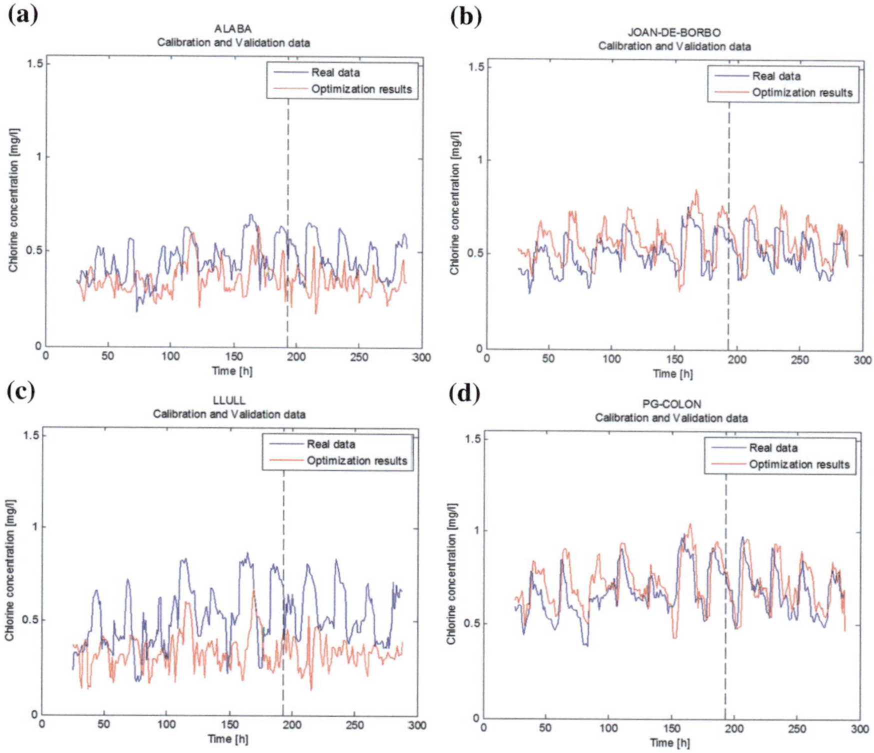

Fig. 8.5 Validation of a two-zone first-order chlorine decay model in pressure level 55th network

8.4.1.6 Case 3: Three Zones

In this case, an extra zone is added to obtain a better model of the chlorine decay through the network in Fig. 8.1. The new zone distribution considered is shown in Fig. 8.6.

As in the previous calibrations corresponding to the previous cases, real chlorine measurements in Cantàbria and Drassanes have been considered as the injected chlorine in the network. Measurements in the four installed sensors have been considered to minimize the cost index in (8.3). The estimated chlorine decay constants after the calibration process for the three different zones are $K_{b_1} = 2.3544\ \mathrm{d}^{-1}$ and $K_{b_2} = 0.56331\ \mathrm{d}^{-1}$ and $K_{b_3} = 0.8686\ \mathrm{d}^{-1}$. The cost value obtained for the calibrated three zones using the first-order chlorine decay model is $J = 0.019871$, and the corresponding average percentage error is $e_v = 13.90\%$ (Fig. 8.7).

Good results have been obtained with the suggested alternative models improving the original model error from 19.41% to 13.90% when using the three-zone model. This leads to consider this better approach instead of the original one, in order to achieve more accurate monitoring of the chlorine evolution through the network. The calibration methodology and the different models proposed should help water companies to obtain a better estimation of the chlorine evolution through a particular

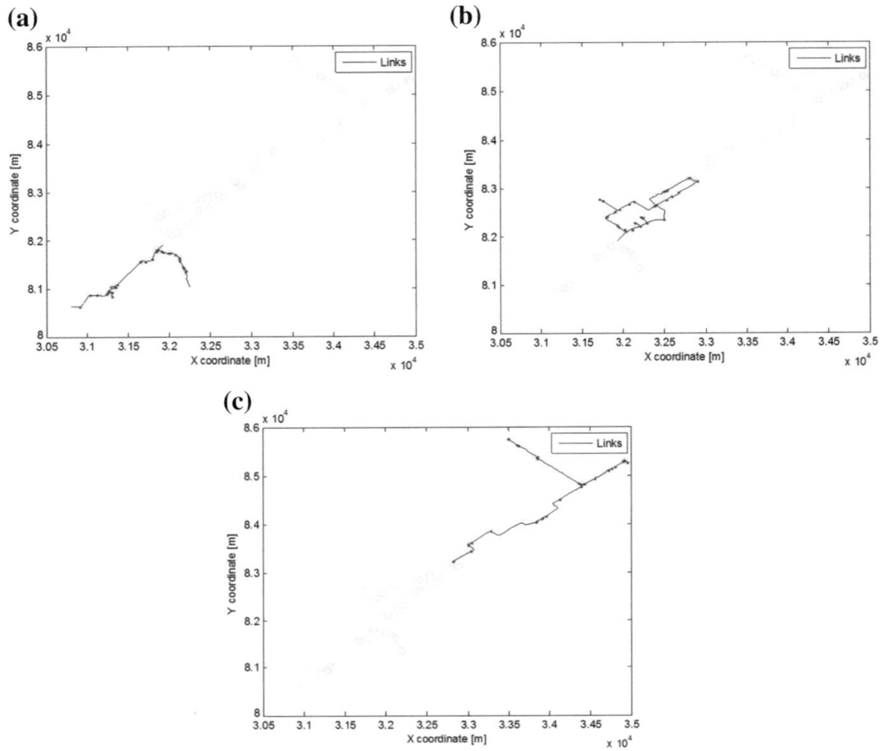

Fig. 8.6 Three zones selected in pressure level 55th network, with a different chlorine decay model

network by simulation and to use this information to perform further monitoring tasks as chlorine event detection in a more reliable way. There are other chlorine decay model approximations which could be taken into account in order to improve the achieved calibration results. This may include different values for the order of the chlorine decay as suggested in [4] or different distributions of the decay zones, e.g., variable zones through time.

8.4.2 Abnormal Quality Detection and Isolation

A DMA of a real water distribution network (Fig. 8.8) was used to verify the proposed method for abnormal water quality detection and location. This DMA, located in Nova Icària area of Barcelona, is included in the 55th pressure level within the city network and has 1996 nodes and 3442 pipes (see Chap. 2). The DMA has two inputs, called Alaba and Llull, and three chlorine sensors, which are also shown in Fig. 8.8.

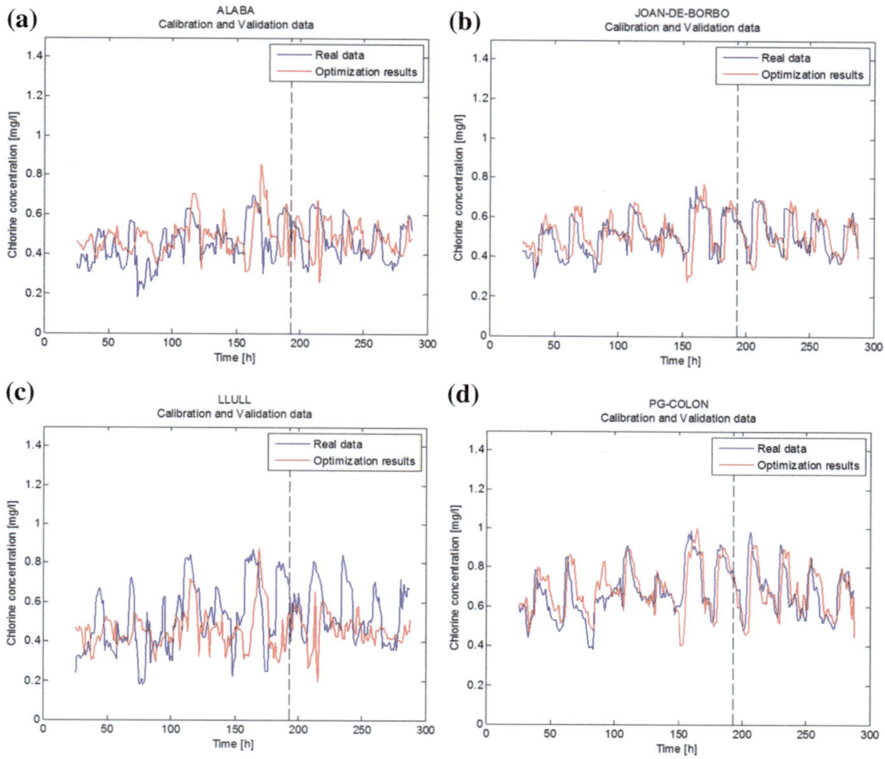

Fig. 8.7 Validation of a three-zone first-order chlorine decay model in pressure level 55th network

Scenarios have been generated using the EPANET simulation software, considering these three sensors already installed in the network (see Fig. 8.8). The events have been generated setting the initial quality and source quality parameters of the injectors Alaba (RE) and Llull (PC3) in Fig. 8.9. The scenarios have been generated changing the K_b parameter at the corresponding links of the network at 08:00 of day 2 (32nd h of the episode). The values of normal/abnormal behaviours on these links are $K_b = -2.306415\ \text{d}^{-1}$ for the normal mode and $K_b = -50\ \text{d}^{-1}$ for the abnormal mode. The scenario has been run for 48 h, and datasets have been created according to the considered scenarios.

Figure 8.10 shows the number of bulk chlorine decay events that may be detected using the sensors located in Fig. 8.8 after 96 h of simulation. It may be seen how, with the sensors already installed, the number of events that may be detected is about 500. The improvement of detection is due to the dynamic of the chlorine concentration within the system. These links are distributed geographically as shown in Fig. 8.11. The constraints in the detectability of events are due to the geographic distribution of the sensors.

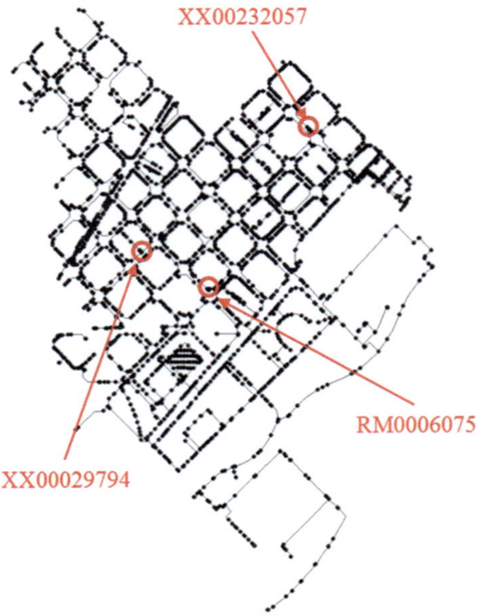

Fig. 8.8 The layout of Nova Icària network with the 3 installed chlorine sensors

Three different links of the network, included in the set of detectable events in Fig. 8.10, are considered to generate the event scenarios. These links labelled

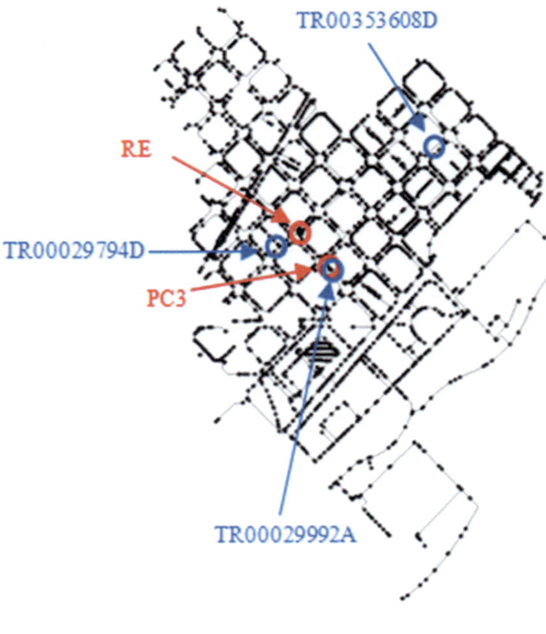

Fig. 8.9 Location of chlorine injectors (*red*) and links with chlorine decay event (*blue*)

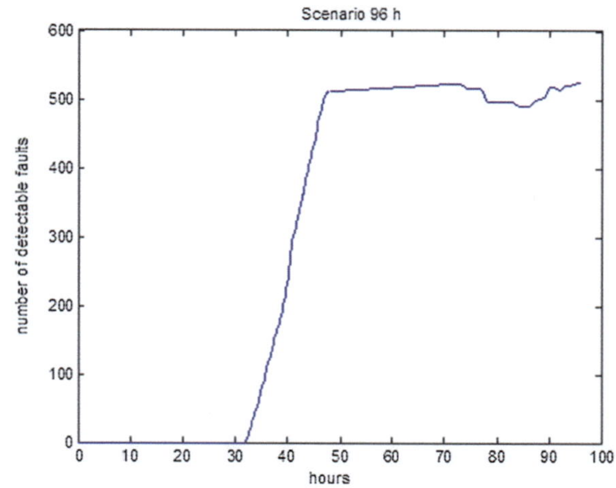

Fig. 8.10 Number of bulk chlorine decay events using 3 preinstalled sensors

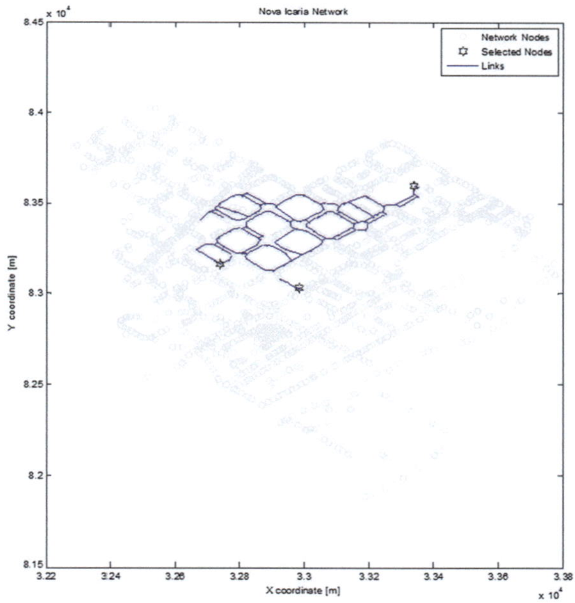

Fig. 8.11 Geographic distribution of events using 3 preinstalled sensors

"TR00353608D", "TR00029992A" and "TR00029794D", respectively, are shown in Fig. 8.9.

The simulated data are obtained using the methodology presented in Sect. 8.3.2. Figures 8.12, 8.13 and 8.14 show the event detection results obtained at hour 48 for every chlorine bulk decay event. In these figures, the link with the actual event is represented with a diamond, and the starred nodes are the potential abnormal nodes suggested by the correlation method. The rest of the nodes are divided in grey-scaled

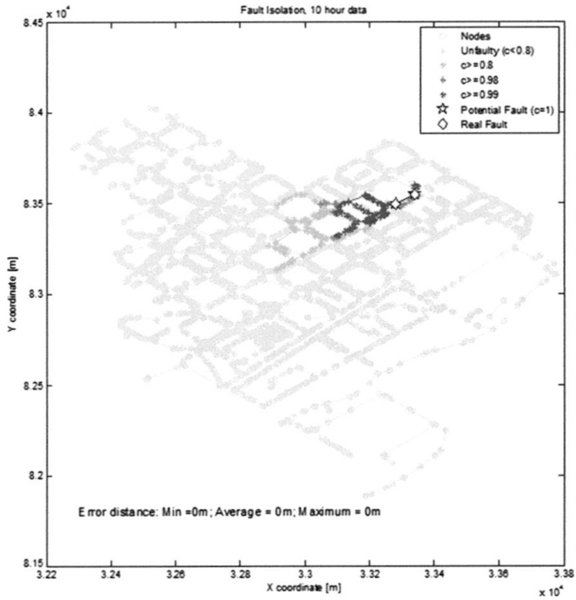

Fig. 8.12 Chlorine bulk decay event detection in link TR00353608D at hour 48

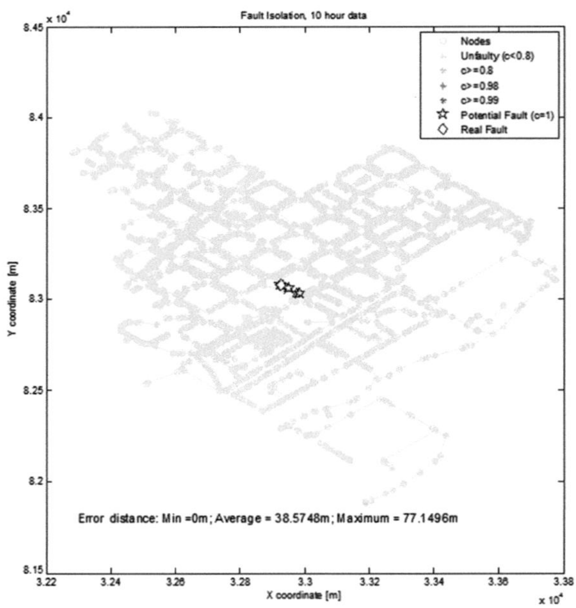

Fig. 8.13 Chlorine bulk decay event detection in link TR00029992A at hour 48

areas, depending on the correlation value they have: the most correlated with the event signature, the darkest the area they are located.

The evolution of the error distance through the period of results' generation (15–24 h) and a detail of the detection for each event are depicted in Figs. 8.15, 8.16 and

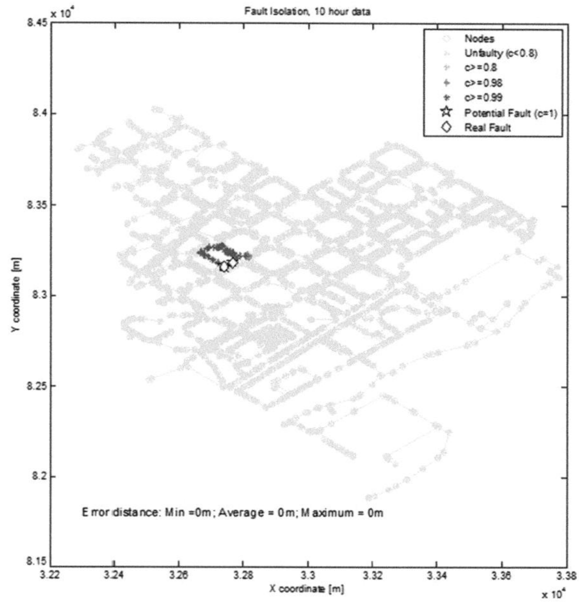

Fig. 8.14 Chlorine bulk decay event detection in link TR00029794D at hour 48

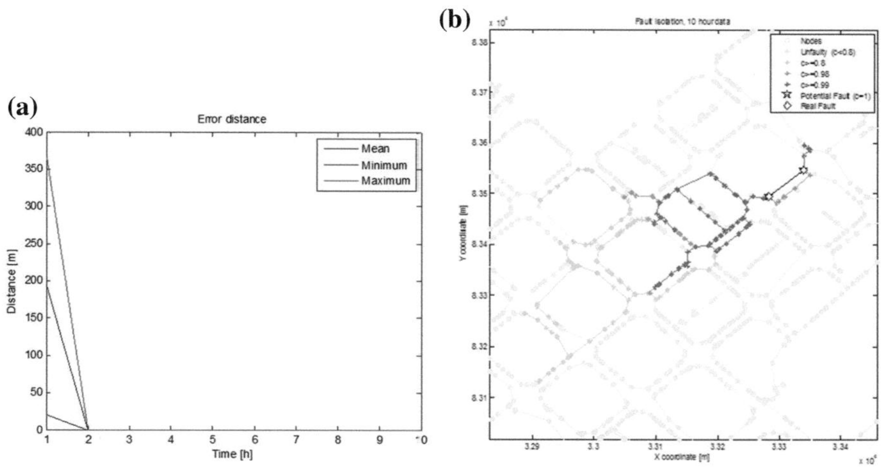

Fig. 8.15 Error distance and geographic detail of chlorine bulk decay event detection for the link TR00353608D at hour 48

8.17. In the latter, the distance between the potential abnormal links (starred) and the actual abnormal link (diamonds) is represented. If more than one potential abnormal link is obtained at a particular time, minimum, maximum and mean distances among potential abnormal links' set and actual abnormal link are represented. For all cases, a proper match between the actual and the identified event has been achieved (Figs. 8.12, 8.13 and 8.14).

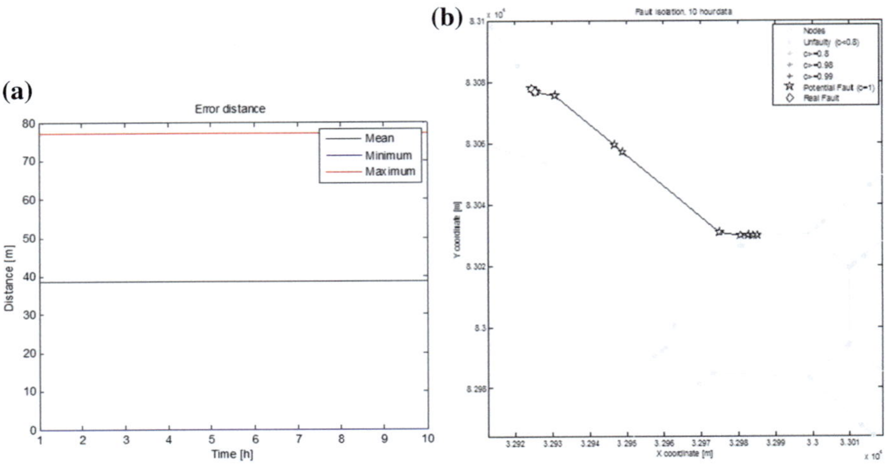

Fig. 8.16 Error distance and geographic detail of chlorine bulk decay event detection for the link TR00029992A at hour 48

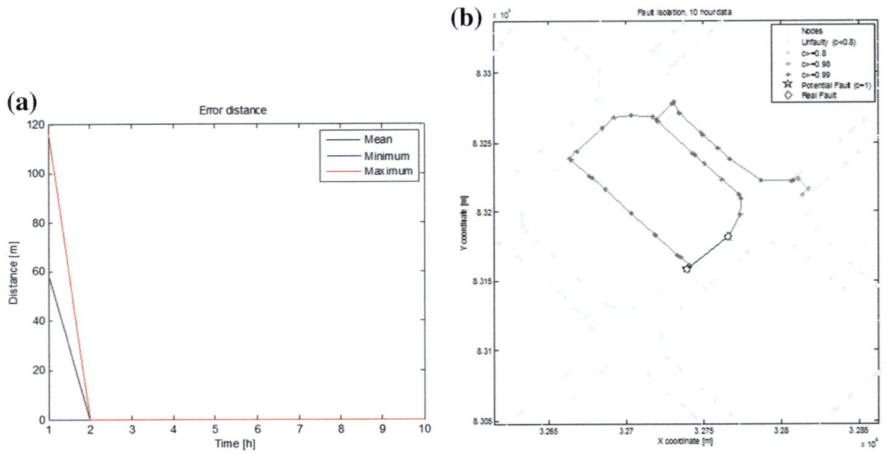

Fig. 8.17 Error distance and geographic detail of chlorine bulk decay event detection for the link TR00029794D at hour 48

8.5 Conclusions

In this chapter, the problem of detection and location of chlorine concentration events has been addressed using a model-based approach. This method is based on the chlorine measurements and chlorine sensitivity analysis of nodes in a DWN. An event location algorithm that correlates online the residuals (generated by comparing the available chlorine measurement with their estimation using a model) with the event sensitivity matrix is used. The proposed algorithm has been applied in simulation to a DMA (Nova Icària) in the Barcelona network using EPANET software. The effectiveness of the method ensures the ability for a rapid response to an abnormal quality event and, consequently, minimizes exposure risks to water consumers. The solution should help water companies to estimate the presence and the location of a bad chlorine concentration through a simple field data analysis.

The model-based water quality event detection and location methodology requires a reliable model that is obtained through the chlorine decay calibration procedure. The decay parameters appear to be highly sensitive to the water origin. Thus, the model calibration process should be previously carried out in any supply system before using it for supervision and monitoring.

References

1. Al-Omari AS, Chaudhry MH (2001) Unsteady-state inverse chlorine modeling in pipe networks. J Hydraul Eng ASCE 127:669–677
2. Blanke M, Kinnaert M, Lunze J, Staroswiecki M (2006) Diagnosis and fault-tolerant control, 2nd edn. Springer
3. Boccelli DL, Tryby ME, Uber JG, Summers RS (2003) A reactive species model for chlorine de-cay and thm formation under rechlorination conditions. Water Res 37:2654–2666
4. Chang EE, Chiang PC, Chao SH, Lin YL (2006) Relationship between chlorine consumption and chlorination by-products formation for model compounds. Chemosphere 64:1196–1203
5. Clark RM (1998) Chlorine demand and tthm formation kinetics: a second-order model. J Environ Eng ASCE, 124:16–24
6. Clark RM, Rossman LA, Wymer LJ (1995) Modeling distribution system water quality: regulatory implications. J Water Res Plan Manag ASCE 121:423–428
7. Gertler J (1998) Fault detection and diagnosis in engineering systems. Marcel Dekker Inc, New York
8. Islam MR, Chandhry MH, Clark RM (1997) Inverse modeling of chlorine concentration in pipe networks under dynamic condition. J Environ Eng 123:1033–1040
9. Laird CD, Biegler LT, van Bloemen Waanders BG, Bartlet RA (2005) Contamination source determination for water networks. J Water Res Plan Manage 131:125–134
10. Liou CP, Kroon JR (1987) Modeling the propagation of waterborne substances in water distribution networks. J Am Water Works Assoc 79:54–58
11. The MathWorks (2009) Genetic algorithm and direct search toolbox. J Water Res Plan Manag ASCE, Version 2.4.1 (R2009a)
12. Munavalli GR, Kumar MS (2003) Water quality parameter estimation in a steady state distribution system. J Water Res Plan Manag ASCE 129:124–134
13. Nejjari F, Perez R, Puig V, Quevedo J, Sarrate R, Cugueró M, Sanz G, Mirats JM (2012) Abnormal quality detection and isolation in water distribution networks using simulation models. Drink Water Eng Sci 5:67–72
14. Munavalli GR, Mohan Kumar MS (2005) Water quality parameter estimation in a distribution system under dynamic state. Water Res 39:4287–4298
15. Rossman LA, Boulos PF, Altman T (1993) Discrete volumeelement method for network water—quality models. J. Water Res Plan Manage ASCE, 119:505–517
16. Rossman Lewis A (2000) EPANET 2 users manual. Environmental Protection Agency, U.S
17. Shang F, Uber JG, Polycarpou MM (2002) Particle backtracking algorithm for water distribution system analysis. J Env Eng 128:441–450
18. Wu ZY (1995) Optimal calibration method for water distribution water quality model. J Env Sci Health Part A, 41:1–16

19. Zeirolf ML, Polycarpou MM, Uber JG (1998) Development and autocalibration of an input-output model of chlorine transport in drinking-water distribution systems. IEEE Trans Control Syst Technol 6:543–553

Chapter 9
Sensor Placement for Monitoring

Ramon Sarrate, Fatiha Nejjari and Joaquim Blesa

9.1 Introduction

As discussed in Chaps. 7 and 8, locating leaks and quality events is of great importance in water networks. Leaks can lead to economic losses while quality events can compromise the safety of the water supply. The location capability for system monitoring highly depends on the set of real-time measurements that are available. Thus, for a water management company, deciding which sensors to install is the key to the success of a monitoring system.

Water loss due to leak in pipelines is one of the main challenges in efficient DWNs. Leaks in DWNs can occur because of damages and defects in pipes, lack of maintenance or uncontrolled increases in pressure. Leaks can cause significant economic losses and must be detected and located as soon as possible to minimize their effects. Continuous improvements in water loss management are being applied, and new technologies are developed to achieve higher levels of efficiency [16].

Methods for locating leaks range from ground-penetrating radar to acoustic listening devices [9]. However, techniques based on locating leaks from pressure/flow monitoring devices allow a more effective and less costly search in situ. The need to identify the location of leaks has promoted the development of several techniques based on the inverse problem and solving it using pressure or flow measurements (see Chap. 7). In the last years, different works that deal with the topic of leak location in DWNs using pressure sensors have been published. Some of these last works tackle with the problem of leak location using the sensitivity matrix [4, 15], which contains

R. Sarrate (✉) · F. Nejjari · J. Blesa
Research Center "Supervision, Safety and Automatic Control" (CS2AC-UPC), Terrassa, Spain
e-mail: ramon.sarrate@upc.edu

J. Blesa
Institut de Robòtica i Informàtica Industrial CSIC-UPC, Barcelona, Spain

V. Puig et al. (eds.), *Real-Time Monitoring and Operational Control of Drinking-Water Systems*, Advances in Industrial Control,
DOI 10.1007/978-3-319-50751-4_9

the information about how leaks affect the different node pressures (see Chap. 7 for a more detailed state of the art in this topic).

These techniques are based on the sensors installed in the network. Ideally, sensor placement should facilitate leak detection and location maximizing the monitoring performance under a given sensor cost budget. In DWNs, only a limited number of sensors can be installed due to these budget constraints. Since improper selections may seriously hamper monitoring performance, the development of sensor placement strategy has become an important research issue in recent years.

The sensor placement problem can be roughly stated as choosing a subset of sensors, from a given candidate sensor location set, such that some monitoring performance is guaranteed or at least maximized. Since installing sensors will involve a cost for the water management company, economic constraints must be additionally taken into account in the choice. Sensor placement entails formulating a combinatorial optimization problem. In such problems, an exhaustive search of the solution is usually not feasible, since its complexity grows exponentially with the number of candidate sensor locations. A DWN may easily involve several thousands of candidate sensor locations, which poses a severe optimization challenge. Thus, the sensor placement methodology should be able to cope with such complexity issues.

Some results devoted to sensor placement for leak monitoring can be found in [10, 14, 20, 21]. All these works use a structural model-based approach and define different leak monitoring specifications to solve the sensor placement problem. A structural model is a coarse model description, based on a bipartite graph, which can be obtained early in the development process, without major engineering efforts. This kind of model is suitable to handle large-scale systems since efficient graph-based tools can be used and does not have numerical problems. Structural analysis is a powerful tool for early determination of fault diagnosis performances [1]. In [20], an algorithm is developed to determine where to install a specific number of pressure sensors in a DMA in order to maximize the capability of detecting and locating leaks. The number of sensors to install is limited in order to satisfy a budgetary constraint requirement. However, in this case, despite using an efficient branch-and-bound search strategy based on a structural model, the approach applicability is still limited to medium-sized networks. To overcome this drawback, the methodology is combined with clustering techniques [19].

On the other hand, optimal pressure sensor placement algorithms based on sensitivity matrix analysis have been developed to determine which pressure sensors have to be installed among hundreds of possible locations in the DWN to carry out an optimal leak location as in [5, 6, 18]. The sensitivity matrix can be obtained by convenient manipulation of model equations as long as leak effects are included in them [3]. Alternatively, it can be obtained by sensitivity analysis through simulation [15]. The elements of this matrix depend on the operating point defined by the heads in reservoirs, the inflow, demand distribution, which is not constant, and the leak magnitudes, which are unknown. In [2], a robustness analysis of the sensor placement problem in DWNs has been addressed. The study has been achieved by optimal sensor placement strategies for different leak magnitudes and DMA operating points and evaluated through a robustness percentage index.

9.2 Problem Statement

9.2.1 Model-Based Fault Diagnosis

As discussed in Chap. 7, model-based leak monitoring relies on the general theory of model-based fault diagnosis [1]. The basic idea behind this approach is the comparison between the observed process behaviour and its estimation using a model. This is performed by means of consistency relations, which can be roughly described as a function of the form

$$h(\mathbf{y}(t), \mathbf{u}(t)) = 0, \tag{9.1}$$

where $\mathbf{y}(t)$ and $\mathbf{u}(t)$ are vectors of known variables, denoting, respectively, process measurements and process control inputs. Function h is obtained from the model and is the basis to generate a residual

$$r(t) = h(\mathbf{y}(t), \mathbf{u}(t)). \tag{9.2}$$

A residual is a temporal signal indicating how close is behaving the process compared with its expected behaviour predicted by the model. In the absence of faults, a residual equals zero. In fact, a threshold-based test is usually implemented in order to cope with noise and model uncertainty effects. Otherwise, when a fault is present, the model is no longer consistent with the observations (known process variables) and the residual exceeds the prefixed threshold.

Detecting faults is possible with only one residual sensitive to all faults. However, fault isolation is usually required rather than just detecting the presence of a fault. The fault isolation task is performed by designing a set of residuals based on several consistency relations. Each residual is sensitive to different faults such that the residual fault signature is unique for each fault. Therefore, distinguishing the actual fault from other faults is possible by looking at the residual fault signature. In case that the faults are leaks, the fault signatures are collected in the sensitivity matrix denoted by $\mathbf{\Omega}$ as explained in Chap. 7 and written as

$$\mathbf{\Omega} = \begin{pmatrix} \frac{\partial r_1}{\partial f_1} & \cdots & \frac{\partial r_1}{\partial f_{n_d}} \\ \vdots & \ddots & \vdots \\ \frac{\partial r_{n_y}}{\partial f_1} & \cdots & \frac{\partial r_{n_y}}{\partial f_{n_d}} \end{pmatrix}, \tag{9.3}$$

where n_y is the number of the available residuals and n_d the number of nodes where a leak might occur. When an element ω_{ij} of $\mathbf{\Omega}$ is close to zero, then residual r_i is weakly sensitive to the leak $f_j \in \mathcal{F}$, being $\mathcal{F}$ the set of leaks that must be monitored, whereas when it diverges from zero then the residual is strongly sensitive to the leak $f_j \in \mathcal{F}$.

In Chap. 7, $\mathbf{\Omega}$ is roughly estimated as follows:

$$\mathbf{\Omega} \simeq \frac{1}{f^0}(\hat{\mathbf{r}}_{f_1}, \ldots, \hat{\mathbf{r}}_{f_{n_d}}), \tag{9.4}$$

where $\hat{\mathbf{r}}_{f_i}$ is the predicted residual considering a leak in node i with magnitude f^0.

Sometimes a binary version of the sensitivity matrix is used. Then, the corresponding binary residuals are usually called structured residuals, whereas in the non-binary matrix they are referred to as directional residuals.

In model-based diagnosis, fault detectability and fault isolability are the main properties that the set of residuals should satisfy [1]. Assuming structured residuals, a fault is detectable if its occurrence can be monitored, whereas a fault $f_i \in \mathcal{F}$ is isolable from a fault $f_j \in \mathcal{F}$ if the occurrence of f_i can be detected independently of the occurrence of f_j.

9.2.2 Optimal Sensor Placement

Sensors measure water physical magnitudes such as pressure, flow, tank level or chlorine concentration. The aim of the sensor placement for monitoring can be roughly stated as the choice of a sensor configuration such that the monitoring performance specification is maximized. Nevertheless, this may lead to a solution involving a large instrumentation cost. A baseline budget is usually assigned to instrumentation by water network companies which should constraint the maximum cost of the sought sensor configuration and consequently will bound the achievable monitoring performance. Thus, water companies rather seek the best monitoring performance that can be achieved by installing the cheapest sensor configuration that satisfies a budget constraint. This chapter focuses on pressure sensor placement for leak monitoring although the methodology could be adapted to quality event monitoring (Chap. 8) or general fault monitoring (Chap. 10).

Let $\boldsymbol{\mathcal{S}}$ be the set of candidate pressure sensors and m_p the maximum number of pressure sensors that can be installed in the water network according to the budget constraint. Just sensor configurations $\mathcal{S} \subseteq \boldsymbol{\mathcal{S}}$ satisfying $|\mathcal{S}| \leq m_p$ will be considered, where $|\mathcal{S}|$ denotes the cardinality of the set $\mathcal{S}$.

The monitoring specification T will be stated based on two fault diagnosis properties: fault detectability and fault isolability. Single fault assumption will hold (i.e., multiple faults will not be considered).

A water network description M is also required to solve the sensor placement problem. Such description will allow the evaluation of leak monitoring specifications for a given pressure sensor configuration. Hence, the sensor placement for fault diagnosis can be formally stated as follows:

GIVEN a candidate pressure sensor set $\boldsymbol{\mathcal{S}}$, a water network description M, a leak set $\mathcal{F}$, a leak monitoring specification T, and a maximum number of pressure sensors m_p.

FIND a pressure sensor configuration $\mathcal{S} \subseteq \boldsymbol{\mathcal{S}}$ such that:

1. $|\mathcal{S}| \leq m_p$
2. T is maximized, and
3. $|\mathcal{S}|$ is minimal.

9.3 Proposed Approach

To solve the sensor placement optimization problem, two alternative methodologies based on two different water network descriptions M and involving a different formulation of the leak monitoring specification T will be investigated. The first approach is based on structural analysis and the second one on sensitivity analysis.

The considered optimization problem is of combinatorial nature, and its complexity critically depends on the cardinality of $\mathcal{S}$. In order to reduce the size and the complexity of this optimization problem, the following two-step hybrid methodology is proposed:

Step 1 Clustering techniques are applied to reduce the initial candidate sensor set $\boldsymbol{\mathcal{S}}$, such that the next step is tractable. At this step, a tentative size n_t for the reduced candidate sensor set is proposed to the clustering analysis. The complexity issues concerning Step 2 should be accommodated through this specification.

Step 2 Given the new candidate sensor set, the optimization problem is solved following either the approach based on structural analysis or the one based on sensitivity analysis.

9.3.1 Clustering Analysis

Given a set of objects $\mathcal{O} = \left\{\mathbf{o}_1, \mathbf{o}_2, \cdots, \mathbf{o}_{n_o}\right\}$ clustering consists in partitioning the n_o objects into ℓ sets $\mathcal{C} = \{\mathcal{C}_1, \mathcal{C}_2, \cdots, \mathcal{C}_\ell\}$ ($\ell \leq n_o$) in such a way that objects in the same group (called cluster) are more similar (in some sense) to each other than those in other groups (clusters).

In this case, the criterion used for determining the similitude between elements (sensors) is the sensitivity pattern of their primary residuals to leaks, i.e., $n_o = n_y$. This information is provided by every row i of the leak sensitivity matrix $\boldsymbol{\Omega}$ defined in (9.4). In this case, a complete sensitivity matrix will be computed. This matrix considers all possible sensors installed in the system, i.e., $n_y = n_d = |\boldsymbol{\mathcal{S}}|$. As proposed in [18], normalized leak sensitivities are considered, i.e., $\mathbf{o}_i = \frac{\boldsymbol{\omega}_{(i,:)}}{\|\boldsymbol{\omega}_{(i,:)}\|}$, $i = 1, \ldots, n_y$, where $\boldsymbol{\omega}_{(i,:)}$ is the ith row vector of matrix $\boldsymbol{\Omega}$ and $\|\boldsymbol{\omega}_{(i,:)}\|$ the Euclidean norm of this vector. Next, applying the ECM (evidential c-means) algorithm defined in [12], a set of ℓ clusters defined by their centroids $\boldsymbol{\mu}_i$ ($i = 1, \ldots, \ell$) and the plausibility matrix

$\mathbf{\Pi}$ $(n_y \times \ell)$ that contains the membership degree of every element to every cluster are obtained. Component π_{ij} of $\mathbf{\Pi}$ contains the plausibility (or the possibility) that object $\mathbf{o}_i$ belongs to cluster $\mathcal{C}_j$. A hard partition can be easily obtained by assigning each object to the cluster with the highest plausibility. The hard cluster membership of the n_y elements can be stored in a vector $\boldsymbol{g}$ whose components are computed as

$$g_i = \arg\max_j \pi_{ij} \quad i = 1, \cdots, n_y. \tag{9.5}$$

Once the set of sensors has been divided into clusters $\mathcal{C}_1, \ldots, \mathcal{C}_\ell$, N representative sensors should be selected of each cluster, setting up the new candidate sensor set of $N \times \ell$ elements $(N \times \ell \leq n_y)$. The number of groups ℓ will be set to the maximum number of installed sensors m_p as long as the validity index provided by the ECM algorithm confirms that this is a suitable number of clusters. Thus, N will be determined by

$$N = \left\lceil \frac{n_t}{m_p} \right\rceil, \tag{9.6}$$

where n_t is the expected cardinality of the reduced candidate sensor set and $\lceil\,\rceil$ denotes the nearest integer in the direction of positive infinity.

Let $\mathbf{pl}_i$ be the plausibility values of the elements of the cluster set $\mathcal{C}_i$, $\mathbf{row}_i$ the row numbers of the sensitivity matrix defined in (9.4) related to the elements of this cluster (sensor numbers) and $\mathbf{modw}_i$ the Euclidean norm of these rows of the sensitivity matrix. Algorithm 9.1 provides the vector $\mathbf{row}_i^0$ with N representative elements (sensors) of the cluster $\mathcal{C}_i$: $\mathbf{row}_i^0(1), \ldots, \mathbf{row}_i^0(N)$. The higher N is, the more representative the elements $\mathbf{row}_i^0$ of the set $\mathcal{C}_i$ are. In this algorithm, in addition to the plausibility values, the Euclidean norm of the sensor sensitivity matrix is taken into account in order to obtain sensor candidates that maximize the leak sensitivity. Once Algorithm 9.1 has been applied to the ℓ clusters, the reduced sensor set is composed by the $N \times \ell$ sensors associated with the obtained variables $\mathbf{row}_i^0$ $i = 1, \ldots, \ell$.

9.3.2 *Structural Analysis Approach*

9.3.2.1 Structural Analysis Framework

The analysis of the model structure has been widely used in the area of model-based fault diagnosis [1]. Therefore, consistent tools exist in order to perform diagnosability analysis and consequently compute the set of detectable and isolable faults.

The structural model is often defined as a bipartite graph $\mathcal{G} = (\mathcal{M}, \mathcal{X}, \mathcal{A})$, where $\mathcal{M}$ is a set of model equations, $\mathcal{X}$ a set of unknown variables and $\mathcal{A}$ a set of edges, such that $(e_i, x_j) \in \mathcal{A}$ as long as $e_i \in \mathcal{M}$ depends on variable $x_j \in \mathcal{X}$. A structural model is a graph representation of the analytical model structure since only the relation

Algorithm 9.1 $\mathbf{row}_i^0$ = N-most-representative($\mathbf{pl}_i$, $\mathbf{row}_i$, $\mathbf{modw}_i$)

```
tempw_i ← modw_i
pl_i^min ← min(pl_i)
pl_i^max ← max(pl_i)
n_i ← length(pl_i)
for j = 1, ..., N do
  for k = 1, ..., n_i do
    if (pl_i(k) < pl_i^min + (j-1)(pl_i^max - pl_i^min)/N) then
      tempw_i(k) ← 0
    end if
  end for
  loc = arg max_k tempw_i(k)
  row_i^0(j) = row_i(loc)
  tempw_i(loc) ← 0
end for
return row_i^0
```

between variables and equations is taken into account, neglecting the mathematical expression of this relation.

Structural modelling is suitable for an early stage of the system design, when the precise model parameters are not known yet, but it is possible to determine which variables are related to each equation. Furthermore, the diagnosis analysis based on structural models is performed by means of graph-based methods which have no numerical problems and are more efficient, in general, than analytical methods. However, due to its simple description, it cannot be ensured that the diagnosis performance obtained from structural models will hold for the real system. Thus, only best case results can be computed.

It is well known that the overdetermined part of the model is the only useful part for system monitoring [1]. The Dulmage–Mendelsohn canonical decomposition [8] is a bipartite graph decomposition that defines a partition on the set of model equations $\mathcal{M}$. It turns out that one of these parts is the overdetermined part of the model and is represented as $\mathcal{M}^+$ (see Chap. 11).

The system fault diagnosis analysis is performed based on the structural model properties. Specifically, fault detectability and isolability are defined as properties of the overdetermined part of the model [11]. First, it is assumed that a single fault $f \in \mathcal{F}$ can only violate one equation (known as *fault equation*), denoted by $e_f \in \mathcal{M}$.

Definition 9.1 A fault $f \in \mathcal{F}$ is (structurally) detectable in a model described by the set of equations $\mathcal{M}$ if

$$e_f \in \mathcal{M}^+. \tag{9.7}$$

Definition 9.2 A fault f_i is (structurally) isolable from f_j in a model described by the set of equations $\mathcal{M}$ if

$$e_{f_i} \in \left(\mathcal{M}\backslash\{\mathrm{e_{f_j}}\}\right)^+. \tag{9.8}$$

Without loss of generality, it is assumed that a sensor $s_i \in \boldsymbol{\mathcal{S}}$ can only measure one single unknown variable $x_i \in \mathcal{X}$. In the structural framework, such sensor will be represented by one single equation denoted as e_s (known as *sensor equation*). Given a set of sensors $\mathcal{S}$, the set of sensor equations is denoted as $\mathcal{M}_\mathcal{S}$. Thus, given a candidate sensor configuration $\mathcal{S}$ and a model $\mathcal{M}$, the complete system model corresponds to $\mathcal{M} \cup \mathcal{M}_\mathcal{S}$.

Let $\mathcal{F}_D(\mathcal{S}) \subseteq \boldsymbol{\mathcal{F}}$ denote the detectable fault set when a sensor configuration $\mathcal{S} \subseteq \boldsymbol{\mathcal{S}}$ is installed in the system. Fault isolability can be characterized in a similar way by means of fault pairs. Let $\mathbb{F} : \boldsymbol{\mathcal{F}} \times \boldsymbol{\mathcal{F}}$ be all fault pair permutations from $\boldsymbol{\mathcal{F}}$, then $\mathcal{F}_I(\mathcal{S}) \subseteq \mathbb{F}$ denotes the set of isolable fault pairs when the sensor configuration $\mathcal{S} \subseteq \boldsymbol{\mathcal{S}}$ is chosen for installation (i.e., $(f_i, f_j) \in \mathcal{F}_I(\mathcal{S})$ means that fault f_i is isolable from f_j when the sensor set $\mathcal{S}$ is installed in the system).

From Definition 9.1, $\mathcal{F}_D(\mathcal{S})$ can be computed as

$$\mathcal{F}_D(\mathcal{S}) = \{f \in \boldsymbol{\mathcal{F}} \mid e_f \in (\mathcal{M}_\mathcal{S} \cup \mathcal{M})^+\}, \tag{9.9}$$

and, from Definition 9.2, $\mathcal{F}_I(\mathcal{S})$ can be computed as

$$\mathcal{F}_I(\mathcal{S}) = \{(f_i, f_j) \in \mathbb{F} \mid e_{f_i} \in (\mathcal{M}_\mathcal{S} \cup (\mathcal{M} \backslash \{e_{f_j}\}))^+\}. \tag{9.10}$$

It is worth noting that testing different sensor configurations involves different sensor equation sets, $\mathcal{M}_\mathcal{S}$, in (9.9) and (9.10) while the other sets remain unchanged.

Definition 9.3 (*Isolability index*) Given a sensor configuration $\mathcal{S} \subseteq \boldsymbol{\mathcal{S}}$, the isolability index is defined as the number of isolable fault pairs, provided the sensors $s \in \mathcal{S}$ are installed, i.e.,

$$I(\mathcal{S}) = |\mathcal{F}_I(\mathcal{S})|. \tag{9.11}$$

9.3.2.2 Optimal Sensor Placement Algorithm

The optimal sensor placement problem stated in Sect. 9.2.2 will be solved under the structural analysis approach. This involves providing a structural model $\mathcal{G}$ as a DWN description M and stating the leak monitoring specification T as follows:

1. All leaks are detectable, i.e., $\mathcal{F}_D(\mathcal{S}) = \boldsymbol{\mathcal{F}}$, according to (9.9).
2. The number of isolable leak pairs is maximized, i.e., the isolability index $I(\mathcal{S})$ is maximized.

Algorithm 9.2 solves the optimal sensor placement problem, by applying a depth-first branch-and-bound search strategy. The search involves building a node tree by recursively calling function $\mathsf{searchOp}_C$, beginning at the root node down to the leaf nodes. Each node corresponds to a sensor configuration ($node.\mathcal{S}$), and child nodes are built by removing sensors from their corresponding parent node. Set $node.\mathcal{R}$ specifies those sensors that are allowed to be removed.

Algorithm 9.2 $\mathcal{S}^* = \mathsf{searchOp}_C(node, \mathcal{S}^*)$

```
childNode.R := node.R
for all s ∈ node.R ordered in decreasing cost do
  childNode.S := node.S\{s}
  childNode.R := childNode.R\{s}
  if |childNode.S\childNode.R| > m_p then
    return S*
  end if
  if I(childNode.S) = I(S*) then
    if |childNode.S\childNode.R| < |S*| then
      if F_D(childNode.S) = 𝓕 then
        if |childNode.S| < |S*| then
          S* := childNode.S {update best solution}
        end if
        S* := searchOp_C(childNode, S*)
      end if
    else
      if I(childNode.S) = I(Node.S) then
        return S*
      end if
    end if
  else
    if I(childNode.S) > I(S*) and
    F_D(childNode.S) = 𝓕 then
      if |childNode.S| ≤ m_p then
        S* := childNode.S {update best solution}
      end if
      S* := searchOp_C(childNode, S*)
    end if
  end if
end for
return S*
```

Throughout the search, the best solution is updated in $\mathcal{S}^*$ whenever a sensor configuration with a higher fault isolability index than the current best one is found, as long as all faults are detectable and the number of sensors does not exceed m_p. The best solution is also updated whenever a smaller sensor configuration is found that has the same isolability index as the current best one.

The search is initialized as follows: $node.\mathcal{S} = node.\mathcal{R} = \boldsymbol{\mathcal{S}}$ and $\mathcal{S}^* = \emptyset$. During the search, only those branches that can be further expanded to a sensor configuration that improves the current solution are explored. Branch exploration is aborted whenever the fault isolability index cannot be improved, a fault is not detectable or the number of sensors cannot satisfy the budget constraint.

9.3.3 Sensitivity Analysis Approach

9.3.3.1 Sensitivity Analysis Framework

Alternatively to the structural method proposed in the previous section, a method that aims at optimizing the performance of the leak location method presented in Chap. 7 is proposed in this section. Considering $\mathbf{r} = [r_1 \cdots r_{n_y}]^T$ be the actual residual vector corresponding to all pressure measurement points $n_y = |\mathcal{S}|$, and $\hat{\mathbf{r}}_{f_j}$ be the column of $\boldsymbol{\Omega}$ corresponding to leak j, the leak location was achieved by solving the problem

$$\arg\max_{j\in\{1,\dots n_d\}} \frac{\mathbf{r}^T \hat{\mathbf{r}}_{f_j}}{\|\mathbf{r}\| \|\hat{\mathbf{r}}_{f_j}\|}. \tag{9.12}$$

Thus, the biggest normalized projection of the actual residual vector on the fault sensitivity space is sought.

The detectable leak set $\mathcal{F}_D$ was defined in terms of structural analysis in (9.9). Next, it will be defined in terms of sensitivity analysis as proposed in [18]. Given a set of sensors $\mathcal{S}$, a set of leaks $\boldsymbol{\mathcal{F}}$ and the corresponding $\boldsymbol{\Omega}$, the set of detectable leaks $\mathcal{F}_D(\mathcal{S})$ is defined as

$$\mathcal{F}_D(\mathcal{S}) = \{f_j \in \boldsymbol{\mathcal{F}} : \exists r_i \in \mathbb{R} : |\omega_{ij}| \geq \epsilon\}, \tag{9.13}$$

where ϵ is a threshold to account for noise and model uncertainty.

Regarding the leak locatability performance, assuming that the leak location is implemented by means of (9.12), a uniform projection angle $\bar{\alpha}$, defined as the average between the residual fault sensitivity vectors for all leak pairs, was proposed in [18]. The resulting sensor locations led to a maximal uniform projection angle $\bar{\alpha}$. In an ideal case, all pairs of leak sensitivity vectors (columns) in $\boldsymbol{\Omega}$ should satisfy this uniform projection angle. This uniform angular separation between leak pairs would allow for a successful leak location method applying (9.12), even when residuals are affected by modelling errors, sensor noise and other uncertainties.

Nevertheless, in a real case, the angle between leak pairs is not uniformly distributed. Some leaks can have similar leak sensitivity vectors, which introduces uncertainty in the leak location results when applying (9.12). This fact can become a critical issue for water network utilities, especially when this uncertainty involves distant leak locations, i.e., two distant leaks that have similar fault sensitivity vectors. So, distances between nodes with a similar fault sensitivity vector should be considered in the optimal sensor placement methodologies. In order to take into account these distances, the following properties are defined.

Definition 9.4 (*Leak expansion set*) Given a leak $f_j \in \boldsymbol{\mathcal{F}}$ and a projection angle threshold α_{th}, the leak expansion set $\boldsymbol{\mathcal{F}}_j^{\alpha_{th}}$ is defined as

$$\boldsymbol{\mathcal{F}}_j^{\alpha_{th}} = \{f_i \in \boldsymbol{\mathcal{F}} : \frac{\hat{\mathbf{r}}_{f_j}^T \hat{\mathbf{r}}_{f_i}}{\|\hat{\mathbf{r}}_{f_j}\| \|\hat{\mathbf{r}}_{f_i}\|} > \cos(\alpha_{th})\}. \tag{9.14}$$

Thus, $\mathcal{F}_j^{\alpha_{th}}$ contains the set of leaks whose correlation with leak f_j is bigger than $\cos(\alpha_{th})$. If $f_i \in \mathcal{F}_j^{\alpha_{th}}$, it follows that $f_j \in \mathcal{F}_i^{\alpha_{th}}$.

Definition 9.5 (*Correlated leak pairs ratio*) Given the leak expansion sets $\mathcal{F}_j^{\alpha_{th}}$ with $j = 1, ..., |\mathcal{F}|$, the correlated leak pairs ratio $\eta^{\alpha_{th}}$ is defined as

$$\eta^{\alpha_{th}} = 100 \frac{\sum_{j=1}^{|\mathcal{F}|} |\mathcal{F}_j^{\alpha_{th}}| - |\mathcal{F}|}{2\binom{|\mathcal{F}|}{2}}. \tag{9.15}$$

Thus, $\eta^{\alpha_{th}}$ provides the percentage of leak pairs from $\mathcal{F}$ whose mutual correlation is bigger than $\cos(\alpha_{th})$.

Definition 9.6 (*Leak node distance matrix*) Given the geographical coordinates of every leak node, the leak node distance matrix $\mathbf{D} \in \mathbb{R}^{|\mathcal{F}| \times |\mathcal{F}|}$ is defined as the matrix whose coefficients d_{ij} are the geographical distance between nodes i and j.

Matrix $\mathbf{D}$ is a symmetric matrix ($d_{ij} = d_{ji}$), with diagonal coefficients equal to zero ($d_{ii} = 0$). This matrix will be used to compute distances in leak expansion sets.

Definition 9.7 (*Worst leak expansion distance*) Given a leak expansion set $\mathcal{F}_j^{\alpha_{th}}$ and the leak node distance matrix $\mathbf{D}$, the worst leak expansion distance $R_j^{\alpha_{th}}$ is defined as

$$R_j^{\alpha_{th}} = \max_{f_i \in \mathcal{F}_j^{\alpha_{th}}} d_{ij}. \tag{9.16}$$

Thus, $R_j^{\alpha_{th}}$ provides the maximum Euclidian distance between the node of leak f_j and the nodes of leaks whose correlation with leak f_j is bigger than $\cos(\alpha_{th})$. This metric is next used to compute the following overall leak location uncertainty index in terms of leak node distances.

Definition 9.8 (*Average worst leak expansion distance*) Given a set of leaks $\mathcal{F}$ and a threshold projection angle α_{th}, leak expansion sets $\mathcal{F}_j^{\alpha_{th}}$ with $j = 1, ..., |\mathcal{F}|$ can be computed applying (9.14). Then, the average worst leak expansion distance can be computed as

$$\bar{R}^{\alpha_{th}} = \frac{1}{|\mathcal{F}|} \sum_{j=1}^{|\mathcal{F}|} R_j^{\alpha_{th}}. \tag{9.17}$$

Thus, $\bar{R}^{\alpha_{th}}$ provides the average of the worst leak expansion distances considering all the possible leaks in $\mathcal{F}$.

As discussed in [13], the greater the threshold α_{th} is, the greater the uncertainty is in terms of leak expansion distance and number of correlated leak pairs. The choice of this threshold should take into account the implementation requirements of the leak location software module, as well as practical issues concerning the water utility maintenance procedures. On the one hand, the leak location software module will have to deal with sensor measurement noise and network modelling uncertainty. Therefore, the bigger the threshold, the better the performance of the leak location procedure. On the other hand, the smaller the leak location result uncertainty, the better for the water utility maintenance department. Indeed, upon the occurrence of a leak, the leak location software module will provide a set of leak node candidates to the maintenance department, which then will undergo leak field testing. Thus, the smaller the leak expansion distance the better, which involves specifying a smaller projection angle threshold. Therefore, a trade-off exists between both criteria.

In order to find a suitable balanced solution, it is expected to find a sensor placement solution suitable for a range of projection angle thresholds as proposed in [13].

Definition 9.9 (*Mean average worst leak expansion distance*) Given a set $\mathcal{A}_{th} = \{\alpha_{th}^1, \dots, \alpha_{th}^{n_\alpha}\}$ that covers a suitable range of projection angle thresholds. Then, the mean average worst leak expansion distance over this set can be computed as

$$\bar{R}^{\overline{\alpha_{th}}} = \frac{1}{|\mathcal{A}_{th}|} \sum_{\alpha_{th} \in \mathcal{A}_{th}} \bar{R}^{\alpha_{th}}, \tag{9.18}$$

where the average worst leak expansion distance $\bar{R}^{\alpha_{th}}$ is computed applying (9.17) to every projection angle threshold in $\mathcal{A}_{th}$.

9.3.3.2 Optimal Sensor Placement Algorithm

The optimal sensor placement problem stated in Sect. 9.2.2 will be solved under the sensitivity analysis approach. This involves providing a node distance matrix $\mathbf{D}$, the fault sensitivity matrix $\mathbf{\Omega}$ and a set of projection angle thresholds $\mathcal{A}_{th}$. Additionaly, the following leak monitoring specification T will be considered:

1. All leaks are detectable, i.e., $\mathcal{F}_D(\mathcal{S}) = \mathcal{F}$ according to (9.13).
2. The mean average worst leak expansion distance $\bar{R}^{\overline{\alpha_{th}}}$ is minimized, i.e., $\frac{1}{\bar{R}^{\overline{\alpha_{th}}}}$ is maximized.

This optimization problem cannot be solved by efficient branch-and-bound search strategies. It is necessary to implement an exhaustive search, i.e., to compute the mean average worst leak expansion distance $\bar{R}^{\overline{\alpha_{th}}}$ by means of (9.18) for all the possible $\frac{n_d!}{(n_d-n_s)!n_s!}$ sensor configurations with $n_s = 1, \dots, m_p$ and choosing the one that provides the best performance (smallest $\bar{R}^{\overline{\alpha_{th}}}$). Alternatively, a genetic algorithm search could be implemented to solve it more efficiently in case that the exhaustive search is not possible because of the size of the network. See [5] for further details.

9.4 Simulations and Results

9.4.1 DMA Case Study

The sensor placement methodology is applied to a DMA located in the Barcelona area (see Fig. 9.1) with 883 nodes and 927 pipes. The network consists of 311 nodes with demand (RM type), 60 terminal nodes with no demand (EC type), 48 nodes hydrants without demand (HI type) and 448 dummy nodes without demand (XX type). Additionally, the network has two inflow inputs modelled as reservoir nodes. The total inflow is distributed using a constant coefficient in each consumption node according to the total demand which is estimated using demand patterns.

Leaks might appear anywhere in the water network. However, due to simulation limitations, leaks are represented in the nodes where the flow balances take place as already discussed in Chap. 7. It is assumed that leaks might only occur at dummy nodes, leading to 448 potential leaks to be detected and located. A similar practical reason is applied when defining the possible location of the network monitoring points. Pressure sensors at RM-type nodes will be used as network monitoring points, leading to 311 candidate sensors that could be chosen for installation. Additionally,

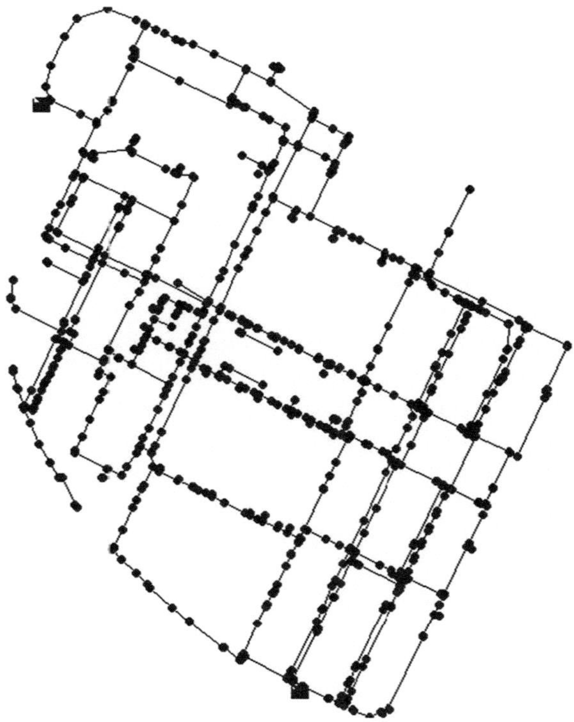

Fig. 9.1 Case study network map

it is also assumed that there is no pressure sensor already installed in the network before solving the sensor placement problem.

The water distribution company has established a maximum budget for investment on instrumentation that makes it possible to install up to five pressure sensors. Hence, up to $m_p = 5$ pressure sensors should be chosen out of 311 such that the leak monitoring specifications are maximized. Despite measuring flow rate could also be useful for leak detection, collecting pressure data is cheaper and easier, and pressure transducers give instantaneous readings, whereas most flow metres do not react instantaneously to flow changes [7].

9.4.2 DMA Network Modelling

The DMA network is represented by a directed graph $\mathcal{G}_N = (\mathcal{V}, \mathcal{J})$ where pipe junctions are nodes, $v \in \mathcal{V}$, and pipes are edges, $j \in \mathcal{J}$. Each node represents, at the same time, a pressure variable and a flow balance equation. Similarly, each edge represents a flow variable and a pipe equation. Therefore, given a node $v \in V$, the following flow balance equation can be derived (see Chap. 3):

$$\sum_{j \in \mathcal{J}_v} q_j = d_v, \tag{9.19}$$

where $\mathcal{J}_v$ represents the set of edges incident to node v, and d_v is the known flow demand associated with node v. Furthermore, given an edge $j \in \mathcal{J}$, the corresponding pipe equation can be deduced (see Chap. 3) as

$$q_j = \operatorname{sgn}(h_{j,1} - h_{j,2})\, c(|h_{j,1} - h_{j,2}|)^{\gamma}, \tag{9.20}$$

where q_j is the flow of edge j, $h_{j,1}$ and $h_{j,2}$ are the heads ($p_{j,1}$ and $p_{j,2}$ pressures plus elevation offsets) at the nodes adjacent to edge $j = (v_{j,1}, v_{j,2})$, and c and γ are parameters modelling physical properties of the pipe, such as length, inside diameter, minor losses and roughness.

Therefore, the DMA model comprises 883 flow balance equations and 927 pipe flow equations that depend on 927 unknown flow variables and 883 unknown pressure variables. The resulting structural model is depicted in Fig. 9.2 in matrix form, where the equation set corresponds to rows and the variable set corresponds to columns. A dot in the (i, j) element indicates that there exists an edge incident to the expression of e_i and variable x_j. Note that the structural model of the DMA network is a just-determined model where all unknown variables can be computed, i.e., the model can be used for system simulation.

Since leaks are assumed to occur in the nodes where a flow balance takes place, flow balance equations corresponding to XX type nodes will be considered as fault equations when applying (9.9)–(9.10).

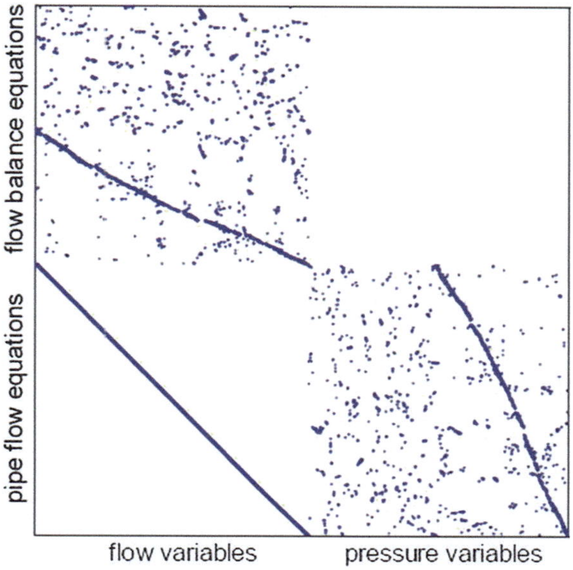

Fig. 9.2 Structural model of the DMA network

A fault sensitivity matrix has also been obtained using the EPANET hydraulic simulator [17]. Leaks are simulated in EPANET through the corresponding emitter coefficient, designed to model fire hydrants/sprinklers, and which can be adapted to provide the desired leak magnitude in the network, according to

$$q_f = C_e P^{P_{exp}}, \tag{9.21}$$

where C_e is the emitter coefficient, q_f is the flow rate, P is the available pressure at the considered node and P_{exp} is the pressure exponent. EPANET permits the value of the emitter coefficient to be specified for individual leak sites, but the pressure exponent can only be specified for the entire network.

Given a set of boundary conditions (such as water demands), EPANET software has been firstly used to estimate the steady-state pressure at the 311 RM-type nodes. Next, 448 leaks have been simulated, and the steady-state pressure has been estimated again in the 311 RM-type nodes. Finally, a 311 $\times$ 448 fault sensitivity matrix has been obtained as the pressure difference between the fault-free case and each faulty situation, according to the procedure described in Sect. 9.2.1. Although the fault sensitivity matrix depends on the leak size, diagnosability properties are robust against this uncertainty. In this chapter, an average leak size of 1.5 l/s has been considered in the simulations.

9.4.3 Clustering Analysis

Recall from Sect. 9.3 that clustering techniques should be applied beforehand in order to accommodate the time complexity of the optimization problem. The methodology described in Sect. 9.3.1 has been applied to the dataset (311 normalized rows of $\boldsymbol{\Omega}$) in order to set up a reduced set of 25 candidate pressure sensors. As it will later be shown in Sect. 9.4.6, $n_t = 25$ is a convenient cardinality for the reduced candidate sensor set. The most time-demanding methodology to solve the sensor placement optimization problem will be the sensitivity analysis approach, requiring a solving time of 27 h. The structural analysis approach will only require 103 min; thus, a bigger candidate sensor set could be accepted. However, since a comparison of both approaches is targeted, the size of the candidate sensor set is decided based on the most time-demanding one.

Firstly, ECM clustering algorithm has been used to classify the dataset into $\ell = 5$ clusters (the same number of clusters as the maximum number of sensors to be installed). Provided the plausibility matrix obtained from the clustering algorithm, a hard partition has been done that assigns each element to its highest plausibility cluster, according to (9.5). Figure 9.3 depicts the 5 network node clusters in different colours, where the closest nodes to the centroid have been highlighted in every cluster.

Next, Algorithm 9.1 has been applied to obtain the N most representative sensors of every cluster, with $N = 5$ given by (9.6). The resulting reduced set $\mathcal{S}'$ with $|\mathcal{S}'| = N \times \ell = 25$ candidate pressure sensor places suggested by the clustering approach is displayed in Fig. 9.4 as blue circles.

Fig. 9.3 Clustering results

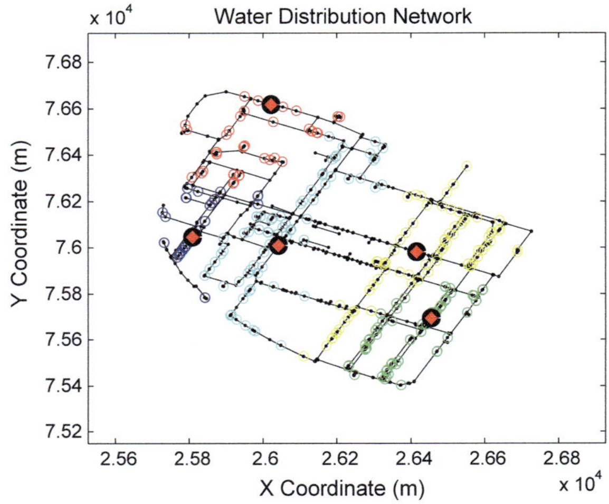

9.4.4 Structural Analysis Approach

The structural analysis approach aims at maximizing the isolability index. To fully isolate all 448 possible leaks, the required isolability index should be $\binom{448}{2} = 100128$. However, according to structural analysis, installing all 311 candidate sensors, the isolability index would just be 100099. Achieving a better performance would require installing more sensors than those provided in the candidate sensor set. So, there is a trade-off between the diagnosis performance and the number of installed sensors. Since the maximum number of sensors to install is 5, the maximum achievable isolability index is expected to be less than 100099.

Algorithm 9.2 is applied to solve the sensor placement problem with $m_p = 5$. The algorithm returns the optimal sensor configuration provided in Fig. 9.4, highlighted as red starred nodes.

With these 5 sensors, all leaks can be detected and the isolability index amounts to 100073, which means that 99.9% of leak pairs are isolable. This is the highest diagnosis performance that can be achieved by a sensor configuration, satisfying the stated budgetary constraint. In this case study, no cheaper sensor configuration can achieve better diagnosis performance. Figure 9.5 provides an evidence to back this assertion up. It displays the highest isolability performance that can be achieved by a sensor configuration of a given size. Remark that, under the structural analysis approach, the isolability performance decreases with the size of the sensor configuration, till

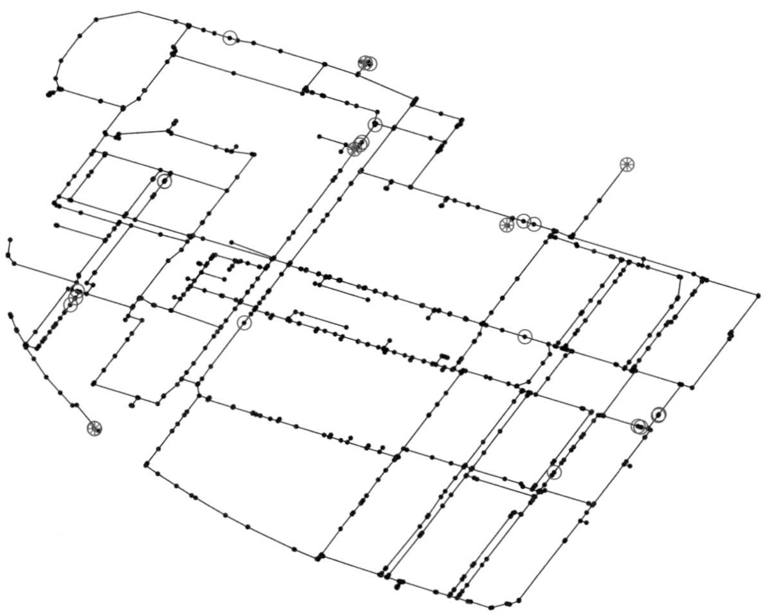

Fig. 9.4 DMA network sensor placement results under structural analysis approach

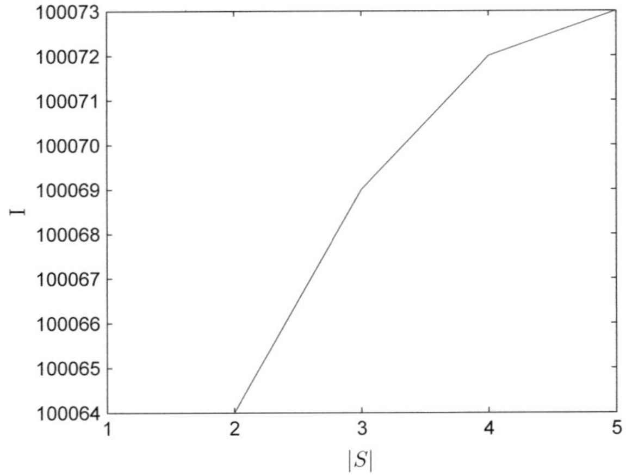

Fig. 9.5 Branch-and-bound search results for several sensor configuration sizes

null when installing just one pressure sensor into the water network. Note however that with one sensor less, the performance just slightly decreases. Thus, savings in the initial budget could be considered by the water distribution company.

9.4.5 Sensitivity Analysis Approach

The sensitivity analysis approach aims at minimizing the mean average worst leak expansion distance, which requires a set of projection angle thresholds to be specified. Recall from Sect. 9.3.3.2 that, in addition to $\mathbf{\Omega}$, a leak node distance matrix and a set of projection angle thresholds should be specified. Matrix $\mathbf{D}$ has been obtained from geographical data contained in the EPANET model, and the following projection angle threshold set has been considered: $\mathcal{A}_{th} = \{10, 20, 30, 40, 50, 60\}$.

As stated in Sect. 9.3.3.2, an exhaustive search is applied to solve the sensor placement problem with $m_p = 5$. The optimal sensor configuration is displayed as red starred nodes in Fig. 9.6.

Installing these 5 pressure sensors, all 448 leaks are detectable, and the mean average worst leak expansion distance is 698.23 m. As in the structural analysis approach, the best diagnosis performance is achieved by a 5-sensor configuration. Figure 9.7 displays the smallest mean average worst leak expansion distance that can be achieved by a sensor configuration of a given size. Remark that, in this case study, the performance index shows a monotonically decreasing trend over values of $|\mathcal{S}|$. Thus, the best leak location performance is achieved with the five pressure sensors displayed in Fig. 9.6. Note again that with fewer sensors, the performance just slightly deteriorates. Thus, savings in the initial budget could be considered by the water distribution company.

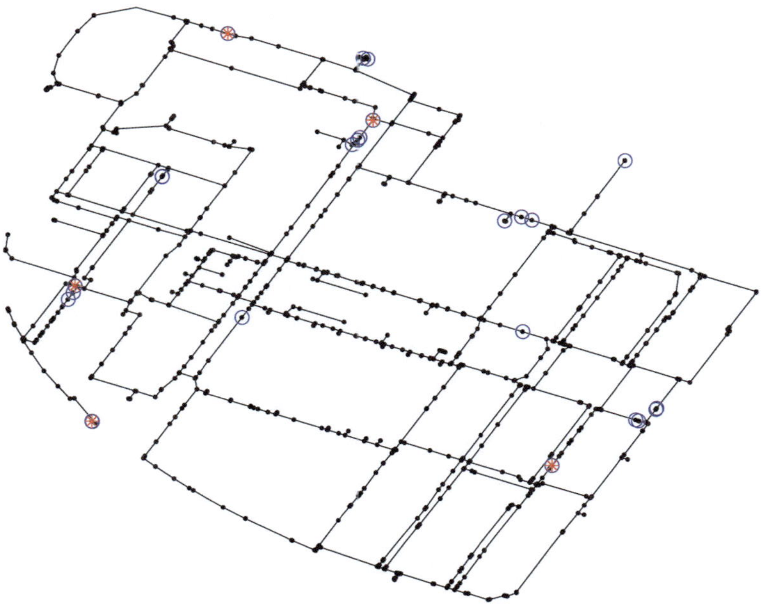

Fig. 9.6 DMA network sensor placement results under sensitivity analysis approach

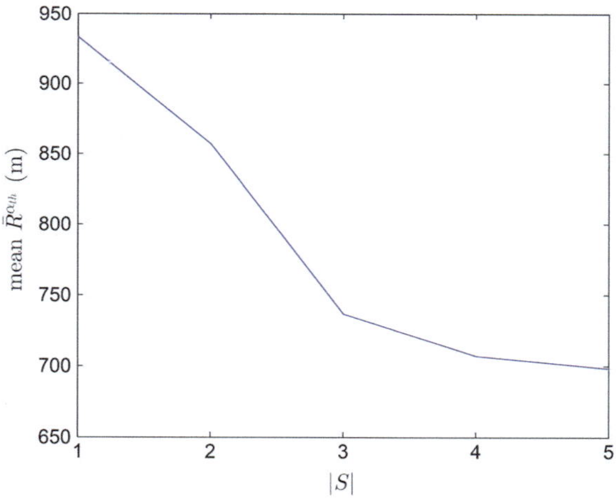

Fig. 9.7 Exhaustive search results for several sensor configuration sizes

9.4.6 Discussion

Regarding the search strategy performance issues, with 25 candidate sensors there are $\binom{25}{5} + \binom{25}{4} + \cdots + \binom{25}{1} = 68405$ possible sensor configurations to check. However, under the structural analysis approach, Algorithm 9.2 with $m_p = 5$ just needs to inspect 4874 sensor configurations. Algorithm 9.2 provides the optimal sensor con-

Table 9.1 Sensor placement results comparison

Approach	$\mathcal{S}$	$I(\mathcal{S})$	$\bar{R}^{\overline{\alpha_{th}}}$ (m)	Computation time
Structural analysis	$\{3, 9, 124, 199, 305\}$	100073	793.4	103 s
Sensitivity analysis	$\{8, 67, 207, 249, 271\}$	100023	698.23	27 h

figuration in 103 min. Tests have been run in an Intel Core i7-4702MQ @ 2.20GHz HP notebook with 16 GB RAM and 64-bit Windows 10.

Alternatively, as an exhaustive search is applied under the sensitivity analysis approach, the leak monitoring performance index must be evaluated for every sensor configuration and every projection angle threshold. Since there are 68405 sensor configurations to check and six projection angle thresholds, the average worst leak expansion distance must be evaluated $68405 \times 6 = 410430$ times. Solving the sensor placement problem under the sensitivity analysis approach takes over 27 h.

Table 9.1 provides a summary of the solution found under both approaches.

As expected, the sensitivity analysis approach provides smaller leak location uncertainty in terms of leak expansion distance at the expense of slightly decreasing the isolability index.

Both approaches provide a different methodology to solve the sensor placement problem stated in Sect. 9.2.2. On the one hand, the structural analysis approach is more efficient since the formulation allows for a branch-and-bound search strategy. However, structural models are a simple description of the network, and only best case results can be computed. This methodology is better suited for an early stage of the network design. On the other hand, the sensitivity analysis approach requires a larger computation time since a highly inefficient exhaustive search is applied. However, the monitoring performance index has a more practical meaning: a leak location uncertainty measure in terms of distance. The search efficiency of the sensitivity analysis approach could be improved by applying other search strategies such as genetic algorithms or simulated annealing at the expense of global optimality.

9.5 Conclusions

This chapter presents two optimal sensor placement strategies for leak location that maximize some monitoring specifications for a DWN. The goal is to characterize and determine a sensor set that guarantees a maximum degree of diagnosability while a budgetary constraint is satisfied. The first methodology is based on a structural model of the water network and a branch-and-bound search, while the second strategy is based on pressure sensitivity matrix analysis and an exhaustive search strategy. To reduce the size and complexity of the optimization problem, the strategies are combined with clustering techniques. The developed strategies are successfully applied to a DMA in the Barcelona water network to decide the best location of pressure sensors for leak monitoring. However, in both approaches, only a suboptimal solution is attained. Thus, the sensor search could be improved using other types of optimization methods that provide some guarantee regarding the solution optimality.

References

1. Blanke M, Kinnaert M, Lunze J, Staroswiecki M (2006) Diagnosis and fault-tolerant control, 2nd edn. Springer
2. Blesa J, Nejjari F, Sarrate R (2016) Robust sensor placement for leak location: analysis and-design. J Hydroinformatics 18(1):136–148
3. Blesa J, Puig V, Saludes J (2012) Robust identification and fault diagnosis based on uncertain multiple input-multiple output linear parameter varying parity equations and zonotopes. J Process Control 22(10):1890–1912
4. Myrna VC, Luis EG, Vicenç P (2012) Extended-horizon analysis of pressure sensitivities for leak detection in water distribution networks. In: 8th IFAC Symposium on Fault Detection, Supervision and Safety of Technical Processes Mexico City. Mexico, pp 570–575
5. Casillas MV, Puig V, Garza-Castaón LE, Rosich A (2013) Optimal sensor placement for leak location in water distribution networks using genetic algorithms. Sensors 13(11):14984–15005
6. Cugueró-Escofet MÀ, Puig V, Quevedo J, Blesa J (2015) Optimal pressure sensor placement for leak localisation using a relaxed isolation index: Application to the Barcelona water network. IFAC-PapersOnLine 48(21):1108–1113 issn: 2405-8963
7. de Schaetzen WBF, Walters GA, Savic DA (2000) Optimal sampling design for model calibration using shortest path, genetic and entropy algorithms. Urban Water 2(2):141–152
8. Dulmage AL, Mendelsohn NS (1958) Covering of bi-partite graph. Canad. J. Math 10:517–534
9. Farley Trow (2003) Losses in water distribution networks. IWA Publishing UK, UK
10. Krysander M (2006) Design and analysis of diagnosis systems using structural analysis. PhD thesis, Linköping Univ., Linköping, Sweden
11. Krysander M, Frisk E (2008) Sensor placement for fault diagnosis. IEEE Trans Syst Man Cybern A, 38(6):1398–1410
12. Masson MH, Denoeux T (2008) Ecm: an evidential version of the fuzzy c-means algorithm. Pattern Recognit 41:1384–1397
13. Nejjari F, Sarrate R, Blesa J (2015) Optimal pressure sensor placement in water distribution networks minimizing leak location uncertainty. Proc Eng 119:953–962
14. Pérez R, Puig V, Pascual J, Peralta A, Landeros E, Jordanas Ll (2009) Pressure sensor distribution for leak detection in Barcelona water distribution network. Water Sci Technol Water Supply 9(6):715–721
15. Pérez R, Puig V, Pascual J, Quevedo J, Landeros E, Peralta A (2011) Methodology for leakage isolation using pressure sensitivity analysis in water distribution networks. Control Eng Pr 19(10):1157–1167
16. Puust R, Kapelan Z, Savic DA, Koppel T (2010) A review of methods for leakage management in pipe networks. Urban Water J 7:25–45
17. Rossman Lewis A (2000) EPANET 2 users manual. Environmental Protection Agency, U.S
18. Sarrate R, Blesa J, Nejjari F (2014) Clustering techniques applied to sensor placement for leak detection and location in water distribution networks. In: 22th Mediterranean Conference on Control and Automation. Palermo, Italy, pp 109–114
19. Sarrate R, Blesa J, Nejjari F, Quevedo J (2014) Sensor placement for leak detection and location in water distribution networks. Water Sci Technol Water Supply 14(5):795–803
20. Sarrate R, Nejjari F, Rosich A (2012) Sensor placement for fault diagnosis performance maximization in distribution networks. In: 20th Mediterranean Conference on Control and Automation. Barcelona, Spain, pp 110–115
21. Travé-Massuyes L, Escobet T, Olive X (2006) Diagnosability analysis based on component supported analytical redundancy relations. IEEE Trans Syst Man Cybern A, 36(6):1146–1160

Chapter 10
Sensor Data Validation and Reconstruction

Joseba Quevedo, Diego Garcia, Vicenç Puig, Jordi Saludes, Miquel Angel Cugueró, Santiago Espin, Jaume Roquet and Fernando Valero

10.1 Introduction

Critical infrastructure systems (CIS), including water, gas or electricity networks, are complex large-scale systems geographically distributed and decentralized with a hierarchical structure. These systems require highly sophisticated supervisory and real-time control schemes to ensure high-performance achievement and maintenance when conditions are non-favourable due to faults (e.g., sensor and/or actuator malfunctions) [29, 47].

In CIS, a telecontrol system is acquiring, storing and validating data gathered from different kind of sensors every given sampling time (e.g., every few minutes) to accurately real-time monitor the whole system. Several problems can occur during the data acquisition process, like those related to the communication system, e.g., communication failure between sensors and data loggers or in the telecontrol system itself. These problems produce lost or corrupted data which may be of great concern in order to have valid historic records. When this is occurring, lost data should be replaced by a set of estimated data, which should be representative of the data lost, since missing data may severely jeopardize further processes needing complete datasets in order to get meaningful conclusions or analysis.

Another common problem in CIS is caused by the unreliability of sensors, which may be affected by faults (e.g., offset, drift or freezing in the measurements) [23, 24, 51]. Unreliable data should be also detected and replaced by forecasted data, since it may be used for system management tasks (e.g., maintenance, planning, investment plans, billing, security and operational control) [45] and/or system fault detection and isolation [36, 37, 44].

J. Quevedo (✉) · D. Garcia · V. Puig · J. Saludes · M.A. Cugueró
Research Center "Supervision, Safety and Automatic Control" (CS2AC-UPC), Terrassa, Spain
e-mail: joseba.quevedo@upc.edu

S. Espin · J. Roquet · F. Valero
ATLL Concessionària de la Generalitat de Catalunya, SA. Sant Joan Despí, Barcelona, Spain

V. Puig et al. (eds.), *Real-Time Monitoring and Operational Control of Drinking-Water Systems*, Advances in Industrial Control,
DOI 10.1007/978-3-319-50751-4_10

In addition to the possible measurement deviations related to the sensor performance itself, the errors can also occur due to heterogeneous reasons such as sensor installation, calibration or electrical problems. Consequently, it is important to provide the data system with procedures that can detect these problems and assist the user in the monitoring and the processing of the incoming data.

Data validation is an essential step to improve data reliability. Traditionally, data validation processes have been developed by manual data analysis, performed by experienced users with the only assistance of basic data analysis and visualization tools [22], which significantly limits the amount of data to be validated [31], and the abnormal situations which may be correctly detected [55]. Furthermore, the volume of real data acquired in CIS is dramatically increasing due to the increment of automated measurement systems allowing their monitoring [7]. Also, real-time operation, paramount in many real applications, makes human data validation even harder to pursuit. In order to cope with this situation and increase the reliability of the data diagnosis system, automatic data validation tools have arisen [26].

The approach presented in this chapter aims to assess the validity of each single sensor measurement by means of a set of tests exploiting not only the model equations (spatial redundancy) but also temporal redundancy, using time series models and a bank of low-level tests (non-model based) aiming to label the data with a certain quality index. The methodology may be applied to alternative systems involving a telemeasured sensor network, such as gas or electrical networks, smart buildings or environmental systems (see, e.g., [12, 13]).

10.2 Problem Statement

The measurements in water systems commonly focus on hydraulic and quality variables, including flow rate in links, pressure in nodes, water level in tanks, water pH, conductivity and turbidity, as well as disinfectant and pollutant concentrations. Each sensor measures a physical quantity and converts it into a signal that can be read by the appropriate instrumentation. Then, the measuring system converts the sensors' signals into values aiming to represent certain real physical quantities. These values, known as raw data, need to be validated before further use in order to assure the reliability of the results derived from their usage.

In this chapter, a methodology is developed for validation and reconstruction of sensors data in a water network, taking into account spatial models (SM) and time series models (TSM) for each flow meter and level meter. Internal models of every component in the local equipment units (e.g., pumps, valves, flows, levels) are also considered. SM take advantage of the relation between different variables in the system (e.g., demand, pump flows and tank levels), while TSM take advantage of the temporal redundancy of the measured variables, by means of Holt–Winters (HW) TSM [28]. Moreover, after the corrupted sensor data are detected, they must be replaced by adequate estimated data using the available temporal/spatial redundancy. The methodology is mainly applied to flow and level meters, since it exploits the temporal redundancy of flow and level data in a water network. An operative software tool implementing the presented methodology which is able to properly handle raw sensor data (including storage, querying and visualization) is also presented. The proposed approach and tool are applied to several subsystems in the Catalonia regional water network using raw data collected from *ATLL Concessionària de la Generalitat de Catalunya, SA* (ATLL), the company managing this water network.

The whole process is considered as two separated stages: data validation and data reconstruction. According to the nature of the available knowledge, different kinds of data validation approaches may be considered, with varying degrees of sophistication. In general, one may distinguish between elementary signal-based ("low-level") methods, and model-based ("high-level") methods (see, e.g., [14, 31, 45]).

Elementary signal-based methods use simple heuristics and limited statistical information of a given sensor [9, 22, 30]. Typically, these methods are based on validating either signal values or signal variations. On the one hand, in the signal value-based approach, data are assessed as valid or invalid according to two different thresholds (high and low), so data are assumed to be invalid when lying outside these threshold values. On the other hand, methods based on signal variations look for high variations (peaks in the curve) and low variations (flat curve) in the signals.

Model-based methods rely on the use of models to check the consistency of sensor data [50]. This consistency check is based on computing the difference between the predicted value from the model and the real value measured by the sensors. Then, this difference (known as residual) is compared with a threshold value (zero in the ideal case). When the residual is bigger than the corresponding threshold, a fault is assumed in the sensor; otherwise, the sensor is assumed to work properly. Moreover, the information of all the available residuals and models allows performing fault isolation in order to discover the faulty sensor. Models are usually derived using either multivariate procedures exploiting the correlation, or analytical relations between several variables, which are sometimes measured at different times (*temporal redundancy*) and/or locations (*spatial redundancy*).

Data validation research applied to water networks can be classified depending on the method use: (a) Time series analysis techniques [3, 4, 6, 11, 27, 40]; (b) Kalman filter-based methods [10, 33, 39]; (c) parity equation methods [5, 20, 42, 46]; (d) pattern recognition methods [54]; and (e) principal component analysis [1, 21, 32].

The output from the data validation process is generally a binary variable indicating whether data are valid or not, or a continuous validity index that can be interpreted as a data confidence index: if the confidence index is too low, data have to be discarded or replaced by forecasted data from a physical or statistical model [38].

High-level data validation methods (model-based) are used not only to validate but also to replace the erroneous data from the faulty sensor by a model prediction. The commonly used data reconstruction methods are: (a) interpolation methods [34]; (b) smoothing methods [31]; (c) data mining technology [16, 25, 53]; and (d) data reconciliation [57].

10.3 Proposed Methodology

This section details the developed procedure, which is divided in two stages (Fig. 10.1): data validation and invalid/missing data reconstruction. The input to this procedure is the raw data vector y_{raw} gathered from the sensors. At the first stage if the data $y_{raw}(k)$ at a certain sample time k is validated, flag v is set to 1 and data $y_{val}(k) = y_{raw}(k)$ is stored in an operational data base (DB) as validated data. Conversely, if data $y_{raw}(k)$ is invalidated, flag v is set to 0 and the data reconstruction process (second stage) is performed to provide a reconstructed estimation $y_{rec}(k)$ of the invalid/missing data $y_{raw}(k)$ to be stored in the DB. The whole procedure is further detailed in the following describing in detail the data validation and reconstruction stages.

10.3.1 Data Validation

The data detection process presented is inspired by the Spanish AENOR-UNE norm 500540 developed for data validation in meteorological stations [52]. The methodology presented here applies a set of consecutive detection tests to a given dataset to finally assign a certain quality level q depending on the number of tests passed. Also, the corresponding tests passed are characterized by a validation vector $\mathbf{l}$. Figure 10.2 presents a schematic view of the whole validation stage. The validation tests include a set of "low-level" tests (Levels 0–3) which check elementary signal properties, and a set of 'high-level' tests (Level 4 and Level 5), which rely on the use of models to check the consistency of the sensor data. The latter models are also used in the reconstruction stage of the potentially invalidated data, as explained in Sect. 10.3.2. If data $y_{raw}(k)$ at a certain sample time k is voided at any validation level, flag v is set to 0 and the data reconstruction process (second stage) is started. Conversely, if data $y_{raw}(k)$ pass all the validation levels, flag v is set to 1 and the data are validated (i.e., $y_{val}(k) = y_{raw}(k)$). In the latter situation, i.e., validated data $y_{raw}(k)$: $q(k) = 6$ and $\mathbf{l}(k) = [111111]^T$ The validation procedure is also detailed in Algorithm 10.1.

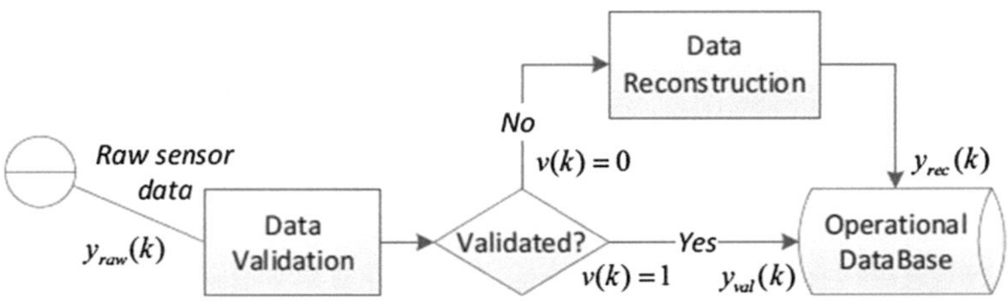

Fig. 10.1 Raw data detection/reconstruction procedure

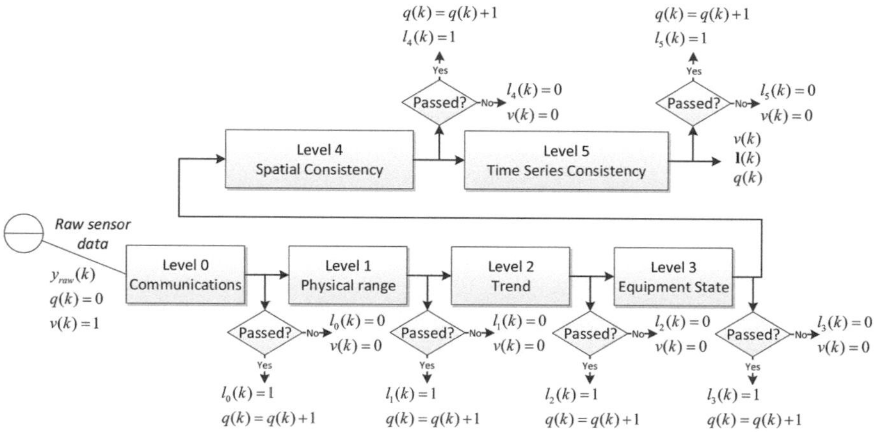

Fig. 10.2 Validation tests diagram

Algorithm 10.1 Data validation

Require: $y_{raw}(k)$
$v(k) = 1$
$q(k) = 0$
for all Validation levels $n = 0, \ldots, 5$ **do**
  Check validation level test n
  if Validation test n passed **then**
    $l_n(k) = 1$
    $q(k) = q(k) + 1$
  else
    $l_n(k) = 0$
    $v(k) = 0$
  end if
end for
if $v(k) = 1$ **then**
  $y_{val}(k) = y_{raw}(k)$
else
  $y_{val}(k) = []$
end if
return : $v(k), l(k), q(k), y_{val}(k)$

An explanation of the low-level tests (elementary signal based) is presented next:

- Level 0 (communications level) checks whether data are properly recorded at a regular sample rate by the acquisition system. If this is not fulfilled, there is some communication problem involving, e.g., the data transmission from the field sensors to the operational database. Hence, this level allows detecting problems in the data acquisition or communication system, which is one of the most common faults in telemeasurement systems in general, as the one considered here.
- Level 1 (physical range limits level) checks whether data are within the physical range of the sensor acquiring the corresponding measurement. The expected range

of the measurements may be obtained from sensor specifications or historical records of the data.

- Level 2 (trend level) checks whether the data derivative, i.e., the magnitude change of the data among consecutive sample times, is within their expected rate. This allows detecting unexpected and possibly undesired sudden changes in the data, e.g., in a water network, tank water level sensors measurements cannot change more than several centimetres per minute.
- Level 3 (equipment state level) allows to check the consistency of the variables in a given equipment unit, i.e., sensor or actuator. For example, in a water network system, in a pipe with a valve and a flow meter installed, there is a relation between the valve state and the flow meter reading.

Level 4: Spatial Model

Level 4 checks the consistency of the data collected by a certain sensor with its spatial model [43], i.e., the correlation between data coming from spatially related sensors. This spatial model is obtained from the physical relations among the network variables. In water networks, this relation is generally obtained from the mass balance model of the element relating the different measured variables involved.

The mass balance expressions for the nth tank is stated as a discrete-time difference equation

$$y_n(k+1) = y_n(k) + \frac{\Delta t}{A_i}\left(\sum_j q_{jn}^{\text{in}}(k) - \sum_h q_{nh}^{\text{out}}(k)\right), \tag{10.1}$$

where $y_n(k)$ is the tank level, A_i is the tank section, $q_{jn}^{\text{in}}(k)$ denotes the manipulated inflows from the jth element to the nth tank, and $q_{nh}^{\text{out}}(k)$ denotes the manipulated outflows from the nth tank to the hth element (which includes the demand flows as outflows).

Moreover, in a water network system nodes are represented as intersections of mains, which mass balance may be expressed as the static equation (see Chap. 3):

$$\sum_i q_i^{in}(k) = \sum_i q_i^{out}(k), \tag{10.2}$$

where similarly to (10.1), $q_i^{in}(k)$ and $q_i^{out}(k)$ correspond to the inflow and outflow of the ith subnet node.

Level 5: Time Series Model

Level 5 checks for temporal consistency of a given sensor measurement, by means of a time series model obtained from sensor historical records under faultless assumption [45]. In contrast to SM level, TSM level only uses information of the considered sensor without needing additional information (e.g., network topology, or extra measurements from the system) to perform the validation, which makes it convenient when there is no such additional information available or the sensors needed by the corresponding SM level are unreliable. At this level, the analysis of the historic measurement records of a certain sensor is used to obtain the corresponding TSM sensor model and to validate the current data acquired by this element.

Usually, the flow in the pipes has a daily repetitive behaviour that can be modelled using a TSM. TSM take advantage of the temporal redundancy of the measured variables. Thus, for each sensor with periodic behaviour, a TS model can be derived:

$$\hat{y}_{ts}(k) = g(y_m(k-1), \ldots, y_m(k-L)), \tag{10.3}$$

where g is the TS model, for data exhibiting a periodicity of L samples.

The aggregate hourly flow model may be built on the basis of a time series modelling approach using ARIMA modelling [6] (see Chap. 6 for more details). A TS analysis is carried out on several daily aggregate series, which consistently showed a daily seasonality, as well as the presence of deterministic periodic components. Once a daily prediction for specific day is obtained, it is distributed along the day hours using a demand pattern. The demand pattern is generated such that the sum of hourly consumes at the end of the day is one, because the sum of the hourly predictions is the daily forecast. Previous works [45] show that several types of demand patterns should be used: one for weekdays, one for Saturdays and one for Sundays/holidays. Then, the hourly prediction is obtained by distributing the daily prediction using the demand pattern.

An alternative widely used method for time series modelling because of its simplicity, low computational and storage requirements and ease of automation, is the HW approach [28, 56] (see Chap. 6 for more details). This method, which was originally created for sales demand forecasting, has been used in a broad range of applications since its appearance. Exponential methods are first introduced in [8], where decreasing series of exponential weights are used. In [56], the former method is extended to include trend and seasonality terms. In [48, 49], multiple (i.e., double and triple) seasonality is explored, expanding the initial single seasonality expression of the former HW method, designed to cope with the sales demands monthly variations across a year period. Further alternative approaches to exponential smoothing forecasting may be found in [18, 35]. Some issues of interest regarding its performance include the effect of the outliers in the forecasting, the consideration of the aforementioned different seasonal periods which may characterize the corresponding time series data sequence to be modelled (e.g., sales demands, water demands) or the consideration of prediction intervals which may provide reliability to the forecast. Regarding outliers, which may be produced by unexpected component behaviours (e.g., sensor malfunctions) these may degrade the performance of the HW method if not accommodated. In [19], this problem is considered and a robust version of the HW method against the outliers is presented, by recursively filtering their effect in the data main stream and applying the standard HW approach to the obtained filtered data. The latter approach is also considered here to provide robustness against the outliers. Moreover, there are different versions of the HW method, e.g., additive or damped trend, additive or multiplicative seasonality, single or multiple seasonality [28]. Here, suitable performance has been attained with the additive single seasonality version, which estimated value is obtained for a forecasting horizon ℓ

$$\hat{x}_{TSM}(k) = \overline{R}(k-\ell) + \ell\overline{G}(k-\ell) + \overline{S}(k-L), \tag{10.4}$$

where $\overline{R}$ is the level estimation removing seasonality

$$\overline{R}(k-\ell) = \alpha(x(k-\ell) - \overline{S}(k-\ell-L)) + (1-\alpha)(\overline{R}(k-\ell-1) + \overline{G}(k-\ell-1)), \tag{10.5}$$

where $\overline{G}$ is the trend estimation

$$\overline{G}(k-\ell) = \beta(\overline{R}(k-\ell) - \overline{R}(k-\ell-1)) + (1-\gamma)\overline{S}(k-\ell-L). \tag{10.6}$$

Moreover, $\overline{S}$ is the seasonal component estimation expressed as

$$\overline{S}(k-\ell) = \gamma(x(k-\ell) - \overline{R}(k-\ell)) + (1-\gamma)\overline{S}(k-\ell-L) \tag{10.7}$$

and L is the season (daily) periodicity, α, β and γ are the HW parameters (level, trend and season smoothing factors, respectively), x is the measured value and $\hat{x}_{TSM}$ is the TSM estimated value. The parameters α, β and γ are in the interval [0,1] and can be estimated from historical data using the least square approach. Hence, analysing the historic records of a certain sensor, a HW TSM model can be obtained and used to estimate missing data of this element when a fault is affecting its readings.

Consistency Tests

The consistency check for the so-called low-level tests is straightforward, since they are based on basic signal-based heuristics. On the other hand, the "high-level" tests are based on the use of models checking for consistency by means of the residuals $r_i(k)$ obtained from the difference between the system measurements and the corresponding SM or TSM expressed in input–output regressor form

$$r_i(k) = x_i(k) - \hat{x}_i(k) = x_i(k) - \phi_i^T(k)\boldsymbol{\theta}_i, \tag{10.8}$$

where $\boldsymbol{\theta}_i$ are the nominal parameters obtained using a training dataset, x_i is the sensor measurement, $\hat{x}_i$ is the model prediction, and $\phi_i(k)$ is the regressor vector of dimensions $n_{\theta_i} \times x_i$ including inputs $(u_i(k), u_i(k-1), u_i(k-2), ...)$ and outputs $(y_i(k), y_i(k-1), y_i(k-2), ...)$. Considering the uncertainty (e.g., modelling errors, noise), the detection test involves checking the condition

$$|r_i(k)| < \tau_i, \tag{10.9}$$

where τ_i is the detection threshold, which can be determined using statistical methods [2] or set-membership approaches [41]. In the case of statistical methods, the noise is assumed to follow a normal distribution with known mean value μ_i and standard deviation σ_i [15]. Then, the threshold of the ith residual can be determined as $\tau_i = \mu_i + 3\sigma_i$, including the 99.7 % of the values of a normal distribution according to the 3-σ rule. On the other hand, when using a set-membership approach the noise is assumed to be unknown but bounded, with a priori known bound. Then, the threshold can be obtained by propagating the uncertainty to the residual computation [41]. Using either one or the other approach, the threshold in (10.9) is determined to include the values of the whole residual distribution in the faultless situation, and hence, it may be used for fault detection purposes. This threshold is also useful to provide prediction interval bounds for the data forecasting process, so test condition (10.9) can be equivalently expressed as follows:

$$x_i(k) \in [\hat{x}_i^{min}(k), \hat{x}_i^{max}(k)], \quad (10.10)$$

where $\hat{x}_i^{max}(k) = \hat{x}_i(k) + \tau_i$ and $\hat{x}_i^{min}(k) = \hat{x}_i(k) - \tau_i$. Condition (10.10) applies both to SM and TSM. These interval bounds (10.10) consider the corresponding model behaviour under faultless conditions including the uncertainty effect, as introduced in the residual bound condition (10.9). Hence, these bounds could alternatively be used in the data validation process, in order to decide whether a data sample at current time instant k is reliable.

10.3.2 Data Reconstruction

As introduced before, when a fault is detected at the validation stage and the corresponding data are voided, a reconstruction process is started until the sensor data are validated again. The output of the data validation process (Fig. 10.1) is used to identify the invalidated data that should be reconstructed. SM, related with Level 4 in Fig. 10.2, and TSM, related with Level 5 in Fig. 10.2, are used for this purpose, depending on the performance of each model. This data reconstruction process is detailed in Algorithm 10.2.

Algorithm 10.2 Data reconstruction

Require: $y_{raw}(k)$, $v(k)$
if $v(k) = 0$ **then**
  Compute $MSE_{SM}(k)$ and $MSE_{TSM}(k)$
  if $MSE_{SM}(k) < MSE_{TSM}(k)$ **then**
    $y_{rec}(k) = \hat{x}_{SM}(k)$
  else
    $y_{rec}(k) = \hat{x}_{TSM}(k)$
  end if
else
  $y_{rec}(k) = []$
end if
return : $y_{rec}(k)$

The models accuracy is measured by the mean square error (MSE) of each model, evaluated in a moving horizon window

$$MSE(k) = \frac{1}{m} \sum_{j=k-m}^{k} e(j)^2, \tag{10.11}$$

where m is the number of data samples considered in the window, $e(j) = x(j) - \hat{x}(j)$ is the error at instant j, $x(j)$ is the measured value at instant j, $\hat{x}(j)$ is the estimated value by the model (SM or TSM, respectively) at instant j and k is the actual time instant. The model having best MSE index before the invalidated data is used to produce the reconstructed sensor signal.

10.3.3 Software for Data Validation and Reconstruction

The architecture of the software framework implemented is depicted in Fig. 10.3. There are two main components: the data management web application and the validation and reconstruction tool [17].

On the one hand, the data management component is a web application focused on collecting and serving time series data, i.e., observations coming from any kind of sensor. It allows authorized users to upload new data, download historical data and visualize data from anywhere using a device with a browser and Internet connection. Thus, this web-based data repository is highly available and provides a solution to the data-driven users to keep centralized data from different projects and sources. It also avoids typical datasets-usage-related drawbacks, e.g., data loss, sparse and

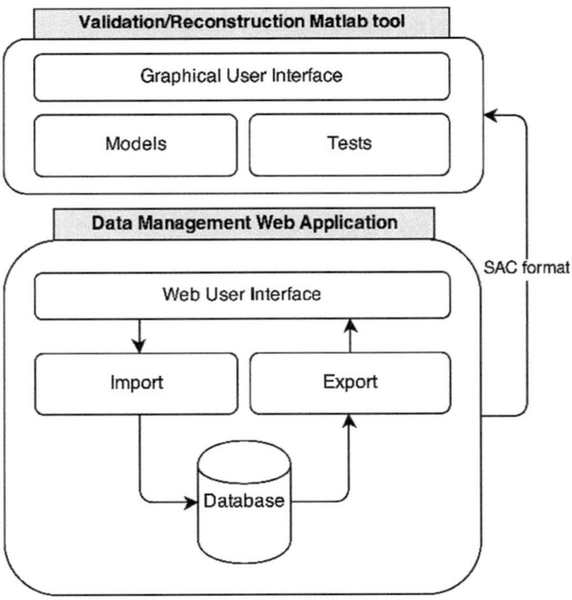

Fig. 10.3 Software architecture diagram

duplicated data locations and e-mails with large datasets between project members. On the other hand, the validation and reconstruction component allows users to apply the methodologies described in Sects. 10.3.1 and 10.3.2 on data provided by the data management web application.

Data Management Web Application

This module provides a user-friendly tool allowing to import and export data so that stored data are available to registered users with read permissions on the dataset. This point is important in order to respect existing confidential agreements: a user must have explicit permission on a dataset to be able to access or visualize it. Only the dataset owner and the administrator can grant read permissions.

People working with data usually need to collect and prepare the raw data (e.g., remove outliers and fill missing data) before being able to apply further analysis (e.g., statistical, exploratory), or even to focus on the real objective of working with the corresponding data. These sort of tasks are time-consuming: there are a many situations when the first two steps introduced take the 80 % of the whole data treatment process time. Thus, this tool provides three services in order to focus the efforts on the data themselves and not on how to collect, obtain and prepare them. These services are the following: the data import service, the data export service and the visualization service.

The data import service handles the data ingestion from different file formats (e.g., CSV, Excel, Access). An import wizard allows the user to specify the input data format, allowing the data to be loaded into the database after being specified. The data export service handles the data extraction. The user can specify the time period to export and the output file format. The current version of the tool allows to download data in CSV, Excel and SAC format.[1] Finally, the data visualization service provides a tool to visually explore the collected data. Hence, the user can plot multiple signals (e.g., time series) to do some exploratory analysis before downloading and to select only the relevant data. The visualization tool allows zooming and panning the time series.

This web application is implemented in two layers, a back end (server layer) that handles the data storage and access with an underlying data model, and a front end (visual layer) to provide a friendly web-based user interface to interact with the three services described before. The back end is developed with the Django[2] web framework connected to a database based on PostgreSQL. The front end is implemented in HTML and JavaScript (see Fig. 10.4). The import and export modules handle the operations of saving and querying data against the PostgreSQL database server.

Validation and Reconstruction MATLAB Tool

The validation and reconstruction methodologies, detailed in this section, have been implemented in a software tool developed in MATLAB. MATLAB is a widely used

[1]SAC format is a binary .mat file containing a defined data structure.

[2]Django is a free open source web framework. Its primary goal is to facilitate the creation of complex, database-driven websites.

Fig. 10.4 Data management web application

numerical computing and programming platform in many research institutions and industrial enterprises, which makes it a convenient prototyping and development framework. This tool includes a graphical user interface (GUI) to configure different modules and to run the validation and reconstruction processes with the configured settings (see Fig. 10.5).

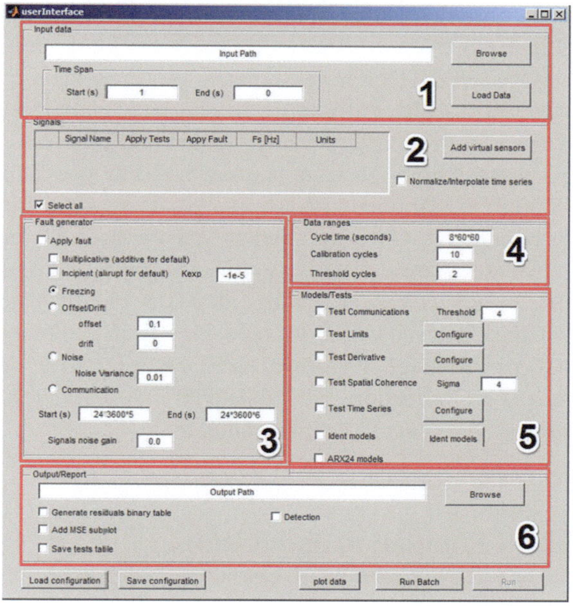

Fig. 10.5 Validation and reconstruction MATLAB tool

Figure 10.5 shows the GUI, composed by six panels. Following the numeration in the figure, each panel has the following purpose:

1. Data input: The user can select the .mat file path in SAC format and load the data into the tool.
2. Signals list: This panel shows the listing of the signals loaded in the previous panel.
3. Fault generator: This module provides a fault generator framework in order to simulate different types of faults; thus, the user can apply a fault to the selected signals. The faults available are freezing, offset, drift, noise and communications.
4. Data ranges: This panel allows the user to indicate the cycle duration of the TSM. For instance, if a signal shows a daily pattern, the cycle duration is 24 h (86400 s). The user can define the number of identification and validation cycles, the rest of the data will be used as testing dataset.
5. Tests and models: This panel lists the tests and the models available. Here, the user can select the tests to apply and configure the required parameters depending on the models and tests selected.
6. Output and reporting: In this panel, the user can enter the path where the results will be recorded and select the reporting options.

As mentioned previously in this section, the input to this tool should be a dataset arranged in the SAC format layout. This input dataset is partitioned in three different subsets (calibration, validation and testing) in order to apply cross-validation to the measurements and validate the generalization of the models. The different subset ranges are defined by the user according to the parameters entered in the panel 4 (Fig. 10.5).

Once all the required parameters are set by the user the process may be executed, which will sequentially apply the presented methodology to the data. This process is divided in three different stages, namely calibration, validation and reconstruction. First, the calibration stage is executed using the calibration and validation datasets in order to learn and estimate the parameters required by the tests and the models to be applied. Once the models and the tests are calibrated, the validation stage runs the sequence of tests in order to validate the testing dataset. Each test applied labels each data $y_{raw}(k)$ with a flag (**l** in Fig. 10.2 and Algorithm 10.1) to indicate whether the test has been fulfilled. Finally, in the reconstruction stage, the model with best performance (i.e., lowest MSE) is selected in order to replace the invalidated data at the validation stage (data with $v = 0$ in Fig. 10.2 and Algorithm 10.1) by its corresponding reconstructed estimation.

10.4 Simulation and Results

This section presents some results obtained with the methodology presented in Sect. 10.3. Realistic situations occurring in the Catalonia regional network have been used.

The Catalonia regional water network managed by ATLL company (Fig. 10.6) supplies water to the Barcelona metropolitan area. Most of the population of the

Fig. 10.6 ATLLs Catalonia regional water network

region (approximately 4.5 million people) is concentrated in this area. This network transports the drinking water from the main water treatment plants (ETAPs), which take the water from two different rivers (Llobregat and Ter), towards the main storing and buffer tanks of 116 municipalities in the Barcelona metropolitan area, using about 1045 km of pipes of up to 3 m diameter. The network is composed by 170 storage tanks, 67 pumps and 212 demand sectors, and is monitored using more than 200 flow meters and 115 tank level sensors by means of a SCADA system with 10 min sample time.

The dataset used to obtain these results is the network's raw data collected by ATLL company, including flow meter measurements, level meter measurements, valve positions and communication system alarms. The time series models considered here are HW TSM.

Figure 10.7 presents a scenario involving the flow meter E6FT00502 CI. In this case, a general communication fault affects all the sensors, a common situation occurring in actual water monitoring systems happening when, e.g., the concentrator (a device collecting data from sensors installed in a particular zone) drops. A SM model is available using the corresponding spatially related sensors data. However, in this particular case the rest of the sensors involved in the SM model are all affected by the same communication fault. Hence, the corresponding measurements are not available for data reconstruction after the communication fault occurs and, consequently, the SM model output cannot be obtained with the available sensor measurements. Alternatively, SM model output might be obtained using TSM estimations of the affected sensors, but this is not considered here. If the use of raw sensor data is required in order to provide the SM estimations, the only available model for reconstruction is

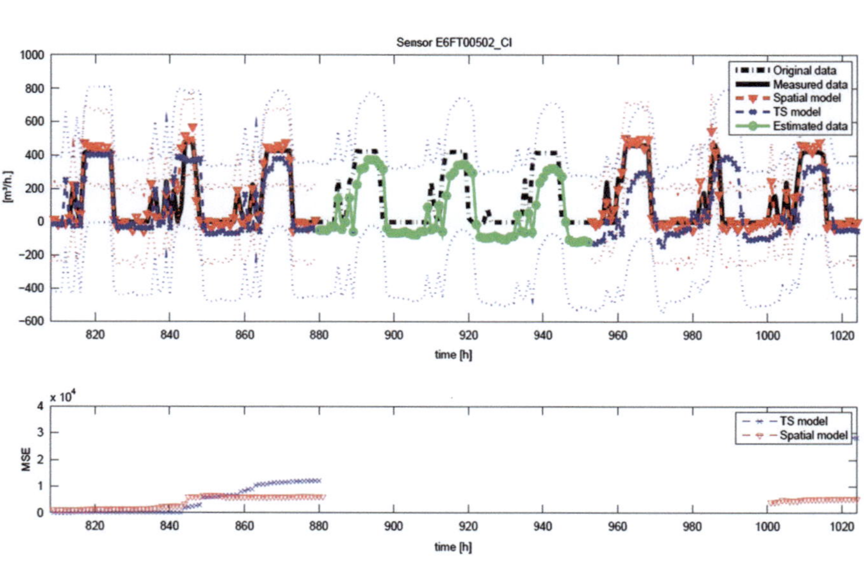

Fig. 10.7 Results of the validation and reconstruction methodology, flow meter E6FT00502 CI

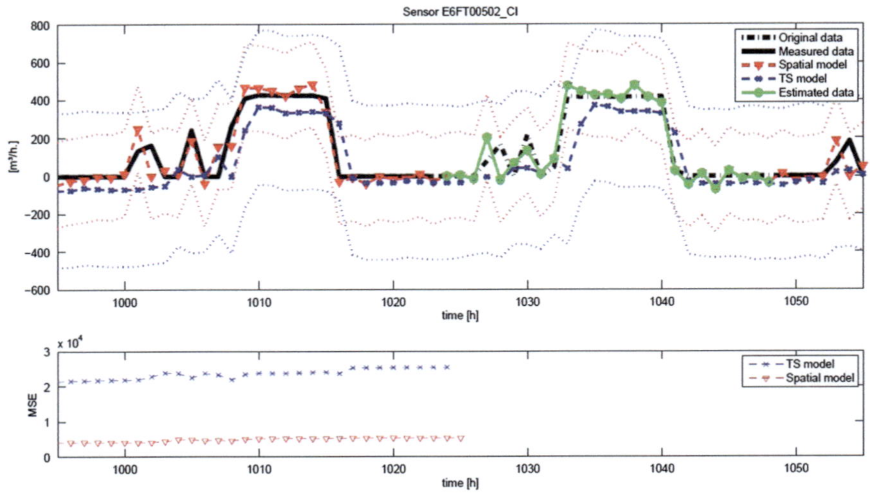

Fig. 10.8 Results of the validation and reconstruction methodology on the flow meter E6FT00502 CI

the TSM, which is finally used for the missing data reconstruction in the scenario considered in Fig. 10.7, based on the limited available information in this particular case.

Figure 10.8 presents a different scenario from the same flow meter, where a communication fault affects the flow meter, which does not transmit data in one-day period (from $t = 1024$ h to $t = 1048$ h). In this particular case, the communication fault only affects the latter sensor; hence, the corresponding SM is available because the spatially related sensors (e.g., D6FT00201 CI) are not affected by this fault. In this scenario, the SM model is used for missing data reconstruction, since it performs better than the corresponding HW TSM model (bottom subplot in Fig. 10.8). It may be noted that the use of the SM assumes that the model input sensor measurement is faultless when the SM is used for data reconstruction. This may be assured since the input model integrity is checked by the methodology presented here at its corresponding stage and, if not verified, the validation test at this stage is not fulfilled.

10.5 Conclusions

In this chapter, a data validation and reconstruction methodology is introduced to overcome the sensor problems arising in CIS, such as water networks. The validation strategy is based on a set of data quality tests that allow to detect potentially erroneous data. Then, a reconstruction scheme is defined using SM and TSM to provide an estimation based on the model having the best fit, also providing prediction intervals for the forecasted reconstructed data. In addition, a software tool is described to provide a homogeneous and accessible database by a user-friendly interface and to apply the methodology presented here. Finally, some results obtained using data from a real network located in the Catalonia area are presented using the software described, showing the ability of the methodology to detect and reconstruct anomalous data. In future steps of this work, the proposed methodology and tool are going to be applied to the water quality sensors.

References

1. Arteaga F, Ferrer A (2002) Dealing with missing data in MSCP: several methods, different interpretations, some examples. J Chemom 16:408–418
2. Basseville M, Nikiforov I (1993) Detection of abrupt changes. Prentice Hall
3. Bennis S, Berrada F, Kang N (1997) Improving single variable and multivariable techniques for estimating missing hydrological data. J Hydrol 191:87–105
4. Bennis S, Kang N (2000) A new methodology for validating historical hydrometric data with redundant measurements. In: Blain WR, Brebbia CA (ed) Hydraulic engineering software VIII. WIT Press
5. Boukhris A, Giuliani S, Mourot G (2001) Rainfall-runoff multi-modelling for sensor fault diagnosis. Control Eng Pract 9(6):659–671
6. Box G, Jenkins G (1970) Time series analysis forecasting and control. Holden-Day
7. Branisavljević N, Kapelan Z, Prodanović D (2011) Improved real-time data anomaly detection using context classification. J Hydroinform 13(3):307–323
8. Brown RG (1959) Statistical forecasting for inventory control. McGraw-Hill
9. Burnell D (2003) Auto-validation of district meter data. In: Maskimovic, Butler, Memon (eds) Advances in water supply management. Swets & Zeitlinger Publish
10. Ciavatta S, Pastres R, Lin Z, Beck MB, Badetti C, Ferrari G (2004) Fault detection in a real-time monitoring network for water quality in the lagoon of Venice (Italy). Water Sci Technol 50(11):51–58
11. Crobeddu E, Bennis S (2006) Data acquisition, validation and forecasting for a combined sewer network. In: Popov V, Kungolos AG, Brebbia CA, Itoh H (eds) Waste management and the environment III. WIT Press
12. Cugueró-Escofet MA, Christodoulou M, Quevedo J, Puig V, García D, Michaelides M (2014) Combining contaminant event diagnosis with data validation/reconstruction: application to smart buildings. In: IEEE 22nd Mediterranean conference on control and automation (MED14), pp 293–298, Palermo, Italy
13. Cugueró-Escofet MA, Quevedo J, Puig V, García D (2014) Inconsistent sensor data detection/correction: application to environmental systems. In: International joint conference on neural networks (IJCNN), pp 84–90
14. Denoeux T, Boudaoud N, Canu S, Dang V, Govaert G, Masson M, Petitrenaud S, Soltani S (1997) High level data fusion methods. Technical report, Université de Technologie de Compiègne
15. Ding SX (2008) Model-based fault diagnosis techniques
16. Elshorbagy A, Simonovic S, Panu U (2002) Estimation of missing streamflow data using principles of chaos theory. J Hydrol 255:123–133
17. García D, Quevedo J, Puig V, Cugueró MA (2014) Sensor data validation and reconstruction in water networks: a methodology and software implementation. In: 9th International conference on critical information infrastructure security
18. Gardner ES (2006) Exponential smoothing: the state of the art—Part II. Int J Forecast 22(4):637–666

19. Gelper S, Fried R, Croux C (2010) Robust forecasting with exponential and holt-winters smoothing. J Forecast 29:285–300
20. Hamioud F, Joannis C, Ragot J (2005) Fault diagnosis for validation of hydrometric data collected from sewer networks. In: 10th International conference on urban drainage, Copenhagen, Denmark
21. Harkat M, Mourot G, Ragot J (2006) An improved PCA scheme for sensor FDI: application to an air quality monitoring network. J Process Control 15(6):625–634
22. Jorgensen HK, Rosenorn S, Madsen H, Mikkelsen S (1998) Quality control of rain data used for urban runoff systems. Water Sci Technol 37(11):113–120
23. Kanakoudis V, Tolikas D (2001) The role of leaks and breaks in water networks: technical and economical solutions. J Water Supply: Res Technol 50(5):301–311
24. Kanakoudis V, Tsitsifli S (2011) Water pipe network reliability assessment using the DAC method. Desalin Water Treat 33(1–3):97–106
25. Kramer M (1991) Nonlinear principal component analysis using autoassociative neural networks. AIChE J 37(2):233–243
26. Lempio G, Podlasly C, Einfalt T (2010) Niklas—automatical quality control of time series data. In: 6th European conference on radar in meteorology and hydrology
27. Lobanova H, Lobanova G (2003) Approach for restoration of missing data, long-term time series and generalization of results. In: Maskimovic, Butler, Memon (eds) Advances in water supply management. Swets & Zeitlinger Publish
28. Makridakis S, Wheelwright SC, Hyndman RJ (1998) Forecasting methods and applications. Wiley
29. Marinaki M, Papageorgiou M (2005) Optimal real-time control of sewer networks. Springer
30. Maul-Kotter B, Einfalt T (1998) Correction and preparation of continuously measured rain gauge data: a standard method in North Rhine-Westphalia. Water Sci Technol 37(11):155–162
31. Mourad M, Bertran-Krajewski J (2002) A method for automatic validation of long time series of data in urban hydrology. Water Sci Technol 45(4–5):263–270
32. Nelson P, Taylor P, MacGregor J (1996) Missing data methods in PCA and PLS: score calculations with incomplete observations. J Chemom Intell Lab Syst 35:45–65
33. Pastres R, Ciavatta S, Solidoro C (2003) The extended kalman filter (EKF) as a tool for the assimilation of high frequency water quality data. Ecol Modell 170(2–3):227–238
34. Olisson G, Nielsen M, Yuan Z, Lynggaard-Jensen A, Steyer J (2005) Instrumentation, control and automation in wastewater systems. Technical report, IWA Publishing
35. Pegels CC (1969) Exponential smoothing: some new variations. Management Science 15(5): 311–315. doi: 10.1287/mnsc.15.5.311
36. Perez R, Puig V, Quevedo J, Nejjari F, Pascual J, Sanz G, Cuguero-Escofet MA, Cembrano M, Landeros E, Peralta A, Mirats J, Meseguer J (2012) Localizacion de fugas en redes de distribucion de agua potable. In: Automatica e instrumentacion, num. 437, pp 36–38
37. Pérez R, Sanz G, Puig V, Quevedo J, Cuguero Escofet MA, Nejjari F, Meseguer J, Cembrano G, Mirats Tur JM, Sarrate R (2014) Leak localization in water networks: a model-based methodology using pressure sensors applied to a real network in Barcelona. In: IEEE control systems magazine, vol. 34, num. 4, pp 24–36. http://hdl.handle.net/2117/23610
38. Petit-Renaud S, Denoeux T (1998) A neuro-fuzzy system for missing data reconstruction. In: IEEE workshop on emerging technologies, intelligent measurement and virtual systems for instrumentation and measurement, Saint-Paul, USA
39. Piatyszek E, Voignier P, Graillot D (2000) Fault detection on a sewer network by a combination of a Kalman filter and a binary sequential probability ratio test. J Hydrol 230(3–4):258–268
40. Prescott S, Ulanicki B (2001) Time series analysis of leakage in water distribution networks. In: Water software systems: theory and applications, vol 2, Baldock, England. Research Studies Press Ltd., pp 17–28
41. Puig V (2010) Fault diagnosis and fault tolerant control using set-membership approaches: application to real case studies. Int J Appl Math Comput Sci 20(4):619–635
42. Puig V, Quevedo J, Figueras J, Riera S, Cembrano G, Salamero M, Wilhelmi G (2003) Fault detection and isolation of rain gauges and limnimeters of Barcelona's sewer system using interval models. In: IFAC SAFEPROCESS, Washington, USA

43. Quevedo J, Blanch J, Puig V, Saludes J, Espin S, Roquet J (2010) Methodology of a data validation and reconstructions tool to improve the reliability of the water network supervision. In: International conference of IWA water loss, Sao Paolo, Brazil
44. Quevedo J, Cugueró MA, Pérez R, Nejjari F, Puig V, Mirats JM (2011) Leakage location in water distribution networks based on correlation measurement of pressure sensors. In: 8th IWA symposium on system analysis and integrated assessment (WATERMATEX 2011), San Sebastian
45. Quevedo J, Puig V, Cembrano G, Blanch J (2010) Validation and reconstruction of flow meter data in the Barcelona water distribution network. Control Eng Pract 18(6):640–651
46. Ragot J, Maquin D (2006) Fault measurement detection in an urban water supply network. J Process Control 16(9):887–902
47. Schutze M, Campisano A, Colas H, Schilling W, Vanrolleghem PA (2004) Real time control of urban wastewater systems—where do we stand today? J Hydrol 299(3–4):335–348
48. Taylor JW (2003) Short-term electricity demand forecasting using double seasonal exponential smoothing. J Oper Res Soc 54(8):799–805
49. Taylor JW (2010) Triple seasonal methods for short-term electricity demand forecasting. J Oper Res 204(1):139–152
50. Tsang KM (2003) Sensor data validation using gray models. ISA Trans 42:9–17
51. Tsitsifli S, Kanakoudis V, Bakouros I (2011) Pipe networks risk assessment based on survival analysis. Water Resour Manage 25(14):3729–3746
52. UNE (2004) Redes de estaciones meteorológicas automáticas: directrices para la validación de registros meteorológicos procedentes de redes de estaciones automáticas: validación en tiempo real. Technical report, AENOR UNE 500540
53. Ustoorikar K, Deo MC (2007) Filling up gaps in wave data with genetic programming. Mar Struct 21(2–3):177–195
54. Valentin N, Denoeux T (2001) A neural network-based software sensor for coagulation control in a water treatment plant. Intell Data Anal 5:23–39
55. Venkatasubramanian V, Rengaswamy R, Kavuri SN (2003) A review of process fault detection and diagnosis: Part II: Qualitative models and search strategies. Comput Chem Eng 27(3):313–326
56. Winters PR (1960) Forecasting sales by exponentially weighted moving averages. Manage Sci 6(52):324–342
57. Yoo CK, Villez K, Lee IB, Van Hulle S, Vanrolleghem PA (2006) Sensor validation and reconciliation for a partial nitrification process. Water Sci Technol 53(4–5):513–521

Chapter 11
Fault Diagnosis

Teresa Escobet, Ramon Sarrate and Ramon Comasolivas

11.1 Introduction

A water transport network (WTN) is a complex large-scale system (LSS) which is composed of a large number of elements with time-varying behaviour, exhibiting numerous operating modes and subject to changes due to external conditions (e.g., weather) and operational constraints. Nowadays, these systems are controlled online and monitored by means of complex computer-based systems [1, 2]. But, in order to take advantage of these expensive infrastructures, it is also necessary to have a highly sophisticated online fault diagnosis (FD) schemes which allow to detect as soon as possible accidental or intentional contamination, sensor and actuator malfunctions (faults) or incorrect operations in order to prevent system malfunctions avoiding catastrophic consequences and ensuring safety and reliable operations. As will see in Chap. 15, online FD is the primary stage of active fault-tolerant control (FTC), but it is also important in the design of health management systems.

The diagnostic process aims at carefully identifying which fault (including hardware faults, software faults and malicious attacks) can be guessed to be the cause of monitored events. In general, when addressing the FD problem, two strategies can be found in the literature: hardware redundancy based on the use of redundancies (adding extra sensors) and software (or analytical) redundancy based on the use of software/intelligent sensors (or model) combining information provided by sensor measurements. In critical infrastructure systems (CIS), hardware redundancy is preferred. However, for large-scale systems, the use of hardware redundancy is quite expensive and increases the number of maintenance and calibration operations. This is the reason why, in CIS applications, systems that allow combining both hardware and analytical redundancy [3] must be developed.

T. Escobet (✉) · R. Sarrate · R. Comasolivas
Research Center "Supervision, Safety and Automatic Control" (CS2AC-UPC), Terrassa, Spain
e-mail: teresa.escobet@upc.edu

V. Puig et al. (eds.), *Real-Time Monitoring and Operational Control of Drinking-Water Systems*, Advances in Industrial Control,
DOI 10.1007/978-3-319-50751-4_11

The problem of FD system based on analytical redundancy has been addressed using quite different techniques, and by many researchers, applying different approaches and assumptions. An overview of techniques in this area is provided by several review papers [4–7]. In this chapter, model-based diagnosis (MBD) is applied, which involves designing residual generators based on structural redundancies in the system.

As remarked in Chap. 9, instrumentation plays a crucial roll in MBD. The performance of MBD depends on the set of measurements that are available in the system. Most of them are provided by sensors installed in the system. Therefore, for a given set of sensors already installed in the system, the maximum MBD performances that can be achieved are bounded. Thus, designing a MBD module ultimately aims at reaching those performance bounds.

In addition to sensors, a WTN consists of three major components: pumps, distribution storage and distribution piping network. Most systems require pumps to supply lift to overcome differences in elevation, and energy losses caused by friction. Pipes may contain flow control devices, such as regulating or pressure-reducing valves. Techniques based on locating leaks in a distribution network is an important issue for the water companies, and this problem has been studied extensively in the literature (see Chap. 7). However, there are less studies that focus on the problem of FD in general. The aim of this chapter is to propose a methodology for detecting component failures in WTNs.

A fault must be understood as an unexpected change in a component or system. Although it may not represent a physical failure or breakdown, it may be due to an erroneous state of hardware or software resulting from failures of components, physical interference from the environment, operator errors or even an incorrect design. An error is the way in which a fault manifests itself, which means the deviation of the system behaviour from the required operation. And, a failure is defined as the inability of a component or system to operate according to its specifications.

A WTN is an inherently continuous system. However, since it is normally controlled by a supervisory system that imposes discrete switching behaviour between several operational modes, the whole system truly behaves as a hybrid dynamical system. Most of the widely used methods for FD for large-scale and hybrid systems rely on the use of hybrid models that integrate continuous and discrete dynamics [8]. Due to its hybrid dynamical nature, a component fault does not always show up as an error. For example, in the case of a WTN with many redundant valves, if a hardware fault causes a valve to stick in a closed position, this fault may not show up until the valve is eventually commanded to open. But, due to physical redundancy the system can operate to perform its required function. And, even in a hybrid dynamical system, an actual error may not be the consequence of a faulty component but that of an operating mode switching.

Traditional FD methods have been mostly developed on a centralized scheme that assumes to have the full system information and its global dynamic model. When dealing with LSS, the centrality assumption usually fails to hold, either because gathering all measurements in one location is not feasible, or because a centralized high-performance computing unit is not available. These difficulties have led to research in fault diagnosis algorithms that operate in either a decentralized or distributed way. Both diagnosis categories are based on the partition of the overall system into subsystems in order to building the local diagnoser [9]. In the decentralized diagnosis, both a central coordination module and a local diagnoser for each subsystem that forms the whole supervision system are running in parallel. On the other hand, in the distributed approach, a set of local diagnosers share information by means of some communication protocol instead of requiring a global coordination process such as in a decentralized approach.

In the literature, most of the works concern fault diagnosis schemes for abrupt faults, which are modelled as instantaneous changes in system behaviour at a given time. However, degradation in system components are often modelled as incipient faults, which are slow drifts in system parameter values over time. In fact, the problem of diagnosing incipient faults is related to the problem of condition-based maintenance which aims at performing actions in the hope of preventing the occurrence of a fault in the future operation of the system. In LSS and, in particular, in WTN where there are many sensors and actuators, the early detection of component degradation should allow to plan maintenance actions reducing maintenance costs and increasing the safety of the system. For instance, the online water quality sensors for field deployment require careful maintenance routines. The incipient fault detection in this components could help in these tasks [10].

In general, a MBD system involves the following tasks [11]. The first task is fault detection, implemented with a set of fault detection tests based on checking errors, which allows deciding whether a fault has occurred, and its occurrence time instant. The second task is fault isolation or localization, which is typically achieved through algorithms that compute a set of possible faulty components. And the third one is fault identification and estimation, which aims at determining the kind of fault and its severity. In general, FD approaches are implemented using a hierarchical structure in which the lowest level consists of a module designed to detect an abnormal behaviour and triggering alarms, and the highest level tries to isolate and localize the problem. A reduced number of approaches include the third one, unless it is necessary for achieving a fault-tolerant control system.

The ability to detect and isolate faults is an important task in order to safeguard the integrity of WTN. To exemplify the FD methodologies in WTN, the Barcelona water transport network is used as the case study.

11.2 Problem Statement

MBD involves using a system model to build a set of consistency tests. In the case of WTN, it is assumed that the process can be described by the following continuous-time nonlinear model:

$$\begin{aligned}\dot{\mathbf{x}} &= \mathbf{g}(\mathbf{x}, \mathbf{u}, \boldsymbol{\theta}) + \mathbf{w},\\ 0 &= \mathbf{f}(\mathbf{x}, \mathbf{u}, \boldsymbol{\theta}) + \boldsymbol{\eta},\\ \mathbf{y} &= \mathbf{h}(\mathbf{x}, \mathbf{u}, \boldsymbol{\theta}) + \boldsymbol{\nu},\end{aligned} \tag{11.1}$$

where $\mathbf{x}$ is the vector of system states, $\mathbf{u}$ is the vector of control actions and $\mathbf{y}$ is the vector of system outputs; $\boldsymbol{\theta}$ is a vector of uncertain parameters; $\mathbf{w}$ and $\boldsymbol{\eta}$ are unmodelled dynamics and disturbances; $\boldsymbol{\nu}$ are measurement noises; $\mathbf{g}$ and $\mathbf{h}$ are the state update and measurement nonlinear functions, respectively; and $\mathbf{f}$ is the nonlinear static relation function.

It should be noted that the controller could generate control actions at discrete-time points than can change the operational models of the plant (e.g., by turning components on and off, changing component parameter values and the set point of regulators [12]). These operating mode changes produce discrete changes in the dynamic models of the system behaviour. Thus, multiple system models are required to analyse their behaviour. Current techniques propose modelling this complex system using a hybrid system model [8, 13, 14] which combines piece-wise linear modelling approaches with a discrete-event model given by an automaton representing the transitions between operation modes, nonlinearities and faulty situations. Each mode could be modelled by (11.1).

The design of a MBD involves building a set of consistency relations that only involves observed variables [15], known as analytical redundancy relations (ARRs). To obtain ARRs for state-space representation such as (11.1), it is necessary to manipulate the model to eliminate the unobserved variables (i.e., the state $\mathbf{x}$).

Given the model defined in (11.1) corresponding to a known operating mode with observed variables $\mathbf{y}$ and $\mathbf{u}$, an ARR is defined as follows:

$$\Psi_i(\mathbf{y}, \mathbf{u}) = 0, \tag{11.2}$$

where Ψ_i is called the residual generator or computational form of residual r_i. The set of residuals, $\mathcal{W}$, can be represented as

$$\mathcal{W} = \{r_i | r_i = \Psi_i(\mathbf{y}, \mathbf{u}), i = 1, \ldots, n_r\}, \tag{11.3}$$

where n_r is the number of residuals.

Then, fault diagnosis is based on monitoring the set of residuals in order to assess the consistency of their corresponding ARRs. The set of inconsistent ARRs is represented by the set of residuals

$$\mathcal{W}^* = \{r_i | r_i = \Psi_i(\mathbf{y}, \mathbf{u}) \neq 0, i = 1, \ldots, n_r\}. \tag{11.4}$$

Fault isolation task starts by obtaining the observed fault signature, where each single fault signal indicator ϕ_i is defined as follows:

$$\phi_i = \begin{cases} 0 & \text{if } r_i \notin \mathcal{W}^*, \\ 1 & \text{if } r_i \in \mathcal{W}^*. \end{cases} \tag{11.5}$$

Typically the interface between fault detection and fault isolation is through a binary codification of the evaluation of every residual; this binary interface could lead to a wrong diagnosis when the residuals present different sensitivities and order/time of activation after the fault appearance [16], and also produce undesirable decision instability (chattering) due to the effect of the noise and uncertainties. In the literature, there are different approaches to deal with this problem. For example, [17] proposed an improved fault diagnosis approach based on the fuzzy evaluation of the residuals that consider not only binary information but also signs/sensitivities as well as the persistence of residual activation. Finally, in [18], the use of the Kramer function [19] is proposed for evaluating the residuals gradient and to compute a fault diagnosis signal.

Fault isolation is based on comparing the history of the fault diagnosis signals with some stored fault patterns based on an extension of the fault signature matrix (with includes other signal properties such as signs, occurrence order and time) and to use a decision logic algorithm for proposing the most probable fault candidate.

11.3 Proposed Approach

11.3.1 Fault Diagnosis Architecture

The proposed approach for online FD consists of two modules, fault detection and fault isolation, as shown in Fig. 11.1. Fault detection module checks the consistency between the observed and the normal system behaviour using a set of analytical redundancy relations (ARRs), which relate the values for measured variables according to a normal operation (fault-free) model of the monitored system. ARRs are derived by performing a structural analysis of the qualitative behaviour of WTN system. Fault detection provides a set of residual indicators and when some inconsistency is detected, the fault isolation mechanism is activated in order to identify the possible fault.

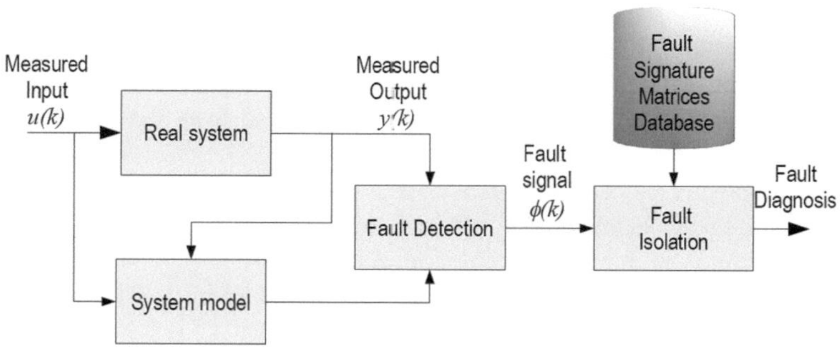

Fig. 11.1 Fault detection and isolation block diagram

Fault isolation module is based on the idea of pattern matching, which consists in comparing the theoretical fault signatures (FS) corresponding to all possible component degradations with the computed observed fault signature obtained from the residuals. In general, FS contain the expected evolution of the residual immediately after that a fault occurs (known also as symptom), which are derived offline from the structural properties of the ARRs. In the proposed approach, FS include both analytic and heuristic information generated to produce distinctive pattern for each particular fault. FS are stored as matrices in the fault diagnosis database and used to match to the trends of online residuals using pattern matching methods. Finally, a decision logic algorithm proposes the most probable fault candidate.

11.3.2 ARR Generation

The first step into the design of a FD system consists of generating a set of ARRs based on a model of the system and the available sensors. When analysing the model of a large-scale system for FD design, the designer is invariably confronted with a dilemma: to use a more accurate model which is harder to manage, or to work with a simpler model which is easier to manipulate but with less confidence. A hierarchy of models with increasing complexity and fidelity is often used for different purposes (control/supervision design, simulation). However, as the number of variables increases, it is worth to start analyzing the system with simple structural models that offer relatively easy ways to identify unsuitable system configurations, causes for lack of desired properties and straightforward remedies.

The structural model of a system is an abstraction of its behavioural model in the sense that only the constraints (i.e., the link between variables and parameters) are considered but not the constraints themselves. Structural analysis aims at the study of system properties using a structural graph [20, 21]. Structural properties of interest in control that can be derived from structural analysis are input/output reachability and observability/controllability.

With regard to fault diagnosis, a structural property that can be derived from structural analysis is diagnosability (i.e., capability to detect and isolate one fault from the others [22]). It also allows determining critical components (i.e., set of system components that are indispensable to satisfy a determined property) or redundant components (i.e., system components which are not critical for the correct functionality of the system, so that they could be subtracted from the system and the satisfaction of the objective would still be achieved) [23]. Structural analysis also allows decomposing a system into subsystems [24] and, as shown in Chap. 9, it can be used to place sensors and actuators for control and supervision [25].

In the structural approach, the behavioural model of a system $\mathcal{M}$ introduced in (11.1) can be seen as a set of n equations, which depend on a set of m variables $\mathcal{Z} = \mathcal{X} \cup \mathcal{O}$, where $\mathcal{X}$ is the set of unknown variables and $\mathcal{O}$ is the set of observed variables, $\mathcal{O} = \{u_i\} \cup \{y_i\}$. A structural model can be formalized as a bipartite graph

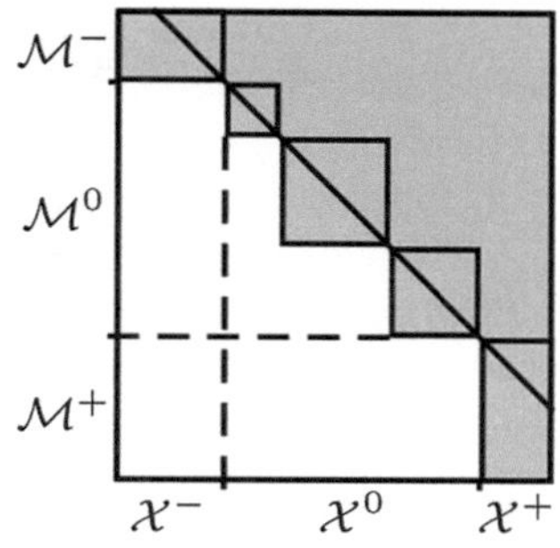

Fig. 11.2 Dulmage–Mendelsohn decomposition of a model

$\mathcal{G} = (\mathcal{M}, \mathcal{Z}, \mathcal{A})$, where $\mathcal{M} \cup \mathcal{Z}$ is the set of vertices and $\mathcal{A}$ is the set of edges, such that $(e_i, z_j) \in \mathcal{A}$ as long as the expression of $e_i \in \mathcal{M}$ depends on variable $z_i \in \mathcal{Z}$.

A structural model is usually represented by a biadjacency matrix that relates equations as rows and variables as columns. An element b_{ij} of the biadjacency matrix is 1 as long as $(e_i, z_j) \in \mathcal{A}$.

The Dulmage–Mendelsohn (DM) decomposition [26] is a well-known theoretical tool in the structural model-based fault diagnosis community. The DM decomposition is usually applied to the structural model $\mathcal{G}_{\mathcal{X}} = (\mathcal{M}, \mathcal{X}, \mathcal{A})$ that relates equations and unknown variables. It defines a partition on the structural model. The biadjacency matrix in Fig. 11.2 shows the Dulmage–Mendelsohn decomposition of $\mathcal{G}_{\mathcal{X}}$. The grey-shaded areas contain ones and zeros, while the white areas only contain zeros. Three main parts of $\mathcal{M}$ can be identified in the partition, namely the under-determined part $\mathcal{M}^-$, the just-determined part $\mathcal{M}^0$ and the overdetermined part $\mathcal{M}^+$. In the overdetermined part, there are more equations than unknown variables, which implies that there exists some degree of redundancy, and this is the part of the model that is useful for process monitoring.

The set of residual generators, $\mathcal{W}$, can be derived following the structural approach, using the algorithms proposed in [20, 27, 28]. In particular, given a set of model equations $\mathcal{M}$, residuals can be obtained from the overdetermined part of the model $\mathcal{M}^+$.

Each residual is obtained from a subset of redundant equations in $\mathcal{M}^+$. The minimal subset of redundant equations that are related to a residual r_i is called minimal structural overdetermined (MSO) set [27] and is represented by $\omega_i \subseteq \mathcal{M}^+$. An ARR can be generated from an MSO by defining a computation sequence. A computation sequence determines how to compute internal unknown variables through a convenient manipulation of MSO equations, and how to check consistency through a redundant equation. A computation sequence can be represented by a directed bipartite graph, where alternate nodes represent variables and equations.

Given a set of model equations $\mathcal{M}$, the set of all possible MSO sets is $\Omega = \{\omega_1, \omega_2, \dots, \omega_r\}$. An efficient algorithm to compute Ω exists [27]. However, it is not always possible to define a computation sequence for an MSO. For instance, solving a certain unknown variable in a nonlinear equation could be a hard task, or even impossible. Specifically, ARRs that depend on a computation sequence that implies the inverse computation of non-invertible functions should be excluded. Another issue concerns differential equations where either integral or mixed causality should be considered [29]. A causal framework was proposed in [30] that allows taking into account these computation issues in the FD system design. In particular, Algorithm 7 in [30] determines the subset $\Omega_c \subseteq \Omega$ of all causally computable MSO sets. These MSO sets are guaranteed to provide a residual using easy back substitution techniques according to their computation sequence. The causal framework requires a causal structural model to be defined through a biadjacency matrix, where

$$b_{ij} = \begin{cases} < empty > & \text{if } (e_i, z_j) \notin \mathcal{A}, \\ \times & \text{if } (e_i, z_j) \in \mathcal{A} \text{ and } z_j \text{ can be computed from } e_i, \\ \Delta & \text{if } (e_i, z_j) \in \mathcal{A} \text{ and } z_j \text{ cannot be computed from } e_i. \end{cases}$$

For illustration proposes, the following Example 1 model is used,

$$\begin{aligned} e_1 &: x_1 + y_1 = 0, \\ e_2 &: \mathbf{g}_1(\dot{x}_1) - x_3 = 0, \\ e_3 &: \mathbf{g}_2(\dot{x}_2) - x_3 = 0, \\ e_4 &: x_2 - y_2 = 0, \\ e_5 &: x_3 - y_3 = 0, \end{aligned}$$

where x_1 and x_2 are the system state variables, x_3 is an unknown variable, and y_1, y_2 and y_3 are systems measurements. For each state-space variables, new constraints capturing the dynamics should be added. In this study case, two more equations should be considered:

$$\begin{aligned} e_6 &: \dot{x}_1 - \frac{dx_1}{dt} = 0, \\ e_7 &: \dot{x}_2 - \frac{dx_2}{dt} = 0. \end{aligned}$$

Table 11.1 Biadjacency matrix for the causal structural model of Example 1

	x_1	$\dot{x}_1$	x_2	$\dot{x}_2$	x_3	y_1	y_2	y_3
e_1	×					×		
e_2		×			×			
e_3				×	×			
e_4			×				×	
e_5					×			×
e_6	×	Δ						
e_7			×	Δ				

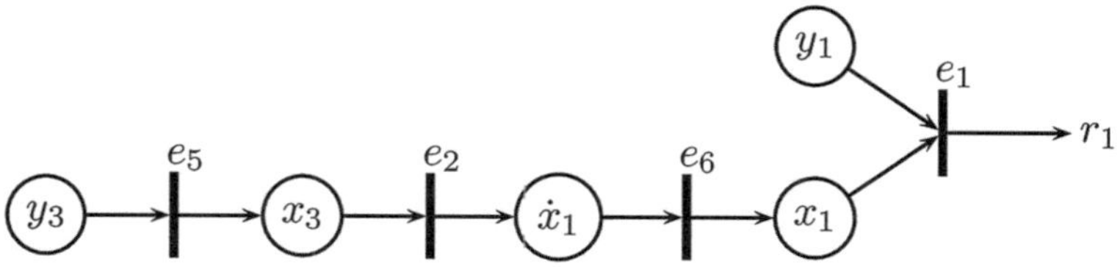

Fig. 11.3 Computation sequence for MSO set ω_1

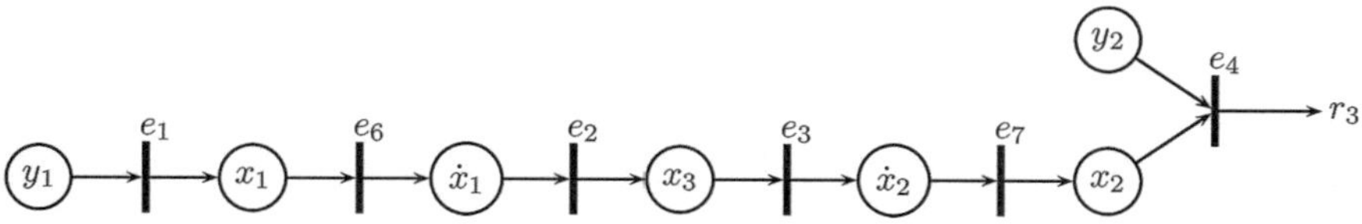

Fig. 11.4 Computation sequence for MSO set ω_3

The causal structural model of Example 1 is represented by the biadjacency matrix shown in Table 11.1.

In this example, the overdetermined part is $\mathcal{M}^+ = \mathcal{M}$. It is assumed integral causality in the differential equations e_6 and e_7 and that functions g_1 and g_2 are fully invertible. Applying the algorithm proposed in [30], two causally computable MSO sets are found, $\Omega = \{\omega_1, \omega_2\}$, where $\omega_1 = \{e_1, e_2, e_5, e_6\}$ and $\omega_2 = \{e_3, e_4, e_5, e_7\}$. Figure 11.3 displays the computation sequence of MSO set ω_1. According to this computation sequence, the computation form of its corresponding residual would be $r_1 = \int g_1^{-1}(y_3)\,dt + y_1$. Note that if function g_1 was not invertible, this MSO would not be causally computable. Similarly, the residual corresponding to MSO set ω_2 would be $r_2 = \int g_2^{-1}(y_3)\,dt - y_2$. Thus, the set of residuals for this example would be $\mathcal{W} = \{r_1, r_2\}$.

An additional MSO set could be obtained when applying the algorithm in [27], $\omega_3 = \{e_1, e_2, e_3, e_4, e_6, e_7\}$, but it would not be causally computable. Figure 11.4 displays the computation sequence of this new MSO set ω_3. According to this computation sequence, the computation form of its corresponding residual would be $r_3 = \int g_2^{-1}(g_1(\frac{dy_1}{dt}))\,dt - y_2$. Since a differentiation operation is involved, the integral causality assumption would be violated.

11.3.3 Fault Detectability and Isolability

Let $\mathcal{F}$ be the set of faults that must be monitored. For example, in a water distribution network, typical faults are pump malfunctioning or valve blocking.

Definition 11.1 (*Detectable fault*) A fault $f \in \mathcal{F}$ is detectable if its occurrence can be observed, i.e., at least one of the residuals $r_i \in \mathcal{W}$ satisfies $r_i \neq 0$.

A detectable fault can also be characterized in the structural analysis framework [31]. Without loss of generality, it is assumed that a single fault $f \in \mathcal{F}$ can only violate one equation, denoted by $e_f \in \mathcal{M}$. Then, a fault $f \in \mathcal{F}$ is *structurally detectable* if there exists at least one MSO $\omega_i \subseteq \Omega_c$ such that

$$e_f \in \omega_i. \tag{11.6}$$

Following Example 1, assume that a fault is defined for the measurement of variable y_1, i.e., its corresponding fault equation would be e_1. Then, residual r_1 would be sensitive to this fault, since $e_1 \in \omega_1$, whereas r_2 not, since $e_1 \notin \omega_2$. Thus, since there exists a residual that is sensitive to it, this fault would be detectable.

Using the set of residual generators (11.3), the fault detection module must check at each time instant whether or not they are consistent with the observations. Under ideal conditions, residuals are zero in the absence of faults and nonzero when a fault is present.

Fault isolation aims at identifying the fault acting on the system in a set of possible faults (or fault hypothesis set) $\mathcal{F} = \{f_1, f_2, \ldots, f_{n_f}\}$. It is assumed that just one fault may be acting on the system at a given time instant. Fault isolation is based on identifying at time instant k those residuals that are consistent and those that are not and implementing a diagnostic reasoning that leads to the isolation of the fault present in the system, assuming that different faults affect different residuals. Thus, unlike for fault detection, where consistency has to be independently evaluated for each residual, a set of residuals is required for fault isolation [11].

Standard fault isolation reasoning exploits the knowledge about the binary relation between the sets $\mathcal{F}$ and $\mathcal{W}$ that is stored in the so called fault signature matrix (FSM). This matrix has as many rows as residuals and as many columns as faults are considered. An element ij of the FSM is equal to 1 if the r_i residual is sensitive to the f_j fault; otherwise, it is equal to 0. A column of the FSM is known as the theoretical fault signature and indicates which residuals are affected by a given fault. A set of faults are isolable if all the columns in the FSM are different.

Definition 11.2 (*Isolable fault*) A fault $f_i \in \mathcal{F}$ is isolable if its occurrence can be uniquely observed through its fault signature.

Following Example 1, assume that a fault is considered for every sensor, i.e., f_1 corresponds to a violation of the expression of e_1, f_2 to e_4 and f_3 to e_5. Then, the FSM corresponding to Example 1 is provided in Table 11.2. Since, all columns are different, all faults are isolable.

Table 11.2 Fault signature matrix corresponding to Example 1

	f_1	f_2	f_3
r_1	1	0	1
r_2	0	1	1

11.3.4 Fault Detection Implementation

However, when building a model of a dynamic process to monitor its behaviour, there is always a mismatch between the modelled and the real behaviour. This is because some effects are neglected in the model, some nonlinearities are linearized in order to simplify the model, some parameters have tolerance when they are compared between several units of the same component, some errors in parameters or in the structure of the model are introduced in the model estimation process, etc. These modelling errors introduce uncertainty in the model and interfere with the fault detection. To properly manage this uncertainty, several authors in the FD [32, 33] and automatic control [34, 35] communities have suggested the use of interval models. Thus, a nominal model plus uncertainty intervals are provided, guaranteeing that all available data from the system in non-faulty scenarios will be included in the interval for the model prediction. Interval methods are quite suitable when additionally to additive uncertainty (noise), modelling uncertainty is present. In particular, no assumption about the noise statistical distribution should be introduced. An alternative to interval-based methods are stochastic models and methods [36].

In the literature, there are different approaches to solve this problem. For example, statistical decision methods [36] can be used when unknown dynamics and measurement noise are stochastically modelled. In many practical situations, this assumption is not realistic, being more natural to assume that disturbances/model errors and measurement noise are bounded and their effect is propagated to the residuals using, for example, interval methods [33].

In the case of models with bounded uncertainties, the ARR defined in (11.2) can be written in discrete time as

$$\mathcal{R}_i(k) = \{\Psi_i(\mathbf{y}(k), \mathbf{u}(k), \boldsymbol{\delta}(k)) | \boldsymbol{\delta}(k) \in \mathcal{D}\}, \tag{11.7}$$

where $\mathcal{D}$ is the interval box $\mathcal{D} = \{\boldsymbol{\delta} \in \mathbb{R}^{n_\delta} | \boldsymbol{\delta}^{min} \leq \boldsymbol{\delta} \leq \boldsymbol{\delta}^{max}\}$ that includes all bounded uncertainties and $\mathcal{R}_i$ being $\mathcal{R}_i \in \left[r_i^{min}, r_i^{max}\right]$

Then, fault detection is formulated as an ARR consistency checking using a set-membership approach [37].

Definition 11.3 (*ARR consistency checking*) Given an ARR described by residual sets as in (11.7) and a sequence of measured inputs $\mathbf{u}(k)$ and outputs $\mathbf{y}(k)$ from the real system, an ARR is consistent with the measurements and the known bounds of uncertain parameters and noise if there exist a set of sequences $\boldsymbol{\delta}(k) \in \mathcal{D}$ which satisfies the ARR mathematical expression.

According to Definition 11.3, a residual $\mathcal{R}_i$ is consistent when zero belongs to the interval bounding it, i.e., $0 \in \mathcal{R}_i(k)$.

Definition 11.4 (*Fault detection*) Given a sequence of observed inputs $\mathbf{u}(k)$ and outputs $\mathbf{y}(k)$ of the real system, a fault is said to be detected at time k if there does not exist a set of sequences $\boldsymbol{\delta}(k) \in \mathcal{D}$ for which some ARR is consistent with.

According to Definition 11.4, a fault is detected when $0 \notin \mathcal{R}_i(k)$. The information provided by the consistency checking procedure is binarized and stored as a fault signal $\phi_i(k)$

$$\phi_i(k) = \begin{cases} 0 & \text{if } 0 \in \mathcal{R}_i(k), \\ 1 & \text{if } 0 \notin \mathcal{R}_i(k). \end{cases} \tag{11.8}$$

In the literature, many algorithms for consistency checking have been proposed, all them with the main objective of trying to eliminate false alarm and providing robust fault indicators. For example, in [38], the fault detection test (11.8) is formulated using the mathematical framework of interval constraint satisfaction problems.

In this implemented approach, the consistency check is performed using the methodology described in [39]. In this case the set $\mathcal{R}_i(k)$ is represented by the bounded interval

$$\mathcal{R}_i(k) \subseteq \left[r_i^{0,min}, r_i^{0,max} \right], \tag{11.9}$$

where $r_i^{0,min} = min(\mathcal{R}_i(k)) - r_i^0$ and $r_i^{0,max} = max(\mathcal{R}_i(k)) - r_i^0$, being r_i^0 the nominal residual. The nominal residual r_i^o is computed according to (11.7) considering the centre of the uncertainty interval δ_o. The fault diagnostic signal is computed using the Kramer function

$$\phi_i^K(k) = \begin{cases} \frac{(r_i^0(k)/r_i^{0,max}(k))^4}{1+(r_i^0(k)/r_i^{0,max}(k))^4} & \text{if } r_i^0(k) \geq 0, \\ -\frac{(r_i^0(k)/r_i^{0,min}(k))^4}{1+(r_i^0(k)/r_i^{0,min}(k))^4} & \text{if } r_i^0(k) < 0. \end{cases} \tag{11.10}$$

When using (11.10), the residuals are normalized to a metric between -1 and 1, $\phi_i^K(k) \in [-1, 1]$, which indicates the degree of satisfaction of (11.8) for every nominal residual, 0 means than r_i is consistent, whereas 1 or -1 means than r_i is inconsistent. Then, the binary fault signal defined in (11.8) is computed by

$$\phi_i(k) = \begin{cases} 0 & \text{if } \left|\phi_i^K(k)\right| \leq 0.5, \\ 1 & \text{if } \left|\phi_i^K(k)\right| > 0.5. \end{cases} \tag{11.11}$$

11.3.5 Fault Isolation Implementation

The standard procedure followed by the FDI community involves finding a matching between the observed fault signature, and one of the theoretical fault signatures. In practice, due to the dynamic nature of a fault signal ϕ_i caused by a given fault f_i and its magnitude, or due to the absence of input signals during system operation, when a fault occurs, an undefined number of residuals affected by the fault can be found inconsistent. For this reason, the fault isolation module proposed in this chapter is an adaptation of the one used in [18] (see Fig. 11.5), and described in more detail in [39].

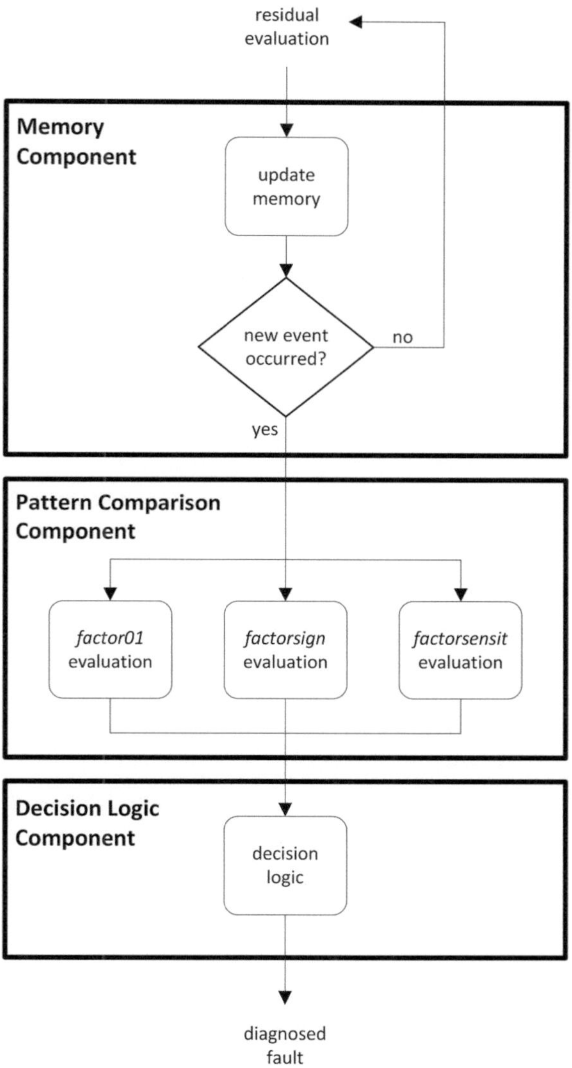

Fig. 11.5 Fault detection and isolation logic scheme

The first component is a memory that stores information about the fault signal occurrence history. It is cyclically updated by the fault detection module. The pattern comparison component compares the memory content with the stored fault patterns. The classical Boolean fault signature matrix concept [11] is generalized since the binary interface is extended to take into account more fault signal properties. The last component represents the decision logic part of the method whose aim is to provide the most probable fault candidate.

11.3.5.1 Memory Component

The memory component consists of a table in which events in the residual history are stored. As soon as the fault detection module detects a new fault signal, i.e., $\phi_i = 1$, the memory is updated by filling out three fields: the occurrence time k_0, the maximum nominal residual $r_i^{0,max}$ and the *sign* of the residual.

The problem of different time instants of fault signal appearance is solved postponing the isolation decision until a prefixed waiting time T_w has elapsed, from the first fault signal appearance.

This T_w could be computed based on the largest transient response [11] from a non-faulty to any faulty situation. However, in the case of a water transport network, which is subject to periodic flow demands, T_w can be also related to the periodic operation time. After this time has elapsed, a diagnosis is provided and the memory component is reset, enabling the diagnosis of future faults. Following [16], inside this diagnosis window, the maximum nominal residual in the memory table $r^o_{i,max}$ is only updated at time k_0 if the current nominal residual is greater than the previous ones.

$$r_i^{0,max} = \max_{k \in [k_0, k_0 + T_w/\Delta t]} |r_i^o(k)|, \tag{11.12}$$

where Δt is the sampling time.

Due to the max-operator, residual values can only increase. Using this strategy, the effect of noise and non-persistent fault indicators are filtered out since just residual peaks are stored.

The memory table makes the residual history accessible for later computation by explicitly storing that data. Using this approach, time aspects of fault isolation can be treated in a straightforward way.

11.3.5.2 Pattern Comparison Component

The pattern comparison component compares the memory content with the stored fault patterns. Fault patterns are organized according to a theoretical FSM. This interpretation assumes that the occurrence of f_j always affect a system and is observable at least in one residual r_i, hypothesis known as fault exoneration and that f_j is the only fault affecting the monitored system. Three different fault signature matrices are considered in the evaluation task: Boolean fault signal activation ($FSM01$), fault signal signs ($FSMsign$) and fault residual sensitivity ($FSMsensit$). Moreover, $FSM01$ coincides with the FSM introduced in Sect. 11.3.3 and can be obtained from structural analysis. The other matrices can be obtained from the analysis of the residual fault sensitivity [39].

In model-based FDI, the fault effects on a residual can be expressed in terms of the residual fault sensitivity that leads to the residual internal form [11]. For example, in the case of two additive faults f_1 and f_2, the internal form of a residual r_1 can be expressed as follows:

$$r_1(k) = S_{f_1}(q^{-1})f_1(k) + S_{f_2}(q^{-1})f_2(k), \tag{11.13}$$

where $S_{f_1}(q^{-1})$ and $S_{f_2}(q^{-1})$ are the residual fault sensitivity transfer functions that characterize the fault effect on the residual, expressed in terms of the shift operator q^{-1}.

Each element of the *FSMsensit* is computed with the following equation:

$$FSMsensit_{ij}(k) = \begin{cases} \frac{S_{r_i f_j}(q^{-1})}{max(\mathcal{R}_i(k))-r_i^0} u_a(k-k_0), & \text{if } r_i^0 \geq 0 \text{ and } k \geq k_0, \\ \frac{S_{r_i f_j}(q^{-1})}{min(\mathcal{R}_i(k))-r_i^0} u_a(k-k_0), & \text{if } r_i^0 < 0 \text{ and } k \geq k_0, \\ 0, & \text{if } S_{r_i f_j} = 0 \text{ or } k < k_0, \end{cases} \tag{11.14}$$

where u_a is a unitary abrupt step input, $S_{r_i f_j}$ is the sensitivity associated with the nominal residual r_i^0 regarding the fault hypothesis f_j, and k_0 is the fault occurrence time instant. As a consequence of the fault residual sensitivity time dependence, $FSMsensit_{ij}$ dynamically evolves since the fault occurrence time instant.

Besides, *FSMsign* matrix can be easily derived from *FSMsensit* (11.14) by applying the following conversion

$$FSMsign_{ij}(k) = \begin{cases} 1 & \text{if } FSMsensit_{ij}(k) > 0, \\ -1 & \text{if } FSMsensit_{ij}(k) < 0, \\ 0 & \text{if } FSMsensit_{ij}(k) = 0. \end{cases} \tag{11.15}$$

In any case, the consistency between the observed sequence of fault signature $\phi_i(k)$ and the theoretical information stored in the FSM matrices for jth-fault hypothesis can be evaluated by computing in real time the *factor* values as follows:

$$factor01_j(k) = \frac{\sum_{i=1}^{n_r} \phi_i(k) FSM01_{ij}(k)}{\sum_{i=1}^{n_r} FSM01_{ij}(k)}, \tag{11.16}$$

$$factorsensit_j(k) = \frac{\left|\sum_{i=1}^{n_r} \phi_i^K(k) FSMsensit_{ij}(k)\right|}{\sum_{i=1}^{n_r} \left|FSMsensit_{ij}(k)\right|}, \tag{11.17}$$

$$factorsign_j(k) = \frac{\left|\sum_{i=1}^{n_r} sign(\phi_i^K(k)) FSMsign_{ij}(k)\right|}{\sum_{i=1}^{n_r} \left|FSMsign_{ij}(k)\right|}, \tag{11.18}$$

where n_r is the number of residuals and k_{ϕ_i} is the occurrence time of the fault signal ϕ_i.

While $factor01_j(k)$ and $factorsign_j(k)$ compute the mapping between residuals and the corresponding FSM matrices, $factorsensit_j(k)$ weighs this mapping considering the sensitivity. Further details on the general rules to obtain these matrices from (11.13) as well as additional fault signature matrix definitions concerning expected fault signal activation time and order can be found in [39].

11.3.5.3 Decision Logic Component

The decision logic algorithm starts when the first residual is activated (i.e., $\phi_i = 1$) and lasts T_w time instants or till all fault hypotheses except one are rejected. Fault hypotheses are rejected when an unexpected activation signal has been observed according to those fault hypotheses. Rejection is based on using the results of $factor01_j(k)$ and $factorsign_j(k)$. If any of these factors is null for a given fault hypothesis, it will be rejected. Every factor represents some kind of a filter, suggesting a set of possible fault hypotheses. At the end of the time window T_w, for each non-rejected fault hypothesis f_j, a fault isolation indicator ρ_j could be determined for instance as follows, although other logic formula could be considered:

$$\rho_j(k) = \max_{p \in [k - T_w/\Delta t, k]} (factor01_j(p) \times |factorsign_j(p)|, |factorsensit_j(p)|). \quad (11.19)$$

A set of fault candidates with their corresponding fault isolation indicator is provided as the final diagnosis result, so that the greatest fault isolation indicator will determine the diagnosed fault. In the case of non-isolable faults, they will all have a high fault isolation indicator.

11.4 Simulations and Results

11.4.1 Barcelona Water Transport Network

The case study used to illustrate the FD methodology proposed in this chapter is tested using a simulator of the Barcelona WTN (described in Sect. 2.4 of Chap. 2).

Without loss of generality, the results presented here are focused on two subsystems, known as Orioles and Cervello (Fig. 11.6). This part of the network includes the following components:

- Tanks: d150SBO, d175LOR, d147SCC, d205CES, d263CES
- Actuators with flow sensors: iStBoi, iOrioles, iStaClmCervello, iCesalpina1, iCesalpina2
- Demands with flow sensors: c150SBO, c175LOR, c147SCC, c205CES, c263CES
- Level sensors: xd150SBO, xd175LOR, xd147SCC, xd205CES, xd263CES

The Barcelona WTN simulator allows activating different kinds of faults in distinct components of the water network. Faults of different nature can be chosen (e.g., freezing, offset, drift, abrupt or incipient) and their magnitude and slope can be adjusted. These generic faults well represent the common hydraulic faults occurring in water networks, e.g., leaks (which may be represented by an abrupt/incipient offset/drift fault), bursts (which may be represented by an abrupt offset fault) or sensor communication faults (which may be represented by an abrupt freezing fault).

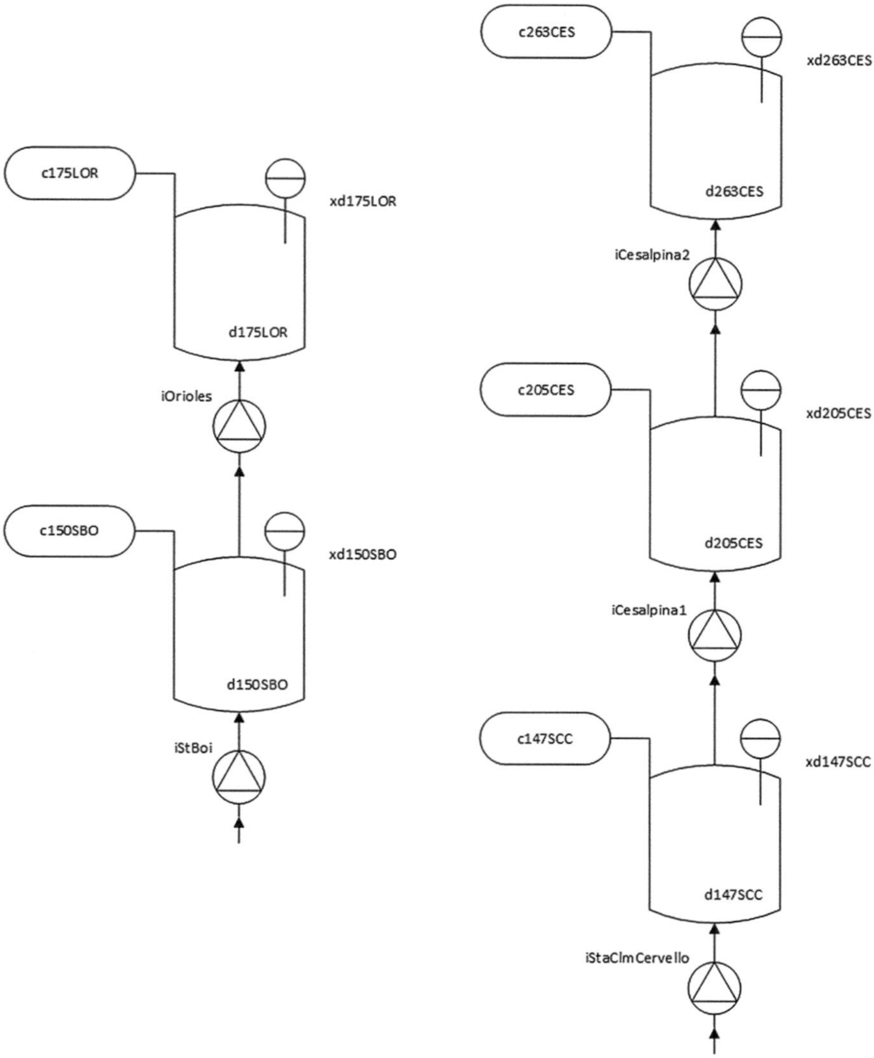

Fig. 11.6 Case study subsystems

The following faults that concern several Barcelona WTN components (see Fig. 11.6) will be considered:

- actuator faults, f_{Pi} for $i = 1 \dots 5$, corresponding to iStBoi, iOrioles, iStaClm-Cervello, iCesalpina1 and iCesalpina2, respectively.
- flow sensor faults, f_{Fi}, corresponding to each actuator.
- level sensor faults, f_{Li}, corresponding to xd150SBO, xd175LOR, xd147SCC, xd205CES and xd263CES, respectively.
- demand sensor faults, f_{Di}, corresponding to c150SBO, c175LOR, c147SCC, c205CES and c263CES, respectively.

11.4.2 WTN Modelling

A WTN can be represented by a directed graph $\mathcal{H} = (\mathcal{N}_{\mathcal{T}} \cup \mathcal{N}_{\mathcal{J}}, \mathcal{E})$ where tanks and pipe junctions are nodes, $\mathcal{N}_{\mathcal{T}} \cup \mathcal{N}_{\mathcal{J}}$, and pipes are edges, $\mathcal{E}$. A dynamic model of a water network can be defined as in (11.1). This general model can be obtained from elementary physical relationships related to nodes and edges in graph $\mathcal{H}$.

Tank node $i \in \mathcal{N}_{\mathcal{T}}$ can be described by a mass balance equation, through function **g** in (11.1) as follows:

$$\dot{x}_i = \frac{1}{A_i}\left(\sum_{q_k \in Q_i} q_k - d_i\right), \tag{11.20}$$

$$x_i^{min} \le x_i \le x_i^{max}, \tag{11.21}$$

where x_i is the water tank level, Q_i is the set of flows q_k corresponding to incident edges $k \in \mathcal{E}$ on node i, d_i is the flow demand, A_i is the tank section, and x_i^{min} and x_i^{max} denote the minimum and the maximum tank level, respectively.

Pipe junction node $j \in \mathcal{N}_{\mathcal{J}}$ can be described by a mass balance equation as well, through function **f** in (11.1) as follows:

$$\sum_{q_k \in Q_j} q_k = 0, \tag{11.22}$$

where Q_j is the set of flows q_k corresponding to incident edges $k \in \mathcal{E}$ on node j.

Pipe flows in a transport network are controlled using actuators (pumps and valves). Let $\mathcal{E}_{\mathcal{P}} \subseteq \mathcal{E}$ denotes the set of pipes that are provided with a pump, and $\mathcal{P}$ denotes the set of pumps. Then, pipe edge $k \in \mathcal{E}_{\mathcal{P}}$ can be described by a pipe equation, through function **f** in (11.1) as well, as follows:

$$q_k - K_l u_l = 0, \tag{11.23}$$

where u_l is the control input and K_l is the gain parameter of pump $l \in \mathcal{P}$.

Additionally, measurements are described through function **h** in (11.1) as follows:

$$y_{Li} = x_i, \tag{11.24}$$

$$y_{Di} = d_i, \tag{11.25}$$

$$y_{Fk} = q_k, \tag{11.26}$$

where y_{Li} corresponds to the ith-tank level, y_{Di} to the ith-demand and y_{Fk} to the kth-pipe flow.

The network subsystem represented in Fig. 11.6 comprises 5 tank nodes, no pipe junction nodes and 5 pipe edges, all provided with a pump. Thus, 5 tank levels, 5 demands and 5 flows are defined. Since all of them are measured, model (11.1) involves up to 25 equations.

11.4.3 ARR Generation

According to Sect. 11.3.2, designing a model-based FD system firstly requires determining the set of ARRs. These will be obtained through structural analysis.

The case study structural model involves 30 equations $\mathcal{M} = \{e_1, \ldots, e_{30}\}$ distributed into

$$
\begin{aligned}
e_1, \ldots, e_5 &: \text{tank node } i \text{ equations, according to (11.20)},\\
e_6, \ldots, e_{10} &: \text{pipe edge } k \text{ equations, according to (11.23)},\\
e_{11}, \ldots, e_{15} &: i\text{th-tank level measurement, according to (11.24)},\\
e_{16}, \ldots, e_{20} &: k\text{th-flow measurement, according to (11.26)},\\
e_{21}, \ldots, e_{25} &: i\text{th-demand measurement, according to (11.25)},\\
e_{26}, \ldots, e_{30} &: \dot{x}_i - \frac{dx_i}{dt} = 0.
\end{aligned}
$$

The set of unknown variables is $\mathcal{X} = \{\dot{x}_i, x_i, q_k, d_i\}$ and the set of known variables is $\mathcal{O} = \{u_l, y_{Li}, y_{Fk}, y_{Di}\}$ for $i = 1, \ldots, 5$, $l = 1, \ldots, 5$ and $k = 1, \ldots, 5$. Table 11.3 provides the biadjacency matrix that represents the structural model of the study case. Notice that, as in the case of Example 1, integral causality is assumed in equations $e_{26}, \ldots, e_{30}$.

Applying the algorithm proposed in [30], 21 causally computable MSO sets are found. For instance, the forth MSO is $\omega_4 = \{e_1, e_7, e_{11}, e_{16}, e_{21}, e_{26}\}$, whose computation sequence is provided in Fig. 11.7. According to this computation sequence, the computation form of its corresponding residual is:

$$
r_4 = y_{L1} - \frac{1}{S_1}\int (y_{F1} - K_2 u_2 - y_{D1})\,dt. \tag{11.27}
$$

Table 11.3 Biadjacency matrix of the case study

	x_1	$\dot{x}_1$	x_2	$\dot{x}_2$	x_3	$\dot{x}_3$	x_4	$\dot{x}_4$	x_5	$\dot{x}_5$	q_1	q_2	q_3	q_4	q_5	d_1	d_2	d_3	d_4	d_5	u_1	u_2	u_3	u_4	u_5	y_{L1}	y_{L2}	y_{L3}	y_{L4}	y_{L5}	y_{F1}	y_{F2}	y_{F3}	y_{F4}	y_{F5}	y_{D1}	y_{D2}	y_{D3}	y_{D4}	y_{D5}
e_1		×									×	×				×																								
e_2				×								×					×																							
e_3						×							×	×				×																						
e_4								×						×	×				×																					
e_5										×					×					×																				
e_6											×										×																			
e_7												×										×																		
e_8													×										×																	
e_9														×										×																
e_{10}															×										×															
e_{11}	×																									×														
e_{12}			×																								×													
e_{13}					×																							×												
e_{14}							×																						×											
e_{15}									×																					×										
e_{16}											×																				×									
e_{17}												×																				×								
e_{18}													×																				×							
e_{19}														×																				×						
e_{20}															×																				×					
e_{21}																×																				×				
e_{22}																	×																				×			
e_{23}																		×																				×		
e_{24}																			×																				×	
e_{25}																				×																				×
e_{26}	×	Δ																																						
e_{27}			×	Δ																																				
e_{28}					×	Δ																																		
e_{29}							×	Δ																																
e_{30}									×	Δ																														

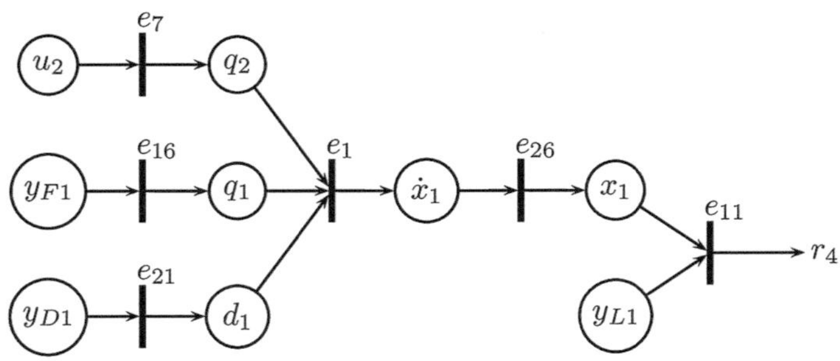

Fig. 11.7 Computation sequence corresponding to ω_4

Note that this residual can be easily computed since integral causality is satisfied. Up to fifteen, other residuals have a similar dynamic nature since their corresponding MSO set also includes an equation in $\{e_{26}, \ldots, e_{30}\}$. The other remaining five residuals have a static nature. For instance, the first MSO is $\omega_1 = \{e_6, e_{16}\}$ and the computation form of its corresponding residual is $r_1 = K_1 u_1 - y_{F1}$.

11.4.4 Structural Fault Detectability and Isolability Analysis

According to the set of MSO determined in the previous section, residual fault sensitivities can be determined. For instance, residual r_4 is sensitive to the set of faults $\{f_{P2}, f_{F1}, f_{L1}, f_{D1}\}$ since their corresponding fault equations satisfy $\{e_7, e_{16}, e_{11}, e_{21}\} \subseteq \omega_4$. Table 11.4 provides the Boolean FSM, which specifies all residual fault sensitivities.

According to Definition 11.1, all faults are detectable since there exists at least one residual that is sensitive to each fault. Regarding fault isolability, according to Definition 11.2 faults f_{Pi} and f_{Fi} are isolable since their corresponding fault signatures are unique. However, fault pairs $\{f_{Li}, f_{di}\}$ cannot be isolated since they have the same fault signature.

11.4.5 FD Implementation

The first step into the implementation of the FD system involves designing a residual generator module that checks system consistency in real time. The residual generator corresponding to the study case includes all 21 residuals determined in Sect. 11.4.4. This module could be implemented by a set of nonlinear difference equations obtained from the discretization of those residuals. One of the residuals that has been implemented is (11.27). Notice that this equation has two terms: measurement of the tank 1 level y_{L1} and an integral term that determines tank 1 level x_{L1} based on some other measurements. The integral term could be discretized using a

Table 11.4 Theoretical fault signature matrix

	f_{P1}	f_{P2}	f_{P3}	f_{P4}	f_{P5}	f_{F1}	f_{F2}	f_{F3}	f_{F4}	f_{F5}	f_{L1}	f_{L2}	f_{L3}	f_{L4}	f_{L5}	f_{D1}	f_{D2}	f_{D3}	f_{D4}	f_{D5}
r_1	1	0	0	0	0	1	0	0	0	0	0	0	0	0	0	0	0	0	0	0
r_2	0	1	0	0	0	0	1	0	0	0	0	0	0	0	0	0	0	0	0	0
r_3	1	1	0	0	0	0	0	0	0	0	1	0	0	0	0	1	0	0	0	0
r_4	0	1	0	0	0	1	0	0	0	0	1	0	0	0	0	1	0	0	0	0
r_5	1	0	0	0	0	0	1	0	0	0	1	0	0	0	0	1	0	0	0	0
r_6	0	0	0	0	0	1	1	0	0	0	1	0	0	0	0	1	0	0	0	0
r_7	0	1	0	0	0	0	0	0	0	0	0	1	0	0	0	0	1	0	0	0
r_8	0	0	0	0	0	0	1	0	0	0	0	1	0	0	0	0	1	0	0	0
r_9	0	0	1	0	0	0	0	1	0	0	0	0	0	0	0	0	0	0	0	0
r_{10}	0	0	0	1	0	0	0	0	1	0	0	0	0	0	0	0	0	0	0	0
r_{11}	0	0	0	0	1	0	0	0	0	1	0	0	0	0	0	0	0	0	0	0
r_{12}	0	0	1	1	0	0	0	0	0	0	0	0	1	0	0	0	0	1	0	0
r_{13}	0	0	0	1	0	0	0	1	0	0	0	0	1	0	0	0	0	1	0	0
r_{14}	0	0	1	0	0	0	0	0	1	0	0	0	1	0	0	0	0	1	0	0
r_{15}	0	0	0	0	0	0	0	1	1	0	0	0	1	0	0	0	0	1	0	0
r_{16}	0	0	0	1	1	0	0	0	0	0	0	0	0	1	0	0	0	0	1	0
r_{17}	0	0	0	0	1	0	0	0	1	0	0	0	0	1	0	0	0	0	1	0
r_{18}	0	0	0	1	0	0	0	0	0	1	0	0	0	1	0	0	0	0	1	0
r_{19}	0	0	0	0	0	0	0	0	1	1	0	0	0	1	0	0	0	0	1	0
r_{20}	0	0	0	0	1	0	0	0	0	0	0	0	0	0	1	0	0	0	0	1
r_{11}	0	0	0	0	0	0	0	0	0	1	0	0	0	0	1	0	0	0	0	1

forward-difference approximation with a sampling time Δt, leading to the following implementation of residual r_4:

$$z_4(k+1) = z_4(k) + \frac{\Delta t}{A_1}(y_{F1}(k) - K_2 u_2(k) - y_{D1}(k)), \tag{11.28}$$

$$\widehat{y}_{L1}(k) = min(z_4(k), y_{F1}^{max}), \tag{11.29}$$

$$r_4(k) = y_{L1}(k) - \widehat{y}_{L1}(k), \tag{11.30}$$

where z and $\widehat{y}_{L1}$ are the estimated value of y_{L1} without and with saturation, respectively, and y_{F1}^{max} is the maximum level of tank 1. The main problem when trying to directly solve (11.28) lies in the modelling uncertainty accounted for in (11.1), which originates in unmodelled dynamics, sensor noise, parameter variations and nonlinearities. In order to implement a robust fault residual generator, instead of computing dynamic residuals as in (11.28), the following nonlinear Luenberger observer structure will be used:

$$z_4(k+1) = (1 - \lambda_4) z_4(k) + \frac{\Delta t}{A_1}(y_{F1}(k) - K_2 u_2(k) - y_{D1}(k)) + \lambda_4 y_{L1}(k), \tag{11.31}$$

where λ_4 is the observer gain.

Thus, the complete set of discretized residuals have been implemented in the state-observer residual form. Recall that the detection test is based on analysing nominal residuals and their corresponding bounds defined in (11.9). Therefore, residuals are computed through the Luenberger observer form using the nominal value of the parameters and the residual bounds are determined experimentally. For example, Fig. 11.8 shows the real-time evolution of r_4^0 under normal operation, which can be bounded by the interval $\mathcal{R}_4(k) \subseteq [-1, 65, +1, 65]$.

The FD system evaluates in real time the fault diagnosis signals (11.10) and the binary fault signals (11.11) based on the nominal residuals computed at each sampling time using the observer-based implementation of the residuals and their bounds.

The second step involves determining the waiting time T_w in the memory component. To decide this time value, the system behaviour must be analysed. Figure 11.9 shows the time evolution of variables y_{L1}, y_{F1} and y_{D1} under normal operation. As in most water transport networks, the flow demand presents a repetitive daily pattern. Thus, $T_w = 13$ h has been chosen, which is greater than half the period.

Next step involves performing consistency analysis by computing *factors* (11.16)–(11.18) in real time, which requires the fault signature matrices to be determined. Matrix *FSM*01 coincides with the FSM determined from structural analysis in Sect. 11.4.4, and matrices *FSMsensit* and *FSMsign* are computed using the residual internal form (11.13). The fault sensitivity of a residual can be derived from the state-observer residual form.

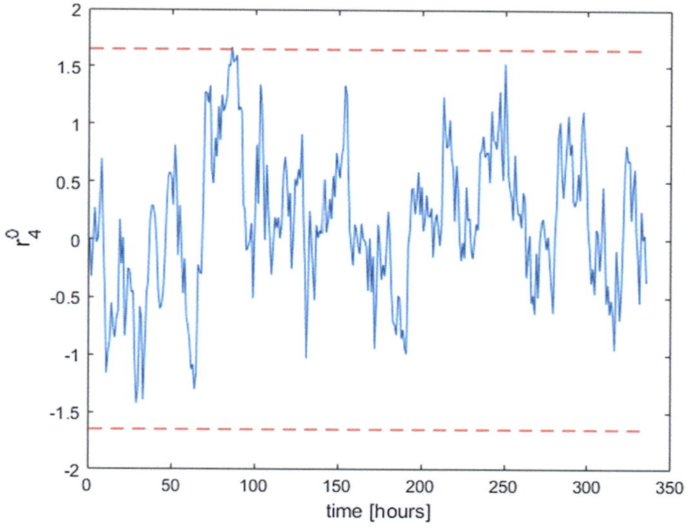

Fig. 11.8 Real-time evolution of r_4^0 under normal operation (*solid blue colour*) and its bounded interval (*dashed red colour*)

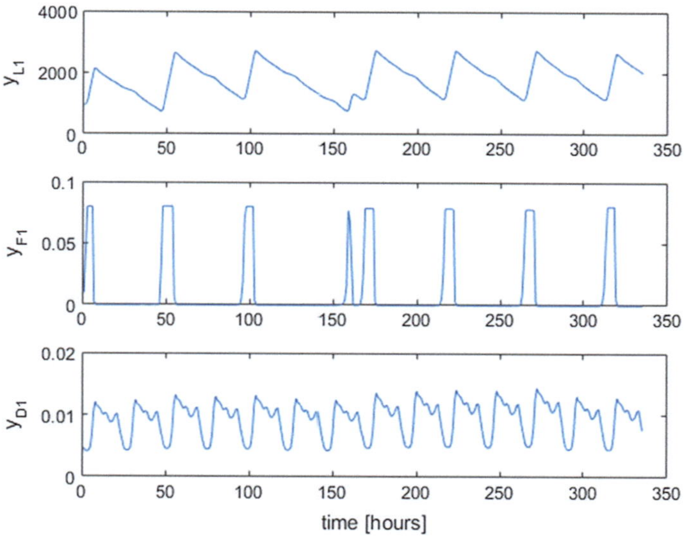

Fig. 11.9 Time evolution of y_{L1}, y_{P1} and y_{D1} under normal operation

For instance, neglecting nonlinearities (i.e., the saturation term), residual r_4 can be rewritten in the input–output form as

$$r_4 = \frac{1-\lambda_4 q^{-1}}{1-(1-\lambda_4)q^{-1}} y_{L_1}(k) + \frac{-(\Delta t/A_1)q^{-1}}{1-(1-\lambda_4)q^{-1}} y_{F1}(k) + \frac{(\Delta t A_1)K_2 q^{-1}}{1-(1-\lambda_4)q^{-1}} u_2(k) + \frac{(\Delta t/A_1)q^{-1}}{1-(1-\lambda_4)q^{-1}} y_{D1}(k), \tag{11.32}$$

where q^{-1} is the delay operator. Therefore, its sensitivity to an additive fault in the sensor measuring the level of tank 1, f_{L1}, and in pump 2, f_{P2}, is

$$S_{r4,f_{L1}} = \frac{1-\lambda_4 q^{-1}}{1-(1-\lambda_4)q^{-1}}, \tag{11.33}$$

$$S_{r4,f_{P2}} = \frac{(\Delta t/A_1)K_2 q^{-1}}{1-(1-\lambda_4)q^{-1}}. \tag{11.34}$$

Once, the *factors* are determined, the final diagnosis results are provided by applying (11.19).

11.4.6 Results

The fault detection and isolation procedure described in Sect. 11.3 has been applied to two simulated scenarios. In both scenarios, a drift fault of distinct magnitude is simulated at 100 h. The first scenario concerns a fault in the iOrioles pump, f_{P2}, whereas a fault in the xd175LOR level sensor, f_{L2}, is simulated in the second scenario.

Regarding the first scenario, Fig. 11.10 shows the time evolution of the nominal residuals r_i^0 and their corresponding binary fault signals ϕ_i for $i = 1, \ldots, 8$ (i.e., the residuals associated with the subsystem on the left in Fig. 11.6, where the fault occurs). An observer gain $\lambda_i = 0.1$ has been chosen for dynamical residuals $r_3, \ldots, r_8$. In order to better appreciate the benefit of considering a waiting time, a $T_w = 1$ h has been firstly chosen. The diagnosis procedure starts when the first fault indicator is activated, which occurs at 113 h (i.e., $k_0 = 113$). Note that ϕ_2, ϕ_3 and ϕ_4 exhibit persistent dynamics (ϕ_2 from 125 h on) whereas ϕ_7 has an intermittent evolution. This intermittent indicator dynamics arises from the periodic nature of the on–off pump commands (see Fig. 11.9).

The periodic dynamics in some residuals may lead to false diagnosis results. In order to overcome these undesired dynamics, a $T_w = 13$ h has been next applied. Figure 11.11 provides the time evolution of the maximum nominal value of the residuals $r_i^{0,max}$ and their corresponding binary fault signals, ϕ_i. Note that in this case non-persistent signals are avoided.

Figure 11.12 shows the time evolution of the subset of *factor* values that differ from zero. The factor subindex, $j = 1, \ldots, 8$, corresponds to the following set of faults $\{f_{P1}, f_{P2}, f_{F1}, f_{F2}, f_{L1}, f_{L2}, f_{D1}, f_{D2}\}$. Note that in this scenario $factor01_j$ and $factorsign_j$ provide the same results.

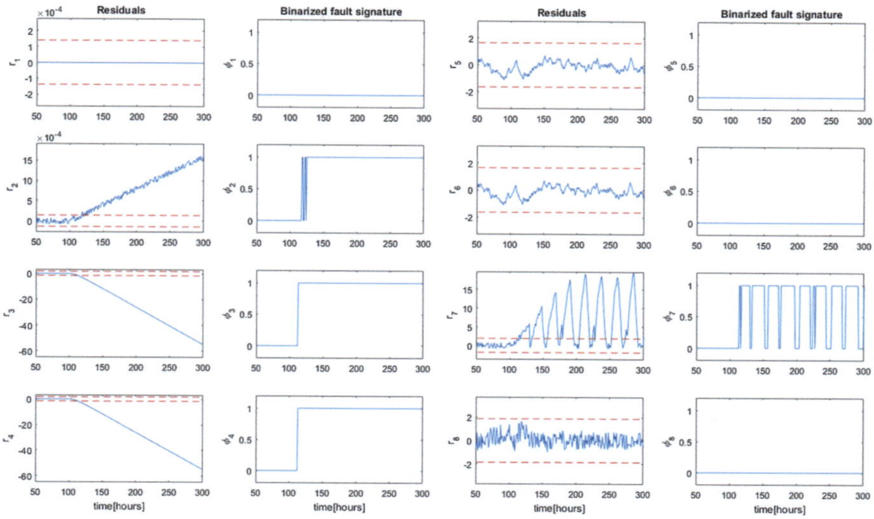

Fig. 11.10 Nominal residuals r_i^0 with their bounded interval (*red-dashed line*) and binary fault signals ϕ_i under a drift fault in iOrioles pump

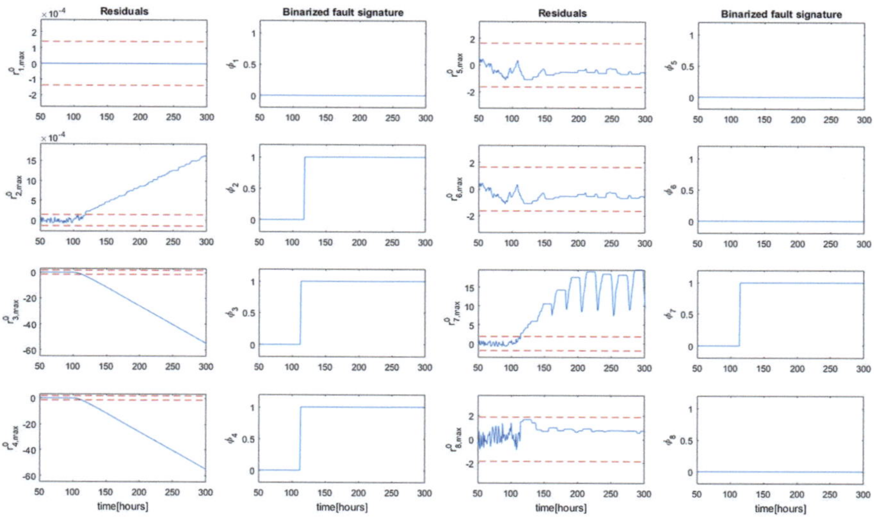

Fig. 11.11 Maximum nominal residuals $r_i^{0,max}$ with their bounded interval (*red-dashed line*) and binary fault signals ϕ_i under a drift fault in iOrioles pump

Finally, the decision logic component provides the diagnosed fault as the one whose fault isolation indicator ρ_j is the highest among all possible fault hypotheses. Figure 11.13 displays the time evolution of the subset of fault isolation indicators ρ_j that differ from zero. The fault isolation subindex, $j = 1, \ldots, 8$, corresponds to the following set of faults $\{f_{P1}, f_{P2}, f_{F1}, f_{F2}, f_{L1}, f_{L2}, f_{D1}, f_{D2}\}$. The fault diagnosis proce-

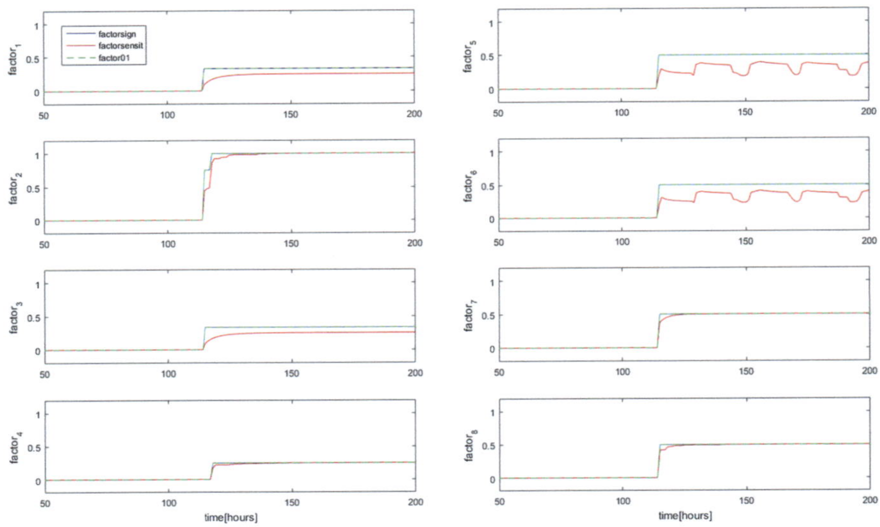

Fig. 11.12 *Factorsign$_j$ (blue line)*, *factorsensit$_j$* (red line) and *factor01$_j$* (*green-dashed line*) signals under a drift fault in iOrioles pump

dure manages to provide the right fault hypothesis: the highest fault isolation indicator corresponds to fault f_{P2} and occurs at 118 h.

Regarding the second scenario, Fig. 11.14 shows the time evolution of the maximum nominal residuals, $r_i^{0,max}$, and their corresponding binary fault signals, ϕ_i, for

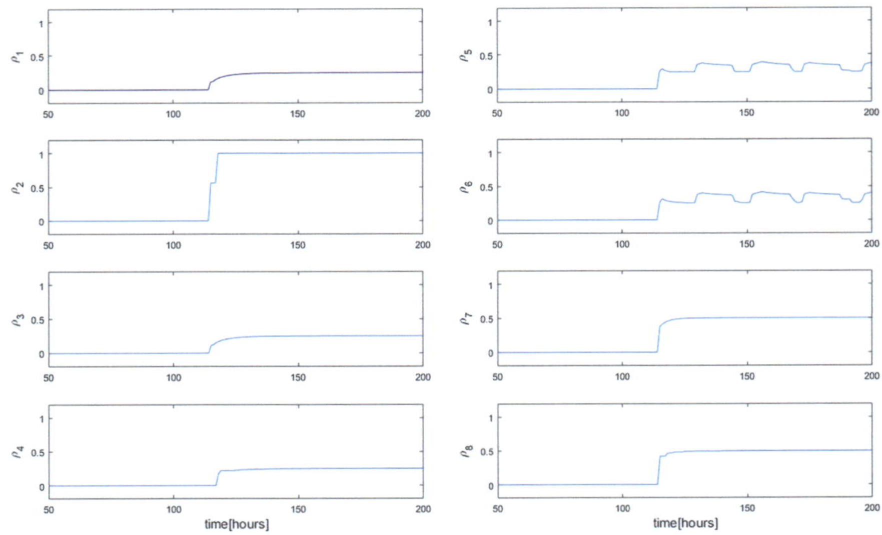

Fig. 11.13 Fault isolation indicators ρ_j under a drift fault in iOrioles pump

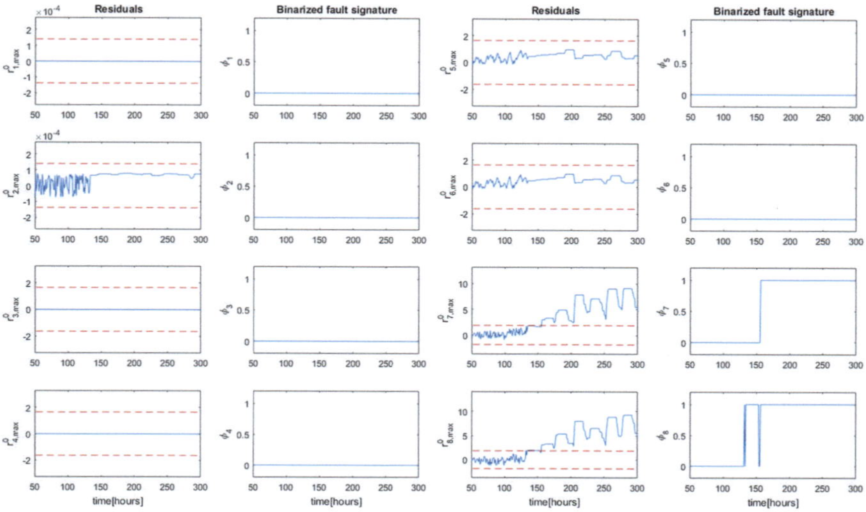

Fig. 11.14 Maximum nominal residuals $r_i^{0,max}$ with their bounded interval (*red-dashed line*) and binary fault signals ϕ_i under a drift fault in the xd175LOR level sensor

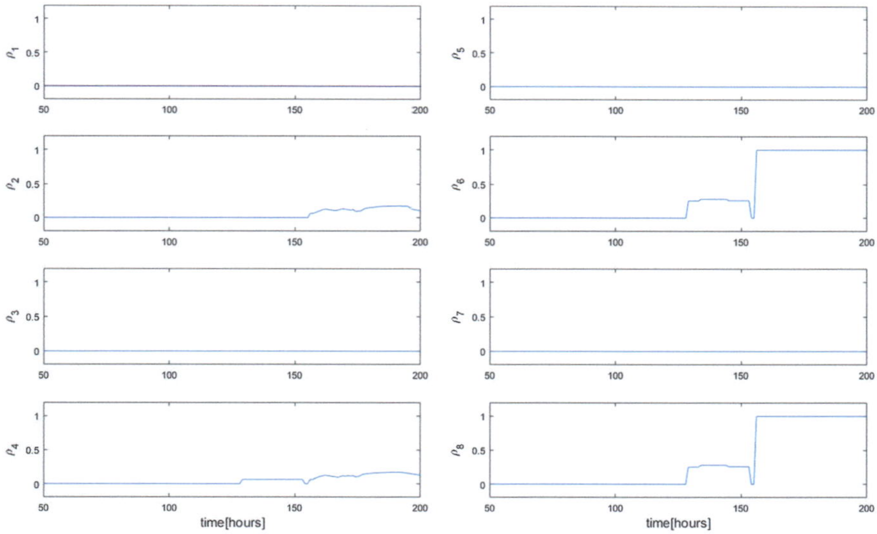

Fig. 11.15 Fault isolation indicators ρ_j under a drift fault in the xd175LOR level sensor

$i = 1, \ldots, 8$ (i.e., the residuals associated with the subsystem on the left in Fig. 11.6, where the fault occurs). Note that in this case binary fault signals are persistent since a waiting time $T_w = 13$ h has been applied, providing fault detection at 129 h.

Figure 11.15 displays the time evolution of the subset of fault isolation indicators ρ_j that differ from zero. In this case, the highest fault isolation indicator corresponds to faults f_{L2} or f_{D2} and occurs at 156 h. As already anticipated in Sect. 11.4.4, the FD system is not able to isolate both faults.

11.5 Conclusions

In WTN, high performance and maintenance are critical factors for enhancing productivity and water quality. To achieve these goals, highly sophisticated and online FD systems able to detect and isolate sensor and actuator malfunctions are required. In this chapter, a methodology has been proposed for developing model-based fault diagnosis for WTN taking into account both theoretical and practical issues. The proposed approach is based on exploiting the analytical redundancy available in the model of the WTN. To exemplify the FD methodologies, a part of the Barcelona WTN was used as a case study.

References

1. Marinaki M, Papageorgiou M (2005) Optimal real-time control of sewer networks. Springer, London, UK
2. Schütze M, Campisano A, Colas H, Schilling W, Vanrolleghem Peter A (2004) Real time control of urban wastewater systems—where do we stand today? J Hydrol 299(3–4):335–348
3. Carrozza G, Cotroneo D, Russo S (2008) Software faults diagnosis in complex ots based safety critical systems. In: Seventh European dependable computing conference, EDCC 2008. IEEE, pp 25–34
4. Ding S (2008) Model-based fault diagnosis techniques: design schemes, algorithms, and tools. Springer Science and Business Media
5. Venkatasubramanian V, Rengaswamy R, Kavuri SN (2003) A review of process fault detection and diagnosis: part ii: qualitative models and search strategies. Comput Chem Eng 27(3):313–326
6. Venkatasubramanian V, Rengaswamy R, Kavuri SN, Yin K (2003) A review of process fault detection and diagnosis: part iii: process history based methods. Comput Chem Eng 27(3):327–346
7. Venkatasubramanian V, Rengaswamy R, Yin K, Kavuri SN (2003) A review of process fault detection and diagnosis: part i: quantitative model-based methods. Comput Chem Eng 27(3):293–311
8. Vento J, Blesa J, Puig V, Sarrate R (2015) Set-membership parity space hybrid system diagnosis. Int J Syst Sci 46(5):790–807
9. Puig V, Ocampo-Martinez CA (2015) Decentralised fault diagnosis of large-scale systems: application to water transport networks. In: 26th DX international workshop on principles of diagnosis, Paris, France, August 2015
10. Escobet T, Puig V, Quevedo J, Garcia D (2014) A methodology for incipient fault detection. In: 2014 IEEE conference on control applications (CCA). IEEE, pp 104–109
11. Gertler J (1998) Fault detection and diagnosis in engineering systems. Marcel Dekker Inc, New York
12. Biswas G, Simon G, Mahadevan N, Narasimhan S, Ramirez J, Karsai G et al (2003) A robust method for hybrid diagnosis of complex systems. In: Proceedings of the 5th symposium on fault detection, supervision and safety for technical processes, pp 1125–1131
13. Bayoudh M, Travé-Massuyes L, Olive X (2009) Active diagnosis of hybrid systems guided by diagnosability properties. IFAC Safeprocess, Barcelona, Spain, pp 1498–1503

14. Bemporad A, Morari M (1999) Control of systems integrating logic, dynamics, and constraints. Automatica 35(3):407–427
15. Cordier MO, Dague P, Lèvy F, Montmain J, Staroswiecki M, Travé-Massuyès L (2004) Conflicts versus analytical redundancy relations: a comparative analysis of the model based diagnosis approach from the artificial intelligence and automatic control perspectives. IEEE Trans Syst Man Cybern B 34(5):2163–2177
16. Combastel C, Gentil S, Rognon JP (2003) Toward a better integration of residual generation and diagnostic decision. In: 5th IFAC symposium on fault detection, supervision and safety of technical processes, Washington, USA, June 2003
17. Ragot J, Maquin D (2006) Fault measurement detection in an urban water supply network. J Process Control 16(9):887–902
18. Puig V, Schmid F, Quevedo J, Pulido B (2005) A new fault diagnosis algorithm that improves the integration of fault detection and isolation. In: 44th IEEE conference on decision and control and European control conference, Sevilla, Spain, December 2005, pp 3809–3814
19. Petti TF, Klein J, Dhurjati PS (1990) Diagnostic model processor: using deep knowledge for process fault diagnosis. AIChE J 36(4):565–575
20. Blanke M, Kinnaert M, Lunze J, Staroswiecki M (2006) Diagnosis and fault-tolerant control, 2nd edition. Springer
21. Staroswiecki M, Cassar JP, Declerck P (2000) A structural framework for the design of FDI system in large scale industrial plants. In: Patton RJ, Frank PM, Clark RN (eds) Issues of fault diagnosis for dynamic systems. Springer, London, UK, pp 245–283
22. Nejjari Fatiha, Pérez Ramon, Escobet Teresa, Travé-Massuyès Louise (2006) Fault diagnosability utilizing quasi-static and structural modelling. Math Comput Mod 45:606–616
23. Staroswiecki M, Hoblos G, Aitouche A (2004) Sensor network design for fault tolerant estimation. Int J Adapt Control Sign Process 18(1):55–72
24. Šiljak DD (1991) Decentralized control of complex systems. Academic Press, Boston
25. Sarrate R, Nejjari F, Rosich A (2012) Model-based optimal sensor placement approaches to fuel cell stack system fault diagnosis. In: 8th IFAC symposium on fault detection, supervision and safety of technical processes, Mexico City, Mexico, pp 96–101
26. Dulmage AL, Mendelsohn NS (1958) Covering of bi-partite graph. Canad J Math 10:517–534
27. Krysander M, Åslund J, Nyberg M (2008) An efficient algorithm for finding minimal overconstrained sub-systems for model-based diagnosis. IEEE Trans Syst Man Cybern A 38(1)
28. Travé-Massuyès L, Escobet T, Olive X (2006) Diagnosability analysis based on component supported analytical redundancy relations. IEEE Trans Syst Man Cybern A 36(6):1146–1160
29. Frisk E, Bregon A, Aslund J, Krysander M, Pulido B, Biswas G (2012) Diagnosability analysis considering causal interpretations for differential constraints. IEEE Trans Syst Man Cybern A 42(5):1216–1229
30. Rosich A, Frisk E, Aslund J, Sarrate R, Nejjari F (2012) Fault diagnosis based on causal computations. IEEE Trans Syst Man Cybern A 42(2):371–381
31. Krysander M, Frisk E (2008) Sensor placement for fault diagnosis. IEEE Trans Syst Man Cybern A 38(6):1398–1410
32. Ploix S, Adrot O, Ragot J (2001) Bounding approach to the diagnosis of uncertain static systems. In: 4th symposium on fault detection, supervision and safety for technical processes, Safeprocess' 2000, vol 1, pp 151–156
33. Puig V, Quevedo J, Escobet T, Nejjari F, de las Heras. S (2008) Passive robust fault detection of dynamic processes using interval models. IEEE Trans Control Syst Technol 16(5):1083–1089
34. Calafiore G, Campi MC, Ghaoui LEI (2002) Identification of reliable predictor models for unknown systems: a data-consistency approach based on learning theory. In: Proceedings of the 15th IFAC world congress
35. El Ghaoui Laurent, Calafiore Giuseppe (2001) Robust filtering for discrete-time systems with bounded noise and parametric uncertainty. IEEE Trans Autom Control 46(7):1084–1089
36. Basseville M, Nikiforov IV (1993) Detection of abrupt changes: theory and application. Prentice-Hall, Upper Saddle River, USA

37. Tornil Sin S, Ocampo Martínez C, Puig V, Escobet T (2012) Robust fault detection of non-linear systems using set-membership state estimation based on constraint satisfaction. Eng Appl Artif Intell 25(1):1–10
38. Tornil-Sin S, Ocampo-Martinez C, Puig V, Escobet T (2014) Robust fault diagnosis of nonlinear systems using interval constraint satisfaction and analytical redundancy relations. IEEE Trans Syst Man Cybern: Syst 44(1):18–29
39. Meseguer J, Puig V, Escobet T (2010) Fault diagnosis using a timed discrete-event approach based on interval observers: application to sewer networks. IEEE Trans Syst Man Cybern A 40(5):900–916

Part III
Real-Time Control

Chapter 12
Model Predictive Control of Water Networks Considering Flow

Gabriela Cembrano, Vicenç Puig, Carlos Ocampo-Martínez, Meritxell Minoves and Ramon Creus

12.1 Introduction

Decision support systems provide useful guidance for operators in complex networks, where resource management best actions are not intuitive. Optimization and optimal control techniques provide an important contribution to a smart management strategy computation for drinking-water networks (DWNs) (see [1–3]). Similarly, problems related to modelling and control of water supply, transport and distribution systems have been an object of important research efforts during the last few years (see, e.g., [4–7].

In general, DWNs contain multiple tanks, pumping stations, valves, water sources (superficial and underground) and sectors of consumer demand. Operational control of DWNs using optimal control techniques has been largely investigated (see [5]). This chapter proposes the use of model predictive control (MPC) techniques to generate flow control strategies in a transport network, delivering water from the drinking-water treatment plants to the consumer areas to meet future demands. Set points for pumps and valves are computed by optimizing a performance index expressing operational goals such as economic cost, safety water storage and smoothness in flow control actions. The main point is to highlight the advantages of using optimization-based control techniques, such as MPC, to improve the performance of a water transport network, taking into account their large-scale nature (in terms of number of dynamic elements and decision variables), the nature of the desired control objectives

G. Cembrano (✉) · C. Ocampo-Martínez
Institut de Robòtica i Informàtica Industrial, CSIC-UPC, Barcelona, Spain
e-mail: gcembrano@iri.upc.edu

M. Minoves · R. Creus
Aguas de Barcelona (AGBAR), Barcelona, Spain

V. Puig
Research Center "Supervision, Safety and Automatic Control" (CS2AC-UPC), Terrassa, Spain

V. Puig et al. (eds.), *Real-Time Monitoring and Operational Control of Drinking-Water Systems*, Advances in Industrial Control,
DOI 10.1007/978-3-319-50751-4_12

and the type and behaviour of the system disturbances (drinking-water demands). The developed control strategies have been tested on the drinking-water transport network of Barcelona.

12.2 Problem Statement

12.2.1 Operational Control of Water Networks

Complex nonlinear models are very useful for offline operations (for instance, calibration and simulation). Detailed mathematical representations such as the pressure-flow models for DWNs allow the simulation of those systems with enough accuracy to observe specific phenomena, useful for design and investment planning. However, for online computation purposes such as those related to global management, a simpler and control-oriented model structure must be conveniently selected. This simplified model includes the following features:

(i) Representativeness of the main network dynamics: It must provide an evaluation of the main representative hydrological/hydraulic variables of the network and their response to control actions at the actuators.
(ii) Simplicity, expandability, flexibility and speed: It must use the simplest approach capable of achieving the given purposes, allowing very easily to expand and/or modify the modelled portion of the network.
(iii) Amenability to online calibration and optimization: This modelling approach must be easily calibrated online using data from the telemetry system and embedded in an optimization problem to achieve the network management objectives.

Figure 12.1, adapted from [8, 9], shows a hierarchical structure for a real-time control (RTC) water system. There, the MPC, as the global control law, determines the references (set points) for the local controllers placed at different elements of the networked system. These references are computed according to the measurements taken from sensors distributed around the network. The management level provides the MPC with its operational objectives, which are reflected in the controller design as the performance indices to be enhanced, which can be either minimized or maximized, depending on the case. Finally, water system control requires the use of a supervisory system to monitor the performance of the different control elements in the network (sensors and actuators) and to take appropriate correcting actions in the case where a malfunction is detected, to achieve a proper fault-tolerant control [10].

In most water networks, the operational control is managed by the operators from the telecontrol centre using a SCADA (supervisory control and data acquisition) system. Operators are in charge of supervising the network status using the telemetry system and providing the set points for the local controllers, which are typically based on PID algorithms. The main goal of the operational control of water networks is

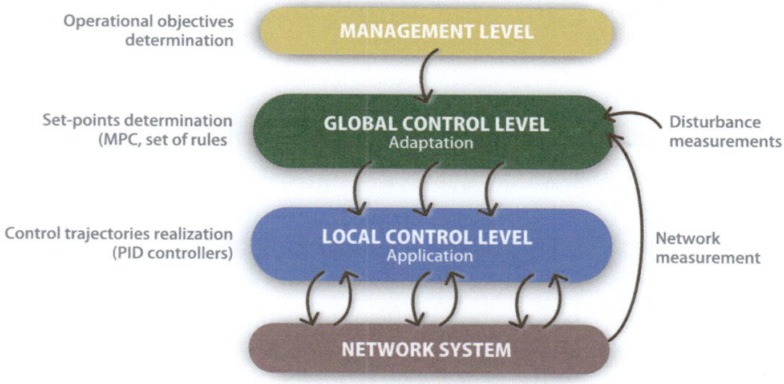

Fig. 12.1 Hierarchical structure for RTC system

to meet the demands at consumer sites, but at the same time with minimum costs of operation and guaranteeing pre-established volumes in tanks (to preserve the satisfaction of future demands) and smooth operation of actuators (valves and pumps) and production plants.

Water consumption in urban areas is usually managed on a daily basis, because water demand generally presents daily patterns and reasonably good hourly 24-h-ahead demand predictions may, in general, be available. Therefore, this horizon is appropriate for evaluating the effects of different control strategies on the water network, with respect to operational goals. However, other horizons may be more appropriate in specific utilities. The approach proposed here is based on demand satisfaction at the transport level, taking into account the supply conditions. For illustration, it uses—but is not restricted to—a 24-h horizon, with hourly sampling. When applied in real-time conditions, the computation of optimal strategies is updated, with new data from the water network, every hour with a sliding 24-h horizon.

At the supply water basin level, strategic planning deals with sustainable use of the water resources, seasonal variations in reservoirs and water levels, etc., so that planning horizon, sampling times and control time steps are usually much longer. In this work, the long-term planning objectives for the supplies are taken into account as bands of admissible requests from the supplies to the transport, production and distribution areas. These admissible bands define bounds on flow from reservoir, aquifer and river sources. Production plant limitations are also used, and these may vary according to weather-related factors, operational schedules and/or breakdowns. The computation of optimal strategies must take into account the dynamics of the complete water system and 24-h-ahead demand forecasts, availability predictions in supply reservoirs and aquifers, defined by long-term planning for sustainable use and predictions of production plant capacity and availability. Moreover, the telemetry system and operational database will provide the current state of the water system.

12.2.2 Operational Control of Water Network Using MPC

Water networks are very complex multi-variable systems. MPC provides suitable techniques to implement the operational control of water systems to improve their performance, since it allows to compute optimal control strategies ahead in time for all the control elements [11, 12]. Moreover, MPC allows taking into account physical and operational constraints, the multi-variable input and output nature, the demand forecasting requirement and complex multi-objective operational goals of water networks. The optimal strategies are computed by optimizing a mathematical function describing the operational goals in a given time horizon and using a representative model of the network dynamics, as well as demand forecasts.

12.3 Proposed Approach

The aim of using MPC techniques for controlling DWNs is to compute, ahead in time, the input actions to achieve the optimal performance of the network according to a given set of control goals. MPC strategies have some important features to deal with complex systems such as DWNs, namely the amenability to include disturbance forecasts, physical constraints and multi-variable system dynamics and objectives in a relatively simple way.

12.3.1 Modelling

Several modelling techniques dealing with DWNs have been presented in the literature (see, e.g., [5, 13]). Here, a control-oriented modelling approach that considers a flow model is outlined, which follows the principles presented by the authors in [6, 14, 15]. The extension to include the pressure model can be found in Chap. 13. A DWN generally contains a set of pressurized pipes, water tanks at different elevation and a number of pumping stations and valves to manage water flows, pressure and elevation in order to supply water to consumers.

The DWN model can be considered as composed of a set of constitutive elements, which are presented and discussed below. Figure 12.2 shows, in a small example, the interconnection of typical constitutive elements.

12.3.1.1 Tanks

Water tanks provide the entire DWN with the storage capacity of drinking water at appropriate elevation levels to ensure adequate water pressure service to consumers. The mass balance expression relating to the stored volume v in the nth tank can be

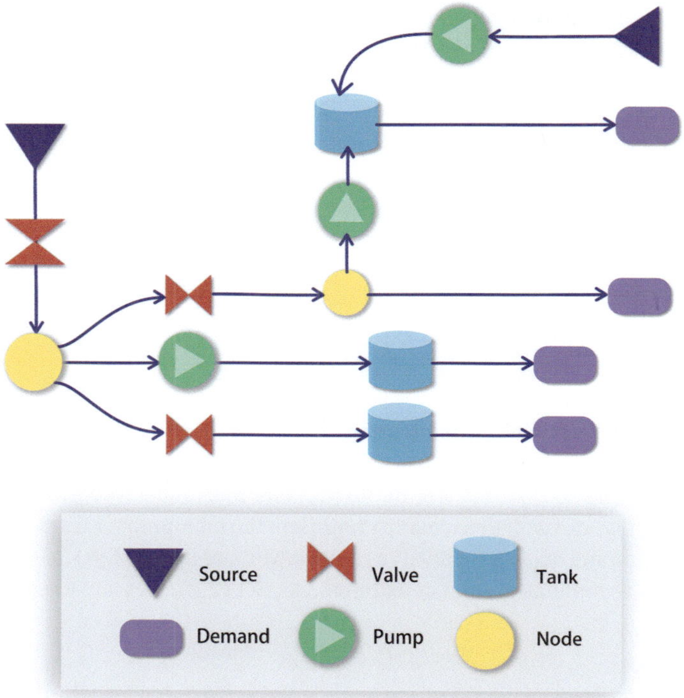

Fig. 12.2 Example of a basic topology of a generic drinking-water transport network. Note that the interaction of the main constitutive elements is shown here: sources' supply water to the system by means of pumps or valves, depending of the nature of the particular source (superficial or underground). Water is moved by using manipulated actuators in order to fill detention tanks and/or supply water to demand sectors

written as the discrete-time difference equation

$$v_n(k+1) = v_n(k) + \Delta t \left(\sum_j q_{\text{in}}^{jn}(k) - \sum_h q_{\text{out}}^{nh}(k) \right), \tag{12.1}$$

where $q_{\text{in}}^{jn}(k)$ denotes the manipulated inflows from the jth element to the nth tank, and $q_{\text{out}}^{nh}(k)$ denotes the manipulated outflows from the nth tank to the hth element (which includes the demand flows as outflows). Moreover, Δt corresponds with the sampling time and k the discrete-time instant. The physical constraint related to the range of admissible storage volume in the nth tank is expressed as

$$v_n^{min} \leq v_n(k) \leq v_n^{max}, \quad \text{for all } k, \tag{12.2}$$

where v_n^{min} and v_n^{max} denote the minimum and the maximum admissible storage capacities, respectively. Notice that $\underline{v}_n$ might correspond with an empty tank; in practice, this value can be set as nonzero in order to maintain an emergency stored volume.

For simplicity, the dynamic behaviour of these elements is described as a function of volume. However, in most cases, the measured variable is the tank water level (by using level sensors), which implies the computation of volume taking into account the tank geometry.

12.3.1.2 Actuators

Two types of control actuators are considered: valves and pumps, or more precisely, complex pumping stations. A pumping station generally contains a number of individual pumps with fixed or variable speed. In practice, it is assumed that the flow through a pumping station is a continuous variable in a range of feasible values. The manipulated flows through the actuators represent the manipulated variables, denoted as q_u. Both pumping stations and valves have lower and upper physical limits, which are taken into account as system constraints. As in (12.2), they are expressed as

$$q_{u_m}^{min} \leq q_{u_m}(k) \leq q_{u_m}^{max}, \quad \text{for all } k, \tag{12.3}$$

where $q_{u_m}^{min}$ and $q_{u_m}^{max}$ denote the minimum and the maximum flow capacity of the mth actuator, respectively. Since this modelling is stated within a supervisory control framework, it is assumed that a *local controller* is available, which ensures that the required flow through the actuator is obtained.

12.3.1.3 Nodes

These elements correspond to the network points where water flows are merged or split. Thus, nodes represent mass balance relations, modelled as equality constraints related to inflows—from other tanks through valves or pumps—and outflows, the latter being not only manipulated flows but also demand flows. The expression of the mass balance in these elements can be written as

$$\sum_j q_{jr}^{\text{in}}(k) = \sum_h q_{rh}^{\text{out}}(k), \tag{12.4}$$

where $q_{jr}^{\text{in}}(k)$ denotes inflows from the jth element to the rth node, and $q_{rh}^{\text{out}}(k)$ denotes outflows from the rth node to the hth element. From now on, node inflows and outflows will be denoted by q^{in} and q^{out}, even if they are manipulated variables (denoted by q_u).

12.3.1.4 Demand Sectors

A demand sector represents the water demand of the network users of a certain physical area. It is considered as a measured disturbance of the system at a given time instant. The demand can be anticipated by forecasting algorithms, which are integrated within the MPC closed-loop architecture. For the case studies in this chapter, the algorithm proposed in [16], among others discussed in Chap. 6, is considered. This algorithm typically uses a two-level scheme composed of

(i) a time series model to represent the daily aggregate flow values and
(ii) a set of different daily flow demand patterns according to the day type to cater for different consumption during the weekend and holiday periods. Every pattern consists of 24 hourly values for each daily pattern.

The algorithm runs in parallel with the MPC algorithm. The daily series of hourly flow predictions are computed as a product of the daily aggregate flow value and the appropriate hourly demand pattern. Regarding the daily demand forecast, its corresponding flow model is built on the basis of an ARIMA time series modelling approach described in [17]. Then, the structure of the daily flow model for each demand sensor may be written as

$$\begin{aligned} y_p(k) = & -b_1 y(k-1) - b_2 y(k-2) - b_3 y(k-3) - b_4 y(k-4) \\ & - b_5 y(k-5) - b_6 y(k-6) - b_7 y(k-7), \end{aligned} \tag{12.5}$$

where the parameters $b_1, \ldots, b_7$ are estimated based on historical data. The 1-h flow model is based on distributing the daily flow prediction provided by the time series model in (12.5) using an hourly flow pattern that takes into account the daily/monthly variation as follows:

$$y_{ph}(k+i) = \frac{y_{pat}(k,i)}{\sum_{j=1}^{24} y_{pat}(k,j)} y_p(k), \qquad i = 1, \ldots, 24, \tag{12.6}$$

where $y_p(k)$ is the predicted flow for the current day k using (12.5), and $y_{pat}(k)$ is the prediction provided considering the flow pattern class corresponding to the current day. Demand patterns are obtained from statistical analysis.

12.3.2 Control-Oriented Model

Considering the set of compositional elements described above, the control-oriented model can be obtained by joining those elements and their corresponding dynamic descriptions. In a general form, the expression which collects all these dynamics can be written as the mapping

$$\mathbf{x}(k+1) = \mathbf{g}(\mathbf{x}(k), \mathbf{u}(k), \mathbf{d}(k)), \tag{12.7}$$

where $\mathbf{x} \in \mathbb{X} \subseteq \mathbb{R}^{n_x}$ corresponds to the system states, $\mathbf{u} \in \mathbb{U} \subseteq \mathbb{R}^{n_u}$ denotes the system inputs (manipulated variables) and $\mathbf{d} \in \mathbb{D} \subseteq \mathbb{R}^{n_d}$ denotes the system disturbances. $\mathbf{g} : \mathbb{R}^{n_x} \times \mathbb{R}^{n_u} \times \mathbb{R}^{n_d} \to \mathbb{R}^{n_x}$ is an arbitrary system state function and $k \in \mathbb{Z}_+$.

In the case of DWN, (12.7) is associated with the set of tank expressions in (12.1). Hence, a control-oriented discrete-time state-space model can be written as [15]

$$\mathbf{x}(k+1) = \mathbf{A}\,\mathbf{x}(k) + \mathbf{B}\,\mathbf{u}(k) + \mathbf{B}_p\,\mathbf{d}(k), \tag{12.8}$$

where, in particular, $\mathbf{x}$ corresponds to the water volumes v of the n_x tanks, $\mathbf{u}$ represents the manipulated flows q_u through the n_u actuators (pumps and valves) and $\mathbf{d}$ corresponds with the vector of n_d water demands (measured disturbances affecting the system). $\mathbf{A}$, $\mathbf{B}$ and $\mathbf{B}_p$ are the system matrices of suitable dimensions. Note that since the system control-oriented model of a DWN does not collect the static dynamics described by DWN nodes in (12.4), then (12.8) can be further rewritten as

$$\mathbf{x}(k+1) = \mathbf{A}\,\mathbf{x}(k) + \mathbf{\Gamma}\,\boldsymbol{\mu}(k), \tag{12.9a}$$

$$\begin{bmatrix} \mathbf{E}_u & \mathbf{E}_d \end{bmatrix} \boldsymbol{\mu}(k) = 0, \tag{12.9b}$$

where $\mathbf{\Gamma} = [\mathbf{B} \;\; \mathbf{B}_p]$, $\boldsymbol{\mu}(k) = [\mathbf{u}(k)^T \;\; \mathbf{d}(k)^T]^T$ and $\mathbf{E}_u$, $\mathbf{E}_d$ are matrices of suitable dimensions. It can be seen that (12.9a) comes from the mass balance in tanks, while (12.9b) comes from the network nodes. Also notice that when all the network flows are manipulated, then $\mathbf{A}$ is an identity matrix of suitable dimensions.

12.3.3 Control Criteria

It is possible to use different control objectives depending on the operational goals considered by the network managers. This section describes the most common control objectives and the resultant multi-objective cost function. Therefore, this chapter considers and discussed the following control objectives [15, 18].

12.3.3.1 Minimization of Water Production and Transport Costs

The main economic costs associated with drinking-water production are due to treatment processes, water acquisition or use costs and, most importantly, to electricity costs associated with pumping. Delivering this drinking water to appropriate pressure levels through the network involves important electricity costs in booster pumping as well as elevation from underground devices. In a specific case, this objective can be mathematically formulated as the minimization of

$$J_1(k) \triangleq (\boldsymbol{\alpha}_1 + \boldsymbol{\alpha}_2(k))^T \mathbf{u}(k), \tag{12.10}$$

where $\boldsymbol{\alpha}_1$ corresponds to a known vector related to water production costs, depending on the selected water source, and $\boldsymbol{\alpha}_2(k)$ is a vector of suitable dimensions associated with the energy pumping costs. Note the k-dependence of $\boldsymbol{\alpha}_2$ since the pumping cost has different values according to the variable electric tariffs along a day.

12.3.3.2 Appropriate Management of Safety Water Storage

The satisfaction of water demands must be fulfilled at all times. However, some risk prevention mechanisms need to be introduced in the tank management so that, additionally, the stored volume is preferably maintained above certain safety value for eventual emergency needs and to guarantee future water availability. Therefore, this objective may be achieved by minimizing the following expression:

$$J_2(k) = \begin{cases} (\mathbf{x}(k) - \mathbf{x}^{\text{safe}})^T \ (\mathbf{x}(k) - \mathbf{x}^{\text{safe}}) & \text{if} \ \ \mathbf{x}(k) \leq \mathbf{x}^{\text{safe}}, \\ 0 & \text{otherwise}, \end{cases} \tag{12.11}$$

where $\mathbf{x}^{\text{safe}}$ is a term which determines the safety volume to be considered for the control law computation. This term might appear as unnecessary given the guarantees of the MPC design but since a trade-off between the other costs and the volumes is present, the controller would tend to keep the lowest possible the tank water volumes. This fact would reduce the safety of the system to handle unexpected extra demands, such as fire extinction, among others.

12.3.3.3 Smoothing of Control Actions

Valves must also operate smoothly in order to avoid big transients in the pressurized pipes. This fact could lead to poor pipe condition. The use of a smooth reference changes also *helps* the lower-level regulator performance. Similarly, water flows requested from treatment plants must have a smooth profile due to plants operational constraints. To obtain such smoothing effect, control signal variation between consecutive time intervals is therefore penalized. The penalty term to be minimized is

$$J_3(k) = \Delta\mathbf{u}(k)^T \ \Delta\mathbf{u}(k), \tag{12.12}$$

where $\Delta\mathbf{u}(k) \triangleq \mathbf{u}(k) - \mathbf{u}(k-1)$.

12.3.3.4 Multi-objective Performance Function

The multi-objective performance function $\mathcal{J}(k)$ that gathers the aforementioned control objectives, either in the case of DWN or SN, can be written as

$$\mathcal{J}(k) = \sum_{j=1}^{\varphi} \gamma_j J_j(k), \tag{12.13}$$

where a set of φ control objectives are considered, and, in turn, a *multi-objective* open-loop optimization problem (OOP) is stated. The prioritization of the control objectives is performed by using the order of the mathematical cost function associated with each objective and also a set of appropriate weights γ_j. These weights are selected offline by means of trial-and-error procedures, taking into account the priority of each objective within the cost function. More sophisticated tuning methodologies for tuning multi-objective control problems based on lexicographic minimizers [19], goal programming [20] or Pareto-front computations [21] may be also considered.

12.3.4 MPC Problem Formulation

Collecting the parts described in previous subsections, the MPC design follows the traditional procedures presented in [11, 12, 22], which involve solving an optimization problem over a prediction horizon H_p, where a cost function is minimized subject to a set of physical and operational constraints. Once the minimization is performed, a vector of H_u control actions over H_p is obtained. Only the first component of that vector is considered and applied to the plant. The procedure is repeated for the next time instant taking into account the feedback measurements coming from the system, following the classic receding horizon strategy.

In general terms, the MPC controller design is based on the solution of a OOP

$$\mathcal{V}(k, H_p) = \min \sum_{i=0}^{H_p} \sum_{j=1}^{\varphi} \gamma_j \, J_j(k + i|k), \tag{12.14}$$

subject to the system model and the physical and operational constraints, where H_p corresponds to the prediction horizon, and index k represents the current time instant, while index i represents the time instant along H_p. Hence, notation $k + i|k$ denotes the time instant $k + i$ given k. Note that (12.14) corresponds with (12.13) over the prediction horizon.

According to the case, the minimum of $\mathcal{V}(k, H_p)$ is achieved by finding a set of optimal variables which generally correspond with the manipulated variables of the system model but that could include further variables of diverse nature. Hence, for a prediction window of length H_p and considering $\mathbf{z} \in \mathbb{R}^{sH_p}$ as the set of s optimization variables for each time instant over H_p, the multi-objective optimization problem can be formulated as

$$\min_{\{\mathbf{z}\in\mathbb{R}^{sH_p}\}} f(\mathbf{z}) \tag{12.15a}$$

subject to

$$H_1(\mathbf{z}) \leq 0, \tag{12.15b}$$

$$H_2(\mathbf{z}) = 0, \tag{12.15c}$$

where $f(\mathbf{z})$ comes from the manipulation of (12.14). Moreover, $H_1(\mathbf{z})$ and $H_2(\mathbf{z})$ are vectors of dimensions $r_i H_p \times 1$ and $r_e H_p \times 1$, respectively, containing the constraint functions. Here, r_i is the number of inequality constraints, and r_e is the number of the problem equality constraints. It can be observed that (12.15b) and (12.15c) gather all problem constraints including those from the system model, the physical restrictions of its variables and the operational and management constraints.

Assuming that the OOP (12.15) is feasible for $\mathbf{z} \in \mathbb{R}^{sH_p}$, there exists an optimal solution given by the sequence

$$\mathbf{z}^* \triangleq \big(\mathbf{z}^*(0|k), \mathbf{z}^*(1|k), \ldots, \mathbf{z}^*(H_p|k)\big) \tag{12.16}$$

and then the receding horizon philosophy sets [12]

$$\mathbf{z}_{\text{MPC}}(\mathbf{x}(k)) \triangleq \mathbf{z}^*(0|k) \tag{12.17}$$

and disregards the computed inputs from $k = 1$ to $k = H_p$, repeating the whole process at the following time step. Equation (12.17) is known as *the MPC law*.

Therefore, the MPC problem formulation in DWNs gives the expressions for each of the problem parts described above. Thus, mapping (12.7) must be replaced by the system modelling in (12.9) when treating a DWN. Finally, constraints in (12.15b) and (12.15c) are conveniently expressed taking into account the type of network and its constitutive components; for example, constraints in (12.9b) must be included when a DWN is considered. Constraints (12.2) and (12.3) are always included. In order to manage the uncertainty of the system disturbances over the prediction horizon, a suitable approach is the stochastic paradigm, which includes explicit models of uncertainty/disturbances in the design of control laws and by transforming hard constraints into probabilistic constraints. As reviewed in [23], the stochastic approach is a classic one in the field of optimization, and a renewed attention has been given to the stochastic programming [24], as a powerful tool for robust control design, leading to the stochastic MPC and specially to the chance-constrained MPC (CC-MPC) [25] (see Chap. 13).

12.4 Simulations and Results

As an application case study to show the performance of the proposed modelling and control approach, some results of its application offline (in simulation) in several real scenarios in the Barcelona WTN are presented. A simulator of this network has been built using MATLAB/Simulink and validated using real data coming from real scenarios (see Figs. 12.10 and 12.11 and the corresponding explanations in Chap. 2). This allows testing the controller against a virtual reality introducing, for example, real demand in the simulator different from the predicted demand used by the controller. The MPC controller was implemented with the PLIO tool presented in [26] that uses GAMS/CONPOPT solver to solve the corresponding optimization problem. This general-purpose decision support tool has been developed to allow the user to implement optimal/predictive control techniques in large-scale drinking-water systems (see Fig. 12.3).

The modelling and predictive control problem solution algorithms are designed for real-time decision support, in connection with a SCADA system. The hydraulic modelling relies on simple, but representative enough, dynamic equations whose parameters are recalibrated online using recursive parameter estimation and real data obtained from sensors in the network. Demand forecast models, based on time series analysis, are also dynamically updated. The real-time calibration using recursive parameter estimation methods contributes to deal with hydraulic uncertainty. This modelling choice, as well as the optimization method selection, allows to deal with very large-scale systems. Another distinguishing feature is its capability to accommodate complex operational goals.

In Fig. 12.4, the evolution of volume at a number of tanks is shown. The simulator output is shown in blue, while red is used for the real data. In some cases, small discrepancies between both volume curves are not associated with modelling errors but with errors in real data due to a faulty sensor. The most important conclusion after

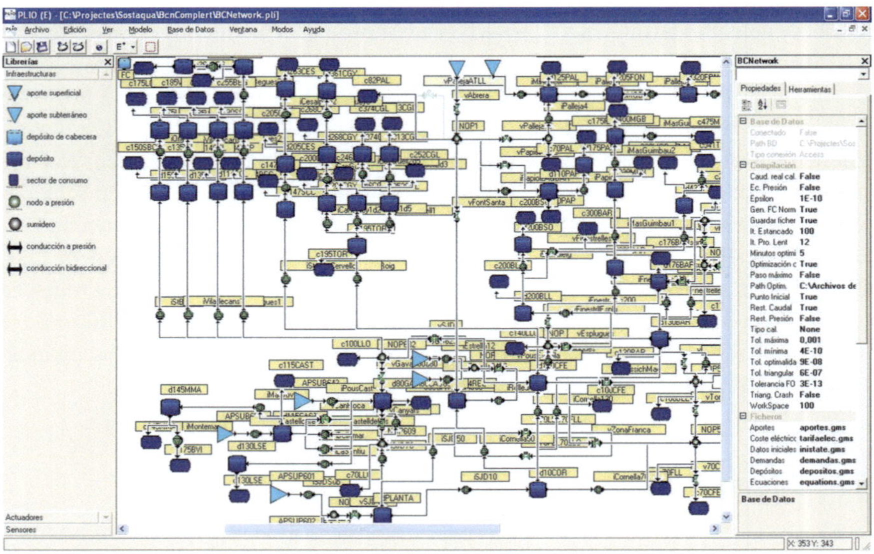

Fig. 12.3 PLIO interface corresponding to the model manager module allows creating/updating the model of the water network in a user friendly way

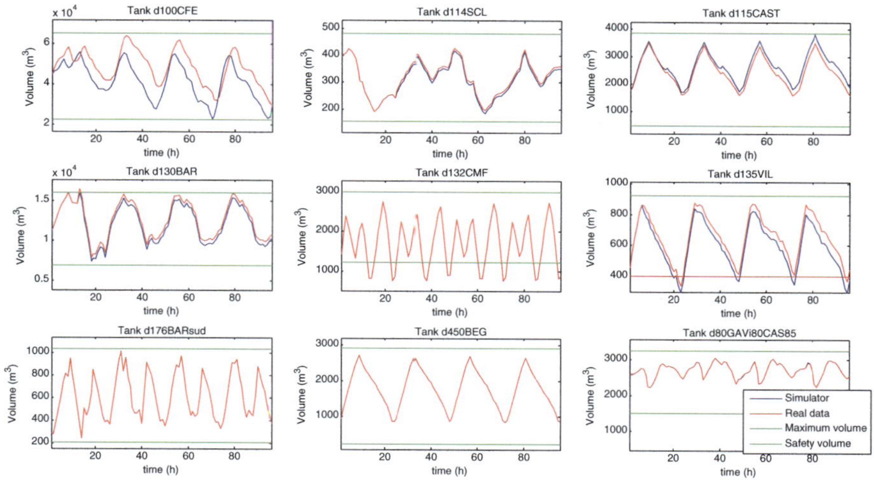

Fig. 12.4 Model validation based on the comparison between real volumes and the simulated ones

this process is that this simulator allows making the model validation process easier. The model has been validated and accepted by Aguas de Barcelona as representative of the network real behaviour.

The Barcelona WTN is organized in different pressure levels. Figure 12.5 presents the several pressure levels in different colours. Each sector will be supplied through a

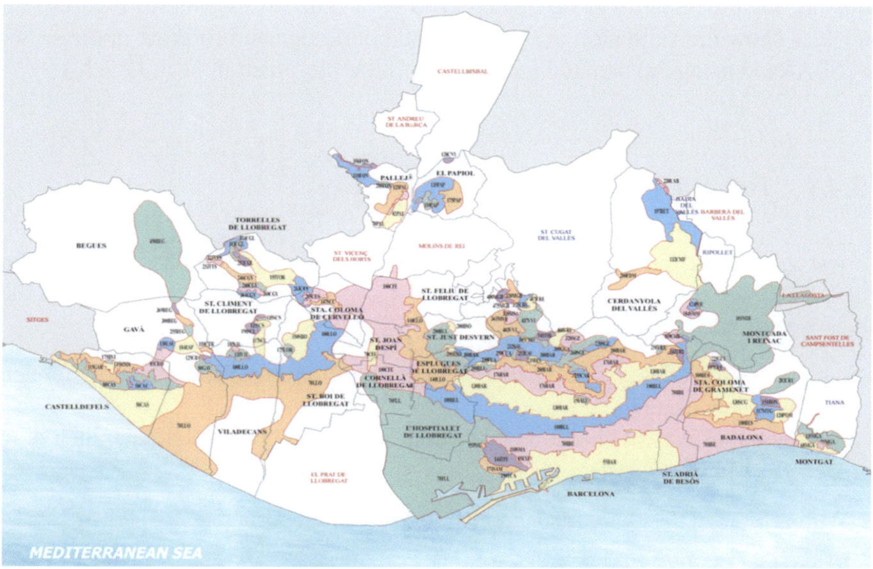

Fig. 12.5 Barcelona water network demand sectors

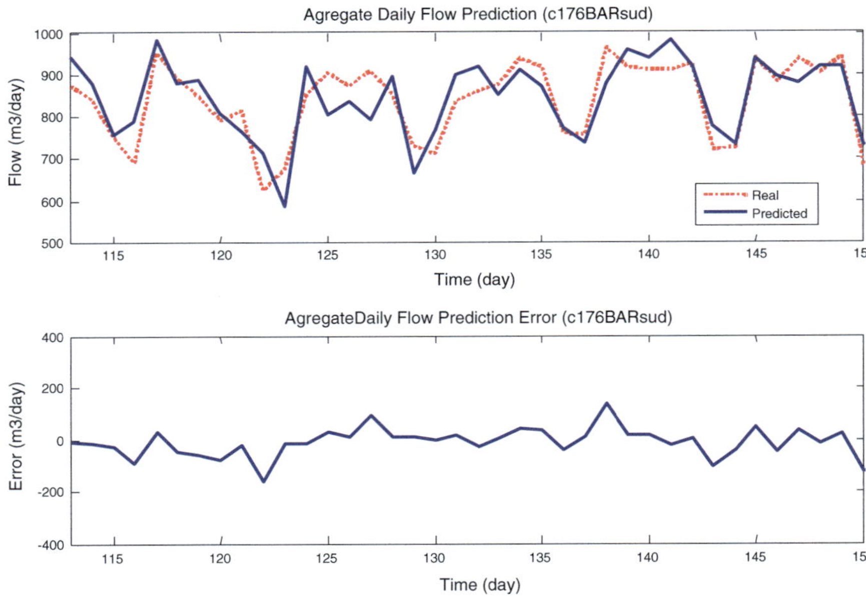

Fig. 12.6 Validation of the aggregate daily demand forecast corresponding to the sector c176BARsud

storage tank. The distribution network that connects each storage tank with individual consumers will not be modelled in detail but will be summarized as an aggregated demand. Each demand will be modelled using a time series pattern. Figures 12.6 and 12.7 show the validation of the daily and hourly demand forecast in the sector c176BARsud using the demand forecast algorithm presented in Sect. 12.3.1.4.

12.4.1 Test Scenarios

To test and adjust the MPC controller, different scenarios have been chosen. The main difference between the selected scenarios is related to source operation. So, the objectives of this study are as follows:

- to compare the effects of the MPC strategies with those of the currently applied control strategies and
- to show the effects of source management in the total operation cost, including electrical and water costs.

With reference to source management, two different scenarios are shown:

- **Scenario 1: Scheduled flow**. In this case, the flow of all sources is fixed to real values obtained from real historical data.

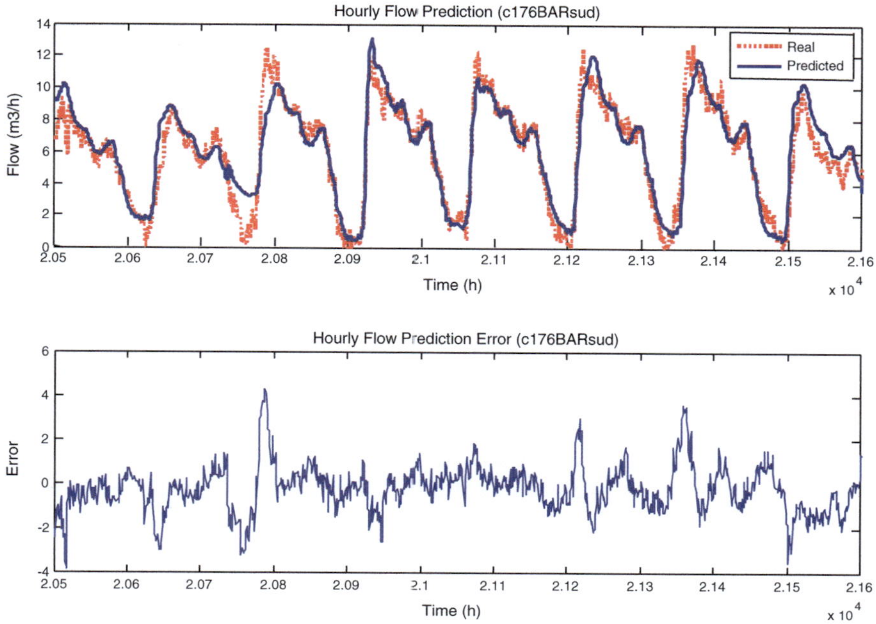

Fig. 12.7 Validation of the hourly demand forecast corresponding to the sector c176BARsud

- **Scenario 2: Flow optimization**. The optimizer calculates the flow to be abducted from each source at each time step, taking into account its operational limits, according to long-term planning.
- **Scenario 3: Fixing main source**. The main source of water is fixed, while the others are optimized.

The parameters taken into account for the calibration of the model are the initial volumes and safety storage volumes in tanks, as well as the objective function weights for each of the operational goals (the economical, safety and smoothness factors). Objective function weights are calibrated by experimentally analysing their effects on the compromise between the operational goals, with historic data. In [21], the authors have explored multi-objective optimization techniques to tune them in a more sophisticated way. Tank initial and safety storage volumes are taken from real historic data of each scenario, in order to make optimization results comparable with current control strategy.

The period in both scenarios is 96 h (4 days), and all of them correspond to the same period, between July 23 and July 26 of 2007. It means that the demand is the same in both scenarios, so they are comparable. To estimate the demand of each sector, the demand forecast method presented in Sect. 12.3.1.4 is used. The total demanded volume for each day is obtained from the total contribution from each source. In Table 12.1, values of volume per day are shown.

Table 12.1 Total input volume for studied days

Date	Total input volume (m^3)	Mean flow (m^3/s)
23/07/2007	633694	7.334
24/07/2007	668136	7.733
25/07/2007	617744	7.150
26/07/2007	627406	7.262
	Mean	7.370

12.4.2 Results and Discussion

In all the test scenarios, the MPC controller computed solutions to meet demands and operational constraints at all times, while optimizing the operational goals. Some illustrative results of the MPC application on the complete Barcelona WTN are presented in this section. For these tests, the same model is used.

12.4.2.1 Scenario 1: Scheduled Flow

In this first scenario, source flows are imposed using real data obtained from Aguas de Barcelona historical database. The interesting point of this scenario is the comparison between MPC control strategy and current control strategy: water sources management is the same in both cases. This scenario is used to show the potential of MPC for minimizing the electrical (pumping) cost. The evolution of source flows is shown in Fig. 12.8.

In Table 12.2, electrical and water cost in percentage of the total cost for the current control strategy are shown. In Table 12.3, costs for the MPC control as an increase or decrease percentage with regard to current control are presented.

Water production cost (acquisition and treatment) represents a value near 70% of the total cost, and there is no variation of this cost in the MPC control because of the fixed sources. With regard to electrical cost, the improvement is between 10 and 25%, which represents a decrease of the total cost between 3 and 8%. To show the differences between the current control and the MPC control, some tank volume and

Table 12.2 Current control strategy costs in percentage

Date	Electricity cost	Water cost	Total cost
23/07/2007	33.13	66.87	100.00
24/07/2007	34.66	65.34	100.00
25/07/2007	32.00	68.00	100.00
26/07/2007	31.29	68.71	100.00

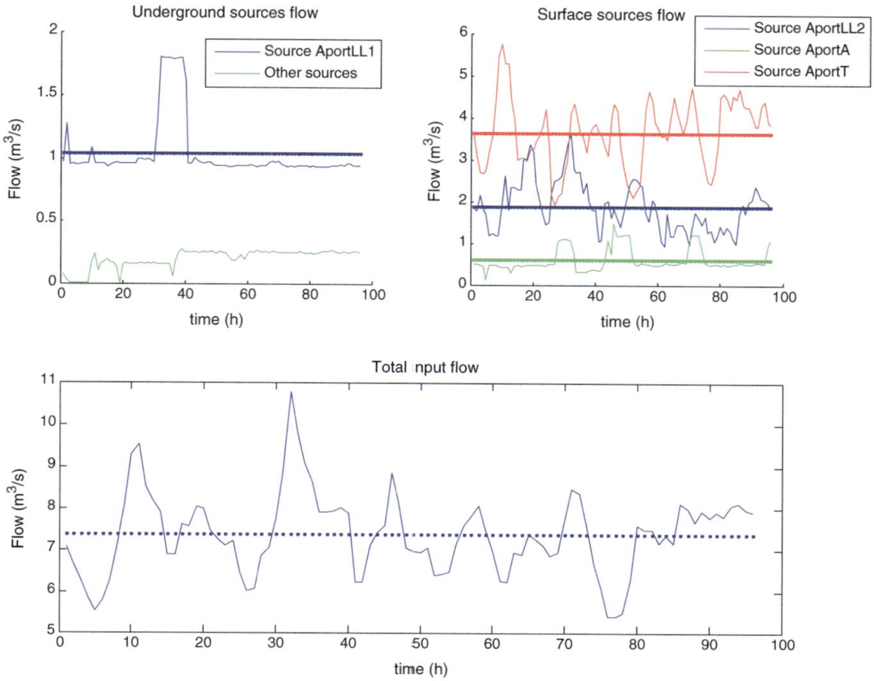

Fig. 12.8 Sources' flow evolution for Scenario 1: scheduled flow

Table 12.3 MPC improvement in percentage for Scenario 1 (scheduled flow) regarding Table 12.2 values

Date	Electricity cost	Water cost	Total cost
23/07/2007	–23.27	+0.00	–7.71
24/07/2007	–10.56	+0.00	–3.66
25/07/2007	–20.61	+0.00	–6.59
26/07/2007	–18.58	+0.00	–5.81

actuator flow plots are shown. In Fig. 12.9, some tank volume evolution can be seen, as well as maximum and security volumes.

The smoothness term is not the only factor with effects on pumps' operation. The electric tariff for each pump is another factor that affects pump operation in order to minimize electrical cost. In Fig. 12.10, the effects of the electricity cost are shown. It can be seen that if it is possible, pumps only run during the cheapest period (e.g., *iPalleja1*). In cases where with a maximum flow during off-peak hours the necessary volume is not reached, pumps must work during other periods. Pump *iFnestrelles200* is an example of this case. Since it is not enough to pump during the cheapest period, this pump is pumping during the medium-cost period too, but with a maximum flow lower than in the cheapest one.

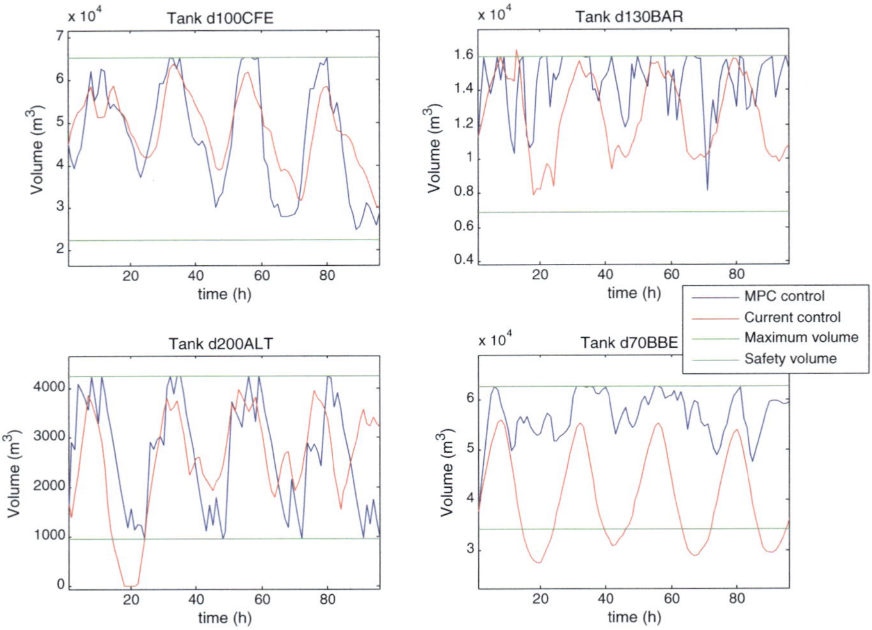

Fig. 12.9 Some tanks' volume evolution: current control and MPC control comparison

12.4.2.2 Scenario 2: Flow Optimization

In this second scenario, the source flows are optimized. It means that the only limitation is the minimum and the maximum flow of actuators in the output of each source. In this case, both electrical and water cost are optimized, so it is expected to obtain a higher improvement in the total cost referring to the Scenario 1, where sources' flow was fixed. This scenario represents a theoretical solution of the water management in the Barcelona WTN. Indeed, the optimization carried out gives total freedom to the different sources, while on a real situation, sources are not unlimited or unrestricted: its availability as well as its future guarantee compromises the total amount of water entering the system from each source. Therefore, the hereby shown results give us an idea of how far flow optimization could go if there were no sources' restrictions. In Fig. 12.11, sources' flow evolution is shown. As it can be seen, Llobregat's mean flow is about 5 m^3/s (which is the maximum possible contribution of this source), while the lack of water necessary to satisfy the total demand is taken from Ter and Abrera. Underground sources' water cost is penalized to avoid its overexploitation.

Electrical and water cost obtained in this scenario is compared with both the current control case and the MPC case of Scenario 1 (scheduled flow). In Tables 12.4 and 12.5, this comparison is shown.

The first point to emphasize is the high water improvement, between 30 and 50%. As shown, it seems that maximizing water taken from Llobregat, water cost is clearly decreased. On the other hand, electrical cost is increased, but the decrease of

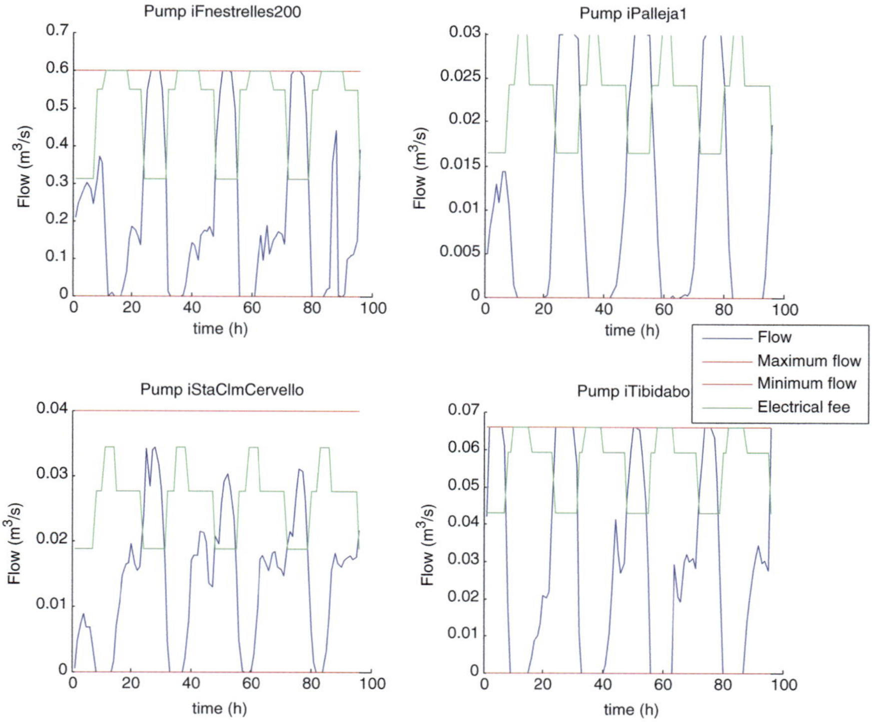

Fig. 12.10 Electrical fee effects on pumps operation

Table 12.4 Scenario 2 improvement with regard to current control case (Table 12.2)

Date	Electricity cost	Water cost	Total cost
23/07/2007	18.92	–50.70	–27.63
24/07/2007	14.04	–32.56	–16.41
25/07/2007	26.29	–43.91	–21.45
26/07/2007	26.09	–44.43	–22.36

Table 12.5 Scenario 2 improvement with regard to Scenario 1 case (scheduled flow)

Date	Electricity cost	Water cost	Total cost
23/07/2007	54.99	–50.70	–21.59
24/07/2007	27.51	–32.56	–13.23
25/07/2007	59.08	–43.91	–15.91
26/07/2007	54.86	–44.43	–17.57

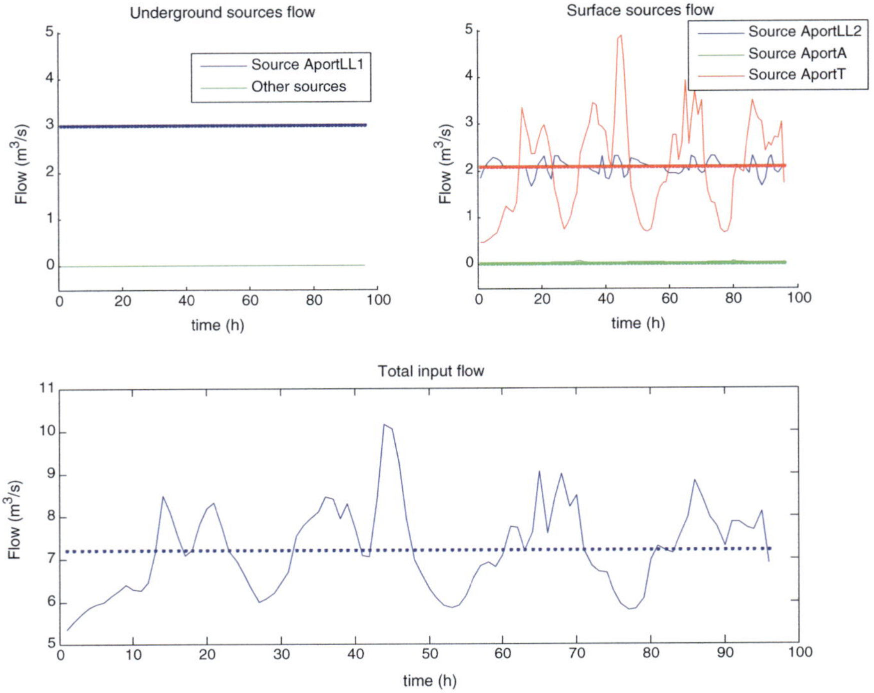

Fig. 12.11 Sources' flow evolution for Scenario 2: flow optimization

the total cost in this second scenario regarding current control case and Scenario 1 is important.

12.4.2.3 Scenario 3: Fixing Main Source

The two main sources of the Barcelona water network are the Llobregat and Ter rivers. Barcelona's average demand is about 7.5 m^3/s. For ecological reasons, Aguas de Barcelona company uses Llobregat source at its maximum capacity in which value depends on the river flow. The rest of flow is supplied by Ter source. From Fig. 12.12, it can be noticed that both sources affect the economic cost in an inverse way. Increasing the amount of water extracted from Llobregat source reduces the water cost while increasing the electrical cost. On the other hand, the Ter source behaves on the opposite sense: increasing the amount of water extracted from this river reduces the electrical cost while augmenting the water cost. The reason for this behaviour is due to a smaller water price in the case of Llobregat. But, since Llobregat source is located close to the sea level, while Ter source is in the upper part of the city, electrical costs will be higher in case of the Llobregat source since more pumping will be required to supply water from this source. In the case when

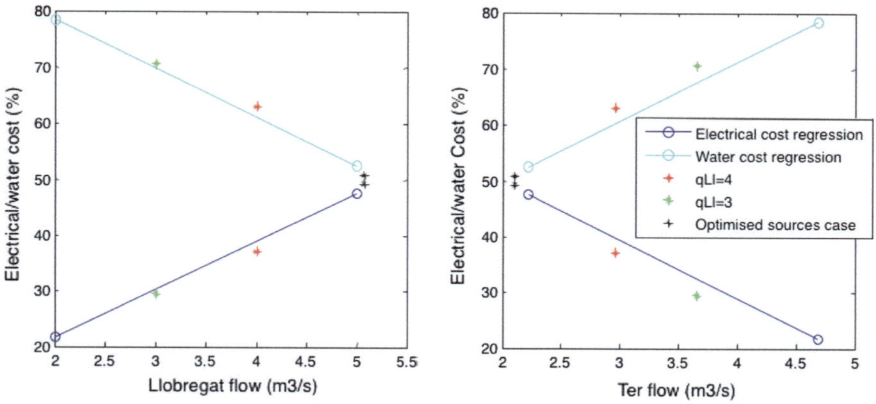

Fig. 12.12 Electrical and water cost when fixing Llobregat source

sources are not fixed, the optimal combination leads to take most of the water from Llobregat source and the remaining from the Ter source.

12.4.3 Complementary Comments

In Table 12.6, a brief summary of results presented is shown, as a mean value of four days of study. The costs of Scenarios 1 and 2 are referred to current control values.

From this table, conclusions that can be emphasized are as follows:

- Maximizing the flow from the source Llobregat to optimize total cost.
- Flow optimization allows higher improvement with regard to fixed real flows because the optimizer can maximize Llobregat's flow contribution if it is possible. Sometimes, it is not possible because of the reasons not related to network characteristics (operational limits of actuators and tanks).
- Ter total cost (only water cost because there is no pump) is higher than the Llobregat one (water and electrical cost associated). This fact, sources' behaviour and results of both test scenarios indicate that:

Table 12.6 Summary of results for scenarios presented

Cost	Current control (%)	Scenario 1 (%)	Scenario 2 (%)
Electrical	32.77	–18.26	+21.34
Water	67.23	0	–42.90
Total	100	–5.94	–21.96

- Reduction of electrical cost involves reduction of the contribution from Llobregat.
- Reduction of water cost involves reduction of Ter source contribution.
- Total cost is minimized by maximizing Llobregat source contribution.

12.5 Conclusions

MPC techniques provide useful tools for generating water management strategies in large and complex water networks, which may be used for decision support, as well as for fully automated control of a water network. This work describes the use of MPC for flow management in a large water system, involving supplies, production plants and water transport into the distribution areas. The chapter presents the application of a unified approach to the water system management including supplies, production, transport and distribution areas. The modelling and predictive control solutions are designed for real-time decision support. The hydraulic modelling relies on simple, but representative, dynamic equations and recursive real-time parameter calibration using updated data from telemetry. Demand predictions are also dynamically updated. The potential of these techniques for real-time control of water supply and distribution has been shown with two representative examples of complex operational situations. The test scenarios are based on real situations which are known to have caused difficulties to operators and, in some cases, severe effects on the service to consumers. The application described in the chapter deals with these scenarios successfully, by producing control strategies that rearrange flows, production plant levels, pumping from underground sources, etc., in a way that demands are met at all times with improved results with respect to management goals. This type of decision support is extremely useful for water system operators in large-scale systems, especially those involving several different water management levels (supply, production, transport and distribution), where the control solutions may not obvious are successfully implemented.

References

1. Tu M, Tsai F, Yeh W (2005) Optimization of water distribution and water quality by hybrid genetic algorithm. J Water Resour Planning Manag 131(6):431–440
2. Nitivattananon V, Sadowski E, Quimpo R (1996) Optimization of water supply system operation. J Water Resour Planning Manage 122(5):374–384
3. Westphal KS, Vogel RM, Kirschen P, Chapra SC (2003) Decision support system for adaptive water supply management. J Water Resour Planning Manag 129(3):165–177
4. Advances in water supply management (2003) Swets and Zeitlinger, The Netherlands
5. Brdys MA, Ulanicki B (1994) Operational control of water systems: structures, algorithms and applications. Prentice Hall

6. Cembrano G, Wells G, Quevedo J, Pérez R, Argelaguet R (2000) Optimal control of a water distribution network in a supervisory control system Control Engineering Practice, 8(10): 1177–1188
7. Butler D, Memon F (2006) Water demand management. IWA Publishing, London
8. Marinaki M, Papageorgiou M (2005) Optimal real-time control of sewer networks. Springer, Secaucus, NJ (USA)
9. Schütze M, Campisanob A, Colas H, Schillingd W, Vanrolleghem P (2004) Real time control of urban wastewater systems: where do we stand today? J Hydrol 299:335–348
10. Blanke M, Kinnaert M, Lunze J, Staroswiecki M (2016) Diagnosis and fault-tolerant control, 3rd edn. Springer, Berlin, Heidelberg
11. Camacho EF, Bordons C (2004) Model predictive control, 2nd edn. Springer, London
12. Maciejowski JM (2002) Predictive control with constraints. Prentice Hall, Essex, England
13. Mays LW (2004) Urban stormwater management tools. McGrawHill Professional Publishing, USA
14. Cembrano G, Quevedo J, Salamero M, Puig V, Figueras J, Martí J (2004) Optimal control of urban drainage systems: a case study. Control Eng Pract 12(1):1–9
15. Ocampo-Martinez C, Puig V, Cembrano G, Creus R, Minoves M (2009) Improving water management efficiency by using optimization-based control strategies: the Barcelona case study. Water Sci Technol: Water Supply 9(5):565–575
16. Quevedo J, Puig V, Cembrano G, Blanch J (2010) Validation and reconstruction of flow meter data in the Barcelona water distribution network. Control Eng Pract 11(6):640–651
17. Quevedo J, Cembrano G, Valls A, Serra J (1998) Computer applications in water supply, volume I, chapter Time series modelling of water demand: a study on short-term and long-term predictions. Research Studies Press, Lechworth, England, pp 268–288
18. Ocampo-Martinez C, Fambrini V, Barcelli D, Puig V (2010) Model predictive control of drinking water networks: a hierarchical and decentralized approach. In: Proceedings of the American control conference, 2010, pp 3951–3956
19. Ocampo-Martinez C, Ingimundarson A, Puig V, Quevedo J (2008) Objective prioritization using lexicographic minimizers for MPC of sewer networks. IEEE Trans Control Syst Technol 16(1):113–121
20. Chen A, Xu X (2012) Goal programming approach to solving network design problem with multiple objectives and demand uncertainty. Expert Syst Appl 39(4):4160–4170
21. Toro R, Ocampo-Martinez C, Logist F, Van Impe J, Puig V (2011) Tuning of predictive controllers for drinking water networked systems. In: Proceedings of 18th IFAC world congress, Milano, Italy, August 2011, pp 14507–14512
22. Rawlings JB, Amrit R. Nonlinear model predictive control
23. Calafiore G, Dabbene F (2006) Probabilistic and randomized methods for design under uncertainty. Springer
24. Shapiro A, Dentcheva D, Ruszczynski A (2009) Lectures on stochastic programming: modeling and theory. Society for industrial and applied mathematics and mathematical programming society, 2009
25. Grosso JM. A robust adaptive model predictive control to enhance the management of drinkingwater networks subject to demand uncertainty and actuators degradation. Master's thesis, Technical University of Catalonia (UPC), Barcelona
26. Cembrano G, Quevedo J, Puig V, Perez R, Figueras J, Verdejo JM, Escaler I, Ramon G, Barnet G, Rodriguez P, Casas M (2011) PLIO: a generic tool for real-time operational predictive optimal control of water networks. Water Sci Technol 64(2):448–459

Chapter 13
Model Predictive Control of Water Networks Considering Flow and Pressure

Ye Wang, Gabriela Cembrano, Vicenç Puig, Maite Urrea, Juli Romera, David Saporta and José Gabriel Valero

13.1 Introduction

As discussed in Chap. 12, WDNs are very complex multi-variable systems since they can contain many tanks, pumping stations, valves, water sources (superficial and underground) and sectors of consumer demand [1]. The main issues associated with such an operational control are the complexity of the network and how to operate the network optimally (that is by using water sources efficiently and minimizing operation costs) while satisfying water demand and quality of service (e.g., minimum pressure) standards. In particular, electrical energy is the main source of operation costs, both for water production and water elevation to adequate pressure levels for consumption, using pumping stations. Currently, depending on the pumping station, different bilateral contracts are established with energy supply companies with a variety of prices and different cost periods (from two periods up to six different price periods per day, depending on working days/weekends, and on seasons). Accordingly, current practice is to pre-allocate the pumping periods of each station when the energy prices arranged by the contract are the lowest possible for that station, guaranteeing that expected demand is satisfied, with the help of intermediate water storage capacity. As in Chap. 12, optimization and optimal control techniques including predictive capabilities provide a suitable framework for the efficient management strategy of WDNs when also pressure constraints are considered, see [1]. In particular, according to [8], MPC is also an appropriate strategy to find optimal flow set-points

Y. Wang · G. Cembrano (✉) · M. Urrea · J. Romera
Institut de Robòtica i Informàtica Industrial, CSIC-UPC, Barcelona, Spain
e-mail: gcembrano@iri.upc.edu

V. Puig
Research Center Supervision, Safety and Automatic Control (CS2AC-UPC), Terrassa, Spain

D. Saporta · J.G. Valero
Aguas de Barcelona (AGBAR), Barcelona, Spain

V. Puig et al. (eds.), *Real-Time Monitoring and Operational Control of Drinking-Water Systems*, Advances in Industrial Control,
DOI 10.1007/978-3-319-50751-4_13

for all the actuators while taking into account hydraulic heads in some specific points (e.g., the entrance of demand sectors) [3, 4]. The optimal strategies are computed by optimizing a cost function describing all the operational objectives in a given prediction horizon subject to a representative model of the network dynamics (including flow and pressure equations), as well as the required constraints on system variables. As discussed in Chap. 12, MPC is very suitable to be used in the global control of networks related to the urban water cycle within a hierarchical control structure [5–8, 12, 13]. In Chap. 12, MPC strategy was successfully applied to a WDN using a control-oriented model that considers only flows, i.e., the pressure/head model of each element in the WDN (including water storage tanks/reservoirs, water demand sectors, pressurized pipes, booster pumps and pressure/flow-controlled valves) is not considered explicitly. But, for certain WDN, in addition to satisfy water demands, it is also necessary to meet the required pressure/head at each water demand sector and particular control points. One possible approach to consider the pressure/head model in the flow-based MPC is presented in [11]. In this work, the nonlinear constraints coming from the flow–head equations are used to update the operational constraints of tanks and actuators by solving a constraint satisfaction problem (CSP) (see [11]), to replace pressure constraints by bounds on flows, before the flow-based linear MPC problem is solved.

This chapter presents a two-layer control scheme, including a nonlinear MPC (NMPC) strategy and a pumping scheduling approach. The NMPC strategy is implemented by solving a nonlinear optimization problem in order to generate flow and pressure set points for the regulatory controllers to transport water from the sources to the consumer areas to meet future demands, optimizing performance indexes associated with operational goals such as economic costs, network safety volumes/pressures and flow/pressure smooth control operations among others. The proposed NMPC strategy is applied to a high-fidelity simulator model of a portion of Barcelona WDN and compared with the current experience-based operational strategies.

13.2 Problem Statement

13.2.1 Control-Oriented Model Including Pressure

The flow equations of WDNs are identical to those described in Chap. 12. However, in some particular configurations (e.g., networks with loops or bidirectional flows depending of the difference of head), these equations are not enough to represent the hydraulics of WDNs, so that additional flow–head equations are required.

As in the case of the flow model presented in previous Chap. 12, the flow–head model can be obtained by the composition of the set of constitutive element models, which are, respectively, presented and discussed below.

Water Storage Tanks

The so-called head at a node in the WDNs is the height to which water would rise in an open-ended vertical pipe installed at the considered point [1]. The head related to the m-th tank with respect to the volume of water stored in the tank, v_m, can be presented as [11]:

$$h_m(k) = \frac{v_m(k)}{S_m} + E_m, \tag{13.1}$$

where v_m is the volume of the m-th tank S_m is the cross-sectional area of the m-th tank and E_m corresponds the m-th tank elevation.

Pipes

Pipes convey water from one place in the network to another. In a pipe, water flow passes directly from one side with a higher head to the other with a lower head. Therefore, the head–flow relationship for a pipe can be described as

$$q_{i,j} = \Phi_{i,j}(h_i - h_j), \tag{13.2}$$

where $\Phi_{i,j}$ is one of the formulas. Typically, the *Hazen-Williams* formula is chosen (see Chap. 3). Hence, the headdrop through a pipe can be calculated as

$$h_i - h_j = R_{i,j} q_{i,j} \left| q_{i,j} \right|^{0.852}, \tag{13.3}$$

where

$$R_{i,j} = \frac{10.67 L_{i,j}}{C_{i,j}^{1.852} D_{i,j}^{4.87}}, \tag{13.4}$$

where $L_{i,j}$, $D_{i,j}$ and $C_{i,j}$ denote the pipe length, diameter and roughness coefficient, respectively.

Actuators

For the nonlinear optimal control problem, the flows across pumps and valves are the control variables. These flows are considered as continuous variables in a range of admissible values, but certain constraints apply to the related head values.

For pumping stations, the following constraints for heads are set as follows:

$$\Delta h_p = h_d - h_s \geq 0, \tag{13.5a}$$

$$h_d \in \left[h_d^{min}, h_d^{max}\right], \tag{13.5b}$$

$$h_s \in \left[h_s^{min}, h_s^{max}\right], \tag{13.5c}$$

where h_d and h_s represent the *suction head* and the *delivery head* of pumping stations, respectively. h_d^{min} and h_s^{min} denote the minima of the suction and delivery heads. h_d^{max} and h_s^{max} denote the maxima of the suction and delivery heads.

For valves, the following constraints for heads are set as follows:

$$\Delta h_v = h_{us} - h_{ds} \geq 0, \tag{13.6a}$$

$$h_{us} \in \left[h_{us}^{min}, h_{us}^{max}\right], \tag{13.6b}$$

$$h_{ds} \in \left[h_{ds}^{min}, h_{ds}^{max}\right], \tag{13.6c}$$

where h_{us} and h_{ds} denote heads of valves in the upstream and downstream, respectively. h_{us}^{min} and h_{ds}^{min} denote the minima of the upstream and downstream heads. h_{us}^{max} and h_{ds}^{max} denote the maxima of the upstream and downstream heads.

13.2.2 Nonlinear MPC Strategy

WDN Mathematical Model

Considering the modelling methodology of each component in the WDN presented above, the control-oriented model of WDNs can be formulated by means of a set of differential algebraic equations (DAEs). The generalized discrete-time DAE model can be written as follows [14]:

$$\mathbf{x}(k+1) = \mathbf{f}\Big(\mathbf{x}(k), \mathbf{z}(k), \mathbf{u}(k), \mathbf{w}(k), \mathbf{d}(k)\Big), \tag{13.7a}$$

$$0 = \mathbf{g}\Big(\mathbf{x}(k), \mathbf{z}(k), \mathbf{u}(k), \mathbf{w}(k), \mathbf{d}(k)\Big), \tag{13.7b}$$

where $\mathbf{x} \in \mathcal{X} \subseteq \mathbb{R}^m$ represents the vector of system states, $\mathbf{z} \in \mathcal{Z} \subseteq \mathbb{R}^z$ represents the vector of algebraic states, $\mathbf{u} \in \mathcal{U} \subseteq \mathbb{R}^n$ denotes the vector of manipulated variables, $\mathbf{w} \in \mathcal{W} \subseteq \mathbb{R}^w$ denotes the vector of non-manipulated variables and $\mathbf{d} \in \mathcal{D} \subseteq \mathbb{R}^d$ corresponds to the vector of system disturbances. $k \in \mathbb{N}$ denotes the time instant. $\mathbf{f}(\cdot)$ and $\mathbf{g}(\cdot)$ are vectors of mapping functions. Moreover, (13.7a) is the discrete-time differential equation describing the system dynamics, while (13.7b) is the discrete-time algebraic equation presenting the static relations of components in the WDN. All the considered variables of the WDN in Sect. 13.2.1 are assigned to control-oriented variables as summarized in Table 13.1. Considering that the tanks are the only elements with dynamics in the WDN, (13.7a) can be explicitly expressed as

$$\mathbf{x}(k+1) = \mathbf{A}x(k) + \mathbf{B}_u u(k) + \mathbf{B}_w w(k) + \mathbf{B}_d d(k), \tag{13.8}$$

where $\mathbf{x}(k)$ represents the vector of hydraulic heads at storage tanks as system states at time instant k, $\mathbf{u}(k)$ corresponds to the vector of the manipulated flows through

Table 13.1 Variable assignments in the control-oriented model of the WDN

Type of variable	Related symbols	Description
System states: $\mathbf{x}$	h_m	Heads at the storage nodes (i.e., storage tanks)
Pseudostates: $\mathbf{z}$	h_d, h_s, h_i, h_j	Heads at the non-storage nodes
Control inputs: $\mathbf{u}$	q_{u_n}	Manipulated flows through actuators (pumps and valves)
Non-control inputs: $\mathbf{w}$	$q_{i,j}$	Non-manipulated flows through pipes
System disturbances: $\mathbf{d}$	d_r	Water demands

actuators (pumps and valves) as control inputs at time instant k, while $\mathbf{w}(k)$ denotes the vector of non-manipulated flows through pipes at time instant k. $\mathbf{d}(k)$ corresponds to the vector of water demands as system disturbances at time instant k. Moreover, $\mathbf{A}$, $\mathbf{B}_u$, $\mathbf{B}_w$ and $\mathbf{B}_d$ are system matrices of appropriate dimensions.

Furthermore, considering the static relations related to flow and head variables in nodes and pipes, the static equation (13.7b) can be explicitly reformulated as follows:

$$0 = \mathbf{E}_u\mathbf{u}(k) + \mathbf{E}_w\mathbf{w}(k) + \mathbf{E}_d\mathbf{d}(k), \tag{13.9a}$$

$$0 = \mathbf{P}_x\mathbf{x}(k) + \mathbf{P}_z\mathbf{z}(k) + \psi\Big(\mathbf{u}(k), \mathbf{w}(k)\Big), \tag{13.9b}$$

where (13.9a) describes the mass balance equations at nodes in the WDN, and (13.9b) expresses the static relations associated with a particular component (e.g., pipe) of the WDN. $\mathbf{E}_u$, $\mathbf{E}_w$, $\mathbf{E}_d$, $\mathbf{P}_x$ and $\mathbf{P}_z$ are system matrices of appropriate dimensions determined by the network topology. Moreover, ψ denotes the vector of nonlinear *Hazen-Williams* mapping functions (13.3).

In general, the control-oriented model of the WDN can be written as

$$\mathbf{x}(k+1) = \mathbf{A}\mathbf{x}(k) + \boldsymbol{\Xi}\boldsymbol{\mu}(k) + \mathbf{B}_d\mathbf{d}(k), \tag{13.10a}$$

$$0 = \boldsymbol{\Theta}\boldsymbol{\mu}(k) + \mathbf{E}_d\mathbf{d}(k), \tag{13.10b}$$

$$0 = \boldsymbol{\Pi}\boldsymbol{\tau}(k) + \psi\Big(\boldsymbol{\mu}(k)\Big), \tag{13.10c}$$

where

$$\begin{aligned}
\boldsymbol{\Xi} &\triangleq \begin{bmatrix}\mathbf{B}_u & \mathbf{B}_w\end{bmatrix}, \\
\boldsymbol{\Theta} &\triangleq \begin{bmatrix}\mathbf{E}_u & \mathbf{E}_w\end{bmatrix}, \\
\boldsymbol{\Pi} &\triangleq \begin{bmatrix}\mathbf{P}_z & \mathbf{P}_x\end{bmatrix}, \\
\boldsymbol{\mu}(k) &\triangleq \begin{bmatrix}\mathbf{u}(k)^T & \mathbf{w}(k)^T\end{bmatrix}^T, \\
\boldsymbol{\tau}(k) &\triangleq \begin{bmatrix}\mathbf{x}(k)^T & \mathbf{z}(k)^T\end{bmatrix}^T.
\end{aligned}$$

Cost Function Settings

As in Chap. 12, the control objectives for the management of WDNs are considered as follows:

1. *Economic*: To provide a reliable water supply with the required pressure minimizing both water production and transport costs.
2. *Safety*: To guarantee the availability of enough water in each storage tank to satisfy its underlying stochastic water demands.
3. *Smoothness*: To operate the WDN under smooth control actions.

The reader is referred to Chap. 12 for the mathematical expressions of these objectives.

System Constraint Settings

Some of the constraints in MPC of water distribution networks correspond to flow balance equations. These are the same as in Chap. 12 and must also hold here. In addition, some additional constraints related to the heads in the network are required. The heads at some certain non-storage nodes are required to be up to some minimum levels as in the case of the water demand sectors. Hence, the following inequality constraint is necessary to be considered:

$$z_i(k) \geq z_i^{\min}, \quad \text{for all } k \text{ and } i \in [1, \ldots, n_h], \tag{13.11}$$

where $z_i^{\min}$ are the required heads at the water demand sectors. Moreover, n_h is the total number of the water demand sectors.

13.2.3 NMPC Formulation

Taking into account the WDN model including the flow and pressure part, a NMPC strategy can be implemented by solving a finite-horizon optimization problem over a prediction horizon H_p. The multi-objective cost function, including the objectives described in Sect. 13.2.2, is minimized subject to the prediction model (13.10), and a set of system constraints is presented above. Thus, the optimization problem behind the NMPC strategy can be formulated as follows:

Problem 13.1 (*NMPC for WDN*)

$$\min_{u^*(k|k),\ldots,u^*(k+H_p-1|k)} J = \sum_{i=0}^{H_p-1} \sum_{j=1}^{\Gamma} \lambda_j J_j\Big(\mathbf{u}(k+i \mid k), \boldsymbol{\xi}(k+i)\Big), \tag{13.12a}$$

subject to

$$\mathbf{x}(k+i+1 \mid k) = \mathbf{A}\mathbf{x}(k+i \mid k) + \boldsymbol{\Xi}\boldsymbol{\mu}(k+i \mid k) + \mathbf{B}_d\mathbf{d}(k+i \mid k), \tag{13.12b}$$

$$0 = \boldsymbol{\Theta}\boldsymbol{\mu}(k+i \mid k) + \mathbf{E}_d\mathbf{d}(k+i \mid k), \tag{13.12c}$$

$$0 = \boldsymbol{\Pi}\boldsymbol{\tau}(k+i \mid k) + \psi\Big(\boldsymbol{\mu}(k+i \mid k)\Big), \tag{13.12d}$$

$$\mathbf{x}^{\min} \leq \mathbf{x}(k+i+1 \mid k) \leq \mathbf{x}^{\max}, \tag{13.12e}$$

$$\mathbf{u}^{\min} \leq \mathbf{u}(k+i \mid k) \leq \mathbf{u}^{\max}, \tag{13.12f}$$

$$\mathbf{z}(k+i \mid k) \geq \mathbf{z}^{min} \tag{13.12g}$$

$$\mathbf{x}(k+i+1 \mid k) \geq \mathbf{x}_s - \boldsymbol{\xi}(k+i), \tag{13.12h}$$

$$\Big(\mathbf{x}(k|k), \mathbf{d}(k|k)\Big) = \Big(\mathbf{x}(k), \mathbf{d}(k)\Big). \tag{13.12i}$$

Since the control-oriented model of the WDN includes the nonlinear relations because of the flow–head equations, the above optimization problem naturally becomes nonlinear. Thus, Problem 13.1 should be solved using nonlinear programming techniques. Assuming that Problem 13.1 is feasible, the sequence of control actions can be expressed as

$$\mathbf{u}^*(k) = \left[\mathbf{u}^*(k \mid k),\, \mathbf{u}^*(k+1 \mid k),\, \ldots,\, \mathbf{u}^*(k+H_p-1 \mid k)\right]. \tag{13.13}$$

And, then by applying the receding-horizon strategy, the optimal control action at time instant k is the first component of the sequence of control actions denoted by

$$\mathbf{u}_{opt}(k) \triangleq \mathbf{u}^*(k \mid k). \tag{13.14}$$

13.3 Proposed Solution

In practice, most of the pumps in WDN are operated in ON–OFF discrete way. Thus, the flows in (13.13) become discrete values and subsequently Problem 13.1 becomes a nonlinear mixed-integer problem. In this section, this nonlinear mixed-integer optimization problem is firstly introduced. Then, in order to avoid solving this complicated optimization problem, a two-layer control strategy is proposed including the NMPC strategy presented above, and the pump scheduling approach is described in the following of this section.

13.3.1 *Mixed-Integer NMPC Including Discrete ON/OFF Pump Scheduling*

Considering that pumps are operated in an ON/OFF discrete manner, Problem 13.1 should be reformulated. By introducing new discrete decision variables that correspond to the ON/OFF operation of the pumps, the mixed-integer NMPC strategy can be implemented by reformulating Problem 13.1 into the following optimization problem:

Problem 13.2 (*Mixed-integer NMPC for WDN including ON/OFF pumps*)

$$\min_{\zeta^*(k|k),\ldots,\zeta^*(k+H_p-1|k)} J = \sum_{i=0}^{H_p-1} \sum_{j=1}^{\Gamma} \lambda_j J_j\Big(\zeta(k+i \mid k), \boldsymbol{\xi}(k+i)\Big), \tag{13.15a}$$

subject to

$$\mathbf{x}(k+i+1 \mid k) = \mathbf{A}\mathbf{x}(k+i \mid k) + \Xi\bar{\mu}(k+i \mid k) + \mathbf{B}_d\mathbf{d}(k+i \mid k), \tag{13.15b}$$

$$0 = \mathbf{\Theta}\bar{\mu}(k+i \mid k) + \mathbf{E}_d\mathbf{d}(k+i \mid k), \quad (13.15c)$$

$$0 = \mathbf{\Pi}\tau(k+i \mid k) + \psi\Big(\bar{\mu}(k+i \mid k)\Big), \quad (13.15d)$$

$$\mathbf{x}^{\min} \leq \mathbf{x}(k+i+1 \mid k) \leq \mathbf{x}^{\max}, \quad (13.15e)$$

$$\mathbf{u}^{\min} \leq \boldsymbol{\zeta}(k+i \mid k)\tilde{\mathbf{u}} \leq \mathbf{u}^{\max}, \quad (13.15f)$$

$$\mathbf{z}(k+i \mid k) \geq \mathbf{z}^{min} \quad (13.15g)$$

$$\mathbf{x}(k+i+1 \mid k) \geq \mathbf{x}_s - \boldsymbol{\xi}(k+i), \quad (13.15h)$$

$$\boldsymbol{\zeta}(k+i) \in [0, 1], \quad (13.15i)$$

$$\Big(\mathbf{x}(k|k), \mathbf{d}(k|k)\Big) = \Big(\mathbf{x}(k), \mathbf{d}(k)\Big), \quad (13.15j)$$

where $\boldsymbol{\zeta}(k+i) \in \{0, 1\}$ represents the vector of binary decision variables that correspond to the ON/OFF pump status at time instant k, where 1 means the ON status of the pump and 0 means the OFF status of the pumps. The ON status pumping flow $\tilde{\mathbf{u}}$ should be estimated based on the pressure conditions that the pumps should establish. Problem 13.1 is considered to operate at hourly basis, which means that optimal hourly flows for actuators are determined. In Problem 13.2, the sampling time should be shorter (at the minutely scale) in order to approximate the optimal continuous flow that would be obtained by solving Problem 13.1 assuming that actuator flows are manipulated in a continuous manner.

Solving a large-scale nonlinear mixed-integer optimization problem with a larger horizon leads Problem 13.1 to be prohibitive from the computational point of view. For this reason, Problem 13.2 is proposed to be replaced by a two-layer optimal control strategy.

13.3.2 Two-Layer Optimal Control Strategy

The two-layer optimal control strategy is presented in Fig. 13.1. In the following, the description of such strategy is provided.

The Upper Layer: NMPC Strategy
In the upper layer, Problem 13.1 is adopted. Note that the NMPC of the WDN is operated on the hourly basis ($\Delta t_u = 3600$ s) over the MPC prediction horizon H_p to find the optimal flow set points. In the lower layer, the sampling time is usually selected as one minute ($\Delta t_l = 60$ s), which means the optimal flow set points at one MPC step decides 60 values using the pump scheduling approach.

Pump Scheduling Approach
In fact, the pumping flow of the i-th pump in the j-th pumping station has been affected by the factors of the suction and delivery heads. Hence, if the heads are given, the real-time pumping flow is located within an interval, which can be formulated as

$$q_{i,j}^{real} \in \left[q_{i,j}^{nom} - \sigma_{i,j},\ \ q_{i,j}^{nom} + \sigma_{i,j}\right], \quad (13.16)$$

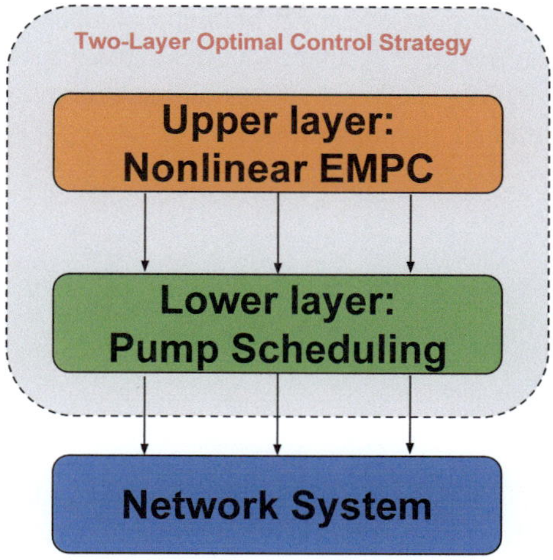

Fig. 13.1 Two-layer optimal control strategy of WDNs

where $q_{i,j}^{nom}$ denotes nominal pumping flows produced through the pumps, and $\sigma_{i,j}$ represents the variance of the pumping flows depending on the uncertainty of the heads in terms of the pumps.

If the actual pumping flow $q_{i,j}^{real}$ is approximated by the nominal pumping flow $q_{i,j}^{nom}$, then the total ON status time can be calculated by

$$T_{i,j}^{ON} = \frac{Q_j^{opt}\Delta t_u}{q_{i,j}^{nom}\Delta t_l}, \tag{13.17}$$

where Q_j^{opt} denotes the optimal hourly flow as the set point of the j-th pumping station and $T_{i,j}^{ON}$ denotes the total ON status time of the pump. In some cases, only one pump cannot provide enough flow to maintain the optimal flow set point. Hence, parallel pumps are set in each pumping station. Therefore, (13.17) can be rewritten as

$$Q_j^{opt}\Delta t_u = \sum_{i=1}^{\chi_j} T_{i,j}^{ON} q_{i,j}^{nom}\Delta t_l, \tag{13.18}$$

where $\chi_j \in [0, \gamma_j] \subset \mathbb{Z}^+$ denotes the number of the opened parallel pumps and γ_j is the total number of the parallel pumps located in the j-th pumping station. Note that parallel pumps are assumed to be opened following a prioritization.

The pump scheduling approach is able to find a sequence value of the number of the ON-status parallel pumps in each pumping station at each sampling time. Hence, the variable χ_j is naturally considered as an admissible time series vector, which can be written as

$$\chi_j = \left[\chi_j(0), \chi_j(1), \ldots, \chi_j(t)\right]^T, \tag{13.19}$$

where $\chi_j(t)$ denotes the number of ON-status parallel pumps at time instant t.

Since q^{real} is unknown prior to run the simulator, $\chi_j(0)$ can be computed in order to identify whether the total nominal pumping flow in j-th pumping station is enough to fulfil the optimal flow found by the upper layer, which can be formulated as follows:

$$\sum_{i=1}^{\chi_j(0)-1} q_{i,j}^{nom} < Q_j^{opt}, \tag{13.20a}$$

$$\sum_{i=1}^{\chi_j(0)} q_{i,j}^{nom} \geq Q_j^{opt}. \tag{13.20b}$$

Considering (13.20), the value of $\chi_j(0)$ at initial time can be determined for the j-th pumping station. Then, the horizon of the pump schedule is within 1 h on the minutely basis. The measured error of the pumping water volume δ_j for the j-th pumping station can be formulated as

$$\delta_j \triangleq V_j^p - V_j^{opt}, \tag{13.21}$$

where V_j^p and V_j^{opt} denote the pumping water volume and optimal water volume for the j-th pumping station, respectively, which are defined as follows:

$$V_j^p \triangleq \int_0^{\Delta t_u} Q_j^{real} dt = \sum_{i=1}^{\chi_j(0)} q_{i,j}^{real} \Delta t_l + \sum_{z=1}^{H_l-1} \sum_{i=1}^{\chi_j(t)} q_{i,j}^{real} \Delta t_l, \tag{13.22a}$$

$$V_j^{opt} \triangleq \int_0^{\Delta t_u} Q_j^{opt} dt = Q_j^{opt} \Delta t_u, \tag{13.22b}$$

where Q_j^{real} denotes the total flow of the j-th pumping station, H_l denotes the calculation horizon for χ_j, which is equation to 60s in this layer. As discussed before, the vector of χ_j along the horizon of H_l can be obtained by solving an optimization problem as follows.

Problem 13.3 (*Pump scheduling approach of the WDN*)

$$\min_{\chi_1^*(t),\ldots,\chi_\Lambda^*(t)} J = \sum_{j=1}^{\Lambda} \|\delta_j\|_2^2, \tag{13.23a}$$

subject to

$$\delta_j = V_j^p - V_j^{opt}, \tag{13.23b}$$

$$V_j^p = \sum_{i=1}^{\chi_j(0)} q_{i,j}^{real} \Delta t_l + \sum_{t=1}^{H_l-1} \sum_{i=1}^{\chi_j(t)} q_{i,j}^{real} \Delta t_l, \tag{13.23c}$$

$$V_j^{opt} = Q_j^{opt}\Delta t_u, \tag{13.23d}$$

$$0 \le \chi_j(t) \le \gamma_j, \tag{13.23e}$$

$$\sum_{i=1}^{\chi_j(0)-1} q_{i,j}^{nom} < Q_j^{opt}, \tag{13.23f}$$

$$\sum_{i=1}^{\chi_j(0)} q_{i,j}^{nom} \ge Q_j^{opt}, \tag{13.23g}$$

$$t \in [1, H_l - 1] \subset \mathbb{Z}^+ \tag{13.23h}$$

where Λ denotes the total number of the pumping stations. Assuming that the Problem 13.3 is also feasible, the optimal sequence of decision variables can be written as

$$\chi^* = \left[\chi_1^{*T}, \chi_2^{*T}, \ldots, \chi_\Lambda^{*\,T}\right]^T, \tag{13.24}$$

where

$$\chi_j^* = \left[\chi_j(0), \chi_j^*(1), \ldots, \chi_j^*(H_l - 1)\right]^T, \quad \text{for all } j \in [1, \Lambda].$$

13.4 Simulation Results

13.4.1 Case Study Description

In order to assess the validity of the proposed approach, a part of the Barcelona transport network (presented in Figs. 2.2, 2.3 and 2.4 of Chap. 2) is used. A “virtual reality” setup has been developed because of the complexity of directly testing on the real network. The selected hydraulic simulation software is *PICCOLO* (developed by Suez company) instead of EPANET, because it is the tool in use in the complete Barcelona WDN and because it has specific advanced functionalities that apply to this case, e.g., in pump and valve modelling.

The simulation online environment used for illustrating the approach proposed in this chapter consists of the *GAMS* model (see [2, 3]) with the third-party software *CONOPT3* as a nonlinear solver, the *PICCOLO* and MATLAB, that is used for the communication between *GAMS* and *PICCOLO* models. The topological graph of the communication is presented in Fig. 13.2. The database (DB) includes the water demands data and electrical tariff data. At each time instant along the simulation scenarios, the data will be selected by means of MATLAB, and then these data will be transferred into *GAMS* and *PICCOLO* models through the MATLAB communication interface.

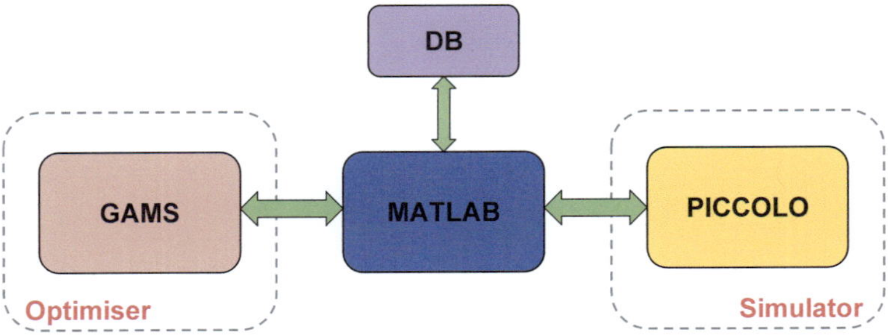

Fig. 13.2 On-line simulation pilot

13.4.2 Results

The obtained control strategies produced by the NMPC meet demands and operational constraints at all times, while optimizing the operational goals when applied to the high-fidelity simulator. Some illustrative results are presented in the following compared with the current heuristic operation obtained from Aguas de Barcelona company that is not optimized. In this chapter, the simulations results are shown by applying the proposed control strategy with the high-fidelity simulator as the virtual real network. To test the NMPC controller implemented, real data corresponding to the period between November 11 and December 2nd, 2013 is used. To show the differences between the current heuristic control and the NMPC control, some tank volume and actuator flow evolutions are shown. In all the graphs, the curves present the current operation (real data) and the result of application of the NMPC strategy in the high-fidelity network simulator (simulation data). In Fig. 13.3, some tank volume evolutions can be seen, as well as maximum and safety volumes. From this figure, it can be observed that the NMPC presents a more regular 24-h cycle than the current control. The tanks are filled at the night when the electricity is cheaper and emptied during the day when the electricity is expensive. In this way, pumps are mainly used when the electricity is cheaper optimizing the performance of the network. The optimal solutions of a selected number of representative pumps and the valves are presented in Figs. 13.4 and 13.5.

Three tests scenarios were chosen based on real data. Each scenario lasts for 120 h (5 days), and the three scenarios span the period from December 12 to December 27, 2013. The obtained control strategies produced by the NMPC and simulated with PICCOLO meet demands and operational constraints at all times. When compared with the actual strategies applied on the period, the NMPC strategy shows significant reductions in energy use, as well as in energy cost, as shown in Table 13.2. Furthermore, in two of the three scenarios, it was also possible to reduce the total amount of water use to meet the same demands.

In Table 13.2, costs achieved with the proposed NMPC when applied to the high-fidelity simulator are compared with those obtained from the current heuristic control strategy. The electrical energy and an associated cost are presented as well as the total

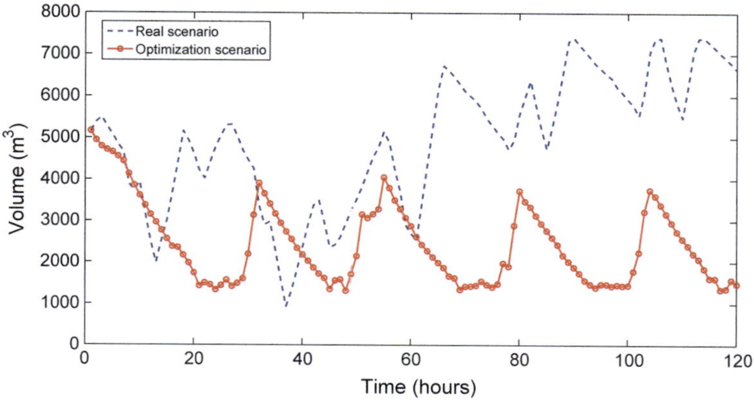

(a) Tank: Finest200

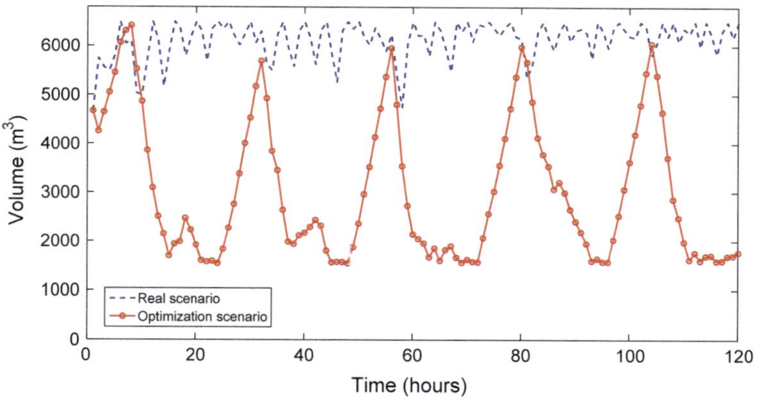

(b) Tank: TrinitatTer200

Fig. 13.3 Results of selected tanks in Barcelona WDN

accumulated energy and electrical cost in the considered days. The cost is given in economical units (e.u.) because of confidential reasons. The electrical cost obtained with the NMPC represents an improvement of almost 19% with respect to the current control in the time period considered. Taking into account the unitary cost of the water in e.u./m^3, the improvement with respect the current heuristic control is between 9 and 15%.

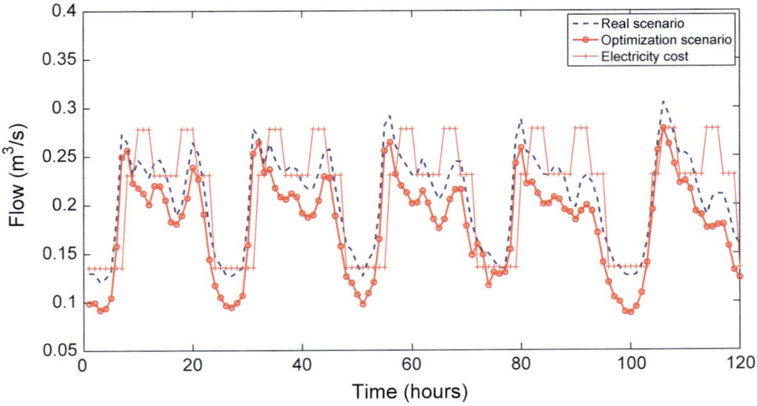

(a) Pump: EspluguesG1234

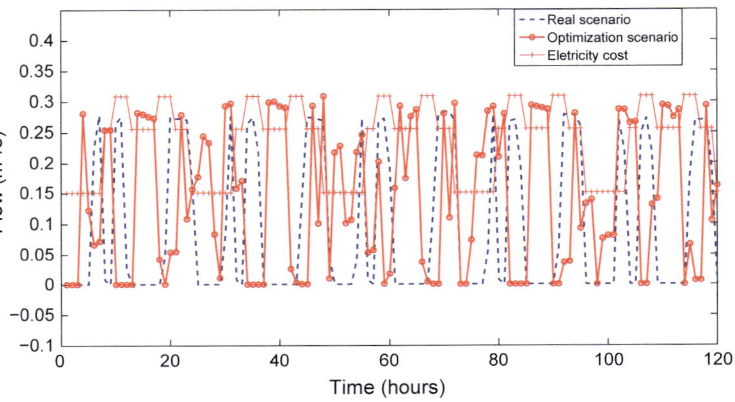

(b) Pump: AlturesG23

Fig. 13.4 Results of selected pumps in Barcelona WDN

13.5 Conclusions

In this chapter, a two-layer NMPC control strategy for WDNs considering both flow and pressure models is proposed. The optimal set points for actuators have been calculated by means of solving a nonlinear optimization problem and subsequently used for deploying the pump scheduling approach considering the ON/OFF operation of the pumps. Throughout a case study based on a part of the Barcelona water transport network, the online simulation results show the feasibility of the proposed control strategy and its economic cost improvement. Compared with current heuristic control strategies, the NMPC strategy of the WDN is able to meet all the demands in the

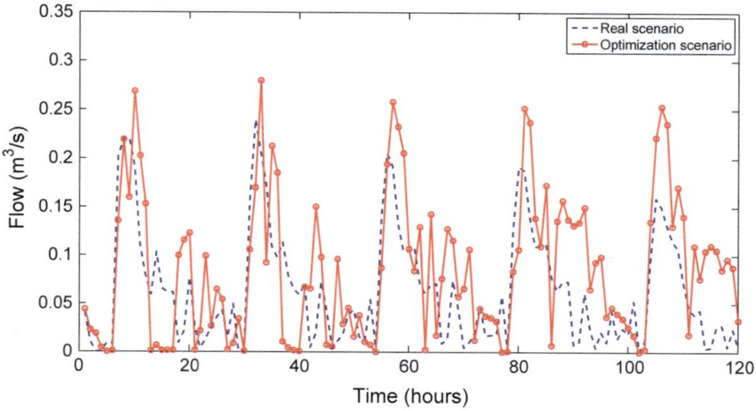

(a) Valve: Bonanova

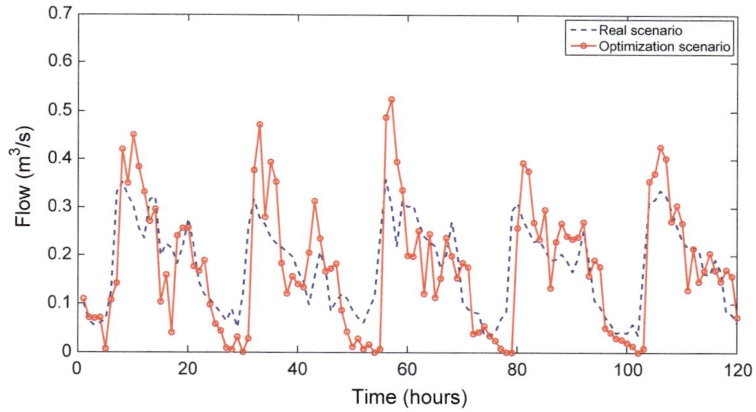

(b) Valve: MinaCiutat

Fig. 13.5 Results of selected valves in Barcelona WDN

considered sectors with their required pressure at the same time at a reduced cost. Hence, the proposed NMPC strategy is considered an adequate control strategy for the operational management of the WDN including both flow and pressure constraints. Furthermore, the lower layer determines a sequence of ON-OFF operations for the pumping stations. For pumps, a logic controller should be used for implementing the strategy determined by the pump scheduler.

Table 13.2 Comparison costs with considered scenarios

Simulation period: 12/12/2013 to 27/12/2013

		Scenario 1 12/12/2013–17/12/2013	Scenario 2 17/12/2013–22/12/2013	Scenario 3 22/12/2013–27/12/2013	Total	Improvement (%)
Total energy consumption (K · Wh)	Real	486	495	440	1422	7.37
	NMPC	473	437	408	1318	
Electrical cost (e.u.)	Real	4659	4699	4280	13640	18.97
	NMPC	3985	3642	3425	11052	
Unitary improvement (e.u./m^3)		9.62%	15.27%	13.74%	12.83%	

References

1. Brdys M, Ulanicki B (1994) Operational control of water systems: structures. Algorithms and applications. Prentice-Hall, Upper Saddle River
2. Brooke A, Kendrick D, Meeraus A, Raman R (2004) GAMS. A user's guide. GAMS Development Corporation, Washington DC, USA
3. Cembrano G, Quevedo J, Puig V, Pérez R, Figueras J, Verdejo JM, Escaler I, Ramón G, Barnet G, Rodríguez P, Casas M (2011) PLIO: a generic tool for real-time operational predictive optimal control of water networks. Water Sci Technol 64(2): 448–459
4. Cembrano G, Wells G, Quevedo J, Perez R, Argelaguet R (2000) Optimal control of a water distribution network in a supervisory control system. Control Eng Pract 8(10): 1177–1188
5. Grosso JM, Ocampo-Martinez C, Puig V, Joseph B (2014) Chance-constrained model predictive control for drinking water networks. J Process Control 24(5): 504–516
6. Grosso JM, Ocampo-Martinez C, Puig V, Limon D, Pereira M (2014) Economic MPC for the management of drinking water networks. In: 2014 European control conference, Strasburg, France, pp 790–795
7. Ocampo-Martinez C, Ingimundarson A, Puig V, Quevedo J (2008) Objective prioritization using lexicographic minimizers for MPCc of sewer networks. IEEE Trans Control Syst Technol 16(1): 113–121
8. Ocampo-Martinez C, Puig V, Cembrano G, Quevedo J (2013) Application of MPC strategies to the management of complex networks of the urban water cycle. IEEE Control Syst Mag 33(1): 15–41
9. Rawlings JB, Mayne DQ (2009) Model predictive control: theory and design. Wis. Nob Hill Pub. cop, Madison
10. Streif S, Kogel M, Bathge T, Findeisen R (2014) Robust nonlinear model predictive control with constraint satisfaction: a relaxation-based approach. In: 19th IFAC world congress, Cape Town, South Africa, pp 11073–11079
11. Sun CC, Puig V, Cembrano G (2016) Combining CSP and MPC for the operational control of water networks. Eng Appl Artif Intell 49: 126–140
12. Wang Y, Ocampo-Martinez C, Puig V (2015) Robust model predictive control based on Gaussian processes: application to drinking water networks. In: 2015 European control conference, Linz, Austria, pp 3292–3297
13. Wang Y, Ocampo-Martinez C, Puig V (2016) Stochastic model predictive control based on Gaussian processes applied to drinking water networks. IET Control Theory Appl 10(8): 947–955

14. Wang Y, Puig V, Cembrano G, Economic MPC with periodic terminal constraints of nonlinear differential-algebraic-equation systems: application to drinking water networks. In: European control conference, Aalborg, Denmark, pp. 1013–1018
15. Wang Y (1), Cembrano G (1), Puig V (2), Urrea M (1), Romera J (1), Saporta D (3), Valero jG (3) (1)Institut de Robotica i Informatica Industrial, CSIC-UPC, Barcelona, Spain (2) Research Center "Monitoring, Safety and Automatic Control" (CS2AC-UPC), Terrassa, Spain (3) Aguas de Barcelona (AGBAR), Barcelona, Spain

Chapter 14
Stochastic Model Predictive Control for Water Transport Networks with Demand Forecast Uncertainty

Juan Manuel Grosso, Carlos Ocampo-Martínez and Vicenç Puig

14.1 Introduction

One of the challenges in water system management is the existence of different sources of uncertainty. The availability of historical data allows to accurately predict the behaviour of the system disturbances over large horizons, but still a meaningful degree of uncertainty is present. In previous chapters, the use of MPC to tackle the complex multi-variable interactions and large-scale nature of drinking-water network control is proposed. There are several examples of MPC applied to water systems in the literature, see, e.g., [2, 7, 10, 16, 29, 30] and references therein.

In a DWN, the main management purpose is the achievement of the highest level of consumer satisfaction and service quality in line with the prevailing regulatory framework, while making best use of available resources. Hence, networks must be reliable and resilient while being subject to constraints and to continuously varying conditions with both deterministic and probabilistic nature. Customer behaviour determines the transport and storage operations within the network, and flow demands can vary in both the long and the short term, often presenting time-based patterns in some applications. Therefore, a better understanding and forecasting of demands will improve both modelling and control of DWNs.

While Chaps. 12 and 13 deal with the uncertainty in the classical way of feed-forward action, this chapter focuses on the way that uncertainty can be faced by using stochastic-based approaches. The simplest way to do this is by ignoring the explicit influence of disturbances or using their expected value as done in the previous chapters. However, dealing with the demand uncertainty explicitly in the control model is expected to produce more robust control strategies. In [12], a reliability-based

J.M. Grosso · C. Ocampo-Martínez (✉)
Institut de Robòtica i Informàtica Industrial, CSIC-UPC, Barcelona, Spain
email: cocampo@iri.upc.edu

V. Puig
Research Center "Supervision, Safety and Automatic Control" (CS2AC-UPC), Terrassa, Spain

V. Puig et al. (eds.), *Real-Time Monitoring and Operational Control of Drinking-Water Systems*, Advances in Industrial Control,
DOI 10.1007/978-3-319-50751-4_14

MPC was proposed to handle demand uncertainty by means of a (heuristic) safety stock allocation policy, which takes into account the short-term demand predictions but without propagating uncertainty along the prediction horizon. As discussed in [5], alternative approaches of MPC for stochastic systems are based on *min-max* MPC, *tube-based* MPC and *stochastic* MPC. The first two consider disturbances to be unmeasured but bounded in a predefined set. The control strategies are conservative, because they consider worst-case demand deviations from their expected value, limiting the control performance. On the other hand, stochastic MPC considers a more realistic description of uncertainty, which leads to less conservative control approaches at the expense of a more complex modelling of the disturbances. The stochastic approach is a mature theory in the field of optimization [3], but a renewed attention has been given to the stochastic programming techniques as powerful tools for control design (see, e.g., [4] and references therein).

From the wide range of stochastic MPC methods, this chapter specializes on *scenario tree-based MPC* (TB-MPC) and *chance-constrained MPC* (CC-MPC). Regarding TB-MPC, see, e.g., [17, 24], uncertainty is addressed by considering simultaneously a set of possible disturbance scenarios modelled as a rooted tree, which branches along the prediction horizon. On the other hand, CC-MPC [28] is a stochastic control strategy that describes robustness in terms of probabilistic (chance) constraints, which require that the probability of violation of any operational requirement or physical constraint is below a prescribed value, representing the notion of reliability or risk of the system. By setting this value properly, the operator/user can trade conservatism against performance. Relevant works that address the CC-MPC approach in water systems can be found in [8, 22] and references therein. Therefore, this chapter is focused on the design and assessment of CC-MPC and TB-MPC controllers for the operational management of transport water networks, which may be described using only flow equations, discussing their advantages and weaknesses in the sense of applicability and performance. The particular case study is related to the Barcelona DWN described in [19] and presented in Chap. 2.

14.2 Problem Formulation

Consider the MPC problem associated with the flow control problem in a DWN (see [20]). In general, a DWN consists in a set of water storage (dynamic) nodes, pipe junction (static) nodes, source nodes and sink nodes, which are interconnected in such a way that the water can be transported from source nodes to sink nodes when demanded. In order to derive a control-oriented model, define the state vector $\mathbf{x} \in \mathbb{R}^{n_x}$ to represent the storage at dynamic nodes. Similarly, define the vector $\mathbf{u} \in \mathbb{R}^{n_u}$ of controlled inputs as the collection of the flow rate through the actuators of the network, and the vector $\mathbf{d} \in \mathbb{R}^{n_d}$ of uncontrolled inputs (demands) as the collection of flow rate required by the consumers at sink nodes. Following flow/mass balance principles, a discrete-time model based on linear difference-algebraic equations can be formulated for a given DWN as follows:

$$
\begin{cases} \mathbf{x}(k+1) = \mathbf{A}\,\mathbf{x}(k) + \mathbf{B}\mathbf{u}(k) + \mathbf{B}_d\mathbf{d}(k), & (14.1a) \\ 0 = \mathbf{E}_u\mathbf{u}(k) + \mathbf{E}_d\mathbf{d}(k), & (14.1b) \end{cases}
$$

where $k \in \mathbb{Z}_+$ is the current time instant and $\mathbf{A}$, $\mathbf{B}$, $\mathbf{B}_d$, $\mathbf{E}_u$ and $\mathbf{E}_d$ are matrices of compatible dimensions dictated by the network topology. Specifically, (14.1a) represents the balance at dynamic nodes while (14.1b) represents the balance at static nodes. The system is subject to state and input constraints considered here in the form of convex polyhedra defined as follows:

$$\mathbf{x}(k) \in \mathcal{X} := \{\mathbf{x} \in \mathbb{R}^{n_x} \mid \mathbf{G}\mathbf{x} \leq \mathbf{g}\}, \tag{14.2a}$$

$$\mathbf{u}(k) \in \mathcal{U} := \{\mathbf{u} \in \mathbb{R}^{n_u} \mid \mathbf{H}\mathbf{u} \leq \mathbf{h}\}, \tag{14.2b}$$

for all k, where $\mathbf{G} \in \mathbb{R}^{r_x \times n_x}$, $\mathbf{g} \in \mathbb{R}^{r_x}$, $\mathbf{H} \in \mathbb{R}^{r_u \times n_u}$ and $\mathbf{h} \in \mathbb{R}^{r_u}$ are matrices/vectors collecting the system constraints, being $r_x \in \mathbb{Z}_+$ and $r_u \in \mathbb{Z}_+$ the number of state and input constraints, respectively.

Regarding the operation of the generalized flow-based networks, the following assumptions are considered in this chapter.

Assumption 14.1 The pair (A, B) is controllable and (14.1b) is reachable,[1] i.e., $m \leq n_u$ with $\text{rank}(E_u) = m$.

Assumption 14.2 The states in $\mathbf{x}$ and the demands in $\mathbf{d}$ are measured at any time instant $k \in \mathbb{Z}_+$. Future demands $d(k+i)$ are unknown for all $i \in \mathbb{Z}_+$ but forecasted information of their first two moments (i.e., expected value and variance) is available for a given prediction horizon $H_p \in \mathbb{Z}_{\geq 1}$.

Assumption 14.3 The realization of demands at any time instant $k \in \mathbb{Z}_+$ can be decomposed as

$$\mathbf{d}(k) = \bar{\mathbf{d}}(k) + \mathbf{e}(k), \tag{14.3}$$

where $\bar{\mathbf{d}} \in \mathbb{R}^{n_d}$ is the vector of expected disturbances, and $\mathbf{e} \in \mathbb{R}^{n_d}$ is the vector of forecasting errors with non-stationary uncertainty and a known (or approximated) quasi-concave probability distribution $\mathcal{D}(0, \Sigma(e_{(j)}(k)))$. The stochastic nature of each jth row of $\mathbf{d}_{(k)}$ is described by $\mathbf{d}_{(j)}(k) \sim \mathcal{D}_i(\bar{\mathbf{d}}_{(j)}(k), \Sigma(\mathbf{e}_{(j)}(k)))$, where $\bar{\mathbf{d}}_{(j)}$ denotes its mean, and $\Sigma(\mathbf{e}_{(j)}(k)$ its variance.

Notice in (14.1b) that a subset of controlled flows are directly related with a subset of uncontrolled flows. Hence, it is clear that $\mathbf{u}$ does not take values in $\mathbb{R}^{n_u}$ but in a linear variety. This latter observation, in addition to Assumptions 14.1 and 14.2,

[1] If $m < n_u$, then multiple solutions exist, so $\mathbf{u}$ should be selected by means of an optimization problem. Equation (14.1b) implies the possible existence of uncontrollable flows $\mathbf{d}$ at the junction nodes. Therefore, a subset of the control inputs will be restricted by the domain of some flow demands.

can be exploited to develop an affine parametrization of control variables in terms of a minimum set of disturbances as shown in [12, Appendix A], mapping control problems to an input space with a smaller decision vector and with less computational burden due to the elimination of the equality constraints. Thus, the system (14.1) can be rewritten as follows:

$$\mathbf{x}(k+1) = \mathbf{A}\mathbf{x}(k) + \tilde{\mathbf{B}}\tilde{\mathbf{u}}(k) + \tilde{\mathbf{B}}_d\mathbf{d}(k), \tag{14.4}$$

and the input constraint (14.2b) replaced with a time-varying restricted set defined as follows:

$$\tilde{\mathcal{U}}(k) := \{\tilde{\mathbf{u}} \in \mathbb{R}^{n_u-m} \mid \mathbf{H}\tilde{\mathbf{P}}\tilde{\mathbf{M}}_1\tilde{\mathbf{u}} \leq \mathbf{h} - \mathbf{H}\tilde{\mathbf{P}}\tilde{\mathbf{M}}_2\mathbf{d}(k)\}. \tag{14.5}$$

Being $\tilde{B} = B\tilde{P}\tilde{M}_1$, $\tilde{B}_d = B\tilde{P}\tilde{M}_2 + B_d$, where $\tilde{P}$, $\tilde{M}_1$ and $\tilde{M}_2$ a control parametritzation matrices ([12], Appendix A). The control goal is considered here as to minimize a convex (possibly multi-objective) stage cost $J(k, \mathbf{x}, \tilde{\mathbf{u}}) : \mathbb{Z}_+ \times \mathcal{X} \times \tilde{\mathcal{U}}(k) \to \mathbb{R}_+$, which might bear any functional relationship to the economics of the system operation. Let $\mathbf{x}(k) \in \mathcal{X}$ be the current state, and let $\mathbf{d}(k)$ be the disturbances. The sequence of disturbances should be known over the considered prediction horizon H_p. The first element of this sequence is measured, while the rest of the elements are estimates of future disturbances computed by an exogenous forecasting system and available at each time instant $k \in \mathbb{Z}_+$. Hence, the MPC controller design is based on the solution of the following finite horizon optimization problem (FHOP):

$$\min_{\tilde{\mathbf{u}}(k)=\{\tilde{u}(k+i|k)\}_{i\in\mathbb{Z}_{[0,H_p-1]}}} \sum_{i=0}^{H_p-1} J(k+i, x(k+i|k), \tilde{u}(k+i|k)), \tag{14.6a}$$

subject to:

$$\mathbf{x}^j(k+i+1|k) = \mathbf{A}\mathbf{x}^j(k+i|k) + \tilde{\mathbf{B}}\tilde{\mathbf{u}}^j(k+i|k) + \tilde{\mathbf{B}}_d\mathbf{d}(k+i), \forall i \in \mathbb{Z}_{[0,H_p-1]} \tag{14.6b}$$

$$\mathbf{x}(k+i|k) \in \mathcal{X}, \qquad \forall i \in \mathbb{Z}_{[1,H_p]} \tag{14.6c}$$

$$\tilde{\mathbf{u}}(k+i|k) \in \tilde{\mathcal{U}}(k+i), \qquad \forall i \in \mathbb{Z}_{[0,H_p-1]} \tag{14.6d}$$

$$\mathbf{x}(k|k) = x(k), \tag{14.6e}$$

Assuming that (14.6) is feasible, i.e., there exists a non-empty solution given by the optimal sequence of control inputs $\tilde{\mathbf{u}}_k^\star = \{\tilde{u}^\star(k+i|k)\}_{i\in\mathbb{Z}_{[0,H_p-1]}}$, then the receding horizon philosophy commands to apply the control action

$$\mathbf{u}(k) := \tilde{\mathbf{u}}^*(k|k), \tag{14.7}$$

and disregards the rest of the sequence of the predicted manipulated variables. At the next time instant k, the FHOP (14.6) is solved again using the current measurements of states and disturbances and the most recent forecast of these latter over the next future horizon.

Due to the stochastic nature of future disturbances, the prediction model (14.6b) involves exogenous additive uncertainty, which might cause that the compliance of state constraints for a given control input cannot be ensured. Therefore, uncertainty has to be represented in such a way that their effect on present decision-making can properly be taken into account. To do so, stochastic modelling based on data analysis, probability distributions, disturbance scenarios, among others, and the use of stochastic programming may allow to establish a trade-off between robustness and performance. In the sequel, two stochastic MPC strategies are proposed for their application to network flow control.

14.3 Chance-Constrained MPC

Since the optimal solution to (14.6) does not always imply feasibility of the real system, it is appropriate to relax the original constraints in (14.6c) with probabilistic statements in the form of the so-called *chance constraints*. In this way, the state constraints are required to be satisfied with a predefined probability to manage the reliability of the system. Considering the form of the state constraint set $\mathcal{X}$, there are two types of chance constraints according to the definitions below.

Definition 14.1 (*Joint chance constraint*) A (linear) state *joint chance constraint* is of the form

$$\mathbb{P}[\mathbf{G}_{(j)}\mathbf{x} \leq \mathbf{g}_{(j)}\,,\ \forall j \in \mathbb{Z}_{[1,r_x]}] \geq 1 - \delta_{\mathbf{x}}, \tag{14.8}$$

where $\mathbb{P}$ denotes the probability operator, $\delta_{\mathbf{x}} \in (0, 1)$ is the *risk acceptability level* of constraint violation for the states, and $\mathbf{G}_{(j)}$ and $\mathbf{g}_{(j)}$ denote the jth row of $\mathbf{G}$ and $\mathbf{g}$, respectively. This requires that all rows j have to be jointly fulfilled with the probability $1 - \delta_{\mathbf{x}}$.

Definition 14.2 (*Individual chance constraint*) A (linear) state *individual chance constraint* is of the form

$$\mathbb{P}[\mathbf{G}_{(j)}\mathbf{x} \leq \mathbf{g}_{(j)}] \geq 1 - \delta_{\mathbf{x},j}, \quad \forall j \in \mathbb{Z}_{[1,r_x]}, \tag{14.9}$$

which requires that each jth row of the inequality has to be fulfilled individually with the respective probability $1 - \delta_{\mathbf{x},j}$, where $\delta_{\mathbf{x},j} \in (0, 1)$.

Both forms of constraints are useful to measure risks; hence, their selection depends on the application. All chance-constrained models require prior knowledge of the acceptable risk $\delta_{\mathbf{x}}$ associated with the constraints. A lower risk acceptability implies a harder constraint. This chapter is concerned with the use of joint chance constraints since they can express better the management of the overall reliability in a DWN. In general, joint chance constraints lack from analytic expressions due to the involved multivariate probability distribution. Nevertheless, sampling-based methods, numeric integration and convex analytic approximations exists (see, e.g., [3] and references therein). Here, (14.8) is approximated following the results in [18, 23] by upper bounding the joint constraint and assuming a uniform distribution of the joint risk among a set of *individual chance constraints* that are later transformed into equivalent deterministic constraints under Assumption 14.4.

Assumption 14.4 Each demand in $\mathbf{d} \in \mathbb{R}^{n_d}$ follows a log-concave univariate distribution, which stochastic description is known.

Given the dynamic model in (14.4), the stochastic nature of the demand vector $\mathbf{d}$ makes the state vector $\mathbf{x} \in \mathbb{R}^{n_x}$ to be also a stochastic variable. Then, let the cumulative distribution function of the constraint be denoted as follows:

$$F_{\mathbf{Gx}}(\mathbf{g}) := \mathbb{P}\left[\left\{\mathbf{G}_{(1)}\mathbf{x} \leq \mathbf{g}_{(1)}, \ldots, \mathbf{G}_{(r_x)}\mathbf{x} \leq \mathbf{g}_{(r_x)}\right\}\right]. \tag{14.10}$$

Defining the events $C_j := \left\{\mathbf{G}_{(j)}\mathbf{x} \leq \mathbf{g}_{(j)}\right\}$ for all $j \in \mathbb{Z}_{[1,r_x]}$, and denoting their complements as $C_j^c := \left\{\mathbf{G}_{(j)}\mathbf{x} > \mathbf{g}_{(j)}\right\}$, then it follows that

$$\begin{aligned} F_{\mathbf{Gx}}(\mathbf{g}) &= \mathbb{P}\left[C_1 \cap \cdots \cap C_{r_x}\right] && (14.11a)\\ &= \mathbb{P}\left[(C_1^c \cup \cdots \cup C_{r_x}^c)^c\right] && (14.11b)\\ &= 1 - \mathbb{P}\left[(C_1^c \cup \cdots \cup C_{r_x}^c)\right] \geq 1 - \delta_{\mathbf{x}}. && (14.11c) \end{aligned}$$

Taking advantage of the *union bound*, the Boole's inequality allows to bound the probability of the second term in the left-hand side of (14.11c), stating that for a countable set of events, the probability that at least one event happens is not higher than the sum of the individual probabilities [23]. This yields

$$\mathbb{P}\left[\bigcup_{j=1}^{r_x} C_j^c\right] \leq \sum_{j=1}^{r_x} \mathbb{P}\left[C_j^c\right]. \tag{14.12}$$

Applying (14.12) to the inequality in (14.11c), it follows that

$$\sum_{j=1}^{r_x} \mathbb{P}\left[C_j^c\right] \leq \delta_x \Leftrightarrow \sum_{j=1}^{r_x} \left(1 - \mathbb{P}\left[C_j\right]\right) \leq \delta_{\mathbf{x}}. \tag{14.13}$$

At this point, a set of constraints arise from the previous result as sufficient conditions to enforce the joint chance constraint (14.8), by allocating the joint risk δ_x in separate individual risks denoted by $\delta_{\mathbf{x},j}$, $j \in \mathbb{Z}_{[1,r_x]}$. These constraints are as follows:

$$\mathbb{P}\big[C_j\big] \geq 1 - \delta_{\mathbf{x},j}, \quad \forall j \in \mathbb{Z}_{[1,r_x]}, \tag{14.14}$$

$$\sum_{j=1}^{r_x} \delta_{\mathbf{x},j} \leq \delta_{\mathbf{x}}, \tag{14.15}$$

$$0 \leq \delta_{\mathbf{x},j} \leq 1, \tag{14.16}$$

where (14.14) forms the set of r_x resultant individual chance constraints, which bounds the probability that each inequality of the receding horizon problem may fail; and (14.15) and (14.16) are conditions imposed to bound the new single risks in such a way that the joint risk bound is not violated. Any solution that satisfies the above constraints is guaranteed to satisfy (14.8). As done in [18], assigning, for example, a fixed and equal value of risk to each individual constraint, i.e., $\delta_{\mathbf{x},j} = \delta_{\mathbf{x}}/r_x$ for all $j \in \mathbb{Z}_{[1,r_x]}$, then (14.15) and (14.16) are satisfied.

Remark 14.1 The single risks $\delta_{\mathbf{x},j}$, $j \in \mathbb{Z}_{[1,r_x]}$, might be considered as new decision variables to be optimized, see, e.g., [21]. This should improve the performance but at the cost of more computational burden due to the greater complexity and dimensionality of the optimization task. Therefore, as generalized flow-based networks are often large-scale systems, the uniform risk allocation policy is adopted to avoid overloading of the optimization problem. ◇

After decomposing the joint constraints into a set of individual constraints, the *deterministic equivalent* of each separate constraint may be used given that the probabilistic statements are not suitable for algebraic solution. Such deterministic equivalents might be obtained following the results in [6]. Assuming a known (or approximated) quasi-concave probabilistic distribution function for the effect of the stochastic disturbance in the dynamic model (14.4), it follows that

$$\begin{aligned}\mathbb{P}\big[\mathbf{G}_{(j)}\mathbf{x}(k+1) \leq \mathbf{g}_{(j)}\big] \geq 1 - \delta_{\mathbf{x},j} &\Leftrightarrow F_{\mathbf{G}_{(j)}\tilde{\mathbf{B}}_d\mathbf{d}(k)}(\mathbf{g}_{(j)} - \mathbf{G}_{(j)}(\mathbf{A}\mathbf{x}(k) + \tilde{\mathbf{B}}\tilde{\mathbf{u}}(k)) \geq 1 - \delta_{\mathbf{x},j} \\ &\Leftrightarrow \mathbf{G}_{(j)}(\mathbf{A}\mathbf{x}(k) + \tilde{\mathbf{B}}\tilde{\mathbf{u}}(k) \leq \mathbf{g}_{(j)} - F^{-1}_{\mathbf{G}_{(j)}\tilde{\mathbf{B}}_d\mathbf{d}(k)}(1 - \delta_{\mathbf{x},j}),\end{aligned} \tag{14.17}$$

for all $j \in \mathbb{Z}_{[1,r_x]}$, where $F_{\mathbf{G}_{(j)}\tilde{\mathbf{B}}_d\mathbf{d}(k)}(\cdot)$ and $F^{-1}_{\mathbf{G}_{(j)}\tilde{\mathbf{B}}_d\mathbf{d}(k)}(\cdot)$ are the cumulative distribution and the left-quantile function of $\mathbf{G}_{(j)}\tilde{\mathbf{B}}_d\mathbf{d}(k)$, respectively. Hence, the original state constraint set $\mathcal{X}$ is contracted by the effect of the r_x deterministic equivalents in (14.17) and replaced by the stochastic feasibility set given by

$$\begin{aligned}\mathcal{X}_{s,k} := \{\mathbf{x}(k) \in \mathbb{R}^{n_x} \; | \exists \tilde{\mathbf{u}}(k) \in \bar{\mathcal{U}}(k), \text{ such that} \\ \mathbf{G}_{(j)}(\mathbf{A}\mathbf{x}(k) + \tilde{\mathbf{B}}\tilde{\mathbf{u}}(k) \leq \mathbf{g}_{(j)} - F^{-1}_{\mathbf{G}_{(j)}\tilde{\mathbf{B}}_d\mathbf{d}(k)}(1 - \delta_{\mathbf{x},j}), \forall j \in \mathbb{Z}_{[1,r_x]}\},\end{aligned}$$

for all $k \in \mathbb{Z}_+$. From convexity of $\mathbf{G}_{(j)}\mathbf{x}(k+1) \leq \mathbf{g}_{(j)}$ and Assumption 14.4, it follows that the set $\mathcal{X}_{s,k}$ is convex when non-empty for all $\delta_{\mathbf{x},j} \in (0, 1)$ in most distribution functions [14]. For some particular distributions, e.g., Gaussian, convexity is retain for $\delta_{\mathbf{x},j} \in (0, 0.5]$.

In this way, the reformulated predictive controller solves the following deterministic equivalent optimization problem for the expectation $\mathbb{E}[\cdot]$ of the cost function in (14.6a):

$$\min_{\tilde{\mathbf{u}}(k)} \sum_{i=0}^{H_p-1} \mathbb{E}[J(k+i, \mathbf{x}(k+i|k), \tilde{\mathbf{u}}(k+i|k))], \tag{14.18a}$$

subject to:

$$\mathbf{x}(k+i+1|k) = \mathbf{A}\mathbf{x}(k+i|k) + \tilde{\mathbf{B}}\tilde{\mathbf{u}}(k+i|k) + \tilde{\mathbf{B}}_d\bar{\mathbf{d}}(k+i), \quad \forall i \in \mathbb{Z}_{[0,H_p-1]}, \tag{14.18b}$$

$$\mathbf{G}_{(j)}(\mathbf{A}\mathbf{x}(k+i|k) + \tilde{\mathbf{B}}\tilde{\mathbf{u}}(k+i|k)) \leq \mathbf{g}_{(j)} - z_{k,j}(\delta_{\mathbf{x}}), \quad \forall i \in \mathbb{Z}_{[0,H_p-1]},\ \forall j \in \mathbb{Z}_{[1,r_x]}, \tag{14.18c}$$

$$\tilde{\mathbf{u}}(k+i|k) \in \tilde{\mathcal{U}}(k+i), \quad \forall i \in \mathbb{Z}_{[0,H_p-1]}, \tag{14.18d}$$

$$\mathbf{x}(k|k) = \mathbf{x}(k), \tag{14.18e}$$

where $\tilde{\mathbf{u}}_k = \{\tilde{\mathbf{u}}(k+i|k)\}_{i\in\mathbb{Z}_{[0,H_p-1]}}$ is the sequence of controlled flows, $\bar{\mathbf{d}}(k+i)$ is the expected future demands computed at time instant $k \in \mathbb{Z}_+$ for i-steps ahead, $i \in \mathbb{Z}_{[0,H_p-1]}$, $n_c \in \mathbb{Z}_{\geq 1}$ is the number of total individual state constraints along the prediction horizon, i.e., $n_c = r_x H_p$ and $z_{k,j}(\delta_{\mathbf{x}}) := F^{-1}_{\mathbf{G}_{(j)}\tilde{\mathbf{B}}_d\mathbf{d}(k+i)}\left(1 - \frac{\delta_{\mathbf{x}}}{n_c}\right)$. Since n_c depends not only on the number of state constraints r_x but also on the value of H_p, the decomposition of the original joint chance constraint within the MPC algorithm could lead to a large number of constraints. This fact reinforces the use of a fixed risk distribution policy for generalized flow-based network control problems, in order to avoid the addition of a large number of new decision variables to be optimized.

Remark 14.2 It turns out that most (not all) probability distribution functions used in different applications, e.g., uniform, Gaussian, logistic, Chi-squared, Gamma, Beta, log-normal, Weibull, Dirichlet, and Wishart, among others, share the property of being log-concave. Then, their corresponding quantile function can be computed offline for a given risk acceptability level and used within the MPC convex optimization problem. ◊

14.4 Tree-Based MPC

The deterministic equivalent CC-MPC proposed before might be still conservative if the probabilistic distributions of the stochastic variables are not well characterized or do not lie in a log-concave form. Therefore, this section presents the TB-MPC strategy that relies on scenario trees to approximate the original problem, dropping Assumption 14.4. The approach followed by the TB-MPC is based on modelling the possible scenarios of the disturbances as a rooted tree (see Fig. 14.1 right). This means that all the scenarios start from the same measured disturbance value. From that point, the scenarios must remain equal until the point in which they diverge from each other, which is called a bifurcation point. Each node of the tree has a unique parent and can have many children. The total number of children at the last stage corresponds to the total number of scenarios. The probability of a scenario is the product of probabilities of each node in that scenario.

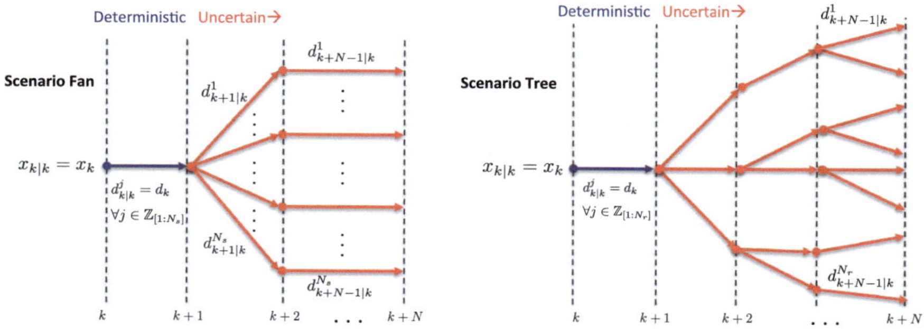

Fig. 14.1 Reduction of a disturbance fan (*left*) of equally probable scenarios into a rooted scenario tree (*right*)

Notice that before a bifurcation point, the evolution followed by the disturbance cannot be anticipated because different evolutions are possible. For this reason, the controller has to calculate control actions that are valid for all the scenarios in the branch. Once the bifurcation point has been reached, the uncertainty is solved and the controller can calculate specific control actions for the scenarios in each of the new branches. Hence, the outcome of TB-MPC is not a single sequence of control actions, but a tree with the same structure of that of the disturbances. As in standard MPC, only the first element of this tree is applied (the root) and the problem is repeated in a receding horizon fashion.

In generalized flow-based networks, the uncertainty is generally introduced by the unpredictable behaviour of consumers. Therefore, a proper demand modelling is required to achieve an acceptable supply service level. For the case study considered in this chapter, the reader is referred to [26], where the authors presented a detailed comparison of different forecasting models. Once a model is selected, it has to be calibrated and then used to generate a large number of possible demand scenarios by Monte Carlo sampling for a given prediction horizon $H_p \in \mathbb{Z}_{\geq 1}$. For the CC-MPC approach, the mean demand path is used, while for the TB-MPC approach, a set of scenarios is selected. The size of this set is here computed following the bound proposed in [27], which takes into account the desired risk acceptability level. A large number of scenarios might improve the robustness of the TB-MPC approach but at the cost of additional computational burden and economic performance losses. Hence, a trade-off must be achieved between performance and computational burden. To this end, a representative subset of scenarios may be chosen using scenario reduction algorithms. In this chapter, the backward reduction algorithm proposed in [13] is used to reduce a specified initial fan of $N_s \in \mathbb{Z}_{\geq 1}$ equally probable scenarios into a rooted tree of $N_r << N_s$ scenarios, where N_r is the number of considered scenarios while N_s is the total number of scenarios (see Fig. 14.1).

The easiest way to understand the optimization problem that has to be solved in TB-MPC is to solve as many instances of Problem (14.6) as the number N_r of considered scenarios, but formally it is a multi-stage stochastic programme and solved as a big optimization for all the scenarios. Due to the increasing uncertainty, it is necessary to include non-anticipativity constraints [25] in the MPC formulation so that the calculated input sequence is always ready to face any possible future bifurcation in the tree. More specifically, if $\mathbf{d}_k^a = \{d^a(k|k), d^a(k+1|k), \ldots, d^a(k+N|k)\}$ and $\mathbf{d}_k^b = \{d^b(k|k), d^b(k+1|k), \ldots, d^b(k+N|k)\}$ are two disturbance sequences corresponding respectively to certain forecast scenarios $a, b \in \mathbb{Z}_{[1,N_r]}$, then the non-anticipativity constraint $\tilde{\mathbf{u}}^a(k+i|k) = \tilde{\mathbf{u}}^b(k+i|k)$ has to be satisfied for any $i \in \mathbb{Z}_{[0,H_p]}$ whenever $d^a(k+i|k) = d^b(k+i|k)$ in order to guarantee that for all $j \in \mathbb{Z}_{[1,N_r]}$ the input sequences $\tilde{\mathbf{u}}^j = \{\tilde{u}^j(k+i|k)\}_{i\in\mathbb{Z}_{[0,H_p-1]}}$ form a tree with the same structure of that of the disturbances.

In this way, the TB-MPC controller has to solve the following optimization problem at each time instant $k \in \mathbb{Z}_+$, accounting for the N_r demand scenarios, each with probability $p_j \in (0, 1]$ satisfying $\sum_{j=1}^{N_r} p_j = 1$:

$$\min_{\tilde{\mathbf{u}}_k^j} \sum_{j=1}^{N_r} p_j \left(\sum_{i=0}^{H_p-1} J(k+i, \mathbf{x}(k+i|k), \tilde{\mathbf{u}}(k+i|k)) \right), \tag{14.19a}$$

subject to:

$$\mathbf{x}^j(k+i+1|k) = \mathbf{A}\mathbf{x}^j(k+i|k) + \tilde{\mathbf{B}}\tilde{\mathbf{u}}^j(k+i|k) + \tilde{\mathbf{B}}_d\mathbf{d}^j(k+i), \quad \forall i \in \mathbb{Z}_{[0,H_p-1]}, \forall j \in \mathbb{Z}_{[1,N_r]}, \tag{14.19b}$$

$$\mathbf{x}^j(k+i+1|k) \in \mathcal{X}, \quad \forall i \in \mathbb{Z}_{[0,H_p-1]}, \forall j \in \mathbb{Z}_{[1,N_r]}, \tag{14.19c}$$

$$\tilde{\mathbf{u}}^j(k+i|k) \in \tilde{\mathcal{U}}^j(k+i), \quad \forall i \in \mathbb{Z}_{[0,H_p-1]}, \forall j \in \mathbb{Z}_{[1,N_r]}, \tag{14.19d}$$

$$\mathbf{x}^J(k|k) = \mathbf{x}(k), \quad \mathbf{d}^j(k|k) = \mathbf{d}(k), \quad \forall j \in \mathbb{Z}_{[1,N_r]}, \tag{14.19e}$$

$$\tilde{\mathbf{u}}^a(k+i|k) = \tilde{\mathbf{u}}^b(k+i|k) \text{ if } \mathbf{d}^a(k+i|k) = d^b(k+i|k), \quad \forall i \in \mathbb{Z}_{[0,H_p-1]}, \forall a, b \in \mathbb{Z}_{[1,N_r]}, \tag{14.19f}$$

where $\tilde{\mathcal{U}}^j(k+i) := \{\tilde{\mathbf{u}}^j \in \mathbb{R}^{n_u-m} \mid \mathbf{H}\tilde{\mathbf{P}}\tilde{\mathbf{M}}_1\tilde{\mathbf{u}}^j \leq \mathbf{h} - \mathbf{H}\tilde{\mathbf{P}}\tilde{\mathbf{M}}_2\mathbf{d}^j(k+i)\}$.

Remark 14.3 The number of scenarios used to build the rooted tree should be determined regarding the computational capacity and the probability of risk that the manager is willing to accept. ◊

14.5 Numerical Results

In this section, the performance of the proposed stochastic MPC approaches is assessed with a case study consisting in a large-scale real system reported in [19], specifically the Barcelona WTN already described in Sect. 2.3 in Chap. 2. The general role of this system is the spatial and temporal reallocation of water resources from both superficial (i.e., rivers) and underground water sources (i.e., wells) to distribution nodes located all over the city. The directed graph of this network can be obtained from the layout shown in Fig. 2.2 of Chap. 2, and its model in the form of (14.1) can be derived by setting the state $\mathbf{x}(k) \in \mathbb{R}^{63}$ as the volume (in m^3) of water stored in tanks at time instant k, the control input $\mathbf{u}(k) \in \mathbb{R}^{114}$ as the flow rate through all network actuators (expressed in m^3/s) and the measured disturbance $\mathbf{d}(k) \in \mathbb{R}^{88}$ as the flow rate of customer demands (expressed in m^3/s). This network is currently managed by Aguas de Barcelona that manages the drinking-water transport and distribution in Barcelona (Spain), and it supplies potable water to the Metropolitan Area of Barcelona (Catalunya, Spain). The main control task for the managers is to economically optimize the network flows while satisfying customer demands. These demands are characterized by patterns of water usage and can be forecasted by different methods (see, e.g., [1, 26]).

The operational goals in the management of the Barcelona DWN are of three kinds, *economic*, *safety* and *smoothness*, and are respectively stated as follows (see Chap. 12. for the mathematical formulation):

1. To provide a reliable water supply in the most economic way, minimizing water production and transport costs.
2. To guarantee the availability of enough water in each reservoir to satisfy its underlying demand, keeping a safety stock to face uncertainties and avoid stock-outs.
3. To operate the DWN under smooth control actions.

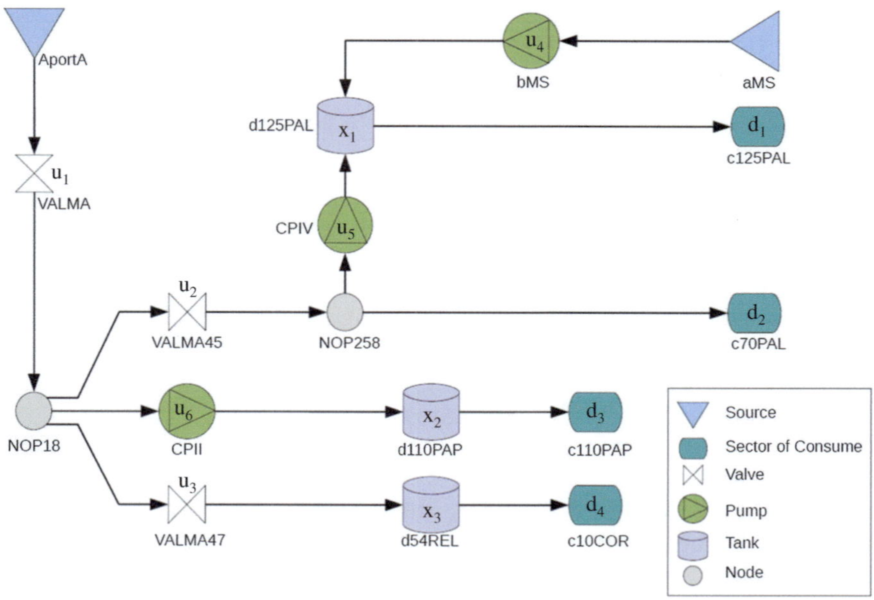

Fig. 14.2 Barcelona DWN small subsystem layout

14.5.1 Performance Comparison on a Small-Scale System

To analyse via simulation the computational burden of each proposed controller, a small portion/subsystem of the complete network is used (see Fig. 14.2 and [9] for details). The DWN is considered as a stochastic constrained system subject to deterministic hard constraints on the control inputs and linear joint chance constraints on the states. The source of uncertainty in the system is assumed to be the forecasting error of the water demands. The stochastic control problem of the DWN is stated as follows:

$$\min_{\tilde{\mathbf{u}}_k} \sum_{i=0}^{H_p-1} \mathbb{E}\left[\lambda_1 J_E(k+i, \mathbf{x}(k+i|k), \tilde{\mathbf{u}}(k+i|k)) + \lambda_2 J_\Delta(\Delta\tilde{\mathbf{u}}(k+i|k))\right], \quad (14.20a)$$

subject to:

$$\mathbf{x}(k+i+1|k) = \mathbf{A}\mathbf{x}(k+i|k) + \tilde{\mathbf{B}}\tilde{\mathbf{u}}(k+i|k) + \tilde{\mathbf{B}}_d\mathbf{d}(k+i|k), \quad (14.20b)$$

$$\mathbb{P}\left[\mathbf{x}(k+i+1|k) \geq \mathbf{x}^{\min}\right] \geq 1 - \frac{\delta_{\mathbf{x}}}{2}, \quad (14.20c)$$

$$\mathbb{P}\left[\mathbf{x}(k+i+1|k) \leq \mathbf{x}^{\max}\right] \geq 1 - \frac{\delta_{\mathbf{x}}}{2}, \quad (14.20d)$$

$$\mathbb{P}\left[\mathbf{x}(k+i+1|k) \geq \mathbf{d}_{\text{net}}(k+i+1|k)\right] \geq 1 - \delta_s, \quad (14.20e)$$

$$\tilde{\mathbf{u}}(k+i|k) \in \tilde{\mathcal{U}}(k+i), \quad (14.20f)$$

$$\mathbf{d}_{\text{net}}(k+i+1|k) = -(\tilde{\mathbf{B}}_{\text{out}}(\tilde{\mathbf{P}}\tilde{\mathbf{M}}_1\tilde{\mathbf{u}}(k+i|k) + \tilde{\mathbf{P}}\tilde{\mathbf{M}}_2\mathbf{d}(k+i|k)) + \tilde{\mathbf{B}}_d\mathbf{d}(k+i+1|k)), \quad (14.20g)$$

$$\Delta\tilde{\mathbf{u}}(k+i|k) = \tilde{\mathbf{u}}(k+i|k) - \tilde{\mathbf{u}}(k+i-1|k), \quad (14.20h)$$

$$(\mathbf{x}(k|k), \tilde{\mathbf{u}}(k-1|k)) = (\mathbf{x}(k), \tilde{\mathbf{u}}(k-1)), \quad (14.20i)$$

for all $i \in \mathbb{Z}_{[0,H_p-1]}$, where $J_E(k+i, \mathbf{x}(k+i|k), \tilde{\mathbf{u}}(k+i|k)) := c_{u,k+i}^\top W_e\, \tilde{\mathbf{u}}(k)\Delta t$ captures the economic costs with $c_{u,k+i} \in \mathbb{R}^{n_u}$ being a known periodically time-varying price of electric tariff, and $J_\Delta(\Delta\tilde{\mathbf{u}}(k+i|k)) := \|\tilde{\mathbf{P}}\tilde{\mathbf{M}}_1\Delta\tilde{\mathbf{u}}(k+i|k) + \tilde{\mathbf{P}}\tilde{\mathbf{M}}_2$ $\Delta\mathbf{d}(k+i|k)\|_{W_{\Delta\tilde{\mathbf{u}}}}^2$ is a control move suppression term aiming to enforce a smooth operation. Moreover, $\delta_{\mathbf{x}}, \delta_s \in (0, 1)$, are the accepted maximum risk levels for the state constraints and the safety constraint (14.20e), respectively. The objectives are traded-off with the scalar weights $\lambda_1, \lambda_2 \in \mathbb{R}_+$, while the elements of the decision vector are prioritized by the weighting matrices $W_e, W_{\Delta\tilde{\mathbf{u}}} \in \mathbb{S}_{++}^m$. The service reliability goal (i.e., demand satisfaction) is enforced by the constraints (14.20e) and (14.20g). In this latter constraint, $\mathbf{d}_{\text{net}}(k+i+1|k) \in \mathbb{R}^{n_d}$ is a vector of net demands above which is desired to keep the reservoirs to avoid stock-outs. The $\tilde{\mathbf{B}}_{\text{out}}(\tilde{\mathbf{P}}\tilde{\mathbf{M}}_1\tilde{\mathbf{u}}(k+i|k) + \tilde{\mathbf{P}}\tilde{\mathbf{M}}_2\mathbf{d}(k+i|k))$ component represents the current prediction step endogenous demand, i.e., the outflow of the tanks caused by water requirements from neighbouring tanks or nodes, and the $\tilde{\mathbf{B}}_d\mathbf{d}(k+i+1|k)$ component denotes the exogenous (customer) demands of tanks for the next prediction step. In the dynamic model (14.4) of the DWN, randomness is directly described by the uncertainty of customer demands, which can be estimated from historical data. In order to solve the above DWN control problem, a tractable safe approximation is derived following Sect. 14.3. The joint chance constraints (14.20c)–(14.20e) are transformed into deterministic equivalent constraints as shown in [11, Appendix B] for the particular case of Gaussian distributions.

The optimization problem associated with the deterministic equivalent CC-MPC for the selected application is stated as follows for a given sequence of forecasted demands denoted by $\bar{\mathbf{d}}_k = \{\mathbf{d}(k+i|k)\}_{i\in\mathbb{Z}_{[0,H_p-1]}}$:

$$\min_{\tilde{\mathbf{u}}_k,\boldsymbol{\xi}_k} \sum_{i=0}^{H_p-1} \mathbb{E}\left[\lambda_1 J_E(k+i,\bar{\mathbf{x}}(k+i|k),\tilde{\mathbf{u}}(k+i|k)) + \lambda_2 J_\Delta(\Delta\tilde{\mathbf{u}}(k+i|k)) + \lambda_3 J_S(\boldsymbol{\xi}(k+i|k))\right], \tag{14.21a}$$

subject to:

$$\bar{\mathbf{x}}(k+i+1|k) = \mathbf{A}\bar{\mathbf{x}}(k+i|k) + \tilde{\mathbf{B}}\tilde{\mathbf{u}}(k+i|k) + \tilde{\mathbf{B}}_d\bar{\mathbf{d}}(k+i|k), \tag{14.21b}$$

$$\bar{\mathbf{x}}_{(j)}(k+i+1|k) \geq \mathbf{x}_{(j)}^{\min} + \Phi^{-1}\left(1-\frac{\delta_{\mathbf{x}}}{2n_x H_p}\right)\Sigma_{x_{(j)}}(k+i+1|k)^{1/2}, \tag{14.21c}$$

$$\bar{\mathbf{x}}_{(j)}(k+i+1|k) \leq \mathbf{x}_{(j)}^{\max} - \Phi^{-1}\left(1-\frac{\delta_{\mathbf{x}}}{2n_x H_p}\right)\Sigma_{x_{(j)}}(k+i+1|k)^{1/2}, \tag{14.21d}$$

$$\begin{aligned}\bar{\mathbf{x}}_{(j)}(k+i+1|k) \geq{}& \bar{\mathbf{d}}_{\text{net}(j)}(k+i+1|k) + \Phi^{-1}\left(1-\frac{\delta_s}{n_x H_p}\right)\Sigma_{d_{\text{net}(j)}}(k+i+1|k)^{1/2} \\ &- \boldsymbol{\xi}_{(j)}(k+i|k),\end{aligned} \tag{14.21e}$$

$$\boldsymbol{\xi}(k+i|k) \geq 0, \tag{14.21f}$$

$$\tilde{\mathbf{u}}(k+i|k) \in \tilde{\mathcal{U}}(k+i), \tag{14.21g}$$

$$\bar{\mathbf{d}}_{\text{net}}(k+i+1|k) = -(\tilde{\mathbf{B}}_{\text{out}}\left(\tilde{\mathbf{P}}\tilde{\mathbf{M}}_1\tilde{\mathbf{u}}(k+i|k) + \tilde{\mathbf{P}}\tilde{\mathbf{M}}_2\bar{\mathbf{d}}(k+i|k)\right) + \tilde{\mathbf{B}}_d\bar{\mathbf{d}}(k+i+1|k)), \tag{14.21h}$$

$$\Delta\tilde{\mathbf{u}}(k+i|k) = \tilde{\mathbf{u}}(k+i|k) - \tilde{\mathbf{u}}(k+i-1|k), \tag{14.21i}$$

$$(\bar{\mathbf{x}}(k|k),\tilde{\mathbf{u}}(k-1|k)) = (\mathbf{x}(k),\tilde{\mathbf{u}}(k-1)), \tag{14.21j}$$

for all $i \in \mathbb{Z}_{[0,N-1]}$ and all $j \in \mathbb{Z}_{[1,n]}$, where $\tilde{\mathbf{u}}_k = \{\tilde{\mathbf{u}}(k+i|k)\}$ and $\boldsymbol{\xi}_k = \{\boldsymbol{\xi}(k+i|k)\}$ are the decision variables. The vectors $\bar{\mathbf{x}}$ and $\bar{\mathbf{d}}$ denote the mean of the random state and demand variables, respectively. Moreover, Φ^{-1} is the left-quantile function of the Gaussian distribution, and $\bar{\mathbf{x}}_{(j)}$ and $\Sigma_{x_{(j)}}$ denote, respectively, the mean and variance of the jth row of the state vector. Notice that Problem (14.21) includes the additional objective $J_S(\boldsymbol{\xi}(k+i|k)) := \|\boldsymbol{\xi}(k+i|k)\|^2_{W_s}$ with $\mathbf{W}_s \succ 0$, and the additional constraint (14.21f), which are related to the safety operational goal. These elements appear due to the safety deterministic equivalent soft constraint (14.21e) introduced with the slack decision variable $\boldsymbol{\xi} \in \mathbb{R}^{n_x}$ to allow the trade-off between safety, economic, and smoothness objectives. Constraints (14.21c) and (14.21d) can be softened in the same way to guarantee recursive feasibility of the optimization problem if uncertainty is too large. For a strongly feasible stochastic MPC approach using closed-loop predictions by means of an affine disturbance parametrization of the control inputs, the reader is referred to [15]. The enforcement of the chance constraints enhances the robustness of the MPC controller by causing an optimal back-off from the nominal deterministic constraints as a risk averse mechanism to face the non-stationary uncertainty involved in the prediction model of the MPC. The states are forced to move away from their limits before the disturbances have chance to cause constraint violation. The $\Phi^{-1}(\cdot)$ terms represent safety factors for each constraint, and specially in (14.21e), it denotes the optimal safety stock of storage tanks. Problem 14.18 may be casted as a second-order cone programming problem. However, state uncertainty is a function of the disturbances only and is not a function of the decision variables of the optimization problem. Therefore, the variance terms in each deterministic equivalent can be forecasted prior to the solution of the optimization problem to include them as known parameters in the MPC formulation. This simplification results in a set of linear constraints, and the optimization remains as a quadratic programming (QP) problem, which can be efficiently solved.

The optimization problem associated with the scenario tree-based MPC approach is stated as follows for all $i \in \mathbb{Z}_{[0,H_p-1]}$ and all $j \in \mathbb{Z}_{[1,N_r]}$:

$$\min_{\tilde{\mathbf{u}}_k^j,\boldsymbol{\xi}_k^j} \sum_{j=1}^{N_r} p_j \sum_{i=0}^{H_p-1} \lambda_1 J_E(k+i, \mathbf{x}^j(k+i|k), \tilde{\mathbf{u}}^j(k+i|k)) + \lambda_2 J_\Delta(\Delta\tilde{\mathbf{u}}^j(k+i|k)) + \lambda_3 J_S(\boldsymbol{\xi}^j(k+i|k)), \tag{14.22a}$$

subject to:

$$\mathbf{x}^j(k+i+1|k) = \mathbf{A}\mathbf{x}^j(k+i|k) + \tilde{\mathbf{B}}\tilde{\mathbf{u}}^j(k+i|k) + \tilde{\mathbf{B}}_d\mathbf{d}^j(k+i|k), \tag{14.22b}$$

$$(\mathbf{x}^j(k+i+1|k), \tilde{\mathbf{u}}^j(k+i|k), \boldsymbol{\xi}^j(k+i|k)) \in \mathcal{X} \times \tilde{\mathcal{U}}^j(k+i) \times \mathcal{R}_+^{n_u}, \tag{14.22c}$$

$$\mathbf{x}^j(k+i+1|k) \geq \mathbf{d}_{\text{net}}^j(k+i+1|k) - \boldsymbol{\xi}^j(k+i|k), \tag{14.22d}$$

$$\mathbf{d}_{\text{net}}(k+i+1|k)^j = -(\tilde{\mathbf{B}}_{\text{out}}\left(\tilde{\mathbf{P}}\tilde{\mathbf{M}}_1\tilde{\mathbf{u}}^j(k+i|k) + \tilde{\mathbf{P}}\tilde{\mathbf{M}}_2\mathbf{d}^j(k+i|k)\right) + \tilde{\mathbf{B}}_d\mathbf{d}_{\text{net}}^j(k+i+1|k), \tag{14.22e}$$

$$\Delta\tilde{\mathbf{u}}^j(k+i|k) = \tilde{\mathbf{u}}^j(k+i|k) - \tilde{\mathbf{u}}^j(k+i-1|k), \tag{14.22f}$$

$$(\mathbf{x}^j(k|k)\mathrm{X}, \tilde{\mathbf{u}}^j(k-1|k), \mathbf{d}^j(k|k)) = (\mathbf{x}(k), \tilde{\mathbf{u}}(k-1), \mathbf{d}(k), \tag{14.22g}$$

$$\tilde{\mathbf{u}}^a(k+i|k) = \tilde{\mathbf{u}}^b(k+i|k) \text{ if } \mathbf{d}^a(k+i|k) = \mathbf{d}^b(k+i|k) \quad \forall a, b \in \mathbb{Z}_{[1,N_r]}. \tag{14.22h}$$

Table 14.1 summarizes the results of applying the deterministic equivalent CC-MPC and the TB-MPC to the aforementioned small example. Simulations have been carried out over a time period of eight days, i.e., $n_s = 192$ hours, with a sampling time of 1 hour. Applied demand scenarios were taken from historical data of the Barcelona DWN. The weights of the multi-objective cost function are $\lambda_1 = 100$, $\lambda_2 = 1$ and $\lambda_3 = 10$. The prediction horizon is selected as $H_p = 24$ h due to the periodicity of demands. The key performance indicators used to assess the aforementioned controllers are defined as follows:

Table 14.1 Assessment of the CC-MPC and TB-MPC applied to the sector model of the DWN case study

	CC-MPC				TB-MPC					
δ_x	KPI_1	KPI_2	KPI_3	KPI_4	KPI_1	KPI_2	KPI_3	KPI_4	N_r	N_s
					58397.14	0	0	0.94	5	
0.3	58535.80	0	0	1.25	58280.69	1	0.51	1.61	10	19
					58279.95	1	4.16	2.37	14	
					58482.14	3	0.18	1.18	7	
0.2	58541.19	0	0	1.21	58903.63	0	0	2.33	14	29
					58452.41	0	0	4.05	21	
					58610.32	0	0	2.57	14	
0.1	58558.29	0	0	1.25	58630.20	0	0	6.65	29	59
					58656.56	1	0.18	13.47	44	
					–	–	–	–	149	
0.01	58612.28	0	0	1.25	–	–	–	–	299	599
					–	–	–	–	449	
					–	–	–	–	1499	
0.001	58667.85	0	0	1.25	–	–	–	–	2999	5999
					–	–	–	–	4499	

$$\mathrm{KPI}_1 \triangleq \frac{24}{n_s+1}\sum_{k=0}^{n_s} \gamma_1 J_E(k, \mathbf{x}(k), \tilde{\mathbf{u}}(k)) + \lambda_2 J_\Delta(\Delta\tilde{\mathbf{u}}(k)) + \lambda_3 J_S(\boldsymbol{\xi}(k)), \quad (14.23a)$$

$$\mathrm{KPI}_2 \triangleq |\left\{k \in \mathbb{Z}_1^{n_s} \mid \mathbf{x}(k) < -\mathbf{B}_d\mathbf{d}(k)\right\}|, \quad (14.23b)$$

$$\mathrm{KPI}_3 \triangleq \sum_{k=1}^{n_s}\sum_{i=1}^{n_x} \max\{0, -\mathbf{B}_{d(i)}\mathbf{d}(k) - \mathbf{x}(k(i))\}, \quad (14.23c)$$

$$\mathrm{KPI}_4 \triangleq \frac{1}{n_s}\sum_{k=1}^{n_s} t(k), \quad (14.23d)$$

where KPI_1 is the average daily multi-objective cost, KPI_2 is the number of time instants where the stored water goes below the demanded volume (for this, $|\cdot|$ denotes the cardinal of a set of elements), KPI_3 is the accumulated volume of water demand that was not satisfied over the simulation horizon, and KPI_4 is the average time in seconds required to solve the MPC problem at each time instant $k \in \mathbb{Z}_{[0,n_s]}$. For the CC-MPC approach, the effect of considering different levels of joint risk acceptability was analysed using $\delta_\mathbf{x} = \{0.3, 0.2, 0.1, 0.01, 0.001\}$ and $\delta_s = \delta_\mathbf{x}$. Regarding the TB-MPC approach, different sizes for the initial set of scenarios were considered, i.e., $N_s = \{19, 28, 59, 599, 5999\}$. The size of this initial set was computed following the bound proposed in [27] taking into account the risk levels involved in the chance constraints. This initial set was reduced later by a factor of 0.25, 0.50 and 0.75 to obtain different rooted trees with N_r scenarios.

As shown in Table 14.1, the different CC-MPC scenarios highlight that reliability and control performance are conflicting objectives; that is, the inclusion of safety mechanisms in the controller increases the reliability of the DWN in terms of demand satisfaction, but also the cost of its operation. The main advantage of the CC-MPC is its formal methodology, which leads to obtain optimal safety constraints that tackle uncertainties and allow to achieve a specified global service level in the DWN. Moreover, the deterministic equivalent CC-MPC robustness is achieved with a low computational burden given that the only extra load (compared with a nominal formulation) is the computation of the stochastic characteristics of disturbances propagated in the prediction horizon. In this way, the deterministic equivalent CC-MPC approach is suitable for real-time control (RTC) of large-scale DWNs. Regarding the TB-MPC approach, numeric results show that considering higher N_s increments the stage cost while reducing the volume of unsatisfied water demand. Nevertheless, this latter observation is not applicable for the different N_r cases within a same N_s. This might be influenced by the quality of the information that remains after the scenario generation and reduction algorithms that affect the robustness of the approach and will be subject of further research. The main drawback of the TB-MPC approach is the solution of the average time and the computational burden. In this case study, the implementation for all cases taking $N_s = \{599, 5999\}$ was not possible due to memory issues. Hence, some simplification assumptions as those used in [17] or parallel computing techniques might be useful.

14.5.2 *Performance Assessment of CC-MPC on a Large-Scale System*

The previous results showed that both CC-MPC and TB-MPC have similar performance under high levels of risk acceptability. Nevertheless, when requiring small risk levels ($\delta_{\mathbf{x}} < 0.1$), CC-MPC retains tractability of the FHOP with low complexity, while the TB-MPC suffers the curse of dimensionality. Therefore, in the following, only the performance of the CC-MPC approach is assessed on the *full model* of the Barcelona DWN. The tuning of the controller parameters is the same as in the previous simulations. In order to further evaluate the proposed CC-MPC scheme, results are compared with the certainty-equivalent MPC approach proposed in [19], which assumes predictions of demands as certain. In these simulations, the CE-MPC strategy has been set up to allow the volume of water in tanks to decrease until the predicted volume of future net demands, which is set as a hard constraint but ignoring the influence of uncertainty. Contrary, the CC-MPC strategy considers and propagates the uncertainty of forecasted demands explicitly in the MPC design and, as a consequence, involves a robust handling of constraints. Again, to analyse the effect of the risk level ($\delta_{\mathbf{x}}$) in this CC-MPC strategy when considering large-scale systems, different scenarios have been simulated for acceptable joint risks of 50, 40, 30, 20, 10, 5 and 1%.

Table 14.2 presents the numeric assessment of the aforementioned controllers through different key performance indicators (KPIs), which are defined as follows:

$$\mathrm{KPI}_E := \frac{1}{n_s+1}\sum_{k=0}^{n_s} c_{u,k}^{\top}\tilde{\mathbf{u}}(k)\Delta t, \tag{14.24a}$$

$$\mathrm{KPI}_{\Delta U} := \frac{1}{n_s+1}\sum_{i=1}^{n_u}\sum_{k=0}^{n_s}\left(\Delta\tilde{\mathbf{u}}_{(i)}(k)\right)^2, \tag{14.24b}$$

$$\mathrm{KPI}_S := \sum_{i=1}^{n_x}\sum_{k=0}^{n_s}\max\left\{0, s_{(i)}(k) - \mathbf{x}_{(i)}(k)\right\}, \tag{14.24c}$$

$$\mathrm{KPI}_D := \sum_{i=1}^{n_x}\sum_{k=0}^{n_s}\max\left\{0, \mathbf{d}_{\mathrm{net}(i)}(k) - \mathbf{x}_{(i)}(k)\right\}, \tag{14.24d}$$

$$\mathrm{KPI}_R := \frac{\sum_{i=1}^{n_x}\sum_{k=1}^{n_s} s_{(i)}(k)}{\sum_{i=1}^{n_x}\sum_{k=1}^{n_s}\mathbf{x}_{(i)}(k)} \times 100\%, \tag{14.24e}$$

$$\mathrm{KPI}_O := t_{\mathrm{opt}}(k), \tag{14.24f}$$

where KPI_E is the average economic performance of the DWN operation, $\mathrm{KPI}_{\Delta U}$ measures the smoothness of the control actions, KPI_S is the amount of water used from safety stocks, KPI_D is the volume of water demand that is not satisfied over the simulation period, KPI_R is the average percentage of safety volume that is contained in the real water volume, and KPI_O determines the difficulty to solve the optimization tasks involved in each strategy accounting $t_{\mathrm{opt}}(k)$ as the average time that takes to solve the corresponding MPC optimization problem. The CE-MPC has been tuned with a safety stock for each tank equal to its net exogenous demand, i.e., $s(k) = d_{\mathrm{net}}(k)$. Therefore, the KPI_S results equal to the KPI_D as should be expected given their definitions. In the case of the CC-MPC, $s(k)$ is equal to the right hand of (14.21e). Regarding the comparison of the KPI_S between the CE-MPC and the CC-MPC, the results present greater values for the CC-MPC cases. This trend is also an expected behaviour given that reducing the risk probability generates a larger back-

Table 14.2 Comparison of the MPC strategies applied to the Barcelona DWN

Controller	KPI_E	$\mathrm{KPI}_{\Delta U}$	KPI_S	KPI_D	KPI_R	KPI_O
CE-MPC	2297.02	2.3586	3.8886	3.8886	19.41	4.82
CC-MPC$_{@50\%}$	2486.40	1.0747	695.54	0	27.79	4.72
CC-MPC$_{@40\%}$	2487.77	1.0767	750.06	0	27.86	4.83
CC-MPC$_{@30\%}$	2489.31	1.0795	819.82	0	27.95	4.79
CC-MPC$_{@20\%}$	2491.61	1.0835	920.36	0	28.07	4.71
CC-MPC$_{@10\%}$	2496.23	1.0964	1101.7	0	28.18	4.70
CC-MPC$_{@5\%}$	2500.52	1.1012	1298.9	0	28.18	4.89
CC-MPC$_{@1\%}$	2509.89	1.1131	1759.4	0	28.43	4.86

off of the demand satisfaction constraint; that is, more safety stock is stored to address demand uncertainty. This latter fact, in addition with the tuning of the multi-objective cost function, leads to higher KPI_S (but lower or null KPI_D) if this is required by the real demand scenario in order to guarantee a service level. It can be observed that the CE-MPC is the cheapest control strategy (lower KPI_E) but the less reliable one given that the certainty equivalence assumption leads to unsatisfying demands (higher KPI_D), especially when the water volume in the tank is close to the expected demand. Thus, the CE-MPC performance represents a strategy for the supply of drinking water with a higher risk of failure. The different CC-MPC scenarios (those of varying the risk acceptability level) have shown that reliability and economic performance are conflicting objectives that have to reach a trade-off; that is, the inclusion of safety mechanisms in the controller increases the reliability of the DWN in terms of demand satisfaction (see Fig. 14.3), but also the economic cost of its operation. The main advantage of the CC-MPC is its formal methodology that leads to obtain optimal dynamic constraints that tackle uncertainties with a minimum cost to achieve also a global service level of the DWN. Table 14.2 shows a smooth degradation of the economic performance under the CC-MPC when varying the risk within a wide range of acceptability levels. Therefore, the CC-MPC approach addressed in this chapter is a suitable mean to compute the proper amount of safety and the proper control actions

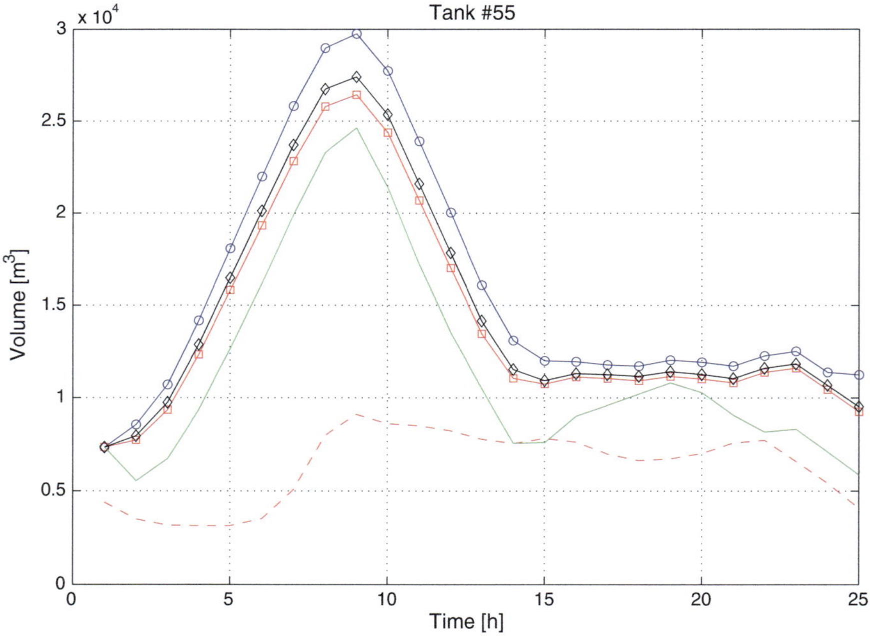

Fig. 14.3 Comparison of the robustness in the management of water storage in a sample of tanks of the Barcelona DWN. (*Blue circle*) CC-MPC$_{1\%}$, (*black diamond*) CC-MPC$_{20\%}$, (*red square*) CC-MPC$_{50\%}$, (*solid green*) CE-MPC, (*dashed red*) net demand

to assure a desired service level. Notice that the computational burden (KPI_O) of the CC-MPC is similar to the CE-MPC given that the complexity of the optimization problem is not altered; that is, the number of constraints and decision variables remain the same. The only extra load that might be added is the computation of the variance of the disturbances propagated in the prediction horizon. Consequently, the CC-MPC approach is suitable for RTC of the Barcelona DWN.

Table 14.3 discloses details of the average production and operational costs related to each strategy. Comparing the CE-MPC controller with the CC-MPC$_{@5\%}$ controller (requiring a reliability of 95%), it can be noticed that the dynamic safety stocks resulting within the stochastic approach might lead to an increase in the operational cost, especially in the electric cost, mainly due to the extra amount of water that is needed to be moved through the network and allocated in tanks to guarantee that the water supply will be feasible with a certain probability for future disturbance realizations.

The conservatism of reformulating the stochastic CC-MPC problem into the tractable deterministic equivalent in (14.21) has been studied in [12]. Table 14.4 shows the conservatism related to approximate constraints (14.20c), (14.20d) and (14.20e), considering different levels of maximum joint risk. It can be observed that conservatism increases when the risk level increases but remains almost constant despite the variation in the number of individual constraints. Hence, the goodness of the approximation using Boole's inequality is not affected, neither by the number of decision variables nor by the prediction horizon. Therefore, the addressed approach is advantageous to be applied to any other DWNs or general flow networks.

Table 14.3 Comparison of daily average economic costs of MPC strategies

Controller	Water cost (e.u./day)	Electric cost (e.u./day)	Daily average cost (e.u./day)
CE-MPC	23015.42	27195.31	50210.73
CC-MPC$_{@5\%}$	22980.34	28514.71	51495.05

e.u.: economic units

Table 14.4 Conservatism of the deterministic equivalent CC-MPC

Joint chance constraint	Number of individual constraints	Joint risk	Conservatism of approximation
		0.001	4.9967×10^{-7}
		0.01	4.9817×10^{-5}
State hard bounds	3024	0.03	4.4539×10^{-4}
		0.05	1.2290×10^{-3}
		0.1	4.8359×10^{-3}
		0.001	4.9950×10^{-7}
		0.01	4.9801×10^{-5}
Safety constraint	1512	0.03	4.4524×10^{-4}
		0.05	1.2286×10^{-3}
		0.1	4.8344×10^{-3}

14.6 Conclusions

In this chapter, two stochastic control approaches have been assessed to deal with the management of generalized flow-based networks. Both the CC-MPC and the TB-MPC approaches focused on robust economic performance under additive disturbances (unbounded and stationary or non-stationary) and avoid relying on heuristic fixed safety volumes such as those used in the CE-MPC or the RB-MPC schemes proposed in Chaps. 12 and 13, what is traduced in better economic performance. According to the results obtained with the considered case study, both techniques showed a relatively similar performance. However, it seems clear that CC-MPC is more appropriate when requiring a low probability of constraint violation, since the use of TB-MPC demands the inclusion of a higher number of scenarios, which may be an issue for the application of the latter to large-scale networks. The analytical approximation of joint chance constraints based on their decomposition into individual chance constraints, these latter bounded by means of the Boole's inequality, has shown to be suitable for large networks regarding that the conservatism involved is not affected neither by the number of the inequalities nor by the prediction horizon of the MPC. The level of resultant back-off is variable and depends on the volatility of the forecasted demand at each prediction step and the suitability of the probabilistic distribution used to model uncertainty. The fact of unbounded disturbances in the system precludes the guarantee of robust feasibility with these schemes. Hence, the approaches proposed in this chapter are based on a service-level guarantee and a probabilistic feasibility. The case study shows that the CC-MPC is suitable for the operational guidance of large-scale networks due to its robustness, flexibility, modest computational requirements, and ability to include risk considerations directly in the decision-making process. Even when the CC-MPC increased the operational costs by around 2.5%, it allowed to improve the service reliability by more than 90% when compared with a CE-MPC setting.

Future research will be directed to incorporate parametric uncertainty and unmeasured disturbances in the model. In addition, future work should include a more detailed study regarding the number of scenarios contained in the tree. Likewise, distributed computation could be used in order to relieve the scaling problems of TB-MPC when the number of scenarios is too high. Moreover, it is of interest to extend the results and develop decentralized/distributed stochastic MPC controllers for large-scale complex flow networks.

References

1. Billings RB, Jones CV (2008) Forecasting urban water demand, 2nd edn. American Water Works Association
2. Biscos C, Mulholland M, Le Lann MV, Buckley CA, Brouckaert CJ (2003) Optimal operation of water distribution networks by predictive control using MINLP. Water SA 29: 393–404

3. Calafiore G, Dabbene F (2006) Probabilistic and randomized methods for design under uncertainty. Springer
4. Calafiore G, Dabbene F, Tempo R (2011) Research on probabilistic methods for control system design. Automatica 47(7): 1279–1293
5. Cannon M, Couchman P, Kouvaritakis B (2007) MPC for stochastic systems. In: Findeisen R, Allgöwer F, Biegler L (eds) Assessment and future directions of nonlinear model predictive control, volume 358 of lecture notes in control and information sciences. Springer, Berlin, pp 255–268
6. Charnes A, Cooper WW (1963) Deterministic equivalents for optimizing and satisfying under chance constraints. Oper Res 11(1): 18–39
7. Congcong S, Puig V, Cembrano G (2014) Temporal multi-level coordination techniques oriented to regional water networks: application to the Catalunya case study. J Hydroinform 16(4): 952–970
8. Geletu A, Klöppel M, Zhang H, Li P (2013) Advances and applications of chance-constrained approaches to systems optimization under uncertainty. Int J Syst Sci 44(7): 1209–1232
9. Grosso JM, Maestre JM, Ocampo-Martinez C, Puig V (2014) On the assessment of tree-based and chance-constrained predictive control approaches applied to drinking water networks. In: Proceedings of 19th IFAC world congress, Cape Town, South Africa, pp 6240–6245
10. Grosso JM, Ocampo-Martinez C, Puig V (2012) A service reliability model predictive control with dynamic safety stocks and actuators health monitoring for drinking water networks. In: Proceedings of 51st IEEE annual conference on decision and control (CDC)
11. Grosso JM, Ocampo-Martinez C, Puig V (2012) A service reliability model predictive control with dynamic safety stocks and actuators health monitoring for drinking water networks. In: Proceedings of 51st IEEE annual conference on decision and control (CDC), Maui, Hawaii, USA, pp 4568–4573
12. Grosso JM, Ocampo-Martinez C, Puig V, Joseph B (2014) Chance-constrained model predictive control for drinking water networks. J Process Control 24(5): 504–516
13. Heitsch H, Römisch W (2009) Scenario tree modeling for multistage stochastic programs. Math Program 118(2): 371–406
14. Kall P, Mayer J (2005) Stochastic linear programming. Number 80 in international series in operations research and management science. Springer, New York, NY
15. Korda M, Gondhalekar R, Cigler J, Oldewurtel F (2011) Strongly feasible stochastic model predictive control. In: Proceedings of 50th IEEE conference on decision and control and European control conference (CDC), pp 1245–1251
16. Leirens S, Zamora C, Negenborn RR, De Schutter B (2010) Coordination in urban water supply networks using distributed model predictive control. In: American control conference (ACC), pp 3957–3962
17. Lucia S, Subramanian S, Engell S (2013) Non-conservative robust nonlinear model predictive control via scenario decomposition. In: Proceedings of 2013 IEEE multi-conference on systems and control (MSC), Hyderabad, India, pp 586–591
18. Nemirovski A, Shapiro A (2006) Convex approximations of chance constrained programs. SIAM J Optim 17(4): 969–996
19. Ocampo-Martinez C, Puig V, Cembrano G, Creus R, Minoves M (2009) Improving water management efficiency by using optimization-based control strategies: the Barcelona case study. Water Sci Technol: Water Supply 9(5): 565–575
20. Ocampo-Martinez C, Puig V, Cembrano G, Quevedo J (2013) Application of predictive control strategies to the management of complex networks in the urban water cycle. IEEE Control Syst 33(1): 15–41
21. Ono M, Williams BC (2008) Iterative risk allocation: a new approach to robust model predictive control with a joint chance constraint. In: Proceedings of 47th IEEE conference on decision and control, Cancun, Mexico, pp 3427–3432
22. Ouarda TBMJ, Labadie JW (2001) Chance-constrained optimal control for multireservoir system optimization and risk analysis. Stoch Environ Res Risk Assess 15: 185–204
23. Prekopa A (1995) Stochastic programming. Kluwer Academic Publishers

24. Raso L, van Overloop PJ, Schwanenberg D (2009) Decisions under uncertainty: use of flexible model predictive control on a drainage canal system. In: Proceedings of the 9th conference on hydroinformatics, Tianjin, China
25. Rockafellar RT, Wets RJ-B (1991) Scenario and policy aggregation in optimization under uncertainty. Math Oper Res 16(1): 119–147
26. Sampathirao AK, Grosso JM, Sopasakis P, Ocampo-Martinez C, Bemporad A, Puig V (2014) Water demand forecasting for the optimal operation of large-scale drinking water networks: the Barcelona case study. In: Proceedings of 19th IFAC world congress, Cape Town, South Africa, pp 10457–10462
27. Schildbach G, Fagiano L, Frei C, Morari M (2013) The scenario approach for stochastic model predictive control with bounds on closed-loop constraint violations. Submitted to Automatica
28. Schwarm AT, Nikolaou M (1999) Chance-constrained model predictive control. AIChE J 45(8): 1743–1752
29. van Overloop PJ, van Clemmens AJ, Strand RJ, Wagemaker RMJ, Bautista E (2010) Real-time implementation of model predictive control on Maricopa-Stanfield irrigation and drainage district's WM canal. J Irrig Drain Eng 136(11): 747–756
30. Zafra-Cabeza A, Maestre JM, Ridao MA, Camacho EF, Sánchez L (2011) A hierarchical distributed model predictive control approach in irrigation canals: a risk mitigation perspective. J Process Control 21(5):787–799

Chapter 15
Fault-Tolerant Model Predictive Control of Water Transport Networks

Vicenç Puig, Carlos Ocampo-Martínez, Deneb Robles and Luis Eduardo Garza-Castañón

15.1 Introduction

WTNs require sophisticated supervisory control strategies to ensure and maintain optimal performance even in faulty conditions. In order to take advantage of these expensive infrastructures, a highly sophisticated real-time control (RTC) scheme is necessary to ensure optimal performance [3, 13]. The RTC scheme in a WTN might be local or global. When control is local, regulation devices only use measurements taken at specific locations. While this control structure is applicable in many simple cases, it may not be the most efficient option for large systems with a highly interconnected and complex sensor and actuator infrastructure. A global control strategy, in contrast, which computes control actions taking into account real-time measurements all through the network, is likely the best way to use infrastructure capacity and all available sensor information. Global RTC deals with the problem of generating control strategies (ahead of time), based on a predictive dynamic model and telemetry readings of the network to optimize operation [13]. The multivariable and large-scale nature of WTNs have led to the use of some variants of MPC as a global control strategy [15], as discussed in previous chapters.

Global RTC of WTNs needs to be operative even in faulty conditions. This problem calls for the use of fault-tolerant control (FTC) mechanisms after a fault is diagnosed so as to avoid the global RTC stopping every time a fault appears. FTC was developed in order to address the growing demand for plant availability [1]. The

V. Puig (✉)
Research Center Supervision, Safety and Automatic Control (CS2AC-UPC), Terrassa, Spain
e-mail: vicenc.puig@upc.edu

C. Ocampo-Martínez
Institut de Robòtica i Informàtica Industrial, CSIC-UPC, Barcelona, Spain

D. Robles · L.E. Garza-Castañón
Technológico de Monterrey, Monterrey, Mexico

V. Puig et al. (eds.), *Real-Time Monitoring and Operational Control of Drinking-Water Systems*, Advances in Industrial Control,
DOI 10.1007/978-3-319-50751-4_15

aim of FTC is to keep a plant fully operative by designing its control system such that system performance can be kept close to desirable levels and stability conditions can be maintained, not only when the system is in nominal conditions, but also in the presence of system component faults; FTC should, at the very least, ensure acceptable degraded performance [11]. Tolerance against faults can be embedded in MPC relatively easily in several different ways, as discussed in [9]:

- Changing the constraints in order to represent the fault effect with the algorithms for actuator faults being especially easy to adapt.
- Modifying the internal plant model used by the MPC in order to reflect fault influence on the plant.
- Relaxing the nominal control objectives in order to reflect system limitations under faulty conditions.

Reviewing the literature, the inclusion of fault tolerance in MPC has already been considered by several authors, including [30], who provides a detailed review of the state of the art in FTC. Camacho et al. [4] provides a general overview on how fault tolerance can be embedded in MPC. The inclusion of fault tolerance in MPC has mainly been addressed by considering practical strategies according to the application domain. For example, [16] described a method for including fault tolerance in MPC for smart grids in order to ensure the proper amount of energy in storage devices and reliable coverage of essential consumer demand. Ocampo-Martinez and Puig [12] applied fault tolerance in MPC to sewage networks considering a hybrid systems framework. Yang and Maciejowski [28] designed a group of predictive controllers to compensate for the fault effects for each component in a wind turbine. More theoretical aspects have also begun to be studied, such as coupling with active fault diagnosis [17] and the use of set-invariance theory [29]. More recent additional objectives for MPC controllers, proposed in [20, 21], have been to preserve system health and reliability, respectively.

The research presented in this chapter is based on three concepts:

- How fault accommodation/reconfiguration strategies were applied in a linear quadratic regulator (LQR) [23].
- The idea that fault configurations should be evaluated before applying FTC strategies [24].
- The idea of using reliability with the FTC design [7].

Starting from these key ideas, it is proposed a new reliable fault-tolerant MPC scheme for application to WTNs. After a fault has occurred, the MPC controller should be redesigned to cope with the fault by considering either a reconfiguration or an accommodation strategy, depending on knowledge available on the fault. Before starting to apply the FTC strategy, whether the MPC controller will be able to continue operating after the fault appears should be evaluated. This is done in two ways: first, a structural analysis is done to determine the level of loss in post-fault controllability; second, a feasibility analysis is done by the optimization problem related to the MPC design so as to consider the fault effect on actuator constraints. By evaluating the admissibility of different actuator-fault configurations (AFCs), critical actuators regarding fault tolerance can be identified considering structural, feasibility, performance and reliability analyses. This has been studied in [19], where only some of the analyses proposed here were considered.

The proposed approach allows a degradation analysis of the system to be performed in terms of performance and reliability. As a result of this analysis, the MPC controller design can be modified, adapting the constraints so as to achieve the best achievable performance with some pre-established level of reliability. The proposed approach was tested in the Barcelona WTN, in an application that also shows that relevant information about critical actuators can be extracted by considering the different analyses proposed.

The main contribution of this chapter is the design of methodologies for the analysis of the influence of faults taking into account reliability features. As discussed, some of the proposed methodologies have been previously documented but not their application in the considered fault tolerance framework, to the best of the knowledge of the authors, after a thorough literature review (a secondary contribution of the chapter).

15.2 Problem Statement

15.2.1 Flow-Based Control-Oriented Model

This chapter considers a general WTN as represented by a digraph $G(\mathcal{V}, \mathcal{E})$ (see [22] for more details), where a set of elements, i.e., n_s sources, n_x storage elements, n_q intersection nodes, and n_d sinks, are represented by $v \in \mathcal{V}$ vertices connected by $a \in \mathcal{E}$ links. Due to the network function, water is transported along the links by n_u flow actuators (i.e., pipes and valves), passing through reservoirs or tanks, from specific origin locations to specific destination locations. The network is subject to several capacity and operational constraints, and to measured stochastic flows to customer sinks as driven by water demand.

Selecting the volume in storage elements as the state variable $\mathbf{x} \in \mathbb{R}^{n_x}$, the flow through the actuators as the manipulated inputs $\mathbf{u} \in \mathbb{R}^{n_u}$ and the demanded flow as *additive* measured disturbances $\mathbf{d} \in \mathbb{R}^{n_d}$, the control-oriented model of the WTN may be described by the following set of linear (or linearized) discrete-time difference-algebraic equations (DAE) for all time instants $k \in \mathbb{Z}_+$:

$$\mathbf{x}(k+1) = \mathbf{A}\mathbf{x}(k) + \mathbf{B}\mathbf{u}(k) + \mathbf{B}_d\mathbf{d}(k), \tag{15.1a}$$

$$0 = \mathbf{E}_u\mathbf{u}(k) + \mathbf{E}_d\mathbf{d}(k), \tag{15.1b}$$

where the difference equation in (15.1a) describes the dynamics of the storage tanks, and the algebraic equation in (15.1b) describes static relations in the network (i.e., mass balance at junction nodes). Moreover, $\mathbf{A}$, $\mathbf{B}$, $\mathbf{B}_d$, $\mathbf{E}_u$ and $\mathbf{E}_d$ are time-invariant matrices of suitable dimensions as dictated by the network topology.

System (5.1) is subject to hard state and input polytopic constraints given by:

$$\mathcal{U} \triangleq \left\{\mathbf{u} \in \mathbb{R}^{n_u} \mid \mathbf{u}^{\min} \leq \mathbf{u} \leq \mathbf{u}^{\max}\right\}, \tag{15.2a}$$

$$\mathcal{X} \triangleq \left\{\mathbf{x} \in \mathbb{R}^{n_x} \mid \mathbf{x}^{\min} \leq \mathbf{x} \leq \mathbf{x}^{\max}\right\}, \tag{15.2b}$$

where $\mathbf{u}^{\min}$, $\mathbf{u}^{\max}$, $\mathbf{x}^{\min}$ and $\mathbf{x}^{\max}$ are the actuator and tank operational limits.

15.2.2 Statement of the Control Problem

The WTN (5.1) is controlled using an MPC law that aims to minimize the operational costs of the WTN as proposed in economic model predictive control (EMPC) [5, 8, 18]. According to [1], the solution of a control problem consists of finding a control law from a given set of *control laws* $\mathbb{U}$, such that the controlled system achieves the *control objectives* $\mathbb{O}$ while its behaviour satisfies a set of *constraints* $\mathbb{C}$. Thus, the solution to the problem is completely defined by the triplet $\langle\mathbb{O}, \mathbb{C}, \mathbb{U}\rangle$. In the case of an MPC, the triplet $\langle\mathbb{O}, \mathbb{C}, \mathbb{U}\rangle$ is defined by

$$\mathbb{O}: \quad \min_{\tilde{\mathbf{x}},\tilde{\mathbf{u}}} J(\tilde{\mathbf{x}}, \tilde{\mathbf{u}}), \tag{15.3a}$$

subject to:

$$\mathbb{C}: \tag{15.3b}$$

$$\mathbf{x}(k+i+1|k) = \mathbf{A}\mathbf{x}(k+i|k) + \mathbf{B}\mathbf{u}(k+i|k) + \mathbf{B}_d\mathbf{d}(k+i|k), \quad \forall i \in \mathbb{Z}_{[0,H_p-1]} \tag{15.3c}$$

$$\mathbf{0} = \mathbf{E}_u\mathbf{u}(k+i|k) + \mathbf{E}_d\mathbf{d}(k+i|k), \quad \forall i \in \mathbb{Z}_{[0,H_p-1]}, \tag{15.3d}$$

$$\mathbf{u}(k+i|k) \in \mathcal{U}, \quad \forall i \in \mathbb{Z}_{[0,H_p-1]}, \tag{15.3e}$$

$$\mathbf{x}(k+i|k) \in \mathcal{X}, \quad \forall i \in \mathbb{Z}_{[1,H_p]}, \tag{15.3f}$$

where

$$\tilde{\mathbf{x}} = \left(\mathbf{x}(1|k), \ldots, \mathbf{x}(H_p|k)\right), \tag{15.4a}$$

$$\tilde{\mathbf{u}} = \left(\mathbf{u}(0|k), \mathbf{u}(1|k), \ldots, \mathbf{u}(H_p-1|k)\right), \tag{15.4b}$$

$$\tilde{\mathbf{d}} = \left(\mathbf{d}(0|k), \mathbf{d}(1|k), \ldots, \mathbf{d}(H_p-1|k)\right) \tag{15.4c}$$

are the state, input and disturbance sequences over H_p, respectively. H_p denotes the prediction horizon used by the MPC controller. The sequence $\tilde{\mathbf{d}}$ comes from a forecasting module based on existing time series techniques (see [15, 26] for more details).

The MPC law belongs to the set $\mathcal{U}$ and is obtained using the *receding horizon philosophy* [9, 18]. This technique consists of solving the optimization problem (15.3a) from the current time instant k to $k + H_p$ using $\mathbf{x}(0|k)$ as the initial condition obtained from measurements (or state estimation) at time k. Only the first value $\mathbf{u}^*(0|k)$ from the optimal input sequence $\tilde{\mathbf{u}}^*$ (which arises from the solution of the optimization problem (15.3a)) is applied to the system. At time $k+1$, in order to compute $\mathbf{u}^*(0|k+1)$ the optimization problem, (15.3a) is solved again from $k+1$ to $k+1+H_p$ (i.e., the time window is shifted), updating initial states $\mathbf{x}(0|k+1)$ from measurements (or state estimation) at time $k+1$. The same procedure is repeated for the following time instants.

The objective function J in (15.3a) collects all the control objectives of the closed-loop system, taking the name *multiobjective cost function*. In general form, (15.3a) can be written as:

$$J(\tilde{\mathbf{x}}, \tilde{\mathbf{u}}) = \sum_{i=0}^{n_J} \sum_{k=0}^{H_p} J_i(k), \tag{15.5}$$

where n_J is the number of objectives and $J_{i,k}$ corresponds to the evaluation of each particular objective i at time k. In the case of WTNs, (15.5) typically includes the objectives presented in Chap. 12.

15.2.3 Inclusion of Fault-Tolerant Capabilities

The control problem $\langle \mathbb{O}, \mathbb{C}, \mathbb{U} \rangle$ described in Sect. 15.2.2 will now be reformulated to consider faults. If an active FTC strategy is considered, there are two main ways to adapt the MPC law to introduce fault tolerance [1]:

1. *System reconfiguration*. This consists of finding a new set of constraints $\mathbb{C}_f(\Theta_f)$, where Θ_f is the set of parameters changed by the faults such that the control problem $\langle \mathbb{O}, \mathbb{C}_f(\Theta_f), \mathbb{U}_f \rangle$ can be solved. This strategy can be applied when the fault detection and isolation (FDI) module does not provide a fault estimation. The faulty components are therefore unplugged by the supervisory system and the control objectives are achieved using non-faulty components. In the case of the actuators, this implies that the model (5.1) used by the MPC controller is modified as follows:

$$\mathbf{x}(k+1) = \mathbf{A}\mathbf{x}(k) + \sum_{i \in I_N} \mathbf{B}_i \mathbf{u}(k,i) + \mathbf{B}_d \mathbf{d}(k), \tag{15.6}$$

$$\mathbf{0} = \sum_{i \in I_N} \mathbf{E}_{u,i} \mathbf{u}(k,i) + \mathbf{E}_d \mathbf{d}(k), \tag{15.7}$$

   where I_N is the subset of non-faulty actuators.
2. *Fault accommodation*. This approach consists of solving the control problem $\langle \mathbb{O}, \hat{\mathbb{C}}_f(\hat{\Theta}_f), \hat{\mathbb{U}}_f \rangle$, where $\hat{\mathbb{C}}_f(\hat{\Theta}_f)$ is an estimate of current system constraints and parameters provided by the FDI module. This strategy can be applied when a change occurs in either system structure or parameters. In this strategy, the control law is modified while the remaining elements within the control loop are kept

unchanged. In the case of the actuators, this requires that the system model 5.1 used by the MPC controller should be modified as follows:

$$\mathbf{x}(k+1) = \mathbf{A}\mathbf{x}(k) + \sum_{i \in I_N} \mathbf{B}_i \mathbf{u}(k,i) + \sum_{i \in I_F} \beta_i(\mathbf{u}(k,i), \theta_i) + \mathbf{B}_d \mathbf{d}(k), \tag{15.8}$$

$$\mathbf{0} = \sum_{i \in I_N} \mathbf{E}_{u,i} \mathbf{u}(k,i) + \sum_{i \in I_F} \varepsilon_i(\mathbf{u}(k,i), \theta_i) + \mathbf{E}_d \mathbf{d}(k), \tag{15.9}$$

where the functions β_i and ε_i and the parameters θ_i should be estimated by the FDI module for actuators belonging to the faulty actuator subset I_F.

Note that, in changing the model (5.1) of the MPC controller using either of the two previous strategies, the controller will consider the effect of the fault in the system model when computing the control action $\mathbf{u}^*(0|k)$. According to [9], this is different from other control laws (e.g., LQR and pole placement), where the control law should be designed offline for the considered set of faults, so as to produce a bank of controllers that should be gain-scheduled online according to the fault features. However, depending on how critical the fault is, the MPC controller will not be able to compute a control input or else the computed control input will not lead to acceptable performance. For this reason, when using an MPC controller, the effect of the fault and the admissibility of the obtained control input needs to be evaluated.

15.3 Proposed Approach

This section describes a series of analyses to assess the fault-tolerant capabilities of the system after a fault has occurred and before applying a reconfiguration or accommodation strategy to achieve fault tolerance.

In case if a fault occurs, then:

- The system might have lost some of the properties required to proceed with system control, or
- That system performance is degraded to an unacceptable level and it is not worth continuing with system control by activating fault-tolerant strategies.

15.3.1 Admissibility Analysis Algorithms

Before starting to apply the FTC strategies described above, it should be evaluated whether the MPC controller will be able to continue operating after fault occurrence. This is done by means of a set of admissibility analysis algorithms, which are based on a structural analysis to determine the loss of post-fault controllability, complemented by a feasibility analysis of the optimization problem related to the MPC design so as

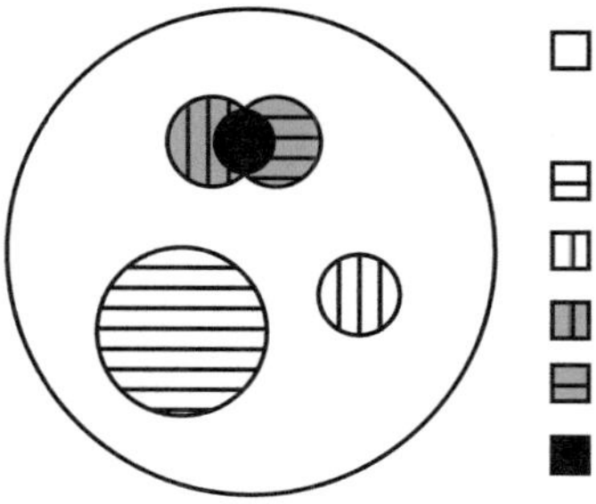

Fig. 15.1 Critical and redundant actuators of the system

to consider the effect of the fault on actuator constraints. Moreover, by evaluating the admissibility of the different AFCs, critical actuators regarding fault tolerance can be identified considering structural, feasibility, performance and reliability analyses.

Let I be the set of system actuators. The different admissibility analysis algorithms consider that the set of all subsets of system actuators is denoted by 2^I. For each subset $K \subseteq I$, corresponding to a given AFC, and using the reconfiguration (or accommodation) approach described in Sect. 15.2.3, the algorithms evaluate whether or not a given system property, denoted by $P(K)$, is satisfied [1]. Thus,

$$P_K = \begin{cases} 1 \text{ if the property is satisfied,} \\ 0 \text{ if the property is not satisfied.} \end{cases} \tag{15.10}$$

This evaluation induces the set of all subsets of I, 2^I, to be partitioned into two classes as follows:

$$2^{I+} = \{K \subseteq I;\, P_K = 1\}, \tag{15.11}$$

$$2^{I-} = \{K \subset I;\, P_K = 0\}. \tag{15.12}$$

The class 2^{I+} contains all the subsets of the actuators for which P_K is satisfied. Thus, the admissibility analysis mainly aims to identify the following (see Fig. 15.1):

- *Critical actuators*, i.e., the set of actuators that are required to satisfy P_K. For every analysis in Fig. 15.2, a set of critical actuators will be identified.
- *Redundant actuators*, i.e., the actuators that are not critical for correct functioning of the system. These may be excluded as P_K will continue to be satisfied.
- *Redundancy degree*, consisting of the number of extra non-critical actuators through which P_K could hold. There are two types of redundancy: *weak* (corresponding to the largest number of sequential faults that can be tolerated in the best case scenario, i.e., while continuing to satisfy P_K) and *strong* (corresponding to the smallest number of sequential faults that can be tolerated in the worst-case scenario).

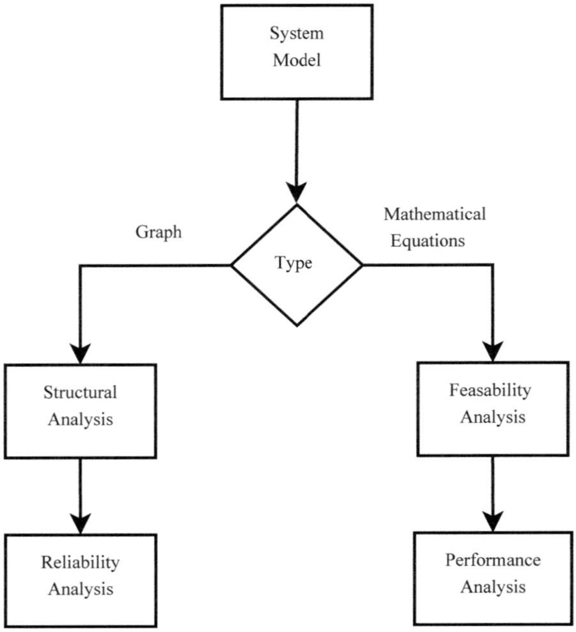

Fig. 15.2 Flow diagram of the proposed actuator fault-tolerant evaluation approach

The approach proposed here consists of a set of analyses based on both the graph and the mathematical model of the system (see Fig. 15.2):

- From the system graph, the *structural analysis* allows to determine whether or not the system with a given AFC is structurally controllable. It does this by checking the existence of at least one path linking demands with sources. At this stage, all possible paths linking demands and sources are also determined. Using this information, the *reliability* of the AFC can also be evaluated.
- From the system mathematical model, a constraint satisfaction problem (CSP) can be formulated that allows a *feasibility analysis* to be performed. This analysis allows the physical capacity of the system to be checked considering constraints in actuators and states (see (15.3a)). Moreover, as a complementary analysis, the *closed-loop performance* based on a given global objective for the AFC can be evaluated.

These two sets of analyses are complementary. When a reconfiguration strategy is used, connectivity between demands and sources may be lost when the faulty actuator is removed (see Sect. 15.2.3). This will affect both controllability and reliability. However, those properties do not take into account the physical limitations of the system actuators. Hence, although connectivity is preserved, the MPC-related optimization problem might lead to an unfeasible solution, due either to the lack of capacity of the remaining actuators or the poor performance of the control loop. This happens when an accommodation strategy is used, since although the connectivity among elements is preserved (the faulty actuator is not removed), the resulting MPC-related optimization problem may be unfeasible or the closed-loop control scheme may perform poorly.

As a result of the application of the methodology presented in Fig. 15.2, it is possible to determine critical actuators as follows (type of analysis in brackets):

- Actuators that are essential to preserving demand–source connectivity (by means of structural controllability analysis).
- Actuators that are indispensable to preserving the capacity to move the desired water volume from sources to meet demands taking into account actuator physical constraints (by means of structural controllability analysis).
- Actuators whose malfunction generates high suboptimality of the considered control objective if the system is maintained in operation after fault detection (by means of performance analysis).
- Actuators whose malfunction does not guarantee reliable operation of the system (by means of reliability analysis).

Figure 15.1 depicts the different types of critical actuators that can be identified applying the sequence of analyses presented in Fig. 15.2. Results for each analysis are considered in subsequent analyses, in such a way that actuators that are considered critical at a given stage of the methodology might not be further considered in the later analyses.

15.3.2 Analyses Based on the System Graph

15.3.2.1 Structural Analysis Algorithm

The structural analysis algorithm copes with connectivity properties of the system without considering the actual value of the model parameters or the limitations of the actuators.[1] This test is used to evaluate the admissibility of a given AFC when the reconfiguration FTC strategy is used, i.e., when an actuator is removed after fault occurrence and the system is controlled by the remaining actuators.

The algorithm starts by determining the digraph[2] $G(\mathcal{V}, \mathcal{E})$ of the model used for the MPC controller. Using the digraph, the *structural controllability* of the system for a given AFC will be evaluated. If this property is preserved after the actuator fails, the AFC is admissible; i.e., it is able to tolerate the fault; otherwise, the AFC is not admissible. To evaluate structural controllability from the system graph, some basic graph theory concepts will be used (see [2] for more details). Using Theorems 15.1 and 15.2, Algorithm 15.9 will perform the structural controllability analysis for a given AFC.

---

[1] See [1] for important definitions related to the topic.

[2] See [22] for details on how to obtain a digraph from the system model.

Algorithm 15.1 Controllability analysis using the structural approach

1: Obtain the digraph $G = (\mathcal{V}, \mathcal{E})$ of the system model used for designing the MPC (related to the optimization problem in (15.3a)) given a particular AFC
2: From the system digraph $G = (\mathcal{V}, \mathcal{E})$, find the reachability matrix Γ
3: **for** each $\mathbf{x}_i \in \mathbb{R}^{n_x}, i = 1, ..., n_x$ **do**
4: **if** $\nexists \mathbf{u}_j \in \mathbb{R}^{n_u}, j = 1, ..., n_u \mid \Gamma_{ij} = 1$ **then**
5: AFC is *non-input-reachable*
6: **else**
7: **if** $s\text{-}rank([\mathbf{A}\ \mathbf{B}]) \neq n$ **then**
8: is *non-structurally controllable*
9: **else**
10: is *structurally controllable*
11: **end if**
12: **end if**
13: **end for**

15.3.3 Analyses Based on the System Mathematical Model

15.3.3.1 Feasibility Analysis Algorithm

To evaluate the admissibility of the control of a given AFC when system constraints (15.2) are considered, it is not possible to use the structural analysis algorithm[3] presented in Sect. 15.3.2.1.

Feasibility in an MPC controller design is a key property to be satisfied before the control action can be computed by solving the optimization problem (15.3a) [9]. In this case, the admissibility evaluation problem for a given AFC can be naturally handled as a CSP. Consequently, the feasibility evaluation of the MPC-related optimization problem (here for a given AFC using the reconfiguration strategy)[4] can be checked using Algorithm 15.10.

15.3.3.2 Performance Analysis Algorithm

The degradation of the control objective in a faulty situation can be quantified by means of maximal loss of efficiency ρ with respect to the objective function in a non-faulty situation J_0. This fact establishes whether or not the control objective degradation after an actuator fault J_f is admissible. Thus, an AFC is admissible regarding performance if the following condition is satisfied: $J_f \leq (1 + \rho)\, J_0$. This condition will enable a performance analysis of the AFC considering the faulty actuator, with either an accommodation or a reconfiguration strategy.

[3]This would also be the case when an accommodation FTC strategy is used, since the actuator would not be removed after the fault but would be operated under the remaining operating range estimated by the FDI module.

[4]If an accommodation strategy is used, then the faulty model used in Algorithm 15.10 should be replaced by the one used in (15.8).

Algorithm 15.2 Feasibility Analysis

1: **for** $k = 1$ to H_p **do**
2: $\quad \mathcal{U}(k-1) \Leftarrow \mathcal{U}$
3: $\quad \mathcal{X}(k) \Leftarrow \mathcal{X}$
4: **end for**
5: $\mathcal{W} \Leftarrow \{\overbrace{\mathbf{x}_1, \mathbf{x}_2, \ldots, \mathbf{x}_{H_p}}^{\tilde{\mathbf{x}}}, \overbrace{\mathbf{u}_1, \mathbf{u}_2, \ldots, \mathbf{u}_{H_p-1}}^{\tilde{\mathbf{u}}}\}$
6: $\mathcal{D} \Leftarrow \{\mathcal{X}_1, \mathcal{X}_2, \ldots, \mathcal{X}_{H_p}, \mathcal{U}_1, \mathcal{U}_2, \ldots, \mathcal{U}_{H_p-1}\}$
7: $\mathcal{Z} \Leftarrow \left\{ \left(\mathbf{x}(k+1) = \mathbf{A}\mathbf{x}(k) + \sum_{i \in I_N} \mathbf{B}_i \mathbf{u}(k,i) + \mathbf{B}_d \mathbf{d}(k), \quad \mathbf{0} = \mathbf{E}_u \mathbf{u}(k) + \mathbf{E}_d \mathbf{d}(k) \right)_{k=0}^{H_p-1} \right\}$
8: $\mathcal{H}_{\mathcal{A}} = (\mathcal{W}, \mathcal{D}, \mathcal{Z})$
9: **if** the CSP $\mathcal{H}_{\mathcal{A}}$ has solution **then**
10: AFC is *admissible*
11: **else**
12: AFC is *non-admissible*
13: **end if**

The procedure for evaluating the performance admissibility of the controller with respect to the fault situation is summarized by Algorithm 15.10, modifying the constraints defined in Step 7 to add a new constraint:

$$\phi_{x_{H_p}} + \sum_{i=0}^{H_p-1} \Phi_i(\mathbf{x}_i, \mathbf{u}_i) \leq (1+\rho)\, J_0. \tag{15.13}$$

Notice that, as in the case of the feasibility analysis, the existence of a solution to the CSP associated with MPC performance evaluation for a given AFC using the reconfiguration strategy[5] can be proved by Algorithm 15.10 but including the new constraint (15.13), which considers the admissibility condition with respect to control performance over the prediction horizon H_p stated in the MPC controller.

15.3.4 Reliability Analysis Algorithm

Reliability is defined as the probability that a given component (or system) will accomplish its intended function during a given period of time and in specific operating conditions and environments [6]. In other words, it is the probability of success in accomplishing a task or achieving a desired property in a process, based on proper operation of components. The main advantages of including a reliability analysis are as follows:

[5] If an accommodation strategy is used, the fault model used in Algorithm 15.10 should be replaced by the one used in (15.8).

- Information on component health is integrated into controller design and improves the life of the system components
- Reliability information on the system can be considered as design criteria to be used in MPC implementation including FTC capabilities
- Essential actuators whose malfunction causes abrupt system reliability decay are identified.

In the case of WTNs, reliability is understood as the ability of the network to provide an efficient water supply to consumers under both normal and abnormal operating conditions. For this reason, reliability is a measure of WTN performance. Reliability in WTNs has already been considered in the literature [14, 25].

When a reconfiguration FTC strategy is used, the reliability of DTWNs can be affected due to the probabilities of success of each of the components in the new configuration. For this case, the admissibility evaluation problem of a given AFC can be handled as composite reliability of the subsystems in the system. In particular, since reliability in DTWNs is related to guaranteed supply to consumers, it can be determined based on all the possible paths linking demands and sources from the network graph already obtained in the structural analysis.

The global reliability of a system, denoted by R_g, generally consists of the decomposition of its subsystems into elementary combinations of serial and parallel subsystems that can be extracted from the matrix containing all paths linking demands and sources [7]:

- Reliability of n_p parallel subsystems is defined as:

$$R_p(k) = 1 - \prod_{i=1}^{n_p}(1 - R_i(k)). \tag{15.14}$$

- Reliability of n_s serial subsystems is defined as:

$$R_s(k) = \prod_{i=1}^{n_s} R_i(k), \tag{15.15}$$

where $R_i(k)$ represents the reliability of the i-th actuator (or subsystem) at time k and where $\gamma_i(k)$ is the failure rate modelled as an exponential distribution

$$R_i(k) = e^{-k\gamma_i(k)}. \tag{15.16}$$

Thus, overall system reliability is given by

$$R_g(k) = \prod_{i=1}^{n_s}(1 - \prod_{i=1}^{n_p}(1 - R_i(k))). \tag{15.17}$$

Algorithm 15.11 shows the reliability evaluation of a given AFC based on computing system reliability. Since the calculation of reliability for each and every AFC could impose a great computational burden, to save time, the path matrix that contains all the possible paths in the system graph is used. This matrix has the following structure:

$$
\begin{array}{c|c|c|c|c|c|}
 & p_1 & p_2 & p_3 & \dots & p_{n_{ph}} \\
\hline
u_1 & 1 & 0 & 1 & \dots & 0 \\
u_2 & 0 & 1 & 1 & \dots & 1 \\
u_3 & 1 & 0 & 0 & \dots & 1 \\
\vdots & \vdots & \vdots & \vdots & \vdots & \vdots \\
u_{n_u} & 0 & 1 & 1 & \dots & 1
\end{array}
\tag{15.18}
$$

where n_{ph} is the number of path, and 1 and 0 indicate the presence and absence, respectively, of an actuator in the path. Each time a component malfunctions, the row assigned to that actuator is withdrawn along with all the paths that make use of it. To evaluate fault tolerance for the rest of the system, the reliability index $R_g(k)$ should be greater than a specific admissibility threshold R_{th} at a given time horizon k_{end}, both defined by the user.

Algorithm 15.3 Reliability analysis

Decompose the system in n_p parallel subsystems and n_s subsystems using the system graph.
for $i = 1$ to n_u **do**
  Evaluate actuator reliability $R_i(k)$ using (15.16).
end for
for $g = 1$ to n_p **do**
  Evaluate reliability of parallel subsystems $R_{p(k)}$ using (15.14) and (15.16).
end for
for $g = 1$ to n_s **do**
  Evaluate reliability of system $R_g(k)$ using (15.17) and the result obtained from the evaluation in (15.14).
end for

15.3.5 *MPC Redesign to Preserve Reliability*

When a fault occurs, the MPC law is modified to cope with the fault, as discussed in Sect. 15.2.3. As explained in [7], the value of the actuator failure rate changes because the control action should be increased in order to compensate for the fault effect. In this case, energy consumption increases and the value of the failure rate also increases due to the actuator load increment. Thus, there is an interplay between maintaining closed-loop performance and reliability. To maintain the desired performance, the relationship between the actuator load increment and reliability can be established.

One of the most commonly used relationships is based on assuming that the actuator failure rate changes with the load through the following exponential law:

$$\gamma_i(k) = \gamma_i^o e^{\beta_i \mathbf{u}_i(k)}, \tag{15.19}$$

where γ_i^o represents the baseline failure rate (nominal failure rate) and $\mathbf{u}_i$ is the control action for the i-th actuator. Parameter β_i is a fixed factor that depends on the actuator characteristics. Thus, the reliability of the actuator can be expressed in terms of its load as follows:

$$R_i(k) = e^{k\gamma_i(k)} = e^{\gamma_i^o e^{k\beta_i \mathbf{u}_i(k)}}. \tag{15.20}$$

Consider that a predefined reliability threshold R_{th} should be maintained until the end of the system mission at time k_{end}. This threshold defines the minimal acceptable reliability value in the degraded fault mode. The aim is to translate this threshold to a load threshold that can be applied to the actuator. This actuator load threshold can be derived from (15.20) as follows:

$$|u_{i,th}| = \frac{1}{\beta_i} \ln \left(\frac{\ln R_{i,th}}{\gamma_i^o k_{end}} \right). \tag{15.21}$$

Hence, the MPC controller (15.3a) can be redesigned by including the following constraint in the i-th actuator control:

$$u_i \in \left[-u_{i,th}, u_{i,th} \right]. \tag{15.22}$$

However, as discussed in [27], this will only preserve the reliability of the i-th actuator. In order to preserve the reliability of the whole WTN, the new actuator constraints (15.22) should be derived taking into account the reliability expression (15.15) and the reliability threshold R_{th} at the end of the MPC prediction horizon H_p. This can be achieved by formulating a CSP problem, such as that reflected in Algorithm 15.12, which considers, as constraints, the reliability of the WTN in (15.15) derived by means of Algorithm 15.11 in terms of the reliability of each actuator, the impact of actuator load (see (15.20)) and the actuator operational constraints defined in (15.3a).

After solving the CSP problem in Algorithm 15.12, to solve the optimization problem associated with the MPC design, the resulting updated actuator constraints are used instead of the actuator operational constraints defined in (15.3a). In this way, it can be guaranteed that the MPC controller computes a control sequence that preserves reliability. There is, of course, a trade-off between reliability and performance. Increasing the reliability threshold R_{th} will imply a reduction in the WTN performance but will extend the life of the remaining actuators.

Algorithm 15.4 MPC redesign to preserve reliability

for $k = 1$ to H_p **do**
  $\mathcal{U}(k-1) \Leftarrow \mathcal{U}$
end for
$\mathcal{W} \Leftarrow \overbrace{\{\mathbf{u}_1, u_2, \ldots, \mathbf{u}_{H_p-1}\}}^{\tilde{\mathbf{u}}}$
$\mathcal{D} \Leftarrow \{\mathcal{U}_1, \mathcal{U}_2, \ldots, \mathcal{U}_{H_p-1}\}$
$\mathcal{Z} \Leftarrow \left\{ \left(R_g(k) = f(R_i(k)),\, R_i(k) = e^{\gamma_i^o e^{k\beta_i |u_i|}},\, i = 1, \ldots, n_u \right)_{k=0}^{H_p-1}, R_g(H_p-1) > R_{th} \right\}$
$\mathcal{H}_{\mathcal{A}} = (\mathcal{W}, \mathcal{D}, \mathcal{Z})$
$\{\mathcal{U}_1, \mathcal{U}_2, \ldots, \mathcal{U}_{H_p-1}\} \Leftarrow$ **solve**$(\mathcal{H}_{\mathcal{A}})$

15.4 Simulations and Results

15.4.1 Case Study Description

The Barcelona WTN, presented in Figs. 2.2, 2.3 and 2.4 of Chap. 2, is used as the case study of this chapter. Figure 15.3 shows the graph derived from this network; the nodes correspond to reservoirs or pipe merging/splitting nodes and the arcs correspond to actuators (valves and pumps). Five of the pumps are used to draw water from underground sources and the remaining pumps satisfy water demand at appropriate pressure levels. The network has 88 main water consumption sectors (for further information regarding the Barcelona WTN, see [15]). Both the demand episode and the calibration set-up of the network are as established by Aguas de Barcelona. The control centre has a telecontrol system for network management. The Barcelona WTN also has some 98 remote stations, which manage about 450 elements in real time, including flow metres, pumps, valves and chlorine-dosing instruments.

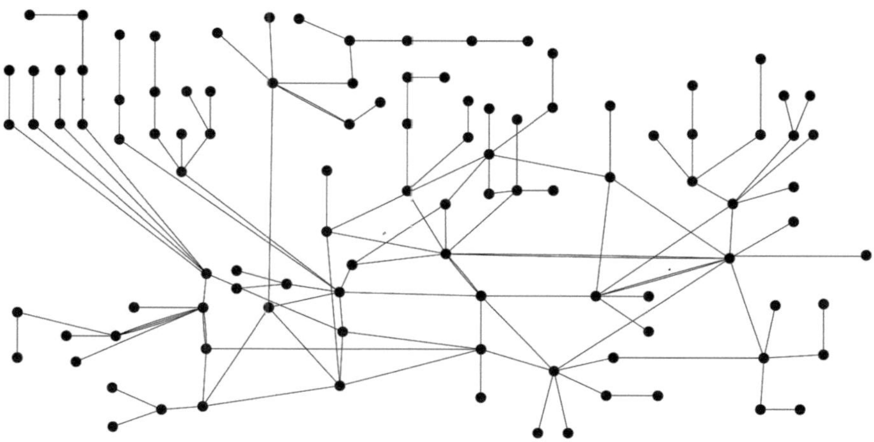

Fig. 15.3 Graph of the Barcelona WTN

The system control objective set for the MPC controller is to minimize operational costs (water transport and production for the entire network) while satisfying water demand for each consumption sector [15]. Thus, recapping on Chap. 12, the economic objective function can be written as follows:

$$J(k) = \sum_{i=0}^{H_p-1} [\alpha_1 + \alpha_2(k)]\mathbf{u}(k), \tag{15.23}$$

which takes into account the water cost α_1 (price of water at source) and the electricity cost $\alpha_2(k)$ (operation of pumps and valves). Note that the time variance of α_2 is due to the fact that pumping costs vary according to the time of day. The prediction horizon H_p is 24 h. No terminal cost is considered in this application.

Demands are imposed as equality constraints in the model (5.1) used by the MPC controller, which, in the case of the WTN, can be expressed in discrete-time state-space form (5.1) using a sampling time $\Delta t = 1$ hour. Moreover, $\mathbf{x} \in \mathcal{X} \subseteq \mathbb{R}^{n_x}$ is the state vector corresponding to the water volumes of the $n_x = 63$ tanks, $\mathbf{u} \in \mathcal{U} \subseteq \mathbb{R}^{n_u}$ represents the vector of manipulated flows through the $n_u = 130$ actuators (pumps and valves), and $\mathbf{d} \in \mathcal{D} \subseteq \mathbb{R}^{n_d}$ corresponds to the vector of the $n_d = 88$ water demands (consumption sectors).

There are 16 nodes in the Barcelona WTN and since demand can be forecasted, these are assumed to be known. Thus, $\mathbf{d}$ is a known vector of non-negative elements that contains the measured disturbances (demands) affecting the system.

15.4.2 Results

Several tests and analyses were performed for the Barcelona WTN case study to illustrate the proposed methodology. Figure 15.2 shows the sequence of tests applied. In this section, all the capabilities of each analysis are explored, while Sect. 15.4.3 describes only the ones necessary for this study. The results were obtained using a 1.5 GHz and 2:00 Gb RAM Intel(R) Core(TM)2 Duo PC. Matlab© and Tomlab were used to perform the simulations.

The structural analysis was carried out using the computed *reachability matrix* and *path computation*, which, as expected, produced equivalent results. However, each technique yielded several additional results that provided important information concerning the operation and behaviour of the WTN. From the reachability analysis, it is possible to determine which states were structurally controllable, while the path computation analysis obtained all possible paths from a source to a destination node as well as, for each path, an approximate operational cost (according to the electricity cost of each element) and a maximal water flow (according to the physical constraints of the actuators). In this stage, critical actuators were located and different approaches were used according to the applied strategy. Although a fault scenario with a faulty actuator at each time instant was considered in both cases, the representation of the malfunction was denoted in different ways. In the reachability analysis, the malfunction was determined from the state-space matrices (a zero value was forced in the position where a connection value previously existed between the state and the actuator failure). In path computation, all paths with the faulty component were extracted from the path matrix (15.18). From this study, the critical actuators for each state and for the whole network could be identified. Note that although the results obtained by both techniques in the structural analysis were similar, the computation time required for the reachability-matrix-based strategy was much higher, at almost 200 times the time consumed by the path computation technique (579 s vs. 3 s).

The feasibility analysis can only be implemented if the previous analysis is first made, since its implementation is based on the path matrix calculation. The result of this analysis was a set of paths that guaranteed that demand was satisfied, taking into account the physical constraints of the network actuators. The cost of maintaining correct network operations was also obtained in this stage. The time consumed by this analysis was 1.57 s.

Performance was computed using the objective function (15.23) and the actuator constraints. The analyses were performed taking into account faulty components and comparing the corresponding performance with the fully operative case (non-faulty system). The computation time needed for this analysis was 8 s. Finally, the reliability analysis showed the level of reliability of each component and path and of the whole network. AFCs were analysed by extracting all paths using the faulty actuator and recomputing the reliability of the WTN. Two rankings were computed: the first one according to demand satisfaction, showing which demands were more likely to be unsatisfied; and the second one according to the most critical actuators, showing how the reliability of the entire network decreased if those actuators were damaged. The computation time in this case was 5 s.

15.4.3 Discussion

Although each of the previous analyses can individually provide a great deal of information about the fault tolerance of a network, linking them up reduces the computational burden. In order to clearly present and easily discuss the proposed methodologies, a smaller portion of the Barcelona WTN (see Fig. 15.4) was used for illustrative purposes.

The first test consisted of locating the critical network actuators by means of a structural analysis. These critical actuators are those without which (outage) path connectivity is lost. The results of this analysis, summarized in Tables 15.1 and 15.2, point to an important number of critical actuators within the network, due to the topology and the way of connecting network elements, as most actuators (valves or pumps) are the only link between tanks and demands. Therefore, if an actuator fails, then the corresponding demand will not be satisfied. Note that the information shown in Tables 15.1 and 15.2 is particularly significant for Aguas de Barcelona

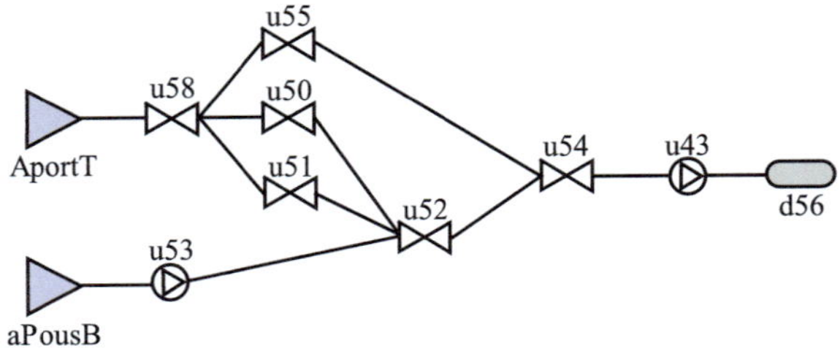

Fig. 15.4 Portion of the WTN related to Demand 56

(the manager of the water infrastructure), since it identifies the critical elements in the network for surveillance/correction policies to be implemented in the event of element damage (fault).

Applying the first test to the network, as depicted in Fig. 15.4, four possible paths were detected. These were:

Path 1: $AportT \rightarrow u58 \rightarrow u50 \rightarrow u52 \rightarrow u54 \rightarrow u43 \rightarrow d56$
Path 2: $AportT \rightarrow u58 \rightarrow u51 \rightarrow u52 \rightarrow u54 \rightarrow u43 \rightarrow d56$
Path 3: $AportT \rightarrow u58 \rightarrow u55 \rightarrow u54 \rightarrow u43 \rightarrow d56$
Path 4: $aPousB \rightarrow u53 \rightarrow u52 \rightarrow u54 \rightarrow u43 \rightarrow d56$

Analysing the structure of the network, as depicted in Fig. 15.4, it can be observed that it contains two critical actuators: 54 and 43. If either of these actuators fails, then Demand 56 will not be satisfied. All the remaining actuators can be considered as redundant actuators.

The second analysis done to the Barcelona WTN was to identify the actuators whose physical constraints limit water transport capacity through a certain path. Note that this analysis did not consider any fault in those actuators. The analysis, performed using Algorithm 15.10, also pinpointed several alternative paths through which water transport is possible (or even mandatory) given the constraints of the paths for supplying demands.

Results for this last analysis considering the whole WTN identified other critical actuators: 26, 52 and 91 (namely *iPalleja4, vBesosMontCerd* and *vGava100a80*). Note that the increase in the number of critical actuators, taking into account their physical constraints, is not significant. For the network in Fig. 15.4, actuator 52 is not a critical element according to the structural controllability property, meaning that connectivity is not lost when this component fails. However, the feasibility analysis determined that this actuator was in fact critical when the actuator physical constraints were considered. Actuator 52 cooperates with a flow of water to satisfy the demand that cannot be satisfied with a flow through a single path.

The third analysis identified the optimal paths to reach a selected destination node without considering the system constraints, i.e., the *structural optimal paths*. This analysis was performed using the structural algorithm, as explained in Sect. 15.3.2.1. For the smaller network, the cost of each path was computed, corresponding to the electricity cost of the actuators for both paths and the cost of water treatment in a

Table 15.1 Structural critical actuators (towards tanks)

No.	Name	No.	Name	No.	Name	No.	Name
122	iAltures	15	iCanGuey2	62	iGuinardera1	30	iPapiol1
10	iBegues1	14	iCanGuey3	60	iGuinardera2	88	iSJD10
6	iBegues2	21	iCanRoig	101	iLaSentiu	7	iStBoi
2	iBegues3	57	iCanRuti	34	iMasGuimbau1	9	iStCliment1
1	iBegues4	37	iCarmel	31	iMasGuimbau2	5	iStCliment2
32	iBellsoleig	43	iCerdMontflorit	100	iMasJove	40	iStGenis1
61	iBonavista	42	iCerdUAB	68	iMntjcStaAmalia	38	iStGenis2
20	iCanGuell1	12	iCesalpina1	69	iMntjcTresPins	13	iStaClmCervello
17	iCanGuell2d3	11	iCesalpina2	3	iOrioles	45	iStaMaMontcada
16	iCanGuell2d5	82	iCornella100	23	iPalleja1	35	iTibidabo
18	iCanGuey1d2	39	iFlorMaig	24	iPalleja2	56	iTorreBaro1
19	iCanGuey1d5	109	iFnestrelles300	27	vPalleja70	65	iTorreoCastell
44	iVallensana1	8	iViladecans1	4	iViladecans2	25	vAbrera
54	vCerdanyola90	63	vMontigala	90	vSJD	59	vTerStaColoma
104	vSJDTot	58	vTer				

Table 15.2 Structural critical actuators (towards demands)

No.	Name	No.	Name	No.	Name	No.	Name
115	vPallejaATLL	116	iPalleja3	117	iMasGuimbau3	118	iVallvidrera
119	vHorta	120	iUAB	121	iVallensana2	122	iBoscVilaro
123	iTorreBaro2	124	iCerdSabadell	125	vBesosStaColoma	126	v117Montigala
127	v70CFE	128	v55BAR	129	iMontemar	130	vAltures

determined source. For paths 1, 2 and 3, the cost was 0.54 e.u.,[6] while for path 4 the cost was 0.77 e.u. This small example would indicate that any of the first three paths is optimal for satisfying Demand 56.

A criterion to decide which of the three paths is optimal for this demand is to calculate the maximum flow of water for each path, which can also be computed in this analysis and is given by the smallest value of the maximum flow of water of the actuators in a given path. In this case, since all paths were restricted to $0.3\,m^3/s$, due to the physical capacities of actuator 43, any of the first three paths is recommended. However, if actuator 43 was not considered, path 1 would be the optimal path as it has a maximum flow of $2.2\,m^3/s$, while in the other paths, actuator 55 is restricted to $0.35\,m^3/s$, and actuators 51 and 52 to $0.8\,m^3/s$.

The fourth analysis consisted of identifying the set of optimal paths including the objective function (15.23) and the system constraints (15.2a)–(15.2b). Path details are not provided here, but the total cost of maintaining the whole DTWN in proper working order and satisfying all its demands was 502.25 e.u. In the case of the network depicted in Fig. 15.4, the optimal path obtained from the fourth analysis was path 4. Although it may appear that, when only Demand 56 is considered without the interconnection of the entire network, the other paths are less costly when the entire network is considered—this is not true. The actuators used in path 4 are also used to satisfy other demands, so sharing components results in an optimal solution.

The fifth test was performance analysis, taking into account the critical actuators already identified in the previous tests, with the difference in costs showing the impact that a single faulty actuator could have on an entire network. Results from this analysis are summarized in Table 15.3. Note that all comparisons took into account an optimal functioning cost (under non-faulty conditions) of 502.25 e.u. Moreover, fault cost denotes the functioning cost under faulty conditions.

Table 15.3 Entire WTN performance analysis

Actuator no.	Faulty cost [e.u.]	Cost overrun (%)
41	514.44	2.43
47	515.94	2.73
74	528.05	5.14
78	557.62	11.03
86	515.08	2.55
89	556.22	10.74
97	510.49	1.64
102	539.87	7.49
103	552.21	9.95

[6]Note that costs are given in *economic units* (e.u.) rather than real units (€) for confidentiality reasons.

According to the analysis of the entire WTN, some actuators did not have a significant impact on the total performing cost (e.g., actuators 28, 29, 33, 64, 71, 80, 81, 85, 87, 94, 107, 108 and 113). However, other actuators such as 78 or 89 significantly increased cost, taking into account daily estimates. These latter actuators are shown in Table 15.3. Degradation in costs obtained with this analysis can be the foundation for the introduction of redundant actuators in the network or an alternative fault tolerance strategy. For the network depicted in Fig. 15.4, the performance analysis shows that the cost of maintaining operations for the network with a fault in any of these actuators does not increase the cost.

The accommodation and reconfiguration strategies presented in Sect. 15.2.3 are now illustrated for the case of a fault in actuator 108 (named *vTerMontcada*), which, according to the previous analysis, is redundant. First the reconfiguration strategy is illustrated. Figure 15.5 presents the volume behaviour of tank 33, which is supplied by two actuators: 73 (*iCornella130*) and 108 (*vTerMontcada*). It can be seen that in a non-faulty situation, the volume of this tank presents a repetitive pattern (filling when pumping is cheaper and emptying otherwise) to satisfy the water demand. However, when a fault occurs (at $k = 50$ h), if the MPC controller is not reconfigured (labelled as fault occurrence in the plots), tank 73 volume drops to zero at $k = 58$ h and demand is not satisfied anymore (unfeasible solution). However, if the MPC controller is reconfigured by removing the faulty actuator 108 from the control model, the tank level is still able to supply the required demand. However, the tank volume decreases with time, indicating that the faulty actuator should be repaired. Figures 15.6 and 15.7, which depict the behaviour of actuators 108 and 73, show that actuator 73

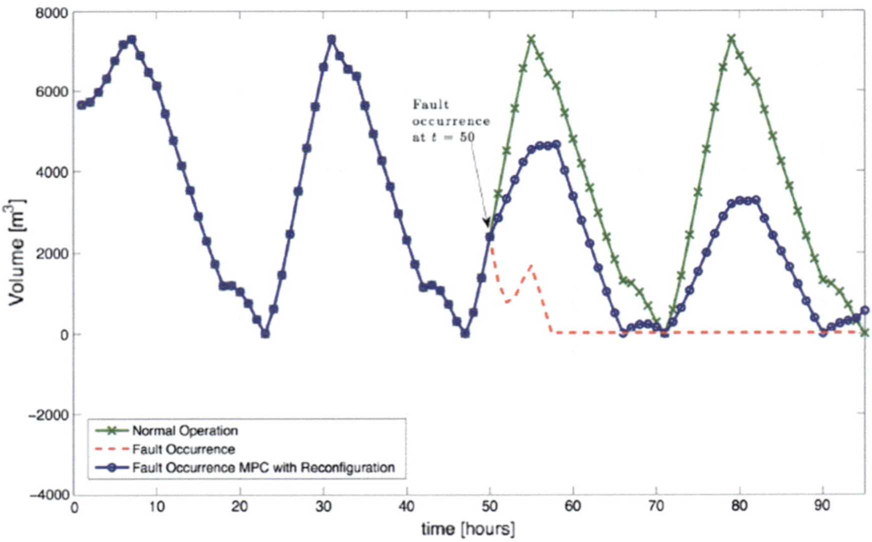

Fig. 15.5 Volume evolution of tank 33 with MPC using reconfiguration

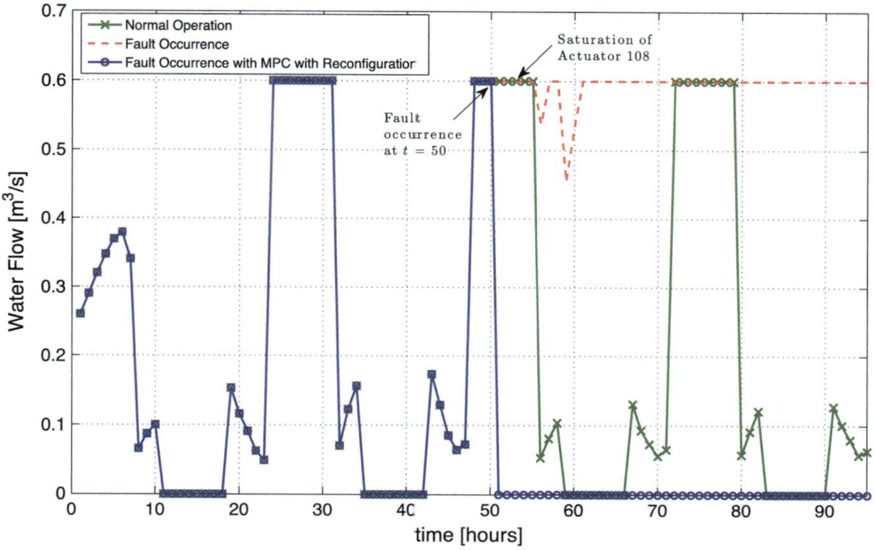

Fig. 15.6 Water flow in actuator 108 with MPC using reconfiguration

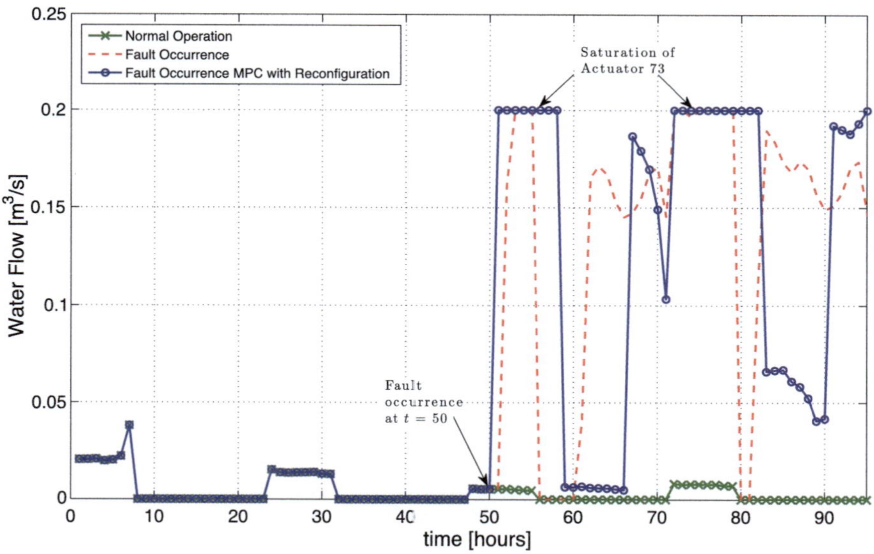

Fig. 15.7 Water flow in actuator 73 with MPC using reconfiguration

starts to deliver more flow in an effort to compensate for the faulty actuator 108 that is removed.

Figures 15.8, 15.9 and 15.10 depict tank 33 volume and actuators 108 and 73 flows when the fault is accommodated by the MPC controller. The fault affecting

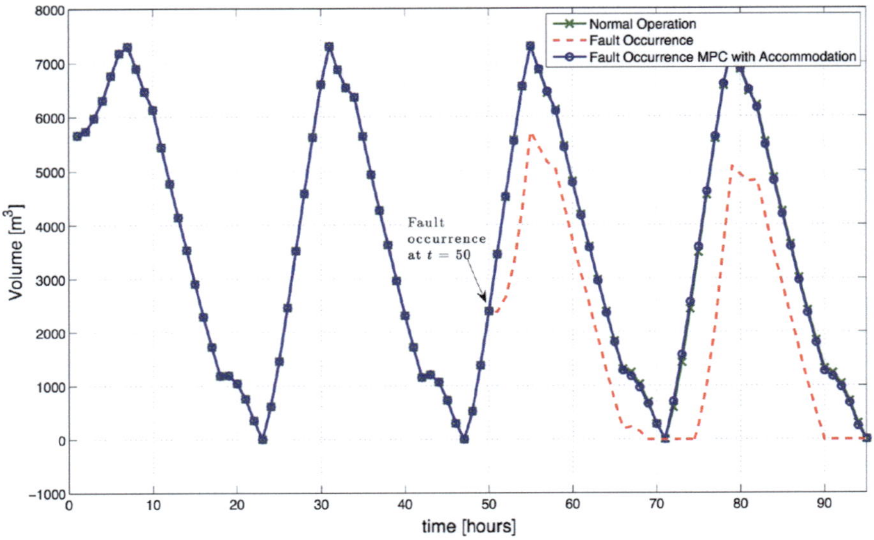

Fig. 15.8 Evolution of volume in tank 33 with MPC using accommodation

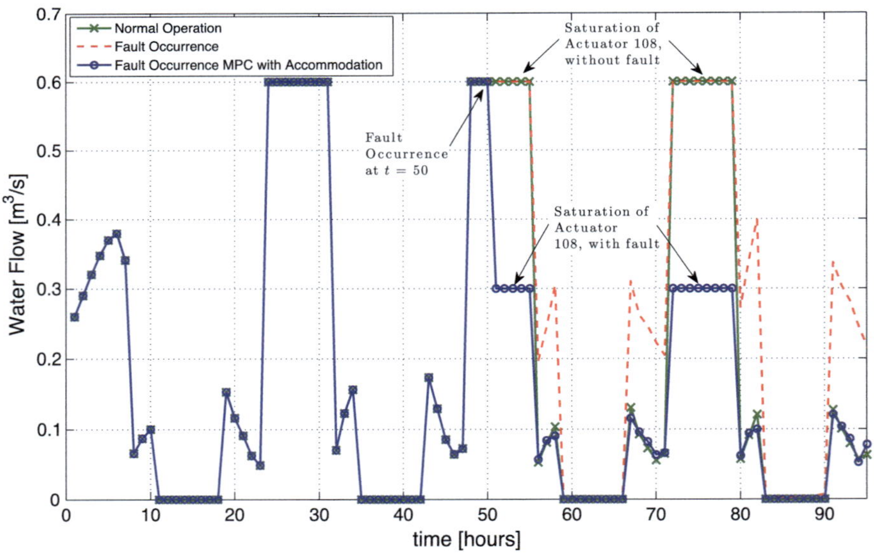

Fig. 15.9 Water flow in actuator 108 with MPC using accommodation

actuator 108 reduces the operating range by 50%. In this case, the faulty actuator is not removed from the control model of the MPC controller; rather, the operating limits of actuator 108 are updated according to the new operating range. Figure 15.8 shows how the volume behaviour of tank 33 in a non-fault situation and when using

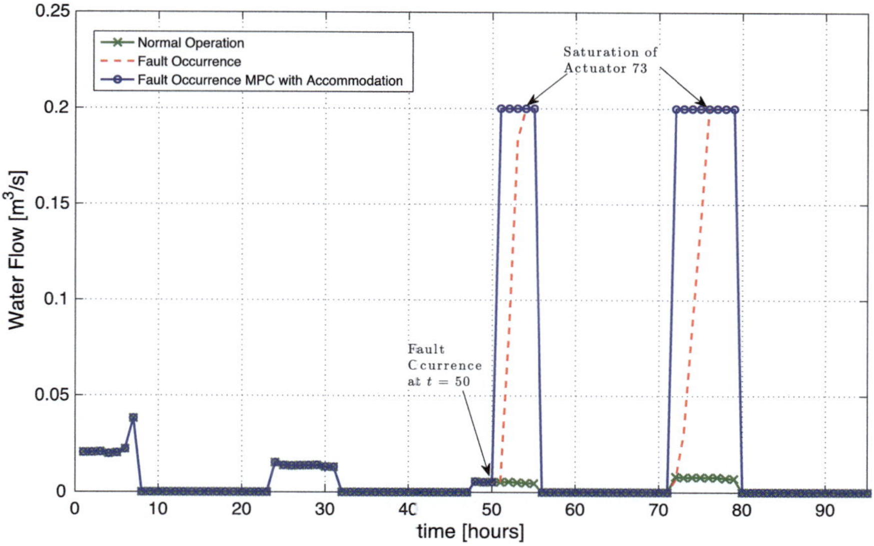

Fig. 15.10 Water flow in actuator 73 with MPC using accommodation

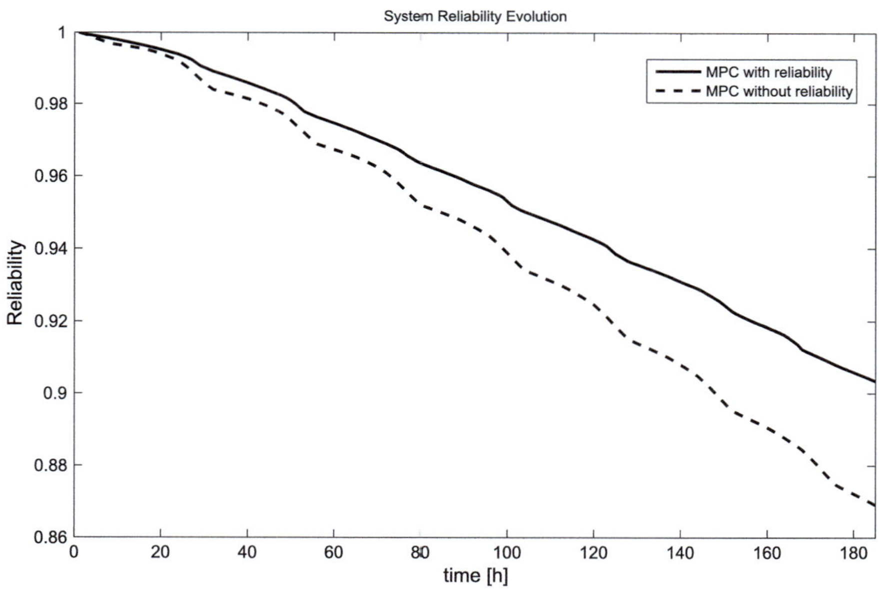

Fig. 15.11 MPC redesign to preserve reliability

accommodation looks exactly the same; in contrast, when the controller is not accommodated, the volume tends to zero and demand is not satisfied.

From Figs. 15.9 and 15.10, it can be seen that the MPC controller compensates for the reduction in the faulty actuator's operating range by increasing use of the non-faulty actuator, thereby compensating for the impact of the fault.

Although the proposed algorithm improves handling of the behaviour of the tank volume and actuator flows, it has computational and financial costs, as implementation of this feature increments computation time by 30 s (12%) and the cost overrun by around 9%.

The reliability analysis also takes into account the results of the previous analysis. The reliability of the entire network considering proper operation is 90.74% successful in satisfying the desired property when the reliability of each component is calculated using (15.16) with $\gamma = 0.0034$ (data obtained [10]). The association between demand satisfaction and reduced reliability when a faulty component exists is shown in Table 15.4.

As shown in Table 15.4, although most faults in actuators do not significantly affect reliability in satisfying demand, some completely override the satisfaction of the desired property. These actuators are critical actuators regarding reliability. The risk of having a malfunction in the system can be better understood when the reliability of the entire network is computed. Examples of critical actuators obtained from this study were actuators 102 and 103 since their malfunction led to a drop of 31.13% in the reliability of the entire network.

The reliability analysis was applied to the network depicted in Fig. 15.4. The reliability of satisfying Demand 56 decreased to 1.33% if actuators 52 and 58 had a fault, highlighting the importance of both these actuators for the operation of this smaller network, and decreased to 0% when actuators 43 and 54 were faulty, reaffirming the fact that these two actuators are critical. Otherwise, reliability remained the same. Regarding the entire WTN, actuator 52 decreased reliability of satisfying the demands in the network by 21.71%, denoting again that it is an important element in system interconnectivity.

Table 15.4 Association between demand satisfaction and reliability

Demand No	Percentage of total demand (%)	Faulty components	R_g in Faulty conditions (%)
69	9.1	128	0
83	4.0069	82, 88, 90, 104	0
70	3.2537	125	0
70	3.2537	58	99.33
70	3.2537	53, 50, 51	99.99
33	1.964	108	99.98
58	1.9407	52, 58	99.33
56	1.6777	52, 58	98.67
64	1.4941	58, 59	0

Critical actuators 43 and 54, when they malfunction, reduced the reliability of the entire system towards zero; in contrast, the fact that other actuators did not affect reliability denotes them to be redundant actuators.

Finally, the MPC redesign approach to preserve the network reliability has been applied to the entire WTN using Algorithm 15.12. Figure 15.11 shows how the reliability of the network evolves in time when this algorithm is used. It can be observed that with the use of Algorithm 15.12, the reliability of the network degrades slowly compared to the case that the reliability is not considered in the MPC design.

15.5 Conclusions

This chapter proposed a reliable fault-tolerant model predictive control strategy for drinking-water transport networks. The proposed approach combines structural, feasibility, performance and reliability analyses. After a fault, the predictive controller is redesigned to cope with the fault by considering either a reconfiguration or an accommodation strategy depending on available knowledge regarding the fault. Before starting to apply the fault-tolerant control strategy, whether the predictive controller will be able to continue operating after the fault appearance needs to be evaluated. This evaluation is performed by means of a structural analysis to determine post-fault loss of controllability, complemented with a feasibility analysis of the optimization problem related to the predictive control design, so as to consider the fault impact on actuator constraints. By evaluating the admissibility of different actuator-fault configurations, critical actuators regarding fault tolerance can be identified. The proposed approach also allows for a degradation analysis of the system in terms of performance and reliability. As a result of this analysis, the predictive controller design can be modified by adapting constraints such that the best achievable performance with some pre-established level of reliability is achieved. The proposed approach, successfully tested on the Barcelona water network, shows that relevant information can be extracted about critical actuators considered in the different analyses. Future research will investigate the impact of uncertainty on the analyses and on the design of the predictive controller including fault-tolerant capabilities.

References

1. Blanke M, Kinnaert M, Lunze J, Staroswiecki M (2016) Diagnosis and fault-tolerant control, 3rd edn. Springer, Berlin
2. Bondy JA, Murty USR (1982) Graph theory with applications. MacMillan Press, Great Britain
3. Brdys MA, Ulanicki B (1994) Operational control of water systems: structures. Algorithms and applications. Prentice Hall International, New York
4. Camacho EF, Alamo T, Muñoz de la Peña D (2010) Fault-tolerant model predictive control. In: IEEE conference on emerging technologies and factory automation (ETFA'2010), pp 1–8

5. Ellis Matthew, Durand Helen, Christofides Panagiotis D (2014) A tutorial review of economic model predictive control methods. J Process Control 24(8):1156–1178
6. Gertsbakh IB (2000) Reliability theory with application to preventive maintenance. Springer
7. Guenab F, Weber P, Theilliol D, Zhang YM (2011) Design of a fault tolerant control system incorporating reliability analysis and dynamic behaviour constraints. Int J Syst Sci 42(1):219–233
8. Limon D, Pereira M, Muñoz de la Peña D, Alamo T, Grosso JM (2014) Single-layer economic model predictive control for periodic operation. J Process Control 8(24):1207–1224 August
9. Maciejowski JM (2002) Predictive control with constraints. Prentice Hall, Great Britain
10. MinEsp (2008) Ntp 417: quantitative risk analysis—reliability of components and implications in preventive maintenance (in spanish only). Technical report, Spain
11. Noura H, Theilliol D, Ponsart JC, Chamssedine A (2009) Fault tolerant control systems: design and practical application. Springer
12. Ocampo-Martinez C, Puig V (2009) Fault-tolerant model predictive control in the hybrid systems framework: application to sewer networks. Int J Adapt Control Signal Process 23(8):757–787
13. Ocampo-Martinez C, Puig V, Cembrano G, Quevedo J (2013) Application of predictive control strategies to the management of complex networks in the urban water cycle. IEEE Control Syst 33(1):15–41
14. Ostfeld A (2001) Reliability analysis of regional water distribution systems. Urban Water 3:253–260
15. Pascual J, Romera J, Puig V, Cembrano G, Creus R, Minoves M (2013) Operational predictive optimal control of Barcelona water transport network. Control Eng Pract 21(8):1020–1034
16. Prodan I, Zico E, Stoican F (2015) Fault tolerant predictive control design for reliable microgrid energy management under uncertainties. Energy 91:20–34
17. Raimondo DM, Marseglia GR, Braatz R, Scott JK (2013) Fault-tolerant model predictive control with active fault isolation. In: 2nd conference on control and fault-tolerant systems (SysTol), pp 444–449, Nice, France
18. Rawlings JB, Angeli D, Bates CN (2012) Fundamentals of economic model predictive control. In: 51st IEEE conference on decision and control, Maui, Hawaii, USA
19. Robles D, Puig V, Ocampo-Martinez C, Garza LE (2012) Methodology for actuator fault tolerance evaluation of linear constrained MPC: application to the Barcelona water network. In: 20th Mediterranean conference on control automation (MED), pp 518–523, July 2012
20. Salazar J, Weber P, Sarrate R, Theilliol D, Nejari F (2015) MPC design based on a DBN reliability model: application to drinking water networks. In: 9th IFAC symposium on fault detection, supervision and safety of technical processes, France, Paris, pp 688–693
21. Sanchez H, Escobet T, Puig V, Odgaard PF (2015) Health-aware model predictive control of wind turbines using fatigue prognosis. In: 9th IFAC symposium on fault detection, supervision and safety of technical processes, France, Paris, pp 1363–1368
22. Šiljak DD (1991) Decentralized control of complex systems. Academic Press
23. Staroswiecki M, Berdjag D (2010) A general fault tolerant linear quadratic control strategy under actuator outages. Int J Syst Sci 41(8):971–985
24. Staroswiecki M, Commault C, Dion JM (2012) Fault tolerance evaluation based on the lattice of system configurations. Int J Adapt Control Signal Process 26:54–72
25. Torii A, Lopez R (2012) Reliability analysis of water distribution networks using the adaptive response surface approach. J Hydraul Eng 138:227–236
26. Wang Y, Ocampo-Martinez C, Puig V (2015) Robust model predictive control based on Gaussian processes: Application to drinking water networks. In: European control conference, Linz, Austria
27. Weber P, Simon C, Theilliol D, Puig V (2012) Fault-tolerant control design for over-actuated system conditioned by reliability: a drinking water network application. In: 8th IFAC symposium on fault detection, supervision and safety for technical processes (Safeprocess12), Mexico City, Mexico

28. Yang X, Maciejowski J (2012) Fault-tolerant model predictive control of a wind turbine benchmark. In: 8th IFAC symposium on fault detection., supervision and safety of technical processes, Mexico City, Mexico, pp 337–342
29. Yetendje A, Seron M, De Doná J (2012) Fault-tolerant model predictive control in the hybrid systems framework: application to sewer networks. Int J Appl Math Comput Sci 22(1):211–223
30. Zhang Y, Jiang J (2008) Bibliographical review on reconfigurable fault-tolerant control systems. Ann Rev Control 32(2):229–252

Chapter 16
Partitioning Approaches for Large-Scale Water Transport Networks

Carlos Ocampo-Martínez and Vicenç Puig

16.1 Introduction

Large-scale systems (LSS) present control theory with new challenges due to the large size of the plant and of its model [13, 22]. The goal to be achieved with control methods for this kind of systems is to obtain a reasonable solution with a reasonable effort in modelling, designing and implementing the controller.

As discussed in previous chapters, MPC has been proved to be suitably applied for the control of LSS as drinking-water networks [3], sewer networks [14], open-flow channel networks [18] or electrical networks [15]. Nevertheless, the main hurdle for MPC control (as any other control technique), when applied to LSS in a centralized way, is the non-scalability. The reason is that a huge control model is needed, being difficult to maintain/update and which needs to be rebuilt on every change of the system configuration, e.g., when some part of the system should be stopped because of maintenance actions or malfunctions. Subsequently, a model change would require re-tuning the centralized controller. It is obvious that the cost of setting up and maintaining the monolithic solution of the control problem is prohibitive. A way of circumventing these issues might be by looking into *decentralized* MPC (DMPC) or *distributed* MPC techniques, where networked local MPC controllers are in charge of controlling part of the entire system. The main difference between distributed and decentralized MPC is that the former *uses* negotiations and re-computations of local control actions within the sampling period to increase the level of cooperation, whereas the latter does not (at the benefit of computation time, but at the cost of optimality).

The industrial success of the traditional centralized MPC (CMPC) drives now a new interest in this old area of distributed control, and distributed MPC has become one of the hottest topics in process control in the early twenty-first century, world-wide. Thus, two research projects (HDMPC [10] and WIDE [24]) are currently being

---

C. Ocampo-Martínez (✉)
Institut de Robòtica i Informàtica Industrial, CSIC-UPC, Barcelona, Spain
e-mail: cocampo@iri.upc.edu

V. Puig
Research Center Supervision, Safety and Automatic Control (CS2AC-UPC), Terrassa, Spain

V. Puig et al. (eds.), *Real-Time Monitoring and Operational Control of Drinking-Water Systems*, Advances in Industrial Control,
DOI 10.1007/978-3-319-50751-4_16

carried out in Europe, both focused on the development of decentralized and distributed MPC techniques. Few works have been recently published in this area; see, e.g., [6, 11, 16, 19, 20, 23], among others.

However, in order to apply decentralized or distributed MPC approaches to LSS, there is a prior problem to be solved: the system decomposition into subsystems. The importance of this issue has already been noticed in classic control books addressing the decentralized control of LSS as [13, 22]. The decomposition of the system into subsystems could be carried out during the modelling of the process by identifying subsystems as parts of the system on the basis of physical insight, intuition or experience. But, when a large-scale complex system with many states, inputs and outputs is considered, it may be difficult, even impossible, to obtain partitions by physical reasoning. A more appealing alternative is to develop systematic methods, which can be used to decompose a given system by extracting information from its structure and representing it as a graph. Then, this structural information can be analysed by using methods coming from graph theory. Consequently, the problem of system decomposition into subsystems leads to the problem of graph partitioning, i.e., the decomposition of graph into subgraphs.

Graph partitioning is an important problem with extensive application in scientific computing [12], optimization, very large-scale integration (VLSI) design [8], task partitioning for parallel processing, control of cascading failures, among others. However, the development of graph partitioning algorithms that allow the decomposition of LSS into subsystems for being used in decentralized or distributed MPC is still very incipient and available methods are quite limited. In [22], a hierarchical LBT decomposition that leads to a input-reachable hierarchy for some particular systems is presented. A more general approach is based on the ε-decomposition method, which is based on decomposing the system in weakly coupled subsystems (see also [22]). The algorithm proceeds sequentially disconnecting the edges of the system graph that are smaller than a prescribed threshold ε and identifying the disconnected subgraph of the resulting graph. The obtained subsystems correspond to the subsystems with mutual coupling smaller or equal than ε. However, the tuning of this parameter is not a trivial issue and only a trial and error approach is currently available.

16.2 Problem Statement

A graph can be defined as an abstract representation of a set of objects from a certain collection, where some pairs of objects are connected by links. The interconnected elements are typically called *vertices* while the connection links are called *edges*. These latter elements may be *directed* (asymmetric) or *undirected* (symmetric) according to their connection features, what makes that the whole graph is directed or undirected as well. It is also possible to distinguish graphs whether or not their vertices and edges are weighted (weighted/unweighted graphs).

Consider a dynamical system represented in general form by the state-space equations

$$\mathbf{x}(k+1) = \mathbf{g}(\mathbf{x}(k), \mathbf{u}(k), \mathbf{d}(k)), \tag{16.1a}$$

$$\mathbf{y} = \mathbf{h}(\mathbf{x}(k), \mathbf{u}(k), \mathbf{d}(k)), \tag{16.1b}$$

where $\mathbf{x}(k) \in \mathbb{R}^{n_x}$ and $\mathbf{x}(k+1) \in \mathbb{R}^{n_x}$ are, respectively, the current and successor system states in discrete time, $\mathbf{u} \in \mathbb{R}^{n_u}$ is the system input, $\mathbf{y} \in \mathbb{R}^{n_y}$ is the system output and $\mathbf{d}\imath\mathbb{R}^{n_d}$ is a bounded process disturbance. Moreover, $\mathbf{g} : \mathbb{R}^{n_x} \times \mathbb{R}^{n_u} \times \mathbb{R}^{n_d} \to \mathbb{R}^{n_x}$ is the states mapping function and $\mathbf{h} : \mathbb{R}^{n_x} \times \mathbb{R}^{n_u} \times \mathbb{R}^{n_d} \to \mathbb{R}^{n_y}$ corresponds with the output mapping function. Suppose now that it is desired to decompose (16.1) into subsystems. With this aim, the graph representation of the system model (16.1) is determined (by using the system topology) and incidence matrix $\mathbf{B}_{ij}$ is then stated, which describes the connections (edges) between the graph vertices (system inputs, outputs and states). Without loss of generality, $\mathbf{B}_{ij}$ and the directionality of the edges are derived from the relation between system equations (rows of $\mathbf{B}_{ij}$) and system variables (columns of I_M), as proposed by [22, 25, 26]. There are alternative matrix representations for a (directed) graph such as the *adjacency matrix* and the *Laplacian matrix* (see [2]), which are related to the matrix representation used in this paper. Once $\mathbf{B}_{ij}$ has been obtained from the system directed graph (digraph), the problem of the decomposition into subsystems can be formulated in terms of partitioning the corresponding graph into subgraphs. Since such partitioning is oriented to the application of a decentralized control strategy (in particular, DMPC), the resultant subgraphs should have the following features (see [13, 22]):

- nearly the same number of vertices;
- few connections between the subgraphs.

These features guarantee that the obtained subgraphs have a similar size which balances computations between subsystem controllers and allows minimizing communications between them. Hence, the problem of graph partitioning can be more formally established as follows:

Problem 16.1 (*Standard Graph Partitioning*) Given a graph $G(\mathcal{V}, \mathcal{E})$, where $\mathcal{V}$ denotes the set of vertices, $\mathcal{E}$ is the set of edges and $M \in \mathbb{Z}_{\geq 1}$, find M subsets $\mathcal{V}_1$, $\mathcal{V}_2, \ldots, \mathcal{V}_M$ of $\mathcal{V}$ such that

1. $\bigcup_{i=1}^{M} \mathcal{V}_i = \mathcal{V}$,
2. $\mathcal{V}_i \cap \mathcal{V}_j = \emptyset$, for $i \in \{1, 2, \ldots, M\}$, $j \in \{1, 2, \ldots, M\}$, $i \neq j$,
3. $|\mathcal{V}_1| \approx |\mathcal{V}_2| \approx \cdots \approx |\mathcal{V}_M|$,
4. the *cut size*, i.e., the number of edges with endpoints in different subsets $\mathcal{V}_i$, is minimized.

Remark 16.1 Defining the *vertex-based weight* of a subset $\mathcal{V}_i$ as

$$\Omega_i \triangleq \sum_{j=1}^{|\mathcal{V}_i|} \omega_i^j, \tag{16.2}$$

where ω_i^j corresponds to the weight of the j-th vertex of the subset $\mathcal{V}_i$, the following condition should be added to Problem 16.1 in the case of *weighted graph partitioning*:

- $\Omega_i \approx \Omega/M$, with $i \in \{1, 2, \ldots, M\}$, where

$$\Omega \triangleq \sum_{i=1}^{M} \Omega_i. \tag{16.3}$$

Remark 16.2 Conditions 3 and 4 of Problem 16.1 are of high interest from the decentralized control point of view since they are related to the degree of interconnection between resultant subsystems and their size balance, respectively.

Graph partitioning is considered as a $\mathcal{NP}$-complete problem [22]. However, it can be solved in polynomial time for $|\mathcal{V}_i| = 2$ (Kernighan-Lin algorithm) [4, 7]. Since this condition is quite restrictive for large-scale graphs, alternatives for graph partitioning based on fundamental heuristics are properly accepted. Two main classes of successful heuristics have evolved over the years, trying to achieve the proper trade-off between partitioning speed and quality. They are the *minimum degree-based* ordering algorithms (MDB) and the *graph partitioning-based* ordering algorithms (GPB) [9].

16.3 Proposed Approaches

16.3.1 Using Graph Theory

This approach consists in proposing a partitioning algorithm, as much automatized as possible, through which a partition of a dynamical system can be found, which allows its decomposition in subsystems. This algorithm requires to represent the dynamical system as a graph, which can be obtained from the system structure [22].

Main Algorithm

The partitioning algorithm proposed in this chapter follows some ideas developed in [9] for graph partitioning purposes. However, some refining steps have been added as well as some of the original procedures have been drastically changed in order to find partitions oriented to split dynamical networked systems. Hence, the different parts/routines of the main proposed algorithm are presented and explained in sections below. The current version of the algorithm is thought to be used offline, i.e., the partitioning of the system is not carried out online. A further improvement could be to adapt the proposed algorithm such that the partitioning could be done online when some structural change of the network occurs. In this way, the potential benefit of using a DMPC approach described in the Introduction could be fully exploited.

Start-up: This procedure requires the definition of the graph, i.e., *the incidence matrix*[1] $\mathbf{B}_{ij}$, which describes the connections between the graph vertices, their directionality and, in some cases, the weight of each edge.

Preliminary partitioning: This procedure performs a preliminary automatic partitioning of the graph as follows. The vertex $v_j \in \mathcal{V}$, for $j \in \{1, 2, \ldots, |\mathcal{V}|\}$, with maximum weight ω is found and defined as the centre of the first subgraph G_1. Then, all vertices connected to this vertex of maximum weight are assigned to G_1. At this point, the set of non-selected vertices is defined as

$$\mathcal{V}_r \triangleq \{v_j \in \mathcal{V} : v_j \notin \mathcal{V}_1\}.$$

This procedure is now repeated for all vertices $v_j \in \mathcal{V}_r$ (now for $j = \{1, 2, \ldots, |\mathcal{V}_r|\}$) until $\mathcal{V}_r$ is empty, after the corresponding updating. This routine highlights the subgraphs of higher connectivity. The resultant subgraphs with just one vertex are merged to the closest subgraph. Once a set of subgraphs $G_i(\mathcal{V}_i, \mathcal{V}_i)$, for $i = 1, 2, \ldots, M$, is obtained, it is possible to determine some useful indexes for the entire graph and each one of the resultant subgraphs. These indexes are as follows:

- $\varphi_i \triangleq |\mathcal{V}_i|$ (from now on called *subgraph internal weight* of G_i);
- ε_i, denoted as the *cut size*[2] of the subgraph G_i (from now on called *subgraph external weight* of G_i);
- $\varphi^{\max} \triangleq \max\limits_i \varphi_i$, for $i = 1, 2, \ldots, M$;
- $\bar{\varphi} \triangleq \frac{1}{M} \sum\limits_{i=1}^{M} \varphi_i$ (arithmetic mean).

Notice that at this stage, the number M of subgraphs is obtained in an automatic way so it is not imposed.

Remark 16.3 Notice that introducing the set $\tilde{\mathcal{E}}_a \subset \mathcal{E}$, defined as the set of edges with endpoints in other subgraphs different to G_a, the representation of subgraphs G_i such that

---

[1]The *incidence matrix* of a directed graph $G(\mathcal{V}, \mathcal{E})$, denoted as $\mathbf{B}_{ij}$, is defined such that

$$\mathbf{B}_{ij} = \begin{cases} -1 & \text{if the edge } z_j \text{ leaves vertex } v_i, \\ 1 & \text{if the edge } z_j \text{ enters vertex } v_i, \\ 0 & \text{otherwise.} \end{cases}$$

This matrix has dimensions $\varphi \times \eta_e$, where φ corresponds with the total number of vertices and η_e denotes de total number of edges [2]. Additionally, the weight of the j-th vertex, denoted as ω^j, for $j = 1, 2, \ldots, \varphi$, where $\varphi \triangleq |\mathcal{V}|$, is computed. The weight ω^j represents the number of edges connected to this vertex. Moreover, ω^j is also known as the *vertex degree* [5].

[2]See Problem 16.1.

$$\bigcup_{i=1}^{M} G_i = G,$$

can be slightly modified to $G_i(\mathcal{V}_i, \mathcal{E}_i, \tilde{\mathcal{E}}_i)$ for completeness purposes. Also notice that $\varepsilon_i \triangleq |\tilde{\mathcal{E}}_i|$.

Uncoarsening—Internal balance: This procedure aims at the reduction of the number of subgraphs, trying to achieve similar internal weights for all of them. This process starts determining the set

$$\mathcal{L} = \{G_i, i = 1, 2, \ldots, m : \varphi_i \leq \bar{\varphi}\}, \tag{16.4}$$

with $m \in \mathbb{Z}_+$ and $m < M$. For each $G_i \in \mathcal{L}$, the set of neighbour[3] subgraphs, denoted as $\mathcal{N}_i$, is determined and expressed as

$$\mathcal{N}_i = \{G_j,\ j = 1, 2, \ldots, h_i : G_j \text{ is neighbour of } G_i\}, \tag{16.5}$$

with $h_i = |\mathcal{L}_i|$. If the condition

$$\varphi_i + \varphi_j \leq \bar{\varphi}, \quad i \in \{1, 2, \ldots, m\},\ j \in \{1, 2, \ldots, h_i\} \tag{16.6}$$

holds for $G_i \in \mathcal{L}$ and $G_j \in \mathcal{N}_i$, then these two subgraphs are merged. If there are two or more subgraphs $G_j \in \mathcal{N}_i$ such that (16.6) holds, the subgraph $G_j \in \mathcal{N}_i$ with minimum internal weight is selected. Once two subgraphs are merged, $\bar{\varphi}$ is updated.

This procedure is iterated until no additional merging was possible. It is considered that the internal balance has been achieved when either

- $\bar{\varphi} \leq \varphi_i \leq \varphi^{\max}$, for $i = 1, 2, \ldots, M$, or
- G_i with $\varphi_i \leq \bar{\varphi}$ cannot be merged with any of its neighbours since the φ associated with the resultant subgraph might be greater than $\varphi^{\max}$.

Refining—External balance: This procedure aims at the reduction of the cut size of the resultant subgraphs. To achieve this goal, define ω_i^j as the degree of the j-th vertex of the i-th subgraph, with $j \in \{1, 2, \ldots, \varphi_i\}$ and $i \in \{1, 2, \ldots, M\}$. From this definition, two indexes can be stated as follows:

- the *vertex internal degree*, denoted as $\hat{\omega}_i^j$, which represents the number of connections of the vertex $v_j \in \mathcal{V}_i$, for $j \in \{1, 2, \ldots, \varphi_i\}$, $i \in \{1, 2, \ldots, M\}$, with other vertices $v_p \in \mathcal{V}_i$, $p \in \{1, 2, \ldots, \varphi_i\}$, $p \neq j$;
- the *vertex external degree*, denoted as $\check{\omega}_i^j$, which represents the number of connections of the vertex $v_j \in \mathcal{V}_i$, for $j \in \{1, 2, \ldots, \varphi_i\}$, $i \in \{1, 2, \ldots, M\}$, with other vertices $v_p \in \mathcal{V}_q$, $p \in \{1, 2, \ldots, \varphi_q\}$, $q \in \{1, 2, \ldots, M\}$, $q \neq i$.

---

[3]Two subgraphs are called *neighbours* if they are contiguous and share edges (see, e.g., [1] among many others).

Algorithm 16.1 Graph partitioning algorithm

$\mathbf{B}_{ij} \leftarrow$ System topology
$G(\mathcal{V}, \mathcal{E}) \leftarrow \mathbf{B}_{ij}$
for $j = 1$ to φ **do**
  Compute ω^j
end for
$\mathcal{V}_r \leftarrow \mathcal{V}, i = 1$
repeat
  Find $v \in \mathcal{V}_r$ with maximum ω
  $\mathcal{V}_i \leftarrow v$ and all its neighbour vertices
  $\mathcal{V}_r \triangleq \mathcal{V} - \left\{ \bigcup_{h=1}^{i} \mathcal{V}_h \right\}$
  $i = i + 1$
until $\mathcal{V}_r = \emptyset$
for $i = 1$ to M **do** {Compute some indexes}
  $\varphi_i \triangleq |\mathcal{V}_i|$ {internal weight}
  $\varepsilon_i \triangleq |\tilde{\mathcal{E}}_i|$ {external weight}
end for
$\varphi^{\max} \triangleq \max_i \varphi_i$
$\bar{\varphi} \triangleq \frac{1}{M} \sum_{i=1}^{M} \varphi_i$ {arithmetic mean}
Compute $\mathcal{L}$ {see (16.4)}
$b_{\text{int}} =$ false {Internal balance}
while $b_{\text{int}} =$ false **do**
  for $i = 1$ to m **do**
    Compute $\mathcal{N}_i$ {see (16.5)}
    for $j = 1$ to h **do**
      if $\varphi_i + \varphi_j \leq \bar{\varphi}$ **then** {see (16.6)}
        $G_* = G_i \cup G_j$
        $G_{\text{new}} \leftarrow G_*$ with minimum φ_*
        Update $\bar{\varphi}$
      end if
    end for
  end for
  Update φ_i
  $b_{\text{ext}} =$ false {External balance}
  while $b_{\text{ext}} =$ false **do**
    for $i = 1$ to M **do**
      for $j = 1$ to φ_i **do**
        Compute $\hat{\omega}_i^j$ and $\breve{\omega}_i^j$
        if $\hat{\omega}_i^j < \breve{\omega}_i^j$ **then**
          Move v^j from G_i to its neighbour
        end if
        Update φ_i, $\bar{\varphi}$, $\varphi^{\max}$
      end for
    end for
    Update all indexes
    Check external balance (nodes)
  end while
  Check internal balance (subgraphs)
end while
return $\mathcal{P}$ {see (16.7)}

Hence, for a given vertex $v_j \in \mathcal{V}_i$, if $\hat{\omega}_i^j < \breve{\omega}_i^j$, then vertex v_j is moved from subgraph $G_i(\mathcal{V}_i, \mathcal{E}_i, \tilde{\mathcal{E}}_i)$ to the subgraph in which most of its edges have their endpoint (like in the AVL tree algorithm [5]). All indexes should be updated for the M subgraphs and the next vertex is analysed. This procedure will last until each subgraph vertex fulfills $\hat{\omega}_i^j \geq \breve{\omega}_i^j$.

The Complete Algorithm: Algorithm 16.1 collects all the procedures/routines mentioned and explained before. Hence, applying this algorithm to the graph associated wit a given dynamical system, the expected result consists of a set of subgraphs which determines a particular system decomposition. This set $\mathcal{P}$ is then defined as

$$\mathcal{P} = \left\{ G_i,\ i = 1, 2, \ldots, M : \bigcup_{i=1}^{M} G_i = G \right\}. \tag{16.7}$$

Auxiliary Routines

Despite Algorithm 16.1 yields an automatic partitioning of a given graph, it does not imply that the resultant set P follows the pre-established requirements stated in Problem 16.1. In this sense, complementary routines can be useful for improving the partitioning process according to the considered application. Additional auxiliary routines could be added such that the generated partitioning takes into account the control performance that would be achieved when used in decentralized or distributed MPC control.

Prefiltering: In general, the resultant solution given by the Algorithm 16.1 is nearly appropriate in terms of $\hat{\omega}$ and $\breve{\omega}$, but it highly depends on the topology and complexity of the graph. For this reason, in order to obtain a better graph partitioning, sometimes it can be useful to make a *Prefiltering* routine, where all the vertexes with $\omega = 1$ are virtually merged to this vertex that shares its unique edge. This procedure creates *supranodes*, which should be properly recognized at the moment of determining the partitioning of the dynamical system from the decomposition of its associated graph. Moreover, doing the manual merging of those vertices reduces the work done by subsequent routines.

Post-filtering: On the other hand, suppose that after partitioning a given graph $G(\mathcal{V}, \mathcal{E})$ by using Algorithm 16.1, all the M resultant subgraphs fulfil

$$\bar{\varphi} \leq \varphi_i \leq \varphi^{\max}, \quad \text{for} \quad i \in \{1, 2, \ldots, M\}. \tag{16.8}$$

However, the following situation could occur. Suppose a subgraph G_a with $\varphi_a \ll \bar{\varphi}$, which is placed next to a subgraph G_b and fulfills (16.8). The merging of subgraphs G_a and G_b, expressed as $G_c \triangleq G_a \cup G_b$, is not allowed since $\varphi_c \geq \varphi_{\max}$. The *post-filtering* routine implements an approximation and a parameterization, i.e., by adding a small tolerance δ, the existence of the resultant subgraph G_c is now allowed since $\varphi_c \leq \varphi^{\max} + \delta$. This relaxation allows to have less subgraphs but with higher complexity and internal weight.

Anti-oscillation: This procedure leads to solve a possible issue when the *refining (external balance)* routine is run. When a vertex is moved from one subgraph to another according to its internal and external degrees, there exists the possibility of doing this movement during an infinite time if there is no specification of routine ending. Therefore, the refining routine is then run within a for loop and the parameter ρ is set as the maximum number of iterations that this procedure is executed. Afterwards, since the resulting set of subgraphs is stored at each iteration $t' \in \mathbb{Z}_+$, $t' = \{1, 2, \ldots, \rho\}$, the configuration of M subgraphs with minor ε_i, for $i = 1, 2, \ldots, M$, can be chosen.

Some Practical Issues

Given that the partitioning algorithm proposed in this chapter is mainly thought for performing decentralized control of LSS, several features could be taken into account to achieve a convenient system partitioning and less complex controller designs. For instance, an additional routine that would restrict the connection of subgraphs with unidirectional edges would be very useful since a pure hierarchical control scheme can be straightforwardly implemented, decreasing the inherent loss of performance of a decentralized control scheme.

16.3.2 Using Masks

The application of DMPC to WTN depends crucially on how the network is decomposed into subsystems. Identifying subsystems is not an easy task in a large-scale network as it involves to find automatically *sufficiently small* sections of the networked plant that are not *too coupled* among them. The partitioning algorithm, proposed in this chapter, aims to obtain this decomposition automatically by identifying clusters of elements that are strongly connected with each other but weakly interconnected with the other clusters, in order to represent the whole network as a set of loosely coupled subsystems [21]. The current version of the algorithm is thought to be used offline, that is the partitioning of the system is static and is not carried out online. A further improvement could be to adapt the proposed algorithm such that the partitioning could be done online when, for instance, some structural change of the network appears.

As a starting point, the partitioning algorithm requires the following information of the WTN:

1. The interconnection structure characterized by the matrix

$$\mathbf{I_c} = \begin{bmatrix} \mathbf{A}_{sp} & \mathbf{B}_{sp} \end{bmatrix}, \tag{16.9a}$$

where

$$\mathbf{A}_{sp} = \begin{bmatrix} \mathbf{A} & 0 \\ 0 & 0 \end{bmatrix}, \quad \mathbf{B}_{sp} = \begin{bmatrix} \mathbf{B} \\ \mathbf{E} \end{bmatrix}, \tag{16.9b}$$

where $\mathbf{A}$ and $\mathbf{B}$ are the system matrices in (12.8), the subscript sp identifies the matrices used for system decomposition and $\mathbf{E} \triangleq [\mathbf{E}_u \quad \mathbf{E}_d]$ is the matrix related to the equality constraints (12.9b). In order to take into account input bounds, new normalized inputs are introduced $\bar{\mathbf{u}} \triangleq \mathbf{u}/\mathbf{u}^{\max}$ so that $\bar{\mathbf{u}} \in [0, 1]$. Thus, new matrices $\bar{\mathbf{B}}$ and $\bar{\mathbf{E}}$ are introduced in (16.9b) to take into account the rescaling. From matrix $\mathbf{I_c}$, the adjacency matrix Ψ of the network graph can be obtained by replacing the nonzero elements by ones, leaving the null elements unchanged.

2. A threshold value ε is used for determining whether a term, which takes into account the actuator capacity (maximum allowable flow) and its usage frequency, has a negligible *effect* on the entire plant. In this way, the less important actuators are filtered out, in order to reduce the coupling degree of the system and identify independent subnetworks.

The partitioning algorithm proceeds by decomposing the matrix $\mathbf{I_c}$ into a set of submatrices, named as *partitions* and denoted by $\mathcal{P}_\varepsilon = \{\mathbf{I}_{\mathbf{c}1}, \ldots, \mathbf{I}_{\mathbf{c}M}\}$. Then, $\mathcal{P}_\varepsilon$ correspond to a set of subgraphs (subsystems) obtaining by deleting the edges corresponding to elements of $\mathbf{I_c}$ with magnitude no larger than ε. That is, the idea behind the partitioning approach is to neglect less important elements (i.e., links) in matrix $\mathbf{I_c}$ such that the resulting $\tilde{\mathbf{I}}_c$ is less coupled. Ideally, $\tilde{\mathbf{I}}_c$ should lead to a permutation matrix $\mathbf{P}$ such that $\mathbf{P}'\tilde{\mathbf{M}}\mathbf{P}$ is block diagonal. This procedure is repeated iteratively by reducing ε until an enough number of partitions is obtained. Algorithm 16.2 summarizes the steps of the proposed partitioning algorithm.

Partitions can be tuned by means of parameter ε of the proposed approach, which makes the user able to attempt matching the desired number and size of subsystems.

Typically, in the first iteration, Algorithm 16.2 neglects a high number of elements of $\mathbf{I_c}$, highly reducing the matrix connectivity degree and obtaining a subsystem decomposition. Then, once the sets of states/inputs relative to each partition are computed, the task of finding a suitable $\mathbf{P}$ that block-diagonalizes the matrix $\mathbf{P}'\tilde{\mathbf{M}}\mathbf{P}$ is a matter of linear algebra implementation. Every subsystem is composed by sets of state and input variables that are linked, meaning that are in the same block in the $\mathbf{P}'\tilde{\mathbf{M}}\mathbf{P}$ diagonal. Let $\mathcal{X}^i$ and $\mathcal{U}^i$ be, respectively, the sets of state and input variables assigned to subsystem i, while $|\mathcal{X}^i|$ and $|\mathcal{U}^i|$ determine the number of variables for each set. A subsystem is created if both numbers are different than zero. All state and input variables that are not assigned to any of the currently created subsystems, i.e., that do not belong to $\mathcal{X}^i$ or $\mathcal{U}^i$, respectively, are available for the next iteration. Otherwise, variables already assigned to a subsystem in the current or in a previous iteration are *masked*[4] to prevent their reassignment to other subsystem.

Then, a new iteration of the algorithm starts by decreasing ε (e.g., halving ε). Algorithm 16.2 iterates until all state variables are assigned to a subsystem. Note that the algorithm may terminate even if some inputs are not assigned to any subsystem, which is due to automatic threshold-based neglecting process. Such issue can be managed by manually including unassigned inputs to proper subsystem following engineering insight.

[4]Consider a variable to be masked when it does not belong to any set since it has already been classified in a previous iteration.

The importance of the mask arises from the structure of the algorithm. In fact, if not excluded, all previously assigned states and inputs would be part of the next iteration partition, introducing couplings and hence increasing the size of the resulting submodels. The aforementioned inclusion easily follows from the decreasing of ε among sequential iterations.

Algorithm 16.2 Automatic partitioning algorithm

```
Initialise masks to a neutral value
Initialise the sets of unassigned variables 𝒳 and 𝒰 with all state and input variables, respectively
Determine the number of unassigned states: N_x = |𝒳|;
Init ε
while N_x > 1 do
  Apply masks to A_sp and B_sp
  I_c = [A_sp B_sp ū]
  For all elements of I_c
  if I_{c i,j} < ε then
    Ĩ_{c i,j} = 0;
  else
    Ĩ_{c i,j} = 1;
  end if
  Find P such that P'M̃P is block diagonal
  Identify parts satisfying N_{x^i} = |𝒳^i| > 0 and N_{u^i} = L(𝒰^i) > 0 and add to previous ones
  Update N_x
  Update masks with updated states and inputs
  Update ε
end while
```

Few remarks on the above algorithm:

1. At any iteration of Algorithm 16.2, the numerical value of ε is a crucial tuning knob of the approach. A guideline is that the larger is the decreasing step, the larger is the size of the obtained subsystems. Ways for automatically determining the step size are a subject of current research.
2. Matrix $\mathbf{E}$ in (16.9b) defines a constraint among actuators that can be easily taken into account if all the actuators belong to the same subsystem. Otherwise, since each controller manipulates every partition independently from the others, negotiations between controllers would be required to guarantee the fulfilment of node constraints.
3. The use of masks to prevent state reassignment avoids that submodels have overlapping states and inputs: if a state variable is used in a model by a controller, no other controller can use it. The main benefit of this choice is the very low level of coupling between partitions, but the price to pay is a potential decrease of closed-loop performance.
4. The current structure of the algorithm is unsuitable to handle state overlaps because it relies on links between elements that present different degree of coupling. Hence, once the stronger couplings are eliminated (using masking), the

weaker ones gain relative importance. State overlaps may be introduced a posteriori based on engineering insight, in order to increase the adherence with respect to the original centralized model. Handling overlapping in an automatic way is also a current research topic.

5. In some cases, even relatively small connections, i.e., capable of carrying a minor amount of water, are very important for demand satisfaction. A way of accounting for such an issue is to perform a simulation using, for instance, a CMPC controller and compute the average percentage of use for each actuator. Thus, this information could be used to weight $\bar{\mathbf{u}}$ component-wise. The main drawback of this approach is the need of (and dependence on) simulation.
6. Note that the proposed algorithm can be customized by setting different importance levels of states vs. inputs, by weighting the related components in $\mathbf{I_c}$ from its statement at (16.9a).
7. The structure of the proposed algorithm suggests that termination is achieved if the ε value is decreased at each iteration. However, at the current status of the development, the algorithm cannot guarantee any property for the resulting partitioning but the assignment of all system state variables to a subsystem.

The decomposition process of matrix $\mathbf{I_c}$ reported here is similar to the one proposed by the ε-decomposition method in [21]. The underlying idea in both cases is to disconnect those actuators corresponding to interconnections with strength smaller than the prescribed ε, identifying the disconnected subsystems. According to [21], there are s different ε-decompositions $\mathcal{P}_\varepsilon$ that can be obtained for different values of ε satisfying

$$\max_{i \neq j} \left| m_{ij} \right| = \varepsilon_1 < \varepsilon_2 < \cdots < \varepsilon_K = 0,$$

with $K \leq dim(\mathbf{I_c})$. Moreover, such decompositions are nested, that is the partitions obtained satisfy: $\mathcal{P}_{\varepsilon_1} \subset \mathcal{P}_{\varepsilon_2} \cdots \mathcal{P}_{\varepsilon_K}$ with $\mathcal{P}_{\varepsilon_1}$ being the finest and $\mathcal{P}_{\varepsilon_k}$ the coarsest. The main novelty of the algorithm presented in this chapter is the matrix normalization taking into account actuator physical/operative limits and the iterative threshold updating that allows one to take into account weaker coupling without being influenced by the stronger ones.

16.4 Simulations and Results

16.4.1 Results Using Masks-Based Approach

Using the partitioning algorithm presented in this section, the aggregate model of the Barcelona WTN is decomposed in three subsystems, as depicted in Fig. 16.1 in different colours. The resultant decomposition follows the scheme shown in Fig. 16.2, where μ_i denotes the i-th vector of shared variables among the subsystems S_j, for $j = 1, \ldots, M$. The subsystems are defined by the following elements:

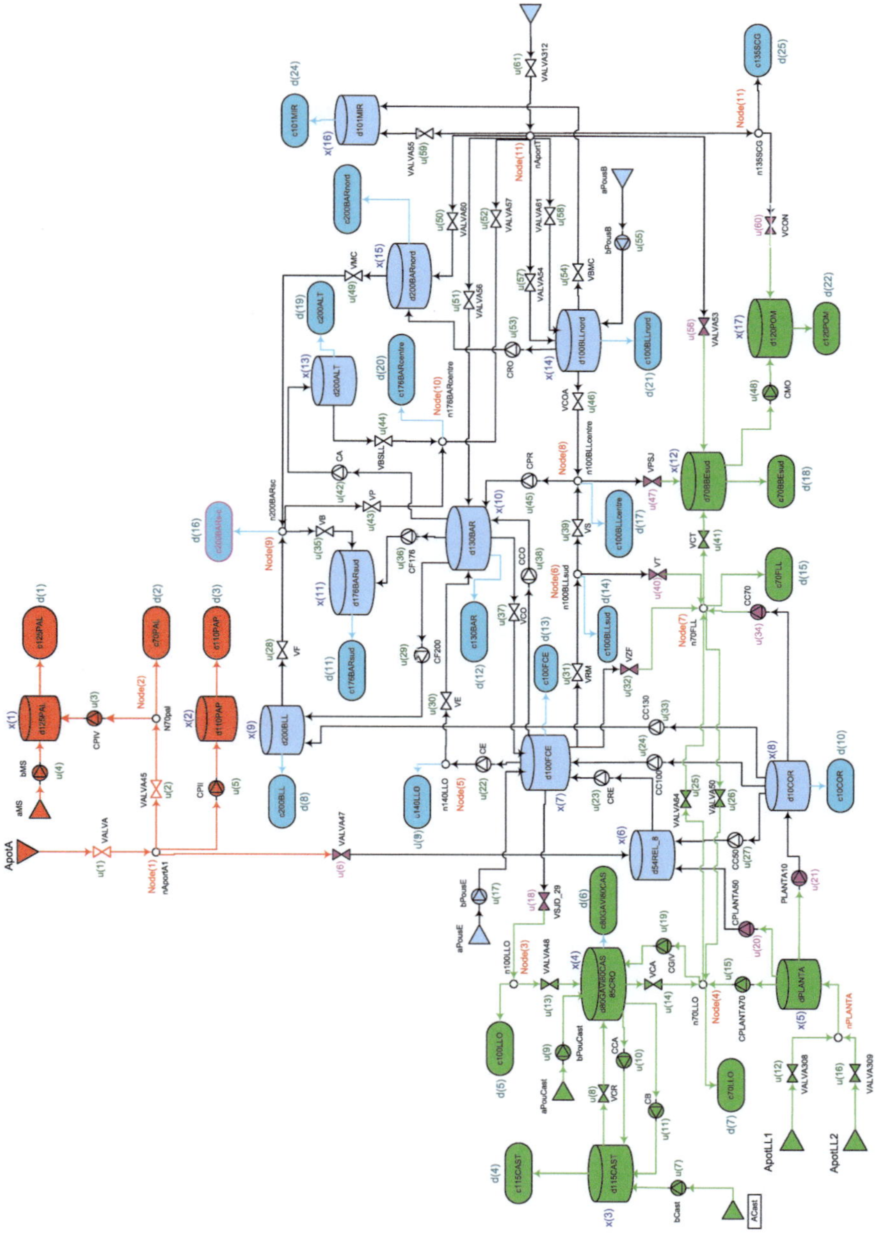

Fig. 16.1 Partition of the Barcelona WTN, aggregate model

- *Subsystem 1:* Composed by tanks x_i, $i \in \{1, 2\}$, inputs u_j, $j \in \{1 : 5\}$, demands d_l, $l \in \{1, 2, 3\}$ and nodes n_q, $q \in \{1, 2\}$. It is represented in Fig. 16.1 with red colour and corresponds to Subsystem S_1 in Fig. 16.2.

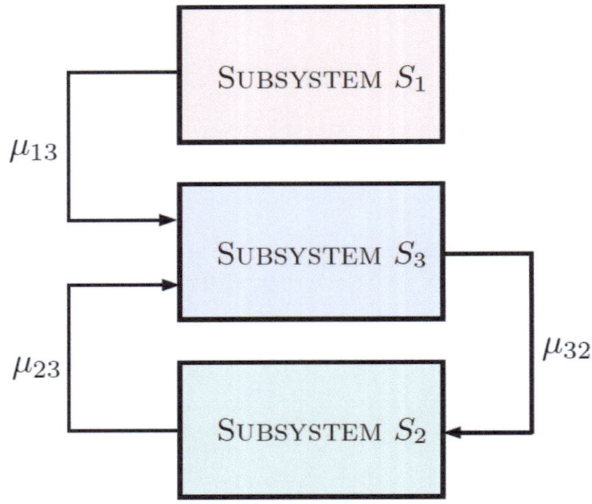

Fig. 16.2 Conceptual scheme of the partitioned Barcelona WTN

Table 16.1 Dimension comparison between the subsystems and the whole network

Elements	Subsystem 1	Subsystem 2	Subsystem 3	Whole model
Tanks	2	5	10	17
Actuators	5	22	34	61
Demands	3	7	15	25
Nodes	2	3	6	11

- *Subsystem 2:* Composed by tanks x_i, $i \in \{3, 4, 5, 12, 17\}$, inputs u_j, $j \in \{7 : 16, 18, 19, 25, 26, 32, 34, 40, 41, 47, 48, 56, 60\}$, demands d_l, $l \in \{4 : 7, 15, 18, 22\}$ and nodes $n_q, q \in \{3, 4, 7\}$. It is represented in Fig. 16.1 with green colour and corresponds to Subsystem S_2 in Fig. 16.2.
- *Subsystem 3:* Composed by tanks x_i, $i \in \{6 : 11, 13 : 16\}$, the inputs u_j, $j \in \{6, 17, 20 : 24, 27 : 31, 33, 35 : 39, 42 : 46, 49 : 55, 57, 58, 59, 61\}$, demands d_l, $l \in \{8 : 14, 16, 17, 19, 20, 21, 23, 24, 25\}$ and nodes n_q, $q \in \{5, 6, 8 : 11\}$. It is represented in Fig. 16.1 with blue colour and corresponds to Subsystem S_3 in Fig. 16.2.

Table 16.1 collects the resultant dimensions for each subsystem and the corresponding comparison with the dimensions of the vectors of variables for the entire aggregate network.

16.4.2 Results using Graph-Theory-Based Approach

This section presents the results of the application of Algorithm 14 for the partitioning of the Barcelona WTN into compositional subsystems [17]. Algorithm 16.1

and auxiliary routines presented in Sect. 16.3.1 have been designed for any system. However, some particular features should be introduced depending on the considered case study and control law in order to obtain a suitable decomposition. More precisely, the graph of the Barcelona WTN has been derived from its mathematical model expressed in the way introduced in Chap. 12, i.e.,

$$\mathbf{x}(k+1) = \mathbf{A}\,\mathbf{x}(k) + \mathbf{B_u}\,\mathbf{u}(k) + \mathbf{B_d}\,\mathbf{d}(k), \tag{16.10a}$$

$$0 = \mathbf{E_u}\,\mathbf{u}(k) + \mathbf{E_d}\,\mathbf{d}(k), \tag{16.10b}$$

under the following considerations:

- every tank, sector of consume, water source and node is considered as a vertex of the graph;
- every pump, valve and link with a sector of consume is considered as a graph edge.

In order to evaluate the partitioning results obtained from the application of Algorithm 16.1 and auxiliary routines to the Barcelona WTN, the following indexes are taken into account additionally to those previously introduced:

- $\varepsilon \triangleq \sum_{i=1}^{M} \varepsilon_i$,
- $\bar{\varepsilon} \triangleq \frac{\varepsilon}{M}$ (arithmetic mean),
- $\sigma_\varphi^2 \triangleq \frac{1}{M} \sum_{i=1}^{M} (\varphi_i - \bar{\varphi})^2$,
- $\sigma_\varepsilon^2 \triangleq \frac{1}{M} \sum_{i=1}^{M} (\varepsilon_i - \bar{\varepsilon})^2$.

Remark 16.4 Notice that although ε is not directly related to the number of shared edges between subgraphs obtained by using Algorithm 16.2, this index gives an indirect idea about their level of interconnection. Recall that the objective of the partitioning algorithm is the minimization of indexes σ_φ^2, ε and ε_i (for $i = 1, 2, \ldots, M$) to obtain a graph decomposition as less interconnected as possible and with similar number of vertices for each subgraph (internal weight).

Table 16.2 summarizes the partitioning results obtained applying Algorithm 16.1 (A1) combined with the auxiliary routine/filters presented in Sect. 16.3.1 performing the following combinations:

Table 16.2 Results for different partitioning approaches

Routine combination	M	$\bar{\varphi}$	$\bar{\varepsilon}$	σ_φ^2	σ_ε^2	ε
1	17	10.59	3.76	53.88	25.32	64
2	13	6.30	4.15	21.39	27.80	54
3	10	8.20	5.10	31.73	32.76	52
4	6	13.67	6.33	**14.88**	25.22	**38**

1. No auxiliary routines are considered.
2. A1 and prefiltering (pre-F) routine only.
3. A1 in addition to pre-F and post-filtering (post-F) routines.
4. A1 in addition to pre-F, post-F and anti-oscillation (AO) routines.

This distinction has been done in order to understand how the proposed routines affect the partitioning results.

Using only the Algorithm 16.1, the resultant partitioning $\mathcal{P}$ is comprised by 17 subgraphs. Many of them are small and cannot be merged since their neighbour subgraphs have internal weights with values quite close to $\bar{\varphi}$ (see Sect. 16.3.1). Moreover, there are several vertices with $\omega = 1$, which correspond to network water sources and demands, leading to unnecessarily difficult algorithm computations due to sizes of the resultant subgraphs (in terms of internal weight). By employing the pre-F routine, the previous problems are fixed and Algorithm 16.1 produces 13 subgraphs (see Table 16.2). Additionally, if the refining routine embedded within Algorithm 16.1 is complemented with the post-F routine, setting $\delta = 2$, a partitioning with ten subgraphs is reached.[5] Finally, if the AO routine is also considered, setting the refining limit to $\rho = 250$, a partitioning with six subgraphs is now reached. According to Table 16.2, this last partitioning (Combination 4) satisfies the minimization of the average of the internal weights for all resultant subgraphs as well as the interconnection degree between subgraph measured through ε. It is important to highlight that the proposed partitioning approach automatically determines the final number of partitions M (six for this case) when the conditions 3 and 4 of Problem 16.1 are fulfilled (see Remark 16.2). The tuning parameters δ and ρ also influence on the obtained value of M.

Notice that each subgraph of the final decomposition corresponds to a subsystem of the Barcelona WTN with the number of elements presented in Table 16.3. Figure 16.3 shows, in different colours, the obtained subsystems of Barcelona WTN.

Moreover, Fig. 16.4 schematically shows the disposition of the resultant subsystems S_i, for $i \in \{1, \ldots, 6\}$, and the sets μ_{ij} of shared links between the network subsystems corresponding to the control inputs u (manipulated flows), whose directionality is defined from S_i to S_j for $j \in \{1, \ldots, 6\}, i \neq j$. Table 16.4 collects the number of control inputs of each set μ_{ij}.

Table 16.3 Dimension comparison of the WTN subsystems

Subsystem	Tanks	Actuators	Demands	Nodes
1	13	36	20	5
2	11	11	11	0
3	13	22	20	3
4	9	16	12	2
5	6	10	8	2
6	15	26	17	3
Total	67	121	88	15

[5] Notice that increasing the parameter δ implies that $\sigma^2_{\varepsilon_i}$ becomes bigger.

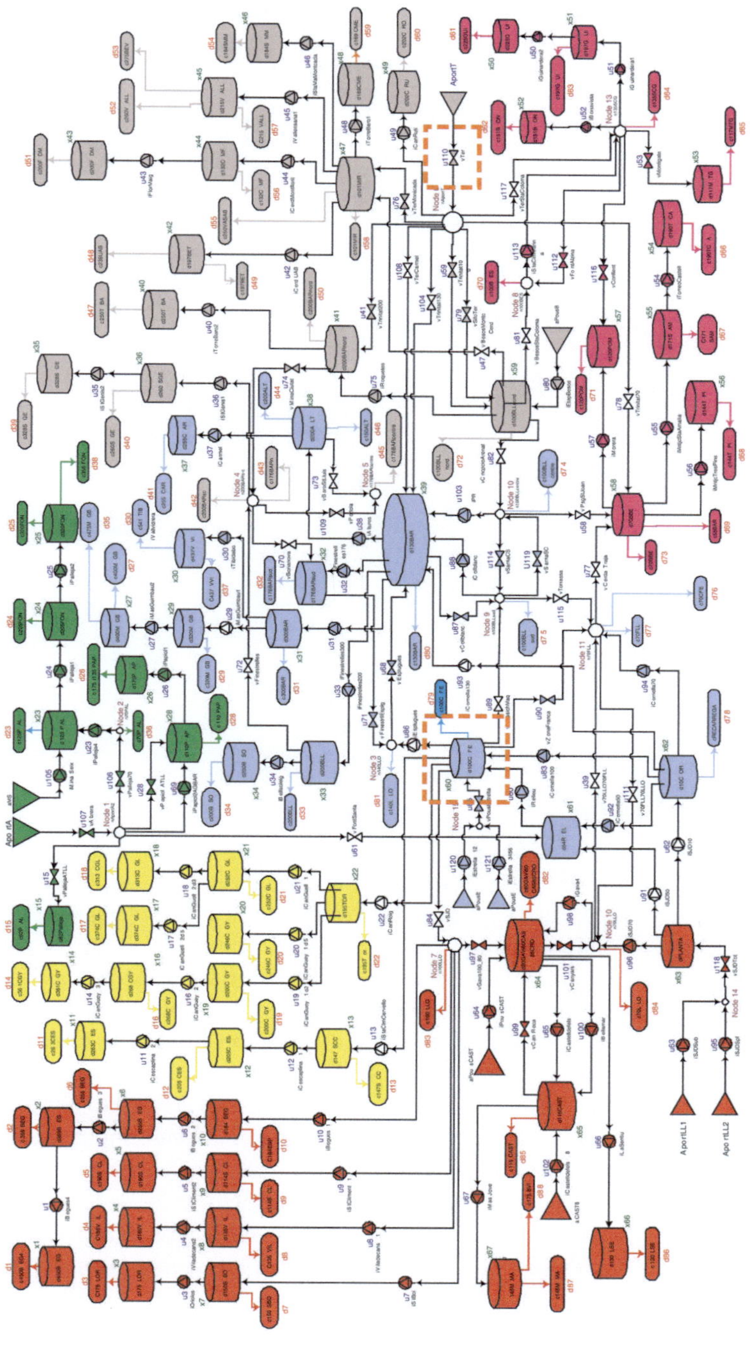

Fig. 16.3 Definitive partition of the Barcelona WTN. The *key* elements are properly featured

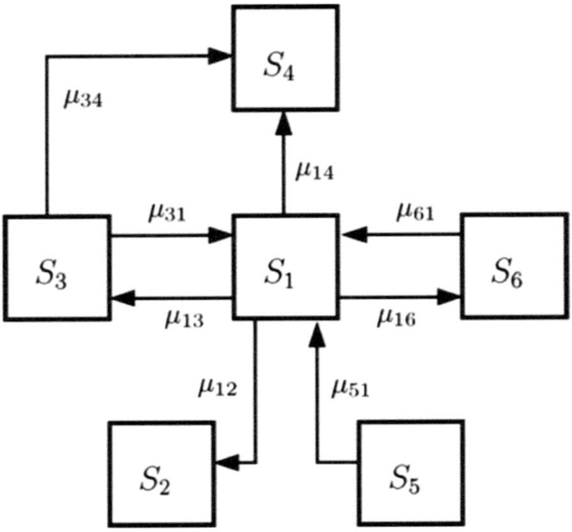

Fig. 16.4 Network subsystems S_i and their sets of shared connections μ_{ij}

Table 16.4 Dimensions of shared links μ_{ij}

Set	μ_{12}	μ_{13}	μ_{14}	μ_{16}	μ_{31}	μ_{34}	μ_{51}	μ_{61}
Number of u's	2	2	2	2	4	3	1	3

16.5 Conclusions

This chapter has proposed two approaches for the automatic partitioning of a WTN into subsystems intended to be applied along with a non-centralized model predictive control strategy. The algorithm transforms the dynamical model of the given system into a graph representation. Once the equivalent graph has been obtained, the problem of graph partitioning is then solved. The resultant partitions are composed of a set of non-overlapping subgraphs such that their sizes, in terms of number of vertices, are similar and the number of edges connecting them is minimal. To achieve this goal, the algorithm applied a set of procedures based on identifying the highly connected subgraphs with balanced number of internal and external connections. Some additional prefiltering and post-filtering routines are also needed to be included to reduce the number of obtained subsystems. The performance of the proposed decomposition approach has been assessed in a real case study based on the Barcelona WTN. A study of the effect of auxiliary routines on the basic partitioning algorithm has also been included showing the benefits of their use.

References

1. Addario-Berry L, Dalal K, Reed B (2008) Degree-constrained subgraphs. Discrete Appl Math 156(7):1168–1174
2. Bondy JA, Murty USR (2008) Graph theory. Graduate series in mathematics, vol 244. Springer
3. Brdys M, Ulanicki B (1994) Operational control of water systems: structures, algorithms and applications. Prentice Hall International, UK
4. Bui TN, Moon BR (1996) Genetic algorithm and graph partitioning. IEEE Trans Comput 45(7):841–855
5. Cormen TH (2001) Introduction to algorithms. The MIT Press
6. Dunbar WB (2007) Distributed receding horizon control of dynamically coupled nonlinear systems. IEEE Trans Autom Control 52(7):1249–1263
7. Dutt S (1993) New faster kernighan-lin-type graph-partitioning algorithms. In: Proceedings of the IEEE/ACM international conference on computer-aided design. IEEE Computer Society Press, pp 370–377
8. Fjallstrom P (1998) Algorithms for graph partitioning: a survey. In: Linkoping electronic articles in computer and information science, vol 3, issue no 10
9. Gupta A (1997) Fast and effective algorithms for graph partitioning and sparse-matrix ordering. IBM J Res Dev 41(1):171–183
10. HD-MPC (2008) Hierarchical and distributed model predictive control of large-scale complex systems. Home page
11. Keviczky T, Borrelli F, Balas GJ (2006) Decentralized receding horizon control for large scale dynamically decoupled systems. Automatica 42(12):2105–2115
12. Li F, Zhang W, Zhang Q (2007) Graphs partitioning strategy for the topology design of industrial network. IET Commun 1(6):1104–1110
13. Lunze J (1992) Feedback control of large-scale systems. Prentice Hall, Great Britain
14. Marinaki M, Papageorgiou M (2005) Optimal real-time control of sewer networks. Springer, Secaucus, NJ (USA)
15. Negenborn RR (2008) Multi-agent model predictive control with applications to power networks. PhD thesis, Delft University of Technology, Delft, The Netherlands
16. Negenborn RR, De Schutter B, Hellendoorn J (2008) Multi-agent model predictive control for transportation networks: serial vs. parallel schemes. Eng Appl Artif Intell 21(3):353–366
17. Ocampo-Martinez C, Bovo S, Puig V (2011) Partitioning approach oriented to the decentralised predictive control of large-scale systems. vol 21, 775–786
18. Van Overloop PJ (2006) Model predictive control on open water systems. Delft University Press, Delft, The Netherlands
19. Rawlings JB, Stewart BT (2008) Coordinating multiple optimization-based controllers: new opportunities and challanges. J Process Control 18(9):839–845
20. Scattolini R (2009) Architectures for distributed and hierarchical model predictive control: a review. J Process Control 19(5):723–731
21. Sezer ME, Siljak DD (1986) Nested ε-decomposition and clustering of complex systems. Automatica 22(3):321–331
22. Šiljak DD (1991) Decentralized control of complex systems. Academic Press
23. Venkat AN, Hiskens IA, Rawlings JB, Wright SJ (2008) Distributed MPC strategies with application to power system automatic generation control. IEEE Trans Control Syst Technol 16(6):1192–1206
24. WIDE (2008) Decentralized and wireless control of large-scale systems. http://ist-wide.dii.unisi.it/
25. Zecevic AI, Šiljak DD (1994) Balanced decompositions of sparse systems for multilevel parallel processing. IEEE Trans Circuits Syst I: Fund Theory Appl 41(3):220–233
26. Zecevic A, Siljak DD (2010) Control of complex systems: Structural constraints and uncertainty. Springer Science & Business Media

Chapter 17
Non-centralized Predictive Control for Drinking-Water Supply Systems

Juan Manuel Grosso, Carlos Ocampo-Martínez and Vicenç Puig

17.1 Introduction

The control schemes proposed in the previous chapters have shown the potential applicability of *centralized* MPC for the economic scheduling control of network flows. Nevertheless, as illustrated with the case study of Barcelona, flow-based networks are generally systems comprised of multiple subsystems and/or large-scale systems. Thus, the centralization of decisions in a single MPC-based agent could be disadvantageous for the reliability of the network operation and the maintenance of the monolithic prediction model. These issues have received a lot of attention from the control research community during the last years. Several *non-centralized* control strategies have been already proposed in the literature, where either large-scale systems are partitioned into subsystems with individual control agents or a plant-wide optimization problem is distributed in a set of smaller optimization problems that are usually coordinated by a master problem. The importance of *system partitioning* and/or *distributed optimization* has already been noticed in classic references addressing the decentralized control of large-scale systems [9, 17] and the decomposition of mathematical programming problems [3]. For distributing the centralized MPC optimization problem, several analytic methods exist, e.g., Dantzig–Wolfe decomposition, Bender's decomposition, and optimality condition decomposition, among other dual or primal decomposition techniques. These analytic decompositions rely strongly on the form of both the constraints and the objective function and are specialized to particular problem structures that might not cover many real

J.M. Grosso · C. Ocampo-Martínez (✉)
Institut de Robòtica i Informàtica Industrial, CSIC-UPC, Barcelona, Spain
e-mail: cocampo@iri.upc.edu

V. Puig
Research Center Supervision, Safety and Automatic Control (CS2AC-UPC), Terrassa, Spain

V. Puig et al. (eds.), *Real-Time Monitoring and Operational Control of Drinking-Water Systems*, Advances in Industrial Control,
DOI 10.1007/978-3-319-50751-4_17

large-scale flow-based networks. Therefore, as discussed in Chap. 16, graph theory is also used to cope with large-scale networks. Basically, the partitioning of a flow-based network consists in choosing subsets of the global variables to be assigned to different local agents that are in charge of controlling individual partitions/subsystems, as described in [5, 6, 11, 14]. This chapter addresses a large-scale network as a system-of-systems instead of analytically decomposing the global optimization problem; the corresponding partitions will be assumed given from now on.

It has been demonstrated in [16] that exchanging only interaction information (even iteratively) among the local controllers is not enough to guarantee closed-loop stability and/or optimal plant-wide performance due to their competitive behaviour. Hence, for economically optimal operation (or to reduce suboptimality) of the network, cooperation between local controllers must be induced. This can be achieved, e.g., by means of cooperative, coordinated or hierarchical MPC schemes, which incorporate negotiation/coordination mechanisms to approach the centralized solution. A crucial issue in all these non-centralized control schemes is that of guaranteeing recursive feasibility of the optimization problem, especially when addressing dynamically coupled subsystems. Among the non-centralized MPC schemes that have been proposed in the literature (see, e.g., [12] and references therein), one important classification criterion is the information exchange between local agents (e.g., predicted trajectories, prices or dual variables), which in general can be either local or global. On the one hand, there are schemes that use local information and iterative communication to improve performance but guaranteeing feasibility mostly only upon convergence to the global optimal solution. To cope with feasibility losses (e.g., due to early termination of the iterative algorithm), other non-iterative distributed MPC schemes consider the shared variables as local disturbances and rely on the design of (possibly over-conservative) robust local controllers, guaranteeing feasibility of the network at the expense of a worse economic performance. On the other hand, there exist several cooperative approaches inspired in [18], which exchange global information and ensure recursive feasibility of the optimization problem (even with non-iterative communication) by using centralized prediction models. Generally, these cooperative schemes converge asymptotically to the central optimum under certain structural assumptions, e.g., sparse couplings.

Most of the available non-centralized MPC algorithms were proposed to control systems operating under a standard (tracking) cost functions, and only few cooperative (iterative) distributed economic MPC schemes have been recently published (see, e.g., [4, 8]). Differently, this chapter proposes a non-iterative multi-layer distributed economic MPC (ML-DMPC) approach for its application to flow-based networks. This approach is based on a temporal and functional decomposition of the centralized economic scheduling-control problem. The architecture of the proposed ML-DMPC controller lies in the class of hierarchical systems [10]. Specifically, the controller comprises two layers that operate at different timescales and interact to fulfil a set $\mathcal{O}$ of desired control objectives. In a top-down hierarchy, the control structure has a centralized coordinator in the upper layer and a set of local distributed MPC controllers in the lower layer. Contrary to the standard coordinated distributed control structures [10], where the local controllers use local information and communicate iteratively only with the coordinator to reconstruct the centralized performance, the proposed ML-DMPC scheme considers non-iterative and hierarchical-like neighbour-to-neighbour communication between the local controllers. The coordinator is used to influence (also non-iteratively) the overall performance through economic *intervention parameters*. The ML-DMPC controller aims to improve the performance of a decentralized MPC strategy (but still being globally suboptimal) and to guarantee the recursive feasibility of the involved tractable distributed algorithm.

17.2 Problem Statement

In Chap. 11, a method to obtain the monolithic state-space model of a given flow-based network graph was described. Once the control-oriented model is stated, it is important to determine the objective of performing the partitioning of the physical network no matter what control strategy is followed. For large-scale network flow problems, the partitioning of the system gains sense from the point of view of modularity of the control architecture and the reduction of computational burden. In any case, the way the network elements are interconnected is a key factor for performing the partitioning and control of the overall network since it determines the type of couplings between subsystems and consequently the complexity and rationality of the control strategy.

In the following sections, the overall system is assumed to be decomposed in a set of $M \in \mathbb{Z}_{\geq 1}$ dynamically coupled non-overlapped subsystems denoted by $\mathcal{S}_i$, $i \in \mathbb{Z}_{[1,M]}$. The number M of subsystems is generally a tuning parameter. In this chapter, a two-stage decomposition is performed. In the first stage, a reachability analysis is used to define a set of subsystems that can be supplied only by one source each. These resultant subsystems are here called *anchored subsystems* and are denoted as $\mathcal{S}_i$, $i \in \mathbb{Z}_{[1,r]}$, where $r \leq M$ is the number of flow sources in the network. The remaining elements of the network are grouped in a subsystem denoted as $\tilde{\mathcal{S}}$, which is supplied by the cross-border outflows of the anchored subsystems. Such flows are considered as *pseudosources* of $\tilde{\mathcal{S}}$. In the second stage of the decomposition, subsystem $\tilde{\mathcal{S}}$ is later subdivided into $M - r$ subsystems by means of the graph-based partitioning algorithm proposed in Chap. 16. This algorithm aims at decomposing $\tilde{\mathcal{S}}$ and its corresponding directed graph into subgraphs, in such a way that all resultant partitions have nearly the same number of vertices and a hierarchical/sequential solution order can be stated. Note that another set of pseudosources may appear after the decomposition of $\tilde{\mathcal{S}}$ and, contrary to the first stage of decomposition, each subsystem may have both entering and leaving cross-border flows depending on the interconnections of the resultant $\mathcal{S}_i$ subsystems, $i \in \mathbb{Z}_{[r+1,M]}$. A sketch of the overall decomposition process is depicted in Fig. 17.1.

Particularly, this chapter considers only input-coupled dynamics and input-coupled constraints. Then, each subsystem can be described by the following discrete-time linear model:

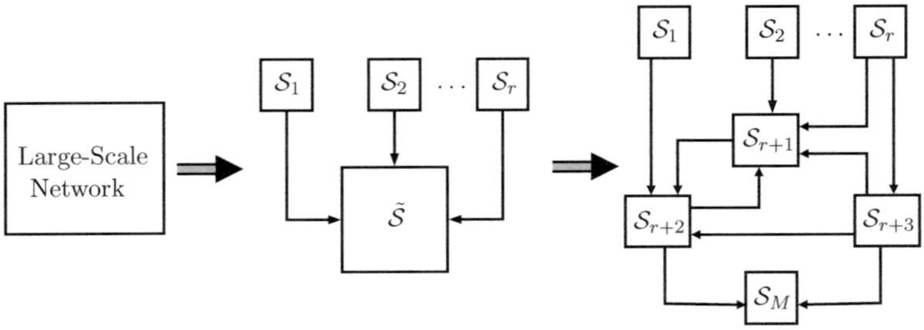

Fig. 17.1 Decomposition of a network with r sources into M subsystems

$$\begin{cases} \mathbf{x}^{[i]}(k+1) = \mathbf{A}_{ii}\mathbf{x}^{[i]}(k) + \mathbf{B}_{ii}\mathbf{u}^{[i]}(k) + \mathbf{B}_{d,ii}\mathbf{d}^{[i]}(k) + \sum_{\substack{j=1 \\ j\neq i}}^{M} \mathbf{B}_{ij}\mathbf{u}^{[j]}(k), & \text{(17.1a)} \\ \mathbf{0} = \mathbf{E}_{u,ii}\mathbf{u}^{[i]}(k) + \mathbf{E}_{d,ii}\mathbf{d}^{[i]}(k) + \sum_{\substack{j=1 \\ j\neq i}}^{M} \mathbf{E}_{u,ij}\mathbf{u}^{[j]}(k), & \text{(17.1b)} \end{cases}$$

for all $k \in \mathbb{Z}_+$ and $i, j \in \mathbb{Z}_{[1,M]}$, where $\mathbf{x}^{[i]} \in \mathbb{R}^{n_{x_i}}$, $\mathbf{u}^{[i]} \in \mathbb{R}^{n_{u_i}}$ and $\mathbf{d}^{[i]} \in \mathbb{R}^{n_{d_i}}$ are, respectively, the local state, input and demand vectors of subsystem $\mathcal{S}_i$, $i \in \mathbb{Z}_{[1,M]}$. Local matrices are given by the topology of each subsystem, with $\mathbf{A}_{ii} = \mathbf{I}_{n_{x_i}}$, $\mathbf{B}_{ii} \in \mathbb{R}^{n_{x_i}\times n_{u_i}}$, $\mathbf{B}_{d,ii} \in \mathbb{R}^{n_{x_i}\times n_{d_i}}$, $\mathbf{B}_{ij} \in \mathbb{R}^{n_{x_i}\times n_{u_j}}$, $\mathbf{E}_{u,ii} \in \mathbb{R}^{q_i\times n_{u_i}}$, $\mathbf{E}_{d,ii} \in \mathbb{R}^{q_i\times n_{d_i}}$ and $\mathbf{E}_{u,ij} \in \mathbb{R}^{q_i\times n_{u_i}}$ for all $i, j \in \mathbb{Z}_{[1,M]}$. The decomposition assures that $\sum_{i=1}^{M} n_{x_i} = n_x$, $\sum_{i=1}^{M} n_{u_i} = n_u$, $\sum_{i=1}^{M} n_{d_i} = n_d$ and $\sum_{i=1}^{M} q_i = q$ for all $n_{x_i}, n_{u_i}, n_{d_i}, q_i \in \mathbb{Z}_{\geq 1}$. Similarly, the global constraint sets $\mathcal{X}, \mathcal{U}$ and $\mathcal{D}$ are decomposed to give place to a set of local constraints defined by:

$$\mathbf{x}^{[j]}(k) \in \mathcal{X}_i := \{\mathbf{x}^{[i]} \in \mathbb{R}^{n_{x_i}} \mid \mathbf{0} \leq \mathbf{x}^{[i]} \leq \mathbf{x}^{[i],\max}\}, \quad \text{(17.2a)}$$

$$\mathbf{u}^{[j]}(k) \in \mathcal{U}_i := \{\mathbf{u}^{[i]} \in \mathbb{R}^{n_{u_i}} \mid \mathbf{0} \leq \mathbf{u}^{[i]} \leq \mathbf{u}^{[i],\max}\}, \quad \text{(17.2b)}$$

$$\mathbf{d}^{[j]}(k) \in \mathcal{D}_i := \{\mathbf{d}^{[i]} \in \mathbb{R}^{p_i} \mid \mathbf{0} \leq \mathbf{d}^{[i]} \leq \mathbf{d}^{[i],\max}\}. \quad \text{(17.2c)}$$

Definition 17.1 (*Neighbour and neighbourhood*) A subsystem $\mathcal{S}_j$ is defined as a *neighbour* of subsystem $\mathcal{S}_i$ if and only if $\mathbf{B}_{ij} \neq \mathbf{0}$ or $\mathbf{E}_{u,ij} \neq \mathbf{0}$, $j \in \mathbb{Z}_{[1,M]}$, $j \neq i$. Hence, the *neighbourhood* of $\mathcal{S}_i$ is defined as $\mathcal{N}_i := \{j \in \mathbb{Z}_{[1,M]} \mid \mathbf{B}_{ij} \neq \mathbf{0} \text{ or } \mathbf{E}_{u,ij} \neq \mathbf{0},\ j \neq i\}$.

Remark 17.1 Note that the overall system model can be obtained by the composition of the above M subsystems, as follows:

$$\begin{cases} \mathbf{x}(k+1) = \mathbf{A}\mathbf{x}(k) + \mathbf{B}\mathbf{u}(k) + \mathbf{B}_d\mathbf{d}(k), \\ \mathbf{0} = \mathbf{E}_u\mathbf{u}(k) + \mathbf{E}_d\mathbf{d}(k), \end{cases}$$

where the vectors and matrices are now a permutation of the original ones, with

$$\mathbf{x}(k) = \begin{bmatrix} \mathbf{x}(k)^{[1]} \\ \vdots \\ \mathbf{x}(k)^{[M]} \end{bmatrix}, \quad \mathbf{u}(k) = \begin{bmatrix} \mathbf{u}(k)^{[1]} \\ \vdots \\ \mathbf{u}(k)^{[M]} \end{bmatrix}, \quad \mathbf{d}(k) = \begin{bmatrix} \mathbf{d}(k)^{[1]} \\ \vdots \\ \mathbf{d}(k)^{[M]} \end{bmatrix}, \tag{17.3}$$

and

$$A = \begin{bmatrix} \mathbf{I}_{n_{x_1}} & \dots & \mathbf{0} \\ \vdots & \ddots & \vdots \\ \mathbf{0} & \dots & \mathbf{I}_{n_{x_M}} \end{bmatrix}, \quad B = \begin{bmatrix} \mathbf{B}_{11} & \dots & \mathbf{B}_{1M} \\ \vdots & \ddots & \vdots \\ \mathbf{B}_{M1} & \dots & \mathbf{B}_{MM} \end{bmatrix},$$

$$\mathbf{B}_d = \begin{bmatrix} \mathbf{B}_{d,ii} & \dots & \mathbf{0} \\ \vdots & \ddots & \vdots \\ \mathbf{0} & \dots & \mathbf{B}_{d,MM} \end{bmatrix}, \quad \mathbf{E}_u = \begin{bmatrix} \mathbf{E}_{u,11} & \dots & \mathbf{E}_{u,1M} \\ \vdots & \ddots & \vdots \\ \mathbf{E}_{u,M1} & \dots & \mathbf{E}_{u,MM} \end{bmatrix},$$

$$\mathbf{E}_d = \begin{bmatrix} \mathbf{E}_{d,ii} & \dots & \mathbf{0} \\ \vdots & \ddots & \vdots \\ \mathbf{0} & \dots & \mathbf{E}_{d,MM} \end{bmatrix}.$$

Moreover, since the dynamic and static nodes of the network were decomposed into M disjoint subsets, it follows that the global constraint sets can be recovered as Cartesian products, i.e.,

$$\mathcal{X} = \prod_{i=1}^{M} \mathcal{X}_i, \quad \mathcal{U} = \prod_{i=1}^{M} \mathcal{U}_i, \quad \mathcal{D} = \prod_{i=1}^{M} \mathcal{D}_i. \tag{17.4}$$

◇

Before getting through the design of the ML-DMPC strategy, the following preliminary assumptions related to the overall system are stated.

Assumption 17.1 All demands have a *periodic* flow request (with period $T \in \mathbb{Z}_{\geq 1}$) that can be supplied by at least one flow source through at least one flow path.[1]

Assumption 17.2 The required control objectives can be grouped in a set $\mathcal{O} = \mathcal{O}_l \cup \mathcal{O}_g$, which is a composition of a set $\mathcal{O}_l$ of *local* control objectives and a set $\mathcal{O}_g$ of *global* control objectives. Moreover, $m_l \triangleq |\mathcal{O}_l|, m_g \triangleq |\mathcal{O}_g|$, and hence $m_l + m_g = |\mathcal{O}|$.

Assumption 17.2 allows to rewrite a centralized general economic stage cost function $J : \mathbb{Z}_+ \times \mathbb{R}^{n_x} \times \mathbb{R}^{n_u} \to \mathbb{R}_+$ in the following form:

[1] A flow path is an ordered sequence of arcs, which may connect sources, intermediate nodes and demands.

$$J(k, \mathbf{x}(k), \mathbf{u}(k) = \sum_{g=1}^{m_g} \lambda_g \ell_g(k, \mathbf{x}(k), \mathbf{u}(k)) + \sum_{l=1}^{m_l} \lambda_l J_l(k, \mathbf{x}(k), \mathbf{u}(k), \tag{17.5}$$

where $\lambda_g, \lambda_l \in \mathbb{R}_+$ are scalar weights that prioritise, within the overall cost function, each global and local control objective, particularly represented by convex functions $J_g : \mathbb{Z}_+ \times \mathbb{R}^{n_x} \times \mathbb{R}^{n_u} \to \mathbb{R}_+$ and $J_l : \mathbb{Z}_+ \times \mathbb{R}^{n_x} \times \mathbb{R}^{n_u} \to \mathbb{R}_+$, respectively. Hence, from (17.1), (17.2) and Remark 17.1, the centralized MPC optimization problem with stage cost (17.5) and prediction horizon N can be rewritten as follows:

$$\min_{\mathbf{u}_k} \sum_{t=0}^{H_p-1} \left(\sum_{g=1}^{m_g} \lambda_g J_g(k, \mathbf{x}(k+t|k), \mathbf{u}(k+t|k)) + \sum_{l=1}^{m_l} \lambda_l J_l(k, \mathbf{x}(k+t|k), \mathbf{u}(k+t|k) \right), \tag{17.6a}$$

subject to:

$$\mathbf{x}^{[i]}(k+t+1|k) = \mathbf{A}_{ii}\mathbf{x}^{[i]}(k+t|k) + \mathbf{B}_{ii}\mathbf{u}^{[i]}(k+t|k) + \mathbf{B}_{d,ii}\mathbf{d}^{[i]}(k+t|k) + \sum_{\substack{j=1 \\ j\neq i}}^{M} \mathbf{B}_{ij}\mathbf{u}^{[j]}(k+t|k), \tag{17.6b}$$

$$\mathbf{0} = \mathbf{E}_{u,ii}\mathbf{u}^{[i]}(k+t|k) + \mathbf{E}_{d,ii}\mathbf{d}^{[i]}(k+t|k) + \sum_{\substack{j=1 \\ j\neq i}}^{M} \mathbf{E}_{u,ij}\mathbf{u}^{[j]}(k+t|k), \tag{17.6c}$$

$$(\mathbf{x}^{[i]}(k+t+1|k), \mathbf{u}^{[i]}(k+t|k)) \in \mathcal{X}_i \times \mathcal{U}_i, \tag{17.6d}$$

$$\mathbf{x}^{[i]}(k|k) = \mathbf{x}^{[i]}(k), \tag{17.6e}$$

for all $i \in \mathbb{Z}_{[1,M]}$ and all $t \in \mathbb{Z}_{[0,H_p-1]}$. The aggregate state and input vectors in the cost function are given by $\mathbf{x}(k+t|k) = (\mathbf{x}^{[1]^T}(k+t|k), \ldots, \mathbf{x}^{[M]^T}(k+t|k))^T$, $mathbfu(k+t|k) = (\mathbf{u}^{[1]^T}(k+t|k), \ldots, \mathbf{u}^{[M]^T}(k+t|k))^T$, respectively. The decision variable is the input sequence $\mathbf{u}_k = \{\mathbf{u}(k+t|k)\}_{t\in\mathbb{Z}_{0,H_p-1}}$.

Thus, the goal of the ML-DMPC approach proposed in this chapter is that of solving (17.6) in a distributed fashion in order to cope with the aforementioned disadvantages of a centralized controller. To do so, a set $\mathcal{C} := \{C_1, \ldots, C_M\}$ of local controllers, their communication network and a coordination mechanism are designed in the following to properly address the effect of couplings between subsystems and to take into account Assumption 17.2.

17.3 Proposed Approach

The whole ML-DMPC set-up consists of the following:

(i) an *upper* layer in charge of achieving the global objectives by solving a centralized optimization problem with a sampling time Δt_1 and
(ii) a lower layer comprising a set of distributed MPC agents that compute the references for the system actuators in order to satisfy the local objectives.

This layer operates with a sampling time Δt_2 ($\Delta t_2 \leq \Delta t_1$). The local controllers solve their associated optimization problem in a hierarchical/sequential fashion and exchange (non-iteratively) in a neighbour-to-neighbour communication strategy the predicted sequence of the inputs affecting the neighbouring subsystems. The upper layer influences the operation of the lower layer by projecting global economic information into the local agents, specifically by modifying the prices/weights of the flow arcs that are shared among the subsystems arising in the lower layer. Figure 17.2 shows the proposed control structure. The ML-DMPC scheme leads to a suboptimal plant-wide performance but with the advantage of a tractable implementation due to a hierarchical-like communication approach that avoids negotiations among local controllers. A formal description of the two optimization layers involved in the ML-DMPC approach and their interaction is given below.

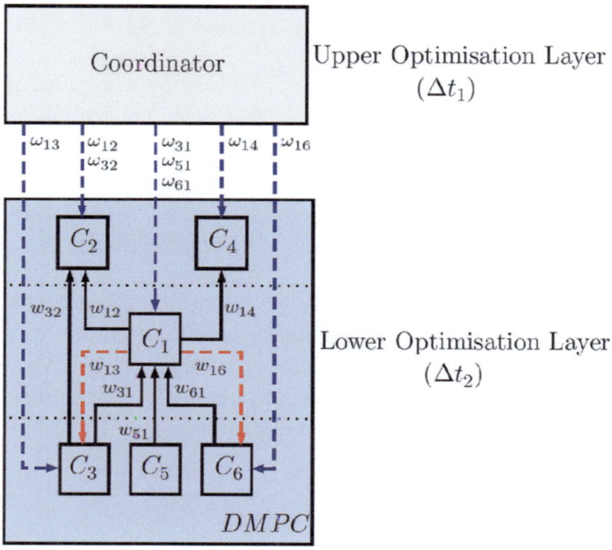

Fig. 17.2 ML-DMPC control architecture

17.3.1 Lower Optimization Layer

Once the network partitioning is performed and the M local models are obtained, it only remains to distribute the original centralized economic MPC problem among the local controllers C_i, considering the given management policies and constraints.

In order to simplify the notation, let rewrite the interaction-oriented local models as rewritten in the following more compact form:

$$\begin{cases} \mathbf{x}^{[i]}(k+1) = \mathbf{A}_{ii}\mathbf{x}^{[i]}(k) + \mathbf{B}_{ii}\mathbf{u}^{[i]}(k) + \mathbf{B}_{d,ii}\mathbf{d}^{[i]}(k) + \tilde{\mathbf{B}}_i\mathbf{w}^{[i]}(k), & (17.7\text{a}) \\ \mathbf{0} = \mathbf{E}_{u,ii}\mathbf{u}^{[i]}(k) + \mathbf{E}_{d,ii}\mathbf{d}^{[i]}(k) + \tilde{\mathbf{E}}_i\mathbf{w}^{[i]}(k), & (17.7\text{b}) \end{cases}$$

for all $i \in \mathbb{Z}_{[1,M]}$, where $\mathbf{w}^{[i]}(k) := (\mathbf{w}_{i_1}^T(k), \ldots, \mathbf{w}_{i_{|\mathcal{N}_i|}}^T(k))^T \in \mathcal{W}_i$ is a vector stacking the flows decided by the controllers of neighbours of subsystem $\mathcal{S}_i$, $\{i_1, \ldots, i_{|\mathcal{N}_i|}\}$ is an ordered sequence of the indices contained in the set $\mathcal{N}_i$ (i.e., $i_1 < \ldots < i_{|\mathcal{N}_i|}$) and $\mathbf{w}_j(k) := \mathbf{T}_{w_j}^T\mathbf{w}^{[j]}(k)$ for all $j \in \mathcal{N}_i$. In the definition of each $\mathbf{w}_j(k)$, the matrix $\mathbf{T}_{w_j} \in \mathbb{R}^{n_{u_j} \times n_{u_{ij}}}$ ($\mathbf{T}_{w_j}^T\mathbf{T}_{w_j} = \mathbf{I}_{n_{u_{ij}}}$) is such that it collects the m_{ij} ($m_{ij} < m_j$) columns of the identity matrix of order n_{u_j}, corresponding to the indices of the rows of $\tilde{\mathbf{u}}^{[j]}(k) \in \mathbb{R}^{n_{u_j}}$ related to the controlled flows decided by the controller C_j and affecting subsystem $\mathcal{S}_i$. Moreover, matrices $\bar{\mathbf{B}}_i$ and $\bar{\mathbf{E}}_i$ are suitably defined to represent the effect of $\mathbf{w}^{[i]}(k)$ on the local state vector $\mathbf{x}^{[i]}(k)$, and the set $\mathcal{W}_i$ is obtained appropriately from $\mathcal{U}_i$. In the sequel, every subsystem $\mathcal{S}_j$ that imposes an outflow $\mathbf{w}_j(k)$ to a subsystem $\mathcal{S}_i$ will be considered as a *virtual demand* of $\mathcal{S}_i$.

Interpretation 1 *At any time instant $k \in \mathbb{Z}_+$ when the controlled flow $\mathbf{u}^{[i]}(k)$ is computed, the controller C_i has knowledge of the state $\mathbf{x}^{[i]}(k)$ and the demands $\mathbf{d}^{[i]}(k)$ and $\mathbf{w}^{[i]}(k)$ imposed by the local and virtual demands, respectively. Future demands $\mathbf{d}^{[i]}(k+t)$ and $\mathbf{w}^{[i]}(k+t)$ might be unknown for all $t \in \mathbb{Z}_{\geq 1}$ and can take arbitrary values in $\mathcal{D}_i$ and $\mathcal{W}_i$, respectively. Nevertheless, the controller C_i has also knowledge of the H_p-step sequences of both the local and virtual demand expectations.*

Each controller C_i will be in charge of deciding only the network flows corresponding to subsystem $\mathcal{S}_i$ by using local and neighbouring information under Interpretation 1. In this chapter, the local problems are defined in such a way that each of them considers a local stage cost function but with a structure similar to the one in (17.5). Specifically, the stage cost function related to each C_i is written as follows:

$$\begin{aligned} J_i(k, \mathbf{x}^{[i]}(k), \mathbf{u}^{[i]}(k)) = &\sum_{g=1}^{m_g} \hat{\lambda}_{g,i}\hat{J}_{g,i}(k, \mathbf{x}^{[i]}(k), \mathbf{u}^{[i]}(k)) \\ &+ \sum_{l=1}^{m_l} \lambda_{l,i}J_{l,i}(k, \mathbf{x}^{[i]}(k), \mathbf{u}^{[i]}(k)), \end{aligned} \quad (17.8)$$

where each $\hat{J}_{g,i}$, $g \in \mathbb{Z}_{[1,m_g]}$, corresponds to the gth global control objective properly expressed and weighted with a suitable $\hat{\lambda}_{g,i} \in \mathbb{R}_+$ in order to influence controllers C_i to improve plant-wide performance. Moreover, each $J_{l,i}$ is assumed to be the corresponding part of the separable local objectives J_l, $l \in \mathbb{Z}_{[1,m_l]}$, related to the subsystem $\mathcal{S}_i$.

For each subsystem $\mathcal{S}_i$, a portion of control *importance* is removed by its neighbours and added to its local uncertainty in a max-min sense due to the local knowledge considered in Interpretation 1. Hence, before fully devising the distributed MPC controllers operating in the lower layer, the following definition (adjusted from [1, Definition 4.1]) is introduced.

Definition 17.2 Denote a given *network decomposition* with $\mathcal{P} = \{\mathcal{S}_i\}_{i \in \mathbb{Z}_{1,M}}$ and let $\mathcal{C}_\infty^{\mathcal{S}_i}$ be the maximal max-min robust control invariant set for subsystem $\mathcal{S}_i$. Then, the *decentralized max-min robust control invariant set* for the overall system

$$\mathbf{x}(k+1) = \mathbf{A}\mathbf{x}(k) + \mathbf{B}\mathbf{u}(k) + \mathbf{B}_d\mathbf{d}(k), \qquad \forall k \in \mathbb{Z}_+ \tag{17.9a}$$
$$\mathbf{0} = \mathbf{E}_u\mathbf{u}(k) + \mathbf{E}_d\mathbf{d}(k), \qquad \forall k \in \mathbb{Z}_+ \tag{17.9b}$$

subject to constraints

$$\mathbf{x}(k) \in \mathcal{X} = \{x \in \mathbb{R}^{n_x} | \mathbf{0} \leq \mathbf{x} \leq \mathbf{x}^{\max}\}, \qquad \forall k \in \mathbb{Z}_+ \tag{17.10a}$$
$$\mathbf{u}(k) \in \mathcal{U} = \{\mathbf{x} \in \mathbb{R}^{n_u} | \mathbf{0} \leq \mathbf{u} \leq \mathbf{u}^{\max}\}, \qquad \forall k \in \mathbb{Z}_+ \tag{17.10b}$$

and decomposed into Δ is given by $\mathcal{C}_\infty^\Delta = \prod_{i=1}^M \mathcal{C}_\infty^{\mathcal{S}_i}$.

For a given network decomposition $\mathcal{P}$ and local sets $\mathcal{X}_i, \mathcal{U}_i, \mathcal{D}_i$ and $\mathcal{W}_i$, $i \in \mathbb{Z}_{[1,M]}$, each maximal max-min robust control invariant set $\mathcal{C}_\infty^{\mathcal{S}_i}$ can be explicitly computed for the overall network.

Note that such sets $\mathcal{C}_\infty^{\mathcal{S}_i}$ may result to be empty for a given $\mathcal{P}$ (consequently $\mathcal{C}_\infty^{\mathcal{P}} = \emptyset$), which implies that there is no guarantee that a decentralized control strategy will be feasibility for all times. In such a case, the sets $\mathcal{U}_i$ (accordingly $\mathcal{W}_i$), $i \in \mathbb{Z}_{[1,M]}$, should be properly modified to make possible the decentralized design of $\mathcal{C}_\infty^{\mathcal{P}}$, see e.g., [1].

Assumption 17.3 The local constraint sets arising for a given network decomposition $\mathcal{P} = \{\mathcal{S}_i\}_{i \in \mathbb{Z}_{1,M}}$ are such that

$$\mathbf{B}_{d,ii}\mathcal{D}_i \oplus \bar{\mathbf{B}}_i\mathcal{W}_i \subseteq -\mathbf{B}_{ii}\mathcal{U}_i \quad \text{and} \quad \mathbf{E}_{d,ii}\mathcal{D}_i \oplus \bar{\mathbf{E}}_i\mathcal{W}_i \subseteq -\mathbf{E}_{u,ii}\mathcal{U}_i,$$

for all $\mathcal{S}_i \in \mathcal{P}$. Hence, $\mathcal{C}_\infty^{\mathcal{S}_i} := \big((\mathcal{X}_i \oplus (-\mathbf{B}_{ii}\mathcal{U}_i)) \ominus \big(\mathbf{B}_{d,ii}\mathcal{D}_i \oplus \bar{\mathbf{E}}_i\mathcal{W}_i\big)\big) \cap \mathcal{X}_i \neq \emptyset$.

Even when Assumption 17.3 holds and $\mathcal{C}_\infty^{\mathcal{P}}$ exists, the algebraic equation (17.7b) for each local model acts as a coupling constraint that forbids the design of non-iterative distributed controllers with parallel solution of the local optimization problems. Thus, the distributed MPC algorithm considered in the lower layer of the proposed ML-DMPC approach involves a non-iterative communication-based MPC design that builds on the hierarchical decentralized MPC approach reported in [13]. The strategy proposed here also follows a hierarchical sequence of solution but considering conditions to deal with the existence of bidirectional complicating flows between neighbour subsystems. The optimization problem to be solved in the lower layer of the ML-DMPC by each local controller C_i, $i \in \mathbb{Z}_{[1,M]}$, with sampling time Δt_2, is defined as follows:

$$\min_{\mathbf{u}_k} \sum_{t=0}^{H_p-1} \left(\sum_{g=1}^{m_g} \hat{\lambda}_{g,i} \hat{J}_{g,i}(k, \mathbf{x}^{[i]}(k+t|k), \mathbf{u}^{[i]}(k+t|k)) + \sum_{l=1}^{m_l} \lambda_{l,i} J_{l,i}(k, \mathbf{x}^{[i]}(k+t|k), \mathbf{u}^{[i]}(k+t|k)) \right), \quad (17.11a)$$

subject to:

$$\mathbf{x}^{[i]}(k+t+1|k) = \mathbf{A}_{ii}\mathbf{x}^{[i]}(k+t|k) + \mathbf{B}_{ii}\mathbf{u}^{[i]}(k+t|k) + \mathbf{B}_{d,ii}\mathbf{d}^{[i]}(k+t|k) + \bar{\mathbf{B}}_i\mathbf{w}^{[i]}(k+t|k), \quad \forall t \in \mathbb{Z}_{[0,H_p-1]} \quad (17.11b)$$

$$\mathbf{0} = \mathbf{E}_{u,ii}\mathbf{u}^{[i]}(k+t|k) + \mathbf{E}_{d,ii}\mathbf{d}^{[i]}(k+t|k) + \bar{\mathbf{E}}_i\mathbf{w}^{[i]}(k+t|k), \quad \forall t \in \mathbb{Z}_{[0,H_p-1]} \quad (17.11c)$$

$$\mathbf{x}^{[i]}(k+1|k) \in \mathcal{C}_{\infty}^{\mathcal{S}_i}, \quad (17.11d)$$

$$\mathbf{x}^{[i]}(k+t|k) \in \mathcal{X}_i, \quad \forall t \in \mathbb{Z}_{[2,H_p]} \quad (17.11e)$$

$$\mathbf{u}^{[i]}(k+t|k) \in \mathcal{U}_i, \quad \forall t \in \mathbb{Z}_{[0,H_p-1]} \quad (17.11f)$$

$$\mathbf{u}_{(r)}^{[i]}(k|k) = u_{(r)}^{[i]\star}(k+1|k-1), \quad \forall r \in \mathcal{I}_u \quad (17.11g)$$

$$\mathbf{x}^{[i]}(k|k) = \mathbf{x}^{[i]}(k), \quad (17.11h)$$

where $\mathcal{I}_u \subset \mathbb{Z}_+$ is a set containing the indices of all the rows of vector $\mathbf{u}^{[i]}(k)$ related to the inputs decided locally by C_i but affecting neighbours whose controllers C_j are located in higher levels of the predefined hierarchy of solution.

Comparing with the algorithms in [13, 15], problem (17.11) has two subtle but important differences:

1. The incorporation of (17.11d) as a robustness constraint that enforces the predicted state to lie within the maximal max-min robust control invariant set at the first prediction step.
2. The incorporation of (17.11f), restricting those components of the first control action that are decided locally but affect neighbouring subsystems whose controllers are located at higher levels of the solution hierarchy.

As demonstrated in [7, Chap. 6] for a min-max interpretation in a standard centralized MPC controller, the robustness constraint (17.11d) leads to a robust strongly feasible MPC algorithm. Nonetheless, this constraint on its own cannot guarantee recursive feasibility of the overall distributed MPC solution sequence, because $\mathcal{C}_\infty^{\mathcal{S}_i}$ is computed under Interpretation 1, which requires that each controller C_i knows at least the first demand value of its local and virtual demands (i.e., $\mathbf{d}^{[i]}(k)$ and $\mathbf{w}^{[i]}(k)$ when solving at k). This requirement is not fulfilled if controllers C_i are allowed to freely optimize their full input vector without considering their effect in the hierarchical sequence of solution of the non-iterative ML-DMPC approach.

To illustrate this observation, assume that a controller C_j optimizes the flow of a complicating arc affecting a subsystem $\mathcal{S}_i$ whose controller C_i has already solved the ith problem in the solution sequence. Then, the trajectory obtained by C_j could be infeasible (specially due to the equality coupling constraint (17.11c)) for $\mathcal{S}_i$ since $\mathbf{w}^{[i]}(k)$ might be changed and C_i does not have the chance to recompute its solution. Hence, constraint (17.11f) is an extra necessary condition to satisfy Interpretation 1 and to maintain feasibility of the overall sequence of local problems.

17.3.2 Upper Optimization Layer

The fulfilment of a global objective from a local point of view often implies information from the entire network, but this is lost when the system partitioning is performed. Therefore, it is necessary to figure out how to induce cooperation among the set of distributed controllers, considering all the control objectives belonging to $\mathcal{O}$ in a suitable way.

One common way to improve overall closed-loop performance of a decentralized/distributed control scheme is to incorporate a supervisor controller or coordinator on top of the local controllers. Two frequently used coordination methods are the *goal coordination* and the *interaction prediction coordination* (cf., [10]). The fundamental idea behind these approaches is to have independent subproblems containing certain coordinating parameters (e.g., Lagrange multipliers, co-state variables and pseudovariables) in addition to the local decision variables. In both coordination methods, duality theory is used as a standard to construct an equivalent two-level problem to the primal (centralized) optimization problem. Within such framework, the coordinating parameters are updated iteratively by the coordinator based on the local solutions until an optimal solution to the overall system is achieved (cf. [3, 10]). Feasibility of these coordinated control strategies is guarantee only upon convergence.

Contrary to the common methods, the upper optimization layer of the ML-DMPC approach proposed in this chapter is not focused on reconstructing the centralized optimal solution in an iterative manner but to improve the economic performance of the local MPC controllers by intervening in their decision process with a low frequency of intervention. Specifically, this upper layer influences the local solutions by computing, in a non-iterative way, the weight $\omega \in \mathbb{R}^{n_\omega}$ (where n_ω is the number of arcs interconnecting the subsystems) related to the shared links between partitions that appear after the selected network decomposition method (see Fig. 17.1). The weights in ω will affect the first term in the local cost function (17.11a) of each controller C_i, $i \in \mathbb{Z}_{[1,M]}$. Therefore, to compute ω, a centralized optimization problem based on a temporal and functional decomposition of the network is stated in the upper layer of the ML-DMPC by considering

(i) a static model of the whole network and
(ii) a cost function that only takes into account the global control objectives associated to the system.

The proposed upper optimization layer works with a sampling time $\Delta t_1 = T$, where $T \in \mathbb{Z}_{\geq 1}$ corresponds to the period of the periodic flow requested by local demands (see Assumption 17.1). Thus, when looking at the volume evolution of storage elements, they show a similar behaviour as the flow to the demands; i.e., volumes might also show a periodic behaviour with period T. For this reason, when modelling the network with sampling time Δt_1, it can be assumed that volumes do not change along the time. From now on, subindex c is used to differentiate the temporal scale of the model in the upper layer to that of the lower layer (e.g., $\mathbf{x}_c(k)$ denotes the state at the coordinator level at time instant k with sampling time Δt_1). Hence, storage nodes behave as static nodes in this layer, and the network dynamic model (17.9a) becomes a stationary model, i.e., $\mathbf{x}_c(k) = \mathbf{A_c}\mathbf{x}_c(k) + \mathbf{B_c}\mathbf{u}_c(k) + \mathbf{B}_{d,c}\mathbf{d}_c(k)$.
The stationary model considered by the coordinator is

$$J_{up}(k, \mathbf{x}_c(k), \mathbf{u}_c(k) := \sum_{g=1}^{m_g} \lambda_{g,c} J_{g,c}(\mathbf{x}_c(k), \mathbf{u}_c(k), \tag{17.12}$$

and the upper layer optimization problem is here proposed to be formulated for a flow-based network as the search of the economically optimal path flows from sources nodes to demand nodes.

Definition 17.3 (*Directed path*) Given a directed graph $\mathcal{G} = (\mathcal{V}, \mathcal{A})$ of a network, a directed path is an ordered sequence of nodes $v_1, v_2, \ldots, v_n$ in which there is an arc (i, j) pointing from each node i in the sequence to its successor node j in the sequence, that is, $\{(v_1, v_2)(v_2, v_3), \ldots, (v_{n-1}, v_n)\}$.

To mathematically and systematically find all flow paths in a given network, this chapter follows the methodology in [2, Appendix A], which exploits the information contained in the node-arc incidence matrix of the network directed graph to construct the path-arc matrix for the given sources and demands. The description of such algorithm is omitted here, and the reader is referred to the aforementioned reference. Once the path-arc matrix is obtained, a constrained optimization problem can be stated to minimize (17.12) in terms of path flows, which are denoted here as $u_p \in \mathbb{R}^{n_p}$ with n_p the number of possible paths.
Hence, the coordinator solves in the upper layer of the ML-DMPC, an optimization problem with the following structure:

$$\min_{u_p} \hat{J}_{up}(\mathbf{x}_c(k), \mathbf{u}_p(k)), \tag{17.13a}$$

subject to:

$$\mathbf{A_p u}_p(k) \leq \mathbf{b}_p(k), \tag{17.13b}$$

$$\mathbf{A}_{eq}\mathbf{u}_p(k) = \mathbf{b}_{eq}(k), \tag{17.13c}$$

where function $\hat{J}_{up}$ is equivalent to (17.12) but properly expressed in terms of the path flows $\mathbf{u}_p(k)$ by using the graph path-arc matrix. Moreover, constraint (17.13b) is used to consider the physical bounds of each actuator involved in each path, while constraint (17.13c) is used to enforce satisfaction of demands $\mathbf{d}_c(k)$. Matrices $\mathbf{A_p}$ and $\mathbf{A}_{eq}$ and vectors $\mathbf{b}_p$ and $\mathbf{b}_{eq}$ are defined accordingly to the considered bounds and balance constraints.

Throughout this chapter, it has been assumed that the flow at each arc of the network is driven by an actuator. Therefore, by using the optimal solution of problem (17.13) and the information contained in the path-arc matrix of the overall network, it is possible to compute the accumulated cost incurred in traversing all the paths that reach the intermediate nodes from which the arcs interconnecting the M subsystems depart. This accumulated cost information, in addition to Assumption 17.1, allows to define the weight ω as a coordinating economic parameter. This weight is used by the coordinator to project, into the cost function of each local controller C_i, the economic impact (from a global point of view) that each subsystem $\mathcal{S}_i$ will suffer when requesting flow from its neighbour subsystems.

In network flow problems, the global objectives are often given as a composition of economic linear cost functions. In this case, the value of ω can be obtained by following Algorithm 17.1.

Note that Assumption 17.1 and the temporal scale selected for the upper layer make (17.13) independent of the state. Furthermore, the weight ω is more an intervention parameter than a coordination variable since the upper layer does not use any feedback information from the local controllers allocated at the lower layer.

17.3.3 ML-DMPC Algorithm

The sharing of information between the two layers of the proposed ML-DMPC depends on the nature and features of each application. For the case considered in this chapter (i.e., periodic demands), the interaction is unidirectional from the upper optimization layer to the lower optimization layer. Once the optimization problem related to the upper layer is solved, the resultant parameters are properly updated for each optimization problem behind each C_i, $i \in \mathbb{Z}_{[1,M]}$. This updating is performed with a periodicity Δt_1 to consider possible changes in the periodic pattern of demands. In fact, if a given application involves an agreement of predefined demands to be satisfied, the optimization problem of the upper layer needs to be executed only once at the beginning of the operation. In general, the computational time that the upper layer spends is quite low with respect to the computational time of the lower layer.

Algorithm 17.1 Computation of the economic intervention parameter ω

1: Compute the path-arc matrix of the network graph, denoted here by $\mathbf{R}_p \in \mathbb{R}^{n_p \times m}$.
2: Define a matrix $\mathbf{C}_p \in \mathbb{R}^{n_p \times m}$ with the same structure of matrix $\mathbf{R}_p$ but containing in each matrix element the unitary flow cost of each actuator in each possible path.
3: Identify all the arcs interconnecting subsystems $\mathcal{S}_i$, $i \in \mathbb{Z}_{[1,M]}$, and denote with $n_{u_s} \in \mathbb{Z}_+$ the number of such arcs, called from now on as complicating arcs.
4: Solve problem 17.13 and identify from the optimal solution all the paths in which each complicating arc participates, and denote by $n_{p_j} \in \mathbb{Z}_+$, $j \in \mathbb{Z}_{[1,n_{u_s}]}$, the numbers of such paths.
5: Define a set of matrices $\mathbf{T}_{s_j} \in \mathbb{R}^{n_p \times n_{p_j}}$, $j \in \mathbb{Z}_{[1,n_{u_s}]}$, each of them collecting the n_{p_j} columns of the identity matrix of order n_p.
6: Define a set of matrices $\mathbf{R}_{p_j} := \mathbf{T}_{s_j}^T \mathbf{R}_p$ and $\mathbf{C}_{p_j} := \mathbf{T}_{s_j}^T \mathbf{C}_p$ for all $j \in \mathbb{Z}_{[1,n_{u_s}]}$.
7: From the sequential order of the directed paths involved in each matrix $\mathbf{R}_{p_j}$, define a set of matrices $\tilde{\mathbf{R}}_{p_j}$ whose elements will be the same as the ones in matrices $\mathbf{R}_{p_j}$ for all the positions related to the sequential arcs that reach the complicating arcs (these latter included) in each path, and zero in those matrix elements related to the successor arcs.
8: Define the vector $\omega := (\omega_1, \ldots, \omega_{n_{u_s}})^T$, with each of its components computed as

$$\omega_j = \frac{\mathbf{1}_{n_{u_s}}^T \left((\mathbf{C}_{p_j} \circ \mathbf{R}_{p_j}) \circ \tilde{\mathbf{R}}_{p_j} \right)^T \mathbf{T}_{s_j}^T \mathbf{u}_p^{\star}(k)}{\left[\mathbf{R}_{p_j}^T \mathbf{T}_{s_j}^T \mathbf{u}_p(k)^{\star} \right]_{(r_j)}}, \quad \forall j \in \mathbb{Z}_{[1,n_{u_s}]}$$

where $\mathbf{1}_{n_{u_s}}$ denotes an all-ones column vector of length n_{u_s}, the operator $(\circ)$ indicates the Hadamard product of matrices, and $[\cdot]_{(r_j)}$ is the r_j row of the vector in the brackets with r_j being the position of the associated jth complicating arc in the input vector $\mathbf{u}_c(k)$. Then, ω_j represents a unitary cost per flow unit.

This fact is due to the difference in the nature of the models handled by each layer and the interactions given by the distributed MPC controllers as well as their amount and disposition within the defined hierarchy. Algorithm 17.2 collects the main steps of the proposed ML-DMPC approach. The computational time spend by the scheme corresponds with the sum of maximum times of each hierarchical level of controllers.

One important property desired in the design of any MPC strategy is recursive feasibility. In the following, it is shown that the proposed ML-DMPC algorithm remains feasible for all times if initial feasibility is assumed. The guarantee of feasibility of the approach is unrelated to optimality of the distributed solution.

Theorem 17.1 *Let Assumptions 17.1–17.3 hold and suppose that an initial feasible solution in Step 1 of Algorithm 17.2 exists. Then, each local MPC problem (17.11) solved in Step 3 of Algorithm 17.2 is robust strongly feasible for each subsystem* $\mathcal{S}_i \in \mathcal{P}$.

Proof The proof is by induction, showing that feasibility at time k implies feasibility at time $k+1$. Let $\mathbf{x}^{[i]}(k)$ be a feasible initial condition for each local problem (17.11) and assume that there exists a pair of feasible (not necessarily optimal) state-input trajectories given by $(\mathbf{x}_k^{[i]}, \mathbf{u}_k^{[i]})$ for each subsystem $\mathcal{S}_i \in \mathcal{P}$.

Algorithm 17.2 Non-iterative Multi-layer Distributed Economic MPC

1: **Initialisation:** Set $k = 0$, establish an arbitrary weight ω in the upper layer and send that information to every local controller C_i, $i \in \mathbb{Z}_{[1,M]}$. For each current local state $\mathbf{x}^{[i]}(k)$ and local demand sequence $\mathbf{d}_k^{[i]} = \{\mathbf{d}^{[i]}(k), \bar{\mathbf{d}}^{[i]}(k+1|k), \ldots, \bar{\mathbf{d}}^{[i]}(k+H_p-1|k)\}$, find for all subsystems $\mathcal{S}_i$ a feasible (not necessarily optimal) pair of state and input sequences $(\mathbf{x}_k^{[i]} = \{\mathbf{x}(k+t|k)\}_{t\in\mathbb{Z}_{[0,H_p]}}, \mathbf{u}_k^{[i]} = \{\mathbf{u}(k+t|k)\}_{t\in\mathbb{Z}_{[0,H_p-1]}})$. Apply $\mathbf{u}^{[i]}(k|k)$ in every subsystem and transmit each $\mathbf{u}_k^{[i]}$ to the controllers of the corresponding neighbours of each $\mathcal{S}_i$.

2: **Collecting of information:** After receiving all the neighbour trajectories $\mathbf{u}_k^{[j]}$, $j \in \mathcal{N}_i$, each controller C_i builds the trajectory $\mathbf{w}_k^{[i]} = \{\mathbf{w}^{[i]}(k+t|k)\}_{\in\mathbb{Z}_{[0,H_p-1]}}$, differencing between shared inputs to be imposed by controllers arranged in higher levels of hierarchy and shared inputs planned by controllers arranged in the same or lower levels of hierarchy. These imposed and planned input trajectories are formed locally as $\mathbf{w}_{a,k}^{[i]} = \{\mathbf{w}_a^{[i]\star}(k|k), \ldots, w_a^{[i]\star}(k+H_p-1|k)\}$ and $\mathbf{w}_{a,k}^{[i]} = \{\mathbf{w}_b^{[i]\star}(k+1|k-1), \ldots, \mathbf{w}_b^{[i]\star}(k+H_p-1|k), \mathbf{w}_b^{[i]\star}(k+1|k-1)\}$, respectively, and it is assumed that $\mathbf{w}^{[i]}(k+t|k) = (\mathbf{w}_a^{[i]\,T}(k), \mathbf{w}_b^{[i]\,T}(k)^T$. At each sampling time, obtain $\mathbf{x}^{[i]}(k)$ and $\mathbf{d}_k^{[i]}$ for each subsystem $\mathcal{S}_i$.

3: **Solution of local problems**: Solve each optimization problem 17.11 following a predefined hierarchical sequence.

4: **Implementation of control action:** Each local controller C_i applies $\kappa_i(\mathbf{x}^{[i]}(k), \mathbf{u}_k^{[i]}, \mathbf{d}_k^{[i]}, \mathbf{w}_k^{[i]}) = \mathbf{u}^{[i]\star}(k|k)$ to the associated subsystem $\mathcal{S}_i$. Transmit each $\mathbf{u}_k^{[i]}$ to the controllers of the corresponding neighbours of each $\mathcal{S}_i$.

5: **Updating of the economic intervention parameter:** If $\lfloor k \rfloor_{\mathcal{P}_1} \in \mathbb{Z}_+$, then solve problem 17.13 for the current $\mathbf{d}_k$ and update ω following Algorithm 17.1. Send the new weight to each local controller C_i. Otherwise, go to step 5.

6: Increment k and go to step 2.

Consider now the hierarchical flow of the solution at the next time instant $k+1$. Since each subsystem applied previously the first control action of the initial feasible trajectory $\mathbf{u}_k^{[i]}$, it follows then that $x_{k+1}^{[i]} = \mathbf{x}^{[i]}(k+1|k)$, and from constraint (17.11d), it holds that $\mathbf{x}^{[i]}(k+1) \in \mathcal{C}_\infty^{\mathcal{S}_i}$ for all $i \in \mathbb{Z}_{[1,M]}$. Since $\mathcal{C}_\infty^{\mathcal{S}_i} \neq \emptyset$ by Assumption 17.3, it follows from the invariance property of $\mathcal{C}_\infty^{\mathcal{S}_i}$ that for all $(\mathbf{x}^{[i]}(k+1), \mathbf{d}_{k+1}^{[i]}, \mathbf{w}_{k+1}^{[i]}) \in \mathcal{C}_\infty^{\mathcal{S}_i} \times \mathcal{D}_i^{H_p} \times \mathcal{W}_i^{H_p}$, there exists a control sequence $\mathbf{u}_{k+1}^{[i]} \in \mathcal{U}_i^{H_p}$ such that the constraints in problem (17.11) are satisfied at time instant $k+1$ for all $i \in \mathbb{Z}_{[1,M]}$.

This claim holds only under Interpretation 1, that is, if and only if each controller C_i knows at least the first demand value of its local and virtual demands ($\mathbf{d}^{[i]}(k+1)$ and $\mathbf{w}^{[i]}(k+1)$ when solving at $k+1$). Such requirement is guaranteed by means of constraint (17.11f), which is feasible by the assumption of existence of any initial feasible trajectory $\mathbf{u}_k$. Therefore, all the local problems solved sequentially by controllers C_i are feasible at $k+1$. Feasibility for all times follows then by induction over k and the assumption of initial feasibility. Consequently, the ML-DMPC approach is strongly feasible, and the claim is proved. □

17.4 Simulations and Results

In order to evaluate the effectiveness of the proposed ML-DMPC approach, the case study related to the model of the Barcelona DWTN is used. In this network, the set $\mathcal{O}_g$ of global control objectives is formed only by the cost function

$$J_E(\mathbf{x}(k), \mathbf{u}(k; \mathbf{c}_u(k), \mathbf{c}_x(k) := \mathbf{c}_u^T(k)\mathbf{W}_e\,\mathbf{u}(k)\Delta t + \mathbf{c}_x^T(k)\mathbf{W}_h\mathbf{x}(k), \tag{17.14a}$$

while the set $\mathcal{O}_l$ of local control objectives is formed by the cost functions

$$J_S(\mathbf{x}(k); \mathbf{s}(k) := \begin{cases} (\mathbf{x}(k) - \mathbf{s}(k)^T\mathbf{W}_s(\mathbf{x}(k) - \mathbf{s}(k) & \text{if } \mathbf{x}(k) \leq \mathbf{s}(k) \\ 0 & \text{otherwise,} \end{cases} \tag{17.14b}$$

$$J_\Delta(\Delta\mathbf{u}(k)) := \Delta\mathbf{u}^T(k)\mathbf{W}_{\Delta u}\,\Delta\mathbf{u}(k). \tag{17.14c}$$

The overall network is assumed to be decomposed in six subsystems ($\mathcal{P} = \{\mathcal{S}_1, \ldots, \mathcal{S}_6\}$), which are non-overlapped, output-decentralized and input-coupled (see Fig. 16.3). The model and constraints of each subsystem $\mathcal{S}_i$ are obtained following Sect. 17.2.

The controller C_i of each subsystem $\mathcal{S}_i$ uses the following local multi-objective stage cost in its optimization problem:

$$J_i(k, \mathbf{x}^{[i]}(k), \mathbf{u}^{[i]}(k) = \hat{\lambda}_{1,i}\hat{J}_{E,i}(\mathbf{x}^{[i]}(k), \mathbf{u}^{[i]}(k; \mathbf{c}_u^{[i]}(k)) + \lambda_{2,i}J_{\Delta,i}(\Delta\mathbf{u}^{[i]}(k)) \\ +\lambda_{3,i}J_{S,i}(\varsigma^{[i]}(k); \mathbf{x}^{[i]}(k), \mathbf{s}^{[i]}(k)),$$

where functions $\hat{J}_{E,i}$, $J_{\Delta,i}$ and $J_{S,i}$ are the local economic, safety and smoothness objectives for subsystems $\mathcal{S}_i$ (see Sect. 17.3.1 for the derivation of the local costs). Moreover, $\hat{\lambda}_{1,i}$, $\lambda_{2,i}$ and $\lambda_{3,i}$ are positive scalar weights to prioritise each objective in the aggregate local cost function.

Each local MPC controller operates with a sampling time $\Delta t_2 = 1$ h and a prediction horizon $H_p = 24$ h. The weight $\hat{\lambda}_{1,i}$ and the internal economic parameters of each function $\hat{\ell}_{E,i}$, $i \in \mathbb{Z}_{[1,6]}$, are modified by the upper optimization layer, placing properly each element of the intervention parameter ω (see Algorithm 17.1) in the local cost of the corresponding complicated arcs. The cost function used in the upper optimization layer is given by

$$J_{up}(k, \mathbf{x}_c(k), \mathbf{u}_c(k)) = J_{E,c}(\mathbf{x}_c(k), \mathbf{u}_c(k)), \tag{17.15}$$

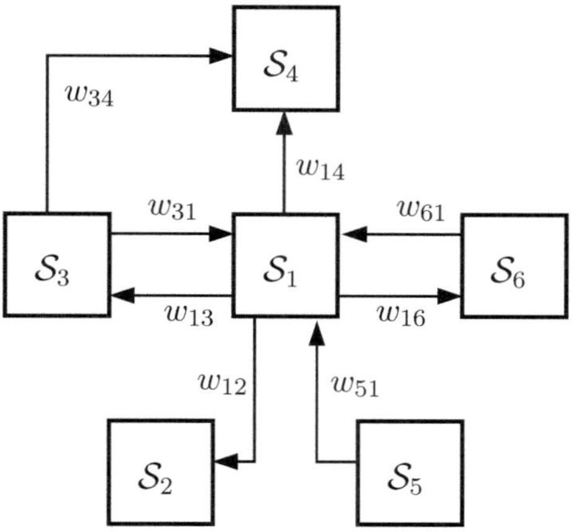

Fig. 17.3 Network subsystems $\mathcal{S}_i$ and their shared connections w_{ij}

which is derived from (17.14a) but expressed in a temporal scale of days (i.e., $\Delta t_1 =$ 24 h).

The constraints and the rest of the parameters involved in the optimization problems (i.e., water demands, economic prices of water end electricity, safety thresholds) are set up according to Chap. 2.

Figure 17.3 shows, in a more compact way, the resulting subsystems and the important couplings between them including their direction. Instead of neglecting the effect of this shared links as classic pure decentralized control schemes do, the ML-DMPC approach applied to the aforementioned case study has the control architecture shown in Fig. 17.2.

The results obtained by applying the ML-DMPC (Algorithm 17.2) are compared with those of applying a centralized MPC (CMPC) approach and a decentralized MPC (DMPC) strategy proposed in [14]. All of the results were obtained for a simulation horizon of 72 h with real data of the network and are summarized in Table 16.2 (Chap. 16) in terms of computational burden and of economic cost as a global management performance indicator. For each MPC approach, the computational time (in seconds) and the water, electric and total cost in economic units (e.u.) are detailed. It can be noticed that an increment of nearly 30% of the total costs of operation occurs when using the one-level hierarchical DMPC strategy reported in [14] with respect to the CMPC baseline. Despite the lower electric costs, the loss of performance in the overall cost is due to the specialized behaviour of local MPC controllers to solve their own optimization problems without knowing the real water supply cost of using shared resources with the neighbours. In contrast, the ML-DMPC outperforms the DMPC results by including the bilevel optimization, which allows to propagate the water cost of sources related with neighbour subsystems to the shared links thanks to the daily centralized control level. With this ML-DMPC approach, the level of suboptimality is acceptable comparing with the CMPC strategy; i.e., total costs are quite similar, but the computational burden is reduced. For this particular application, the computational time of the three approaches is able to satisfy the real-time constraint

Table 17.1 Performance comparisons

Index	CMPC	DMPC	ML-DMPC
Water cost	93.01	205.55	97.11
Electric cost	90.31	34.58	87.53
Total cost	183.33	240.13	184.65
CPU time	1143	537	540

since the control sampling time is 1 h. Thus, the main motivation for using ML-DMPC is the scalability and easy adaptability of the submodels if network changes, as well as the modularity of the control policy that leads to face some malfunction/fault without stopping the overall supervisory MPC strategy (Table 17.1).

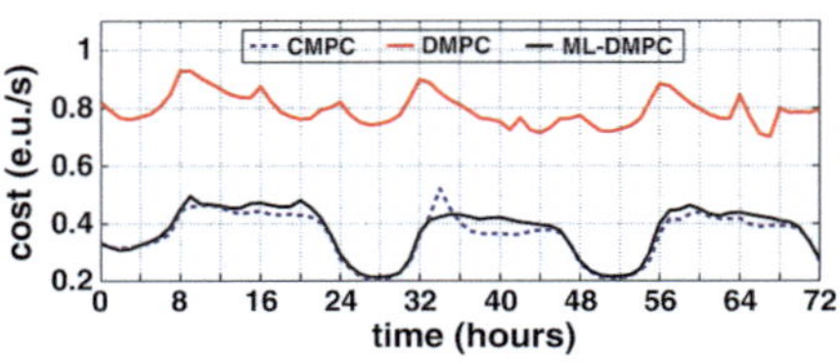

(a) Water cost

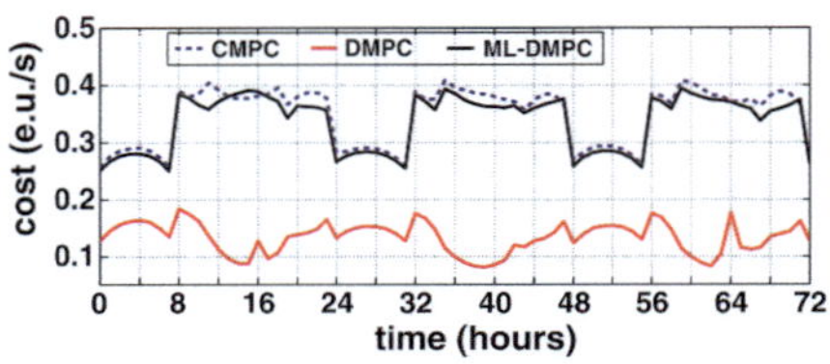

(b) Electric cost

Fig. 17.4 Economic costs of the three MPC strategies

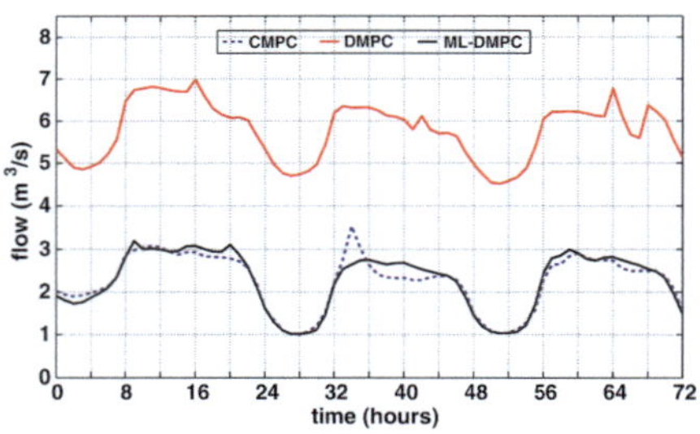

(a) Ter source

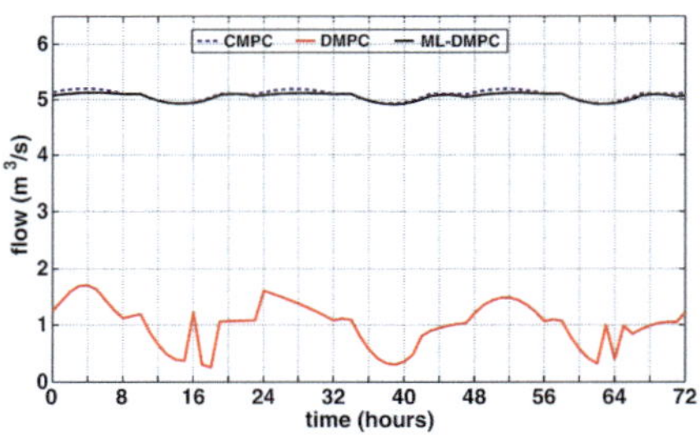

(b) Llobregat source

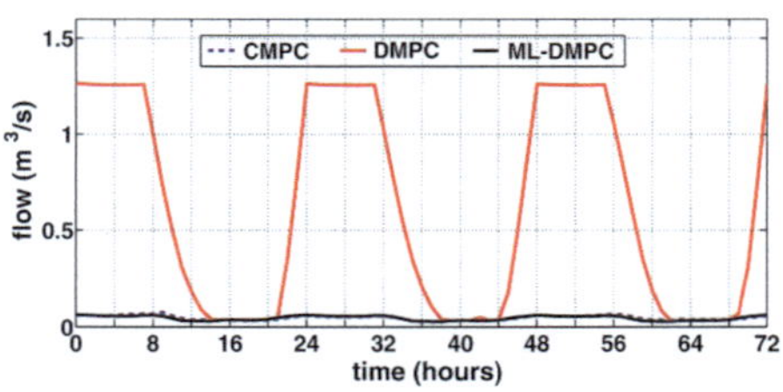

(c) Abrera source

Fig. 17.5 Total flow per water source in the Barcelona DWTN

Due to the difference of price between water sources and the impact of electric costs on the overall economic performance, the CMPC and ML-DMPC strategies decide to use more water from the Llobregat source despite the consequent pumping of more water through the network (see Fig. 17.5), but achieving a lower total cost, while the hierarchical DMPC decides to exploit in each subsystem their own water source (which could be expensive) and minimize the pumping operation cost. Figure 17.4 shows in detail the evolution of water cost and electric cost, respectively. These results confirm the improvement obtained by including an upper optimization layer to coordinate the local MPCs and face the lack of communication when solving their problems in a tractable way.

17.5 Conclusions

This chapter proposed a non-iterative multi-layer distributed economic MPC approach for large-scale flow-based networks. The control architecture consists in two optimization layers. The upper layer, working with a larger timescale, is in charge of improving the global performance (in general related to an optimal economic cost) by influencing a set of distributed MPC controllers by means of an intervention economic parameter. These distributed controllers are hierarchically arranged in a lower optimization layer and are in charge of determining the set point of the flow actuators to satisfy the local management/control objectives. The system decomposition is based on graph partitioning theory. Results obtained on selected simulation scenarios have shown the effectiveness of the control strategy in terms of system modularity, reduced computational burden and, at the same time, reduced loss of performance in contrast to a CMPC strategy and a hierarchical-like DMPC strategy. Additionally, it has been proved that the proposed approach results in a strongly feasible distributed MPC algorithm. For clarity of presentation, in Algorithm 17.2 it was required that each subsystem calculates its input trajectory at each time step in a hierarchical and sequential order. However, the algorithm works in the same way if non-neighbouring systems located in the same level of hierarchy solve their problems in parallel. Future work will be focused on finding stability conditions under the framework of economic MPC and also on improving the mechanism of coordination to avoid the requirement of plant-wide information in the upper layer of the ML-DMPC approach.

References

1. Barić M, Borrelli F (2010) Decentralized robust control invariance for a network of integrators. In: Proceedings of the American control conference (ACC), Baltimore, MD, USA, June 2010, pp 3939–3944
2. Cheng WC, Hsu NS, Cheng WM, Yeh W (2009) A flow path model for regional water distribution optimization. Water Resour Res 45:1–12

3. Conejo AJ, Castillo E, Minguez R, Garcia-Bertrand R (eds) (2006) Decomposition techniques in mathematical programming. Springer, Berlin
4. Driessen PAAA, Hermans RM, van den Bosch PPJ (2012) Distributed economic model predictive control of networks in competitive environments. In: Proceedings of the IEEE conference on decision and control (CDC), Maui, HI, USA, December 2012, pp 266–271
5. Jilg M, Stursberg O (2013) Optimized distributed control and topology design for hierarchically interconnected systems. In: Proceedings of the European control conference (ECC), July 2013, pp 4340–4346
6. Kamelian S, Salahshoor K (2015) A novel graph-based partitioning algorithm for large-scale dynamical systems. Int J Syst Sci 46(2):227–245
7. Kerrigan EC (2000) Robust constraint satisfaction: invariant sets and predictive control. PhD thesis, Department of Engineering, University of Cambridge, UK, November 2000
8. Lee J, Angeli D (2014) Cooperative economic model predictive control for linear systems with convex objectives. Eur J Control 20(3):141–151
9. Lunze J (1992) Feedback control of large-scale systems. Prentice Hall, Great Britain
10. Mesarović MD, Macko D, Takahara Y (1970) Theory of hierarchical multilevel systems. Academic Press, New York
11. Motee N, Sayyar-Rodsari B (2003) Optimal partitioning in distributed model predictive control. Proc Am Control Conf (ACC) 6:5300–5305
12. Negenborn RR, Maestre JM (2014) Distributed model predictive control: an overview and roadmap of future research opportunities. IEEE Control Syst Mag 34(4):87–97
13. Ocampo-Martinez C, Barcelli D, Puig V, Bemporad A
14. Ocampo-Martinez C, Bovo S, Puig V (2011) Partitioning approach oriented to the decentralised predictive control of large-scale systems. J Process Control 21(5):775–786. Special issue on Hierarchical and distributed model predictive control
15. Ocampo-Martinez C, Puig V, Grosso JM, Montes-de-Oca S (2014) Multi-layer decentralized MPC of large-scale networked systems. In: Distributed Model predictive control made easy. Springer, pp 495–515
16. Rawlings JB, Stewart BT (2008) Coordinating multiple optimization-based controllers: new opportunities and challanges. J Process Control 18(9):839–845
17. Šiljak DD (1991) Decentralized control of complex systems. Academic Press
18. Stewart BT, Venkat AN, Rawlings JB, Wright SJ, Pannocchia G (2010) Cooperative distributed model predictive control. Syst Control Lett 59(8):460–469

Part IV
Future Trends

Chapter 18
Data-Driven Evolutionary-Game-Based Control for Drinking-Water Networks

Julián Barreiro-Gomez, Gerardo Riaño-Briceño, Carlos Ocampo-Martínez and Nicanor Quijano

18.1 Introduction

Around 663 million people had no access to safe drinking water in 2015, and around 2.4 billion people live without adequate sanitation according to [25]. This situation has impacts on the economy of the society according to the Millennium Summit of 2000, on which the United Nations agreed the Millennium Development Goals (MDG). One of the biggest concerns of the MDG, due to the rapid population growth and industrialization, is to guarantee the access to drinking water, achieving a proper management of the available water resources. Hence, it becomes essential to overcome the lack of drinking water for achieving sustainable development including in both social and economic aspects, poverty reduction and equity, and also sustainable environmental services [13].

Over the last decade, several optimization-based control strategies have been proposed to manage efficiently drinking water and to solve resource allocation problems in water applications. For instance, in [10] a nonlinear multi-objective optimization procedure has been proposed to manage water flows and reserves in drinking water transport networks (DWTNs), considering the uncertainty of climate and global change development, using an integrated approach, i.e., modelling the drinking-water system, the climate, and the society as a whole. However, this solution implies to consider a lot of variables and constraints which increase the complexity of the optimization problem. Likewise, optimization-based strategies such as model predictive

J. Barreiro-Gomez · C. Ocampo-Martínez (✉)
Institut de Robòtica i Informàtica Industrial, CSIC-UPC, Barcelona, Spain
e-mail: cocampo@iri.upc.edu

G. Riaño-Briceño · N. Quijano
Departmento de Ingenería Eléctrica y Electrónica, Universidad de los Andes, Bogotá, Colombia

V. Puig et al. (eds.), *Real-Time Monitoring and Operational Control of Drinking-Water Systems*, Advances in Industrial Control,
DOI 10.1007/978-3-319-50751-4_18

control (MPC) have been designed for this kind of systems, considering the uncertainty of demand patterns as in [26] and minimizing operational costs and shortage events [9].

Another approach to address the DWTN control design is the use of population dynamics taking advantage of their stability properties and the close relationship between the solution in a population game (Nash equilibrium) and the unique maximizer of a constrained convex optimization problem [23]. Recently, game theory has been used in the solution of engineering problems [1, 14, 17, 24] and for the solution of optimization problems [15, 16]. Furthermore, in this chapter, the population game approach is presented as a powerful tool for the design of data-driven controllers. More precisely, two different directions in the design of data-driven population-games-based controllers are treated in this work. First, the DWTN is controlled by making a partitioning into subsystems that satisfy specific conditions, and a resource allocation problem is solved at each partition. This approach generates a decentralized control scheme since the local controllers neither communicate to each other nor exchange information among them. Secondly, it is proposed the design of data-driven controllers by minimizing a cost function and considering flow-balance constraints. Under this approach, the network is divided into subsystems according to the established constraints over the control inputs, which constitutes a distributed scheme due to the existing intersection among the different subsystems.

The presented contents in this chapter are a compilation of the theory proposed in previous works [3–6, 21]. However, some new case studies are incorporated as well as new simulation results.

Notation

Although this book follows an unified notation and in order to facilitate the reading of this chapter, some additional notation is introduced. The subindex is associated to a node of a graph or to a strategy in a game. On the other hand, the superindex refers to a population. For instance, the subindex i in u_i, $\mathcal{P}_i$, u_i^p or f_i refers either to a node in a graph or to a strategy, and the superindex p in m^p, $\mathbf{u}^p$, u_i^p or n^p indicates a population. Also it should be clear that the superindex is not an operational number, i.e., n^3 refers to population three but $n^3 \neq nnn$. We use bold font for column vectors and matrices, e.g., $\mathbf{u}$, and $\mathbf{H}$; and non-bold style is used for scalar numbers, e.g., n^p. Calligraphy style is used for sets, e.g., $\mathcal{S}$. The column vector with n unitary entries is denoted by $\mathbf{1}_n$, and the column vector with null entries and suitable dimension is denoted by $\mathbf{0}$. The identity matrix with dimension $n \times n$ is denoted by $\mathbb{I}_n$. The cardinality of a set $\mathcal{S}$ is denoted by $|\mathcal{S}|$. The continuous time is denoted by t, and it is mostly omitted throughout the manuscript in order to simplify the notation. Finally, $\mathbb{R}_{\geq 0}$ represents the set of all non-negative real numbers, and $\mathbb{Z}_{>0}$ represents the set of positive integer numbers.

18.2 Problem Statement

18.2.1 First Data-Driven Perspective

In the proposed DWTN model for the design of the population dynamics-based controllers, which is composed by several storage tanks, the flow direction is unique since it is assumed that the pressure head at upstream tanks of the network is always higher than the pressure head at downstream tanks. This consideration is common in DWTNs that have been designed for places where the topography is steep and the slope is descending. Due to this assumption, it is possible to distinguish between source and receptor tanks, taking into account that the former ones are always upstream and directly linked to the latter ones.

Consider then a simple DWTN composed by n receptor tanks, and only one source tank as shown in Fig. 18.1. This topology is known as *branched* [19], which means that there are no loops in the network due to the fact that several outflows might go out from a single source tank, but no several inflows come into a single receptor. Let $\mathcal{S} = \{1, ..., n\}$ be the set of receptor tanks in the *branched* subsystem. The volume of the tank $i \in \mathcal{S}$ is denoted by $x_i \in \mathbb{R}_{\geq 0}$, its maximum volume is denoted by $x_i^{\max} \in \mathbb{R}_{\geq 0}$ and its inflows and outflows are given by $q_i^{in} \in \mathbb{R}_{\geq 0}$ and $q_i^{out} \in \mathbb{R}_{\geq 0}$, respectively. Hence, the vector of all the tank volumes is denoted $\mathbf{x} \in \mathbb{R}^n_{\geq 0}$, and the vector of maximum volumes is $\mathbf{x}^{\max} \in \mathbb{R}^n_{\geq 0}$. The parameter $u_i \in [0, 1]$ determines the setting of the input valve in the *ith* tank, $K_i > 0$ scales the outflow, and it can be considered as a volume-flow conversion factor or the discharge coefficient of the tank. Moreover, the system is affected by perturbations that are related to daily demand patterns.

The control objective consists in avoiding shortages throughout the system, i.e., to avoid that the current volume of the tank x_i runs out, not supplying the demand, for all $i \in \mathcal{S}$. To achieve this objective, it is proposed to do an allocation of the available resource stored in the n tanks, i.e., to distribute the current available volume given by $x_i^{\max} - x_i$ in an optimal way by controlling the inflows q_i^{in}, for all $i \in \mathcal{S}$. For instance, considering the hypothetical situation in which one tank is completely filled and another tank is empty, more priority should be assigned to the inflow of the empty tank rather than the inflow assigned to the filled one, in order to prevent shortages.

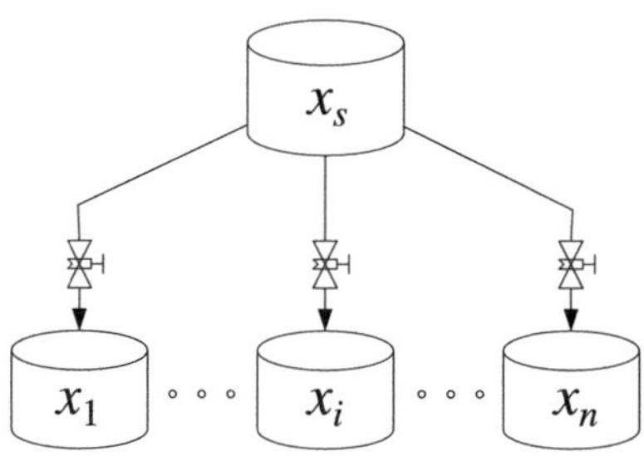

Fig. 18.1 Branched topology with n receptor tanks, and one source tank whose volume is denoted by v_s. The source tank is upstream of the receptor tanks

For each subsystem, the topology of interest is given by different receptor tanks and one source. The entire control system for the DWTN is composed by π local controllers that do not communicate with each other and which operate independently in parallel, i.e., all local controllers may operate their corresponding control inputs at once.

18.2.2 Second Data-Driven Perspective

This section presents the design of a controller without considering the model of the system, but just by considering the fact that the error within a tank (i.e., the difference between the safety value and the current volume) can be reduced as the control action is increased. In order to design a data-driven controller based on the proposed methodology, it is defined a cost function corresponding to the desired behaviour of the system. In this particular case, a volume error at each tank is considered.

The controller is designed through an optimization problem minimizing economical costs, the volume error with respect to the safety storage term and variations in the control actions. The economical costs are given by $(\boldsymbol{\alpha}_1^p + \boldsymbol{\alpha}_2^p(k))^\top \mathbf{u}^p(k)$, where $\boldsymbol{\alpha}_1^p$ is a constant vector defining the energy costs, and $\boldsymbol{\alpha}_2^p$ is a time varying vector determining the water costs. The volume error is given by $\mathbf{x}_s^p - \mathbf{x}^p$, where $\mathbf{x}_s^p$ is the safety storage imposed by the company in charge of the system management. Finally, the $\Delta\mathbf{u}^p(k)^\top \Delta\mathbf{u}^p(k)$ corresponds to the smooth operation cost.

These objectives are minimized subject to constraints of mass balance and physical constraints of actuators. To this end, new variables $\tilde{\mathbf{x}}_s \in \mathbb{R}^{n_u}$ of safety values, and $\tilde{\mathbf{x}} \in \mathbb{R}^{n_u}$ composed of tank volumes, are introduced. Notice that the dimension of the new vectors of volumes corresponds to the dimension of the vector of control actions, i.e., $\tilde{\mathbf{x}}_s, \tilde{\mathbf{x}}, \mathbf{u} \in \mathbb{R}^{n_u}$. The scalar $\tilde{x}_i$ denotes the volume corresponding to the tank whose inflow is given by u_i, and null in case that u_i is not an inflow for any tank. The safety volume $\tilde{x}_{s,i}$ corresponds to the safety volume of the tank whose inflow is given by u_i, and null otherwise. Briefly, $\tilde{x}_i = x_j$, and $\tilde{x}_{s,i} = x_{s,j}$ if u_i is the inflow of the jth tank, and null if u_i is not an inflow for any tank. Notice that the constraints over the system states (i.e., tanks volumes) may not be considered since this approach does not use a Control-Oriented Model (COM). The following optimization problem only depends on measured state values (volumes) and decision variables (control inputs):

$$\begin{aligned}
\underset{\mathbf{u}^p}{\text{maximize}} \quad & V(\mathbf{u}^p(k)) = -\gamma_1(\boldsymbol{\alpha}_1^p + \boldsymbol{\alpha}_2^p(k))^\top \mathbf{u}^p(k) \\
& -\gamma_2(\tilde{\mathbf{x}}_s^p - \tilde{\mathbf{x}}^p(k))^\top \text{diag}(\mathbf{u}^p(k))\,(\tilde{\mathbf{x}}_s^p - \tilde{\mathbf{x}}^p(k)) - \gamma_3 \Delta\mathbf{u}^p(k)^\top \Delta\mathbf{u}^p(k), \\
\text{subject to} \quad & \mathbf{E}_u^p \mathbf{u}^p(k) = -\mathbf{E}_d^p \mathbf{d}^p(k), \\
& \begin{bmatrix} \mathbb{I}_{n_u} \\ -\mathbb{I}_{n_u} \end{bmatrix} \mathbf{u}^p(k) \leq \begin{bmatrix} \mathbf{u}^{p,\max} \\ -\mathbf{u}^{p,\min} \end{bmatrix}.
\end{aligned} \tag{18.1}$$

18.3 Proposed Approach

18.3.1 Population-Games Approach: First Data-Driven Perspective

In this section, a detailed description of the population dynamics-based controller is done, taking into consideration that it is presented for the case of a single partition or subsystem. As it was stated before, the control approach is conceived from an analogy between the population dynamics framework and the DWTN model (see Table 18.1). In order to make clearer the analogy, it is worth to understand the process of transport between a source tank and the final user.

First of all, storage tanks receive water from treatment plants and/or natural water bodies (e.g., aquifers and reservoirs). Then, this water is redistributed among several storage tanks, which are located close to the final user. For instance, these can be placed in houses to prevent shortage when there is a lack of the resource. Consumers use the water that is available for them into the closest tank. In order to match supply and demand, the utility has the possibility to manipulate the amount of water that is deposited into receptor tanks through valves.

Considering this process, one can notice that the control problem is reduced to a resource allocation problem, in which the system can be seen as the population of a game. The population is composed by water or flow units, which summed all up form a mass (outflow). When the population mass reaches a point in which the flow diverges, it has the possibility to select one of the n paths (strategies) that lead to one of the receptor tanks in $\mathcal{S}$. The mass is going to select certain strategy based on the maximization of its wealth, which is defined by a fitness function.

Now that the analogy has been exposed, consider the branched DWTN with $n \in \mathbb{Z}_{>0}$ receptor tanks (strategies). The total flow through the system (population mass) is denoted by $Q \in \mathbb{R}_{\geq 0}$, which corresponds to the outflow of the source tank. Each flow unit is assigned to an inflow of one of the receptor tanks.

The scalar $u_i \in \mathbb{R}_{\geq 0}$ is the proportion of flow units assigned to each flow associated to the tank $i \in \mathcal{S}$ as a percentage, i.e., the inflow for the *ith* tank is given by $u_i Q$. The

Table 18.1 Equivalence between population dynamics and DWTN

	Population dynamics	DWTN
$\mathcal{P}$	Population	System
i	Strategy	Receptor tanks
m	Population mass	Total outflow source tank
q	Agents	Flow units
u_i	Proportion of agents	Proportion of flow
$\mathbf{u}$	Strategic distribution	Flow distribution in receptor tanks
f_i	Fitness of a strategy	Available volume capacity

vector $\mathbf{u} \in \mathbb{R}^n_{\geq 0}$ is the flow proportion distribution involving the n tanks according to the topology. The set of the possible distributions of flow is given by a simplex

$$\Delta = \left\{ \mathbf{u} \in \mathbb{R}^n_{\geq 0} : \sum_{i \in \mathcal{S}} u_i = 1 \right\},$$

and the tangent space of the set of possible distributions of flow is defined as

$$\mathrm{T}\Delta = \left\{ \mathbf{z} \in \mathbb{R}^n : \sum_{i \in \mathcal{S}} z_i = 0 \right\}.$$

Each flow unit is assigned to each tank $i \in \mathcal{S}$ depending on the current volume capacity, which is described by a function $f_i(\mathbf{u})$. Therefore, less inflow is assigned to those tanks close to be filled up.

The design of the population dynamics-based controllers are given by the proper selection of the fitness functions that define the incentives for the proportion of agents to choose a particular strategy. The proper selection of the fitness functions is further discussed below, and it depends on how the water is distributed in a DWTN with branched topology. Furthermore, it is necessary that the fitness functions satisfy conditions to obtain a class of population game known as stable game [11].

Definition 18.1 The game $\mathbf{F}(\mathbf{u})$ is stable if the Jacobian matrix $\mathbf{J} = \mathrm{D}\mathbf{F}(\mathbf{u})$ is negative semi-definite with respect to the tangent space $\mathrm{T}\Delta$ [11], i.e.,

$$\mathbf{z}^\top \mathbf{J}\, \mathbf{z} \leq 0, \quad \text{for all} \;\; \mathbf{z} \in \mathrm{T}\Delta, \;\; \mathbf{u} \in \Delta.$$

Then, it implies that a game is stable if the fitness functions are decreasing with respect to the proportion of agents.

Notice that for the branched topology, the fitness functions can be selected decreasing with respect to the current volume, e.g., the error with respect to the maximum capacity volume as in [20] (see Fig. 18.2a). When a proportion of agents is increased, it is expected that the corresponding volume increases (see Fig. 18.2b). Consequently, due to the fact that fitness functions are increasing with respect to the volume, the fitness function decreases with respect to the proportion of agents (necessary condition for a stable game).

The Distributed Replicator Dynamics

The results presented on this chapter are obtained using the replicator dynamics [23] in order to find a solution to the resource allocation problem. The solution, in which no agent has incentives to switch from one strategy to another one [23], is determined in terms of a *Nash equilibrium*,[1] which can be found when the dynamics converge,

[1] $\mathbf{u}^* \in \Delta$ is a Nash equilibrium if each used strategy entails the maximum benefit for the proportion of agents selecting it, i.e., the set of Nash equilibria is given by $\{\mathbf{u}^* \in \Delta : u_i^* > 0 \Rightarrow f_i(\mathbf{u}^*) \geq f_j(\mathbf{u}^*)\}$, for all $i, j \in \mathcal{S}$ [23].

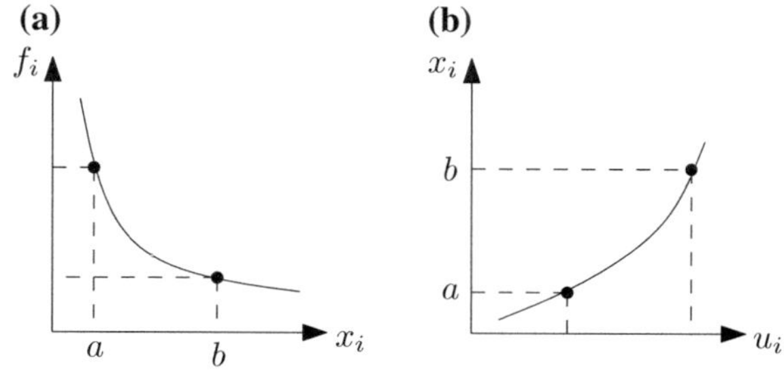

Fig. 18.2 Proper selection of fitness functions for divergence topology (**a**) and (**b**). Correspondence is as follows: **a** decreasing fitness function with respect to volume. **b** increasing relation existing between proportion of agents and volume for divergence topology

and is denoted by $\mathbf{u}^* \in \Delta$. The replicator dynamics are of interest in this work since they share gradient properties studied in [22], and because of their passivity properties studied in [3]. However, the replicator dynamics require full information (i.e., all the tanks (strategies) need information about the states of the others in order to evolve).

Since the problem is handled using a distributed control approach, it is necessary to use the distributed replicator dynamics, which were deduced in [2] from a local revision protocol that only needs partial information. Due to the fact that only local information is needed, then there is an undirected non-complete connected graph describing the interactions among agents. It is denoted by $\mathcal{G} = (\mathcal{V}, \mathcal{E})$, where $\mathcal{V}$ is the set of nodes, which represents the tanks, and $\mathcal{E} \subset \{(i,j) : i,j \in \mathcal{V}\}$ is the set of links representing the information sharing within the system. Furthermore, the set of neighbours of the node $i \in \mathcal{V}$ is given by $\mathcal{N}_i = \{j : (i,j) \in \mathcal{E}\}$. Notice that $i \notin \mathcal{N}_i$, and that $\mathcal{N}_i \neq \emptyset$, for all $i \in \mathcal{V}$ since $\mathcal{G}$ is connected.

The distributed replicator dynamics are given by

$$\dot{u}_i = u_i \left(f_i(\mathbf{u}) \sum_{j \in \mathcal{N}_i} u_j - \sum_{j \in \mathcal{N}_i} u_j f_j(\mathbf{u}) \right), \quad \text{for all } i \in \mathcal{S}.$$

Now that the distributed replicator dynamics have been defined, consider a population composed by a large and finite number of agents. Agents in the population have incentives to select the tank outflows (e.g., in a general control system, the error is an incentive for the controller to apply more energy to the system and then correct the states to achieve the desired values). The incentives, associated to rewarding that the proportion of agents u_i receives, for selecting the tank $i \in \mathcal{S}$, are given by a fitness function $f_i(\mathbf{u})$ whose mapping is $f_i : \Delta \to \mathbb{R}$. Moreover, the vector of all the fitness functions is denoted by $\mathbf{F} = [f_1 \ \cdots \ f_n]^\top$ with mapping $\mathbf{F} : \Delta \mapsto \mathbb{R}^n$.

The solution of the population game is given by the condition $f_i = f_j$, for all $i,j \in \mathcal{S}$. In order to control the *case of flow divergence* topology, it is proposed the following fitness function

$$f_i = -\left(\frac{1}{e_i + \varepsilon} \right), \quad \text{for all } i \in \mathcal{S}, \tag{18.2}$$

with

$$e_i = 1 - \frac{x_i + s_i + \gamma}{x_i^{\max} + \gamma}, \quad \text{for all } i \in \mathcal{S},$$

where $s_i \in \mathbb{R}_{\geq 0}$ is the shortage volume, i.e., the volume that is demanded but cannot be supplied by the *ith* tank, $\gamma \in \mathbb{R}_{>0}$ is a constant that ensures $0 \leq e_i \leq 1$ and $\varepsilon \in \mathbb{R}_{>0}$ is a small factor that prevents the indetermination of f_i when $e_i = 0$. Moreover, the proposed fitness function for the strategy $i \in \mathcal{S}$, only depends on the volume v_i and the proportion of agents u_i, making it suitable to apply in this case where only local information is available.

All the valves, defining the inflow of the receptor tanks in a partition, are established by the vector $\mathbf{u} \in \mathbb{R}^n_{\geq 0}$. These settings in the output gates affect the behaviour of the tank volumes, i.e., $\mathbf{x} \in \mathbb{R}^n_{\geq 0}$. Then, the variation of the tank volumes modifies the fitness function (18.2), affecting the control actions over the output valves $\mathbf{u} \in \mathbb{R}^n_{\geq 0}$.

18.3.2 Population-Games Approach: Second Data-Driven Perspective

Consider a society whose topology is represented by an undirected non-complete connected graph denoted by $\mathcal{G} = (\mathcal{V}, \mathcal{E})$, where $\mathcal{V}$ denotes the set of nodes of the graph $\mathcal{G}$. These nodes represent the set of n available strategies in a social game denoted by $\mathcal{S} = \{1, \ldots, n\}$. Besides, the set $\mathcal{E} \subset \{(i,j) : i,j \in \mathcal{V}\}$ denotes the edges of the graph $\mathcal{G}$ that determines the possible interactions among social strategies.

The graph $\mathcal{G}$ is divided into $\pi \in \mathbb{Z}_{>0}$ subcomplete graphs known as cliques [7]. Additionally, each clique represents a population within the society. The set $\mathcal{P} = \{1, \ldots, \pi\}$ denotes the collection of the π populations, and the set of cliques is denoted by $\mathcal{C} = \{\mathcal{C}^p : p \in \mathcal{P}\}$. The clique corresponding to the population $p \in \mathcal{P}$ is a graph given by $\mathcal{C}^p = (\mathcal{V}^p, \mathcal{E}^p)$, where the set $\mathcal{V}^p$ represents the n^p available strategies in a population game, which are denoted by $\mathcal{S}^p = \{i : i \in \mathcal{V}^p\}$. On the other hand, $\mathcal{E}^p = \{(i,j) : i,j \in \mathcal{V}^p\}$ is the set of all the possible links in $\mathcal{C}^p$ determining full interaction among the population strategies.

In this work, it is assumed that the set of cliques is already known, i.e., the number of cliques π, the set of vertices $\mathcal{V}^p$ and the set of edges $\mathcal{E}^p$ for all $p \in \mathcal{P}$ are known. Although if it is desired to obtain the optimal set of cliques,[2] there are several methods to find them, e.g., the Bron Kerbosh algorithm [12], or the maximum clique problem using replicator dynamics as shown in [7]. Once the optimal set of cliques $\mathcal{C}$ has been identified, it is possible to find *redundant links*. A link $(i,j) \in \mathcal{E}$ is redundant if $(i,j) \notin \tilde{\mathcal{E}}$, i.e., $(i,j) \notin \mathcal{E}^p$, for all $p \in \mathcal{P}$.

[2]The minimum amount of cliques π such that $\bigcup_{p\in\mathcal{P}} \mathcal{V}^p = \mathcal{V}$, and the minimum amount of links $|\tilde{\mathcal{E}}|$, where $\tilde{\mathcal{E}} = \bigcup_{p\in\mathcal{P}} \mathcal{E}^p \subseteq \mathcal{E}$ such that the graph $\tilde{\mathcal{G}} = (\mathcal{V}, \tilde{\mathcal{E}})$ is connected.

Then, the number of cliques that contain a node $i \in \mathcal{V}$, denoted by $G(i)$, is defined as follows:

$$G(i) = \sum_{p \in \mathcal{P}} g(i, p),$$

and

$$g(i, p) = \begin{cases} 1 & \text{if } i \in \mathcal{V}^p \\ 0 & \text{otherwise.} \end{cases}$$

Due to the fact that the graph $\mathcal{G}$ is a non-complete and connected, then all cliques share at least one node with another clique, which is known as an intersection node. The set $\mathcal{I}^p = \{i \in \mathcal{V}^p : G(i) > 1\}$ collects all the intersection nodes in a population $p \in \mathcal{P}$. Moreover, the set of intersection nodes in the graph $\mathcal{G}$ is given by $\mathcal{I} = \bigcup_{p \in \mathcal{P}} \mathcal{I}^p$.

Furthermore, all the populations $p \in \mathcal{P}$ such that a node $i \in \mathcal{V}$ belongs to the set of nodes $\mathcal{V}^p$ are collected in a set denoted by $\mathcal{P}_i$. The set of all the populations that includes a node $i \in \mathcal{V}$ is given by $\mathcal{P}_i = \{p : i \in \mathcal{V}^p\}$, where $\mathcal{P}_i \subseteq \mathcal{P}$.

The scalar $u_i \in \mathbb{R}_{\geq 0}$ is the proportion of agents in the society selecting the strategy $i \in \mathcal{S}$. Similarly, the scalar $u_i^p \in \mathbb{R}_{\geq 0}$ is the proportion of agents selecting the strategy $i \in \mathcal{S}^p$ in the population $p \in \mathcal{P}$. Moreover, the distribution of agents throughout the available strategies in the society and populations is known as the social strategic distribution and the population strategic distribution denoted by $\mathbf{u} \in \mathbb{R}^n_{\geq 0}$ and $\mathbf{u}^p \in \mathbb{R}^{n^p}_{\geq 0}$, respectively.

The set of possible social strategic distributions is given by a simplex denoted by Δ, which is a constant set, i.e., $\Delta = \left\{\mathbf{u} \in \mathbb{R}^n_{\geq 0} : \sum_{i \in \mathcal{S}} u_i = m\right\}$, where $m \in \mathbb{R}_{>0}$ is the constant mass of agents in the society. Similarly, the set of possible strategic distributions of the population $p \in \mathcal{P}$ is given by a non-constant simplex defined as $\Delta^p = \left\{\mathbf{u}^p \in \mathbb{R}^{n^p}_{\geq 0} : \sum_{i \in \mathcal{S}^p} u_i = m^p\right\}$, where $m^p \in \mathbb{R}_{>0}$ corresponds to the mass of agents in the population $p \in \mathcal{P}$. Furthermore, there is a relationship between the social proportions and the population proportions given by

$$u_i = \frac{1}{G(i)} \sum_{p \in \mathcal{P}_i} u_i^p. \tag{18.3}$$

Notice that if it is considered that $u_i^p = 0$ for all $i \notin \mathcal{V}^p$, then (18.3) can be written as

$$u_i = \frac{1}{G(i)} \sum_{p \in \mathcal{P}} u_i^p. \tag{18.4}$$

The fitness functions take a social or population strategic distribution and return the payoff that a proportion of agents playing a certain strategy receives. Let $f_i : \Delta \mapsto \mathbb{R}$ be the mapping of the fitness function for the proportion of agents playing the strategy $i \in \mathcal{S}$, and $f_i^p : \Delta^p \mapsto \mathbb{R}$ be the mapping of the fitness function for the proportion of agents playing the strategy $i \in \mathcal{S}^p$ in the population $p \in \mathcal{P}$. The fitness corresponding to a strategy $i \in \mathcal{S}$ is the same as the fitness for a strategy $j \in \mathcal{S}^p$ for all $p \in \mathcal{P}$ if $i = j$. Consequently, for all $i \in \mathcal{S}^p$ and for all $p \in \mathcal{P}_i$,

$$f_i(\mathbf{u}) = f_i^p(\mathbf{u}^p), \quad \text{if} \quad u_i = u_i^p. \tag{18.5}$$

The vector of the fitness functions for a society is given by $\mathbf{F} = [f_1 \quad \ldots \quad f_n]^\top \in \mathbb{R}^n$. The social average fitness is denoted by $\bar{f}$, where $\bar{f} = (\mathbf{u}^\top \mathbf{F})/m$. Similarly, the vector of fitness functions for a population $p \in \mathcal{P}$ is given by $\mathbf{F}^p \in \mathbb{R}^{n^p}$, whose fitness functions are associated to the strategies $\mathcal{S}^p$. The average fitness for a population $p \in \mathcal{P}$ is denoted by $\bar{f}^p = (\mathbf{u}^{p\top} \mathbf{F}^p)/m^p$. There is a relationship between the population masses and the social mass given by

$$m = \sum_{p \in \mathcal{P}} m^p - \sum_{i \in \mathcal{S}} (G(i) - 1) u_i. \tag{18.6}$$

The framework of this paper is given by the assumptions stated next.

Assumption 18.1 The game $\mathbf{F}$ is a full potential game [23], i.e., there is a continuously differentiable function $V(\mathbf{u})$, known as the potential function, satisfying

$$\frac{\partial V(\mathbf{u})}{\partial u_i} = f_i(\mathbf{u}), \quad \text{for all} \quad i \in \mathcal{S}, \text{ and } \mathbf{u} \in \Delta.$$

Assumption 18.2 Fitness functions depend only on strategies on which there is connection, i.e., each node requires only available information given by the graph topology.

Assumption 18.3 The proportion of agents playing the strategies corresponding to intersection nodes are strictly positive for all the time, i.e., $u_i^p > 0$ for all $i \in \mathcal{I}$, and for all $p \in \mathcal{P}$ (i.e., there is not extinction of the intersection population). This also implies that population masses are strictly positive, i.e., $m^p > 0$, for all $p \in \mathcal{P}$, since the population masses are composed of proportion of agents within populations.

Assumption 18.4 The game $D\mathbf{F}$ is a stable game [11], i.e., the Jacobian matrix $D\mathbf{F}(\mathbf{u})$ is negative semi-definite with respect to the tangent space $T\Delta$ (see Definition 18.1).

The features of the potential function $V(\mathbf{u})$ determine whether the full potential game $\mathbf{F}$ is stable, as shown in Lemma 18.1.

Lemma 18.1 *If $V(\mathbf{u})$ is twice continuously differentiable and concave, then the full potential game $\mathbf{F}$ is a stable game.*

The objective for the society is to converge to a Nash equilibrium[3] of the game $\mathbf{F}$ denoted by $\mathbf{u}^* \in \Delta$. In order to achieve this objective, there is a game at each

[3] $\mathbf{u}^* \in \Delta$ is a Nash equilibrium if each used strategy entails the maximum benefit for the proportion of agents selecting it, i.e., the set of Nash equilibria is given by $\{\mathbf{u}^* \in \Delta : u_i^* > 0 \Rightarrow f_i(\mathbf{u}^*) \geq f_j(\mathbf{u}^*)\}$, for all $i, j \in \mathcal{S}$ [23].

population $p \in \mathcal{P}$ converging to a Nash equilibrium of the game $\mathbf{F}^p$ denoted by $\mathbf{u}^{p*} \in \Delta^p$, and the intersection nodes $i \in \mathcal{I}$ allow a mass interchange among the different populations.

Population Dynamics and Mass Dynamics

A game is solved for each population with constraints given by the population masses m^p, which vary dynamically. Dynamics associated to each population are shown in (18.7a). There are π different dynamics of this form, one for each clique $\mathcal{C}^p$ for all $p \in \mathcal{P}$, i.e.,

$$\dot{u}_i^p = u_i^p \left(f_i^p - \bar{f}^p - \phi^p\right), \quad \text{for all } i \in \mathcal{S}^p, \tag{18.7a}$$

$$\phi^p = \beta \left(\frac{1}{m^p} \sum_{j \in \mathcal{S}^p} u_j^p - 1 \right), \tag{18.7b}$$

where β is the convergence factor for the whole system that takes a positive and finite value. Notice that, when $\phi^p = 0$ (i.e., $\mathbf{u}^p \in \Delta^p$), then (18.7a) becomes the classical replicator dynamics equation [27].

On the other hand, there are as many mass dynamics as intersection nodes in the graph, i.e., one for each $i \in \mathcal{I}$. The dynamics for population masses m^p are given by

$$\dot{m}_i^p = m_i^p \left(u_i - u_i^p\right), \quad \text{for all } p \in \mathcal{P}_i, \tag{18.8}$$

Equation (18.8) describes the movements of agents among populations through intersection nodes for the case in which there is no social mass constraint [6]. There might be alternative possibilities in the selection of the mass dynamics (18.8). However, the requirements that should be satisfied are as follows: (i) the dynamics satisfy the communication constraints established by the graph $\mathcal{G}$, and (ii) dynamics converge to the equilibrium point given by $u_i = u_i^p$, for all $p \in \mathcal{P}_i$.

There is a relationship between m_i^p, for all $i \in \mathcal{I}^p$, and the population masses m^p given by

$$m^p = \frac{1}{|\mathcal{I}^p|} \sum_{i \in \mathcal{I}^p} m_i^p, \quad \text{for all } p \in \mathcal{P}. \tag{18.9}$$

For the mass dynamics at intersection nodes in (18.8), the vector of masses and the vector of states associated to an intersection node $i \in \mathcal{I}$ are defined next. The masses vector is denoted by $\mathbf{m}_i = [m_i^{p_1} \ \ldots \ m_i^{p_{G(i)}}]^\top \in \mathbb{R}^{G(i)}$, where $p_1, \ldots, p_{G(i)} \in \mathcal{P}_i$; and the vector of population states is $\mathbf{u}_i = [u_i^{p_1} \ \ldots \ u_i^{p_{G(i)}}]^\top \in \mathbb{R}^{G(i)}$, where $p_1, \ldots, p_{G(i)} \in \mathcal{P}_i$; both vectors $\mathbf{m}_i$, and $\mathbf{u}_i$ for all $i \in \mathcal{I}$. Notice that, $\mathbf{m}_i \neq m_i$ and $\mathbf{u}_i \neq u_i$.

Finally, the dynamical system can be forced to converge to a Nash equilibrium $\mathbf{u}^*$ such that $\mathbf{F}(\mathbf{u}^*) = \nabla V(\mathbf{u}^*)$ converges to a desired value $f_i(r)$ for an $i \in \mathcal{I}$, where r is a known value (e.g., a reference). Modifying the relationship between the states in (18.4) by adding the reference r, the following new relationship is obtained:

$$u_i = \frac{1}{G(i)+1}\left(\sum_{p\in\mathcal{P}} u_i^p + r\right),$$

where $u_i^p = 0$, if $i \notin \mathcal{V}^p$. Using this modification, by (18.8), u_i tends to r. This makes $\bar{f}$ to converge to the desired value $f_i(r)$, for only one $i \in \mathcal{I}$.

Optimization Problems

The presented population dynamics with time-variant mass may be implemented to solve different constrained optimization problem forms. First, it is presented a population game without social mass constraint but with the positiveness over the proportion of agents. Afterwards, the population-games approach is presented to solve a constrained optimization problem with several constraints over the proportion of agents.

First, consider optimization problems without social mass constraint. This problem only demands the positiveness of optimization variables. From a mass dynamics perspective, it implies a variation of the social mass arbitrarily. The problem is stated as follows:

$$\begin{aligned} &\underset{\mathbf{u}}{\text{maximize}} && V(\mathbf{u}) \\ &\text{subject to} && \mathbf{u} \in \mathbb{R}^n_{\geq 0}, \end{aligned}$$

where $V : \mathbb{R}^n_{\geq 0} \mapsto \mathbb{R}$, and V is continuously differentiable and concave. Also, it is supposed that the solution point of this problem is an interior point. The solution for the optimization problem with one constraint is found by $\mathbf{F}(\mathbf{u}) = \nabla V(\mathbf{u}) = 0$, since $V(\mathbf{u})$ is concave and by the fact that it is known that the maximum point is an interior point. Therefore, the desired value for the average fitness is $f_i(r) = 0$, and it is enough to find the correct value for reference r and any intersection $i \in \mathcal{I}$.

Secondly, consider optimization problems with multiple constraints over agents proportions. Suppose that there is a strategic interaction with more than one constraint, e.g., different constraints over the proportion of agents. It is desired that the total amount of certain groups of proportions of agents are constant. This problem is stated as

$$\begin{aligned} &\underset{\mathbf{u}}{\text{maximize}} && V(\mathbf{u}) \\ &\text{subject to} && \mathbf{H}\mathbf{u} = \mathbf{h}, \text{ and } \mathbf{u} \in \mathbb{R}^n_{\geq 0}, \end{aligned} \tag{18.10}$$

where $\mathbf{u} \in \mathbb{R}^n_{\geq 0}$, $V : \mathbb{R}^n_{\geq 0} \mapsto \mathbb{R}$, and V is concave and continuously differentiable. Moreover, $\mathbf{H} \in \mathbb{R}^{L\times n}$ since there are L constraints and n decision variables, and $\mathbf{h} \in \mathbb{R}^L$. For this optimization problem, $\boldsymbol{\mu}$ is the Lagrange multiplier vector. The Lagrange function $l : \mathbb{R}^n \times \mathbb{R}^L \mapsto \mathbb{R}$ is

$$l(\mathbf{u}, \boldsymbol{\mu}) = V(\mathbf{u}) + \boldsymbol{\mu}^\top(\mathbf{H}\mathbf{u} - \mathbf{h}). \tag{18.11}$$

Moreover, $\nabla_{\mathbf{u}} l(\mathbf{u}, \boldsymbol{\mu}) = \nabla f(\mathbf{u}) + \mathbf{H}^\top \boldsymbol{\mu}$, and $-\nabla_{\mu} l(\mathbf{u}, \boldsymbol{\mu}) = -\mathbf{Hu} + \mathbf{h}$. The Lagrange condition is used to find possible extreme points in the objective function, in which $\nabla_{\mathbf{u}} l(\mathbf{u}, \boldsymbol{\mu}) = \mathbf{0}$, $\nabla_{\mu} l(\mathbf{u}, \boldsymbol{\mu}) = \mathbf{0}$ [8].

Consequently, fitness functions for each node are chosen to be defined as $\mathbf{F}(\mathbf{u}) = \nabla_{\mathbf{u}} l(\mathbf{u}, \boldsymbol{\mu})$, and $\mathbf{F}(\boldsymbol{\mu}) = \nabla_{\mu} l(\mathbf{u}, \boldsymbol{\mu})$. This problem is solved by using a reference r as it was explained in Sect. 18.3.2 in order to force a convergence value for the fitness functions associated to the social states and the Lagrange multipliers. In order to use the population and the mass dynamics, it is necessary that the games are stable according to Assumption 18.3.

Lemma 18.2 *If* $V(\mathbf{u})$ *is twice continuously differentiable and concave, and the constraints have the form* $\mathbf{Hu} = \mathbf{h}$*, then the games* $\mathbf{F}(\mathbf{u}) = \nabla_{\mathbf{u}} l(\mathbf{u}, \boldsymbol{\mu})$ *and* $\mathbf{F}(\boldsymbol{\mu}) = \nabla_{\mu} l(\mathbf{u}, \boldsymbol{\mu})$ *are stable.*

18.4 Simulations and Results

18.4.1 Case Study: First Data-Driven Perspective

In the design of the proposed decentralized controller, it is necessary to make a partitioning of the DWTN into different subsystems. Each subsystem must correspond to a case of flow divergence (i.e., each subsystem must be of the form shown in Fig. 18.1). In order to clarify the partitioning process in a typical branched DWTN, an arbitrary DWTN is presented in Fig. 18.3. At this general example, it is possible to identify that the whole system is composed of three partitions or subsystems.

When performing the partitioning, it is possible to find some tanks that are a source and also a receptor for different subsystems in the DWTN (this is typical when the topology is *branched*). For instance, in the partitioning presented in Fig. 18.3, the grey tanks are receptors for the partition 1, and source tanks for the partitions 2, and 3.

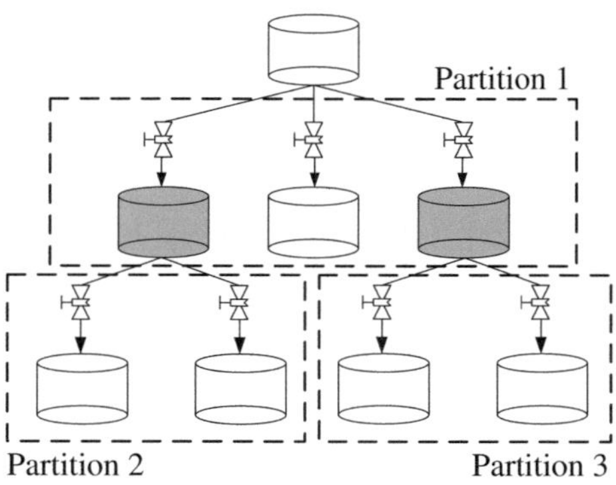

Fig. 18.3 Partitions over a branched topology. Some tanks are source and receptor in different partitions (*gray tanks*)

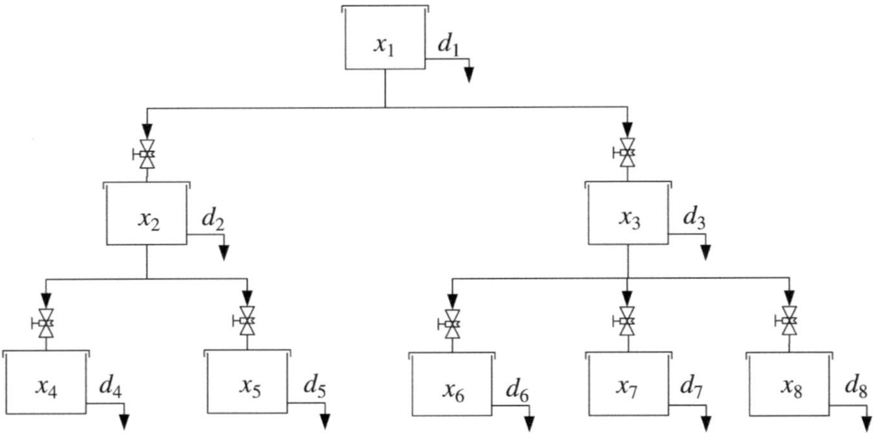

Fig. 18.4 Case study with eight tanks in a branched DWTN

Table 18.2 Maximum volumes and scale factors of the tanks in the DWTN

Tank i	$x_i^{\max}$ (m^3)	K_i (1/ms)
1	2.0	0.123
2	1.1	0.160
3	2.0	0.326
4	0.5	0.599
5	2.6	0.660
6	0.2	0.632
7	2.0	0.255
8	3.5	0.427

A DWTN composed by eight tanks is controlled (see Fig. 18.4), for an scenario in which shortages are produced due to the fact that the network is only operating with water stored in the main upstream tank. The system is a branched DWTN whose topology is mainly divergent, so it can be partitioned in three main subsystems; all independently controlled by a distributed replicator-dynamics-based controller. The maximum storage capacity and the scale factors of each tank are presented in Table 18.2. Since the system is branched and the divergence topology prevails, it is possible to divide the system into three partitions, as it was described before. The first one is composed by tanks 2 and 3, the second by tanks 4 and 5 and the third by tanks 6, 7 and 8. Each partition receives the flow from a source tank, which is distributed in different proportions, depending on the setting of the input valves of each tank of the partition.

Each tank attends a different demand pattern along the day denoted by d_i. The tanks with volumes x_5 and x_8 supply a constant demand pattern of 4.5×10^{-3} l/s, while the others, denominated as inactive tanks (i.e., tanks 2, 3, 4, 6 and 7) are just operating to store water, not attending any demand pattern. When there is not

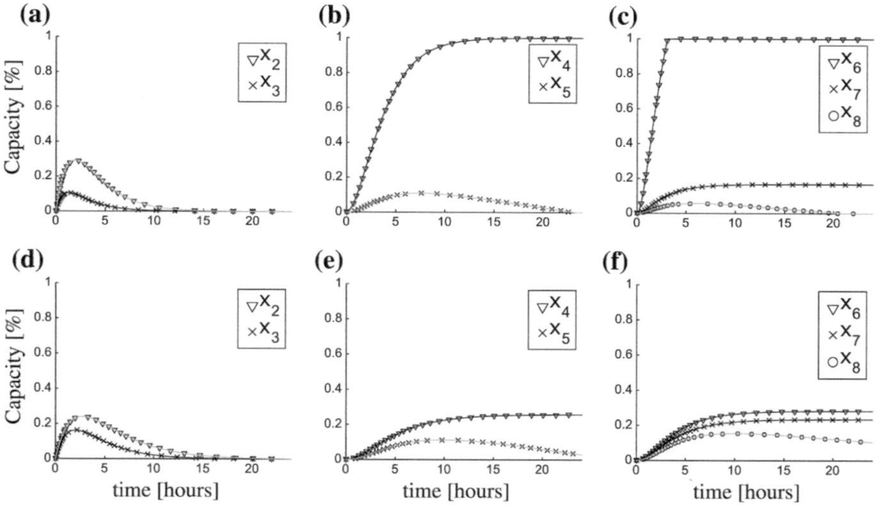

Fig. 18.5 Evolution of the used capacity in storage tanks; **a**, **b**, **c** capacity of tanks when no control strategy is applied; and **d**, **e**, **f** capacity of tanks when the evolutionary-game-based strategy with the replicator dynamics is applied

a control strategy, the flow is divided equally, and shortage of 26 m^3 is produced because the distribution of flows is inefficient, as shown in Figs. 18.5a, 18.5b and 18.5c.

When the control strategy is applied, then the priority is given to the tanks that supply the demand, and inactive tanks become less filled up since they are not attending any demand pattern. Thereby, no shortages are produced, the demand is fully supplied and the distribution of flows is more efficient, in comparison to the case with no control. This is because all the tanks keep some volume stored on them at the end of the day, while in the other case, tanks 5 and 8 are completely empty.

It has been shown that the proposed decentralized population dynamics-based control is efficient in terms of a better distribution of drinking water throughout the DWTN, avoiding shortages. The partitioning proposed methodology allows to design the decentralized controller by using different local controllers with a lower computational burden with respect to a centralized controller.

18.4.2 Case Study: Second Data-Driven Perspective

Consider the case study presented in Fig. 18.6, which corresponds with the aggregate model of the Barcelona drinking water network presented in Fig. 2.2. For this system, consider $\tilde{\mathbf{x}} = [\tilde{x}_1 \;\; \tilde{x}_2 \;\; \dots \;\; \tilde{x}_{61}]^\top$, $\tilde{\mathbf{x}}_s = [\tilde{x}_{s,1} \;\; \tilde{x}_{s,2} \;\; \dots \;\; \tilde{x}_{s,61}]^\top$ and $\mathbf{u} = [u_1 \;\; u_2 \;\; \dots \;\; u_{61}]^\top$ according to the explanation presented in Sect. 18.2.2.

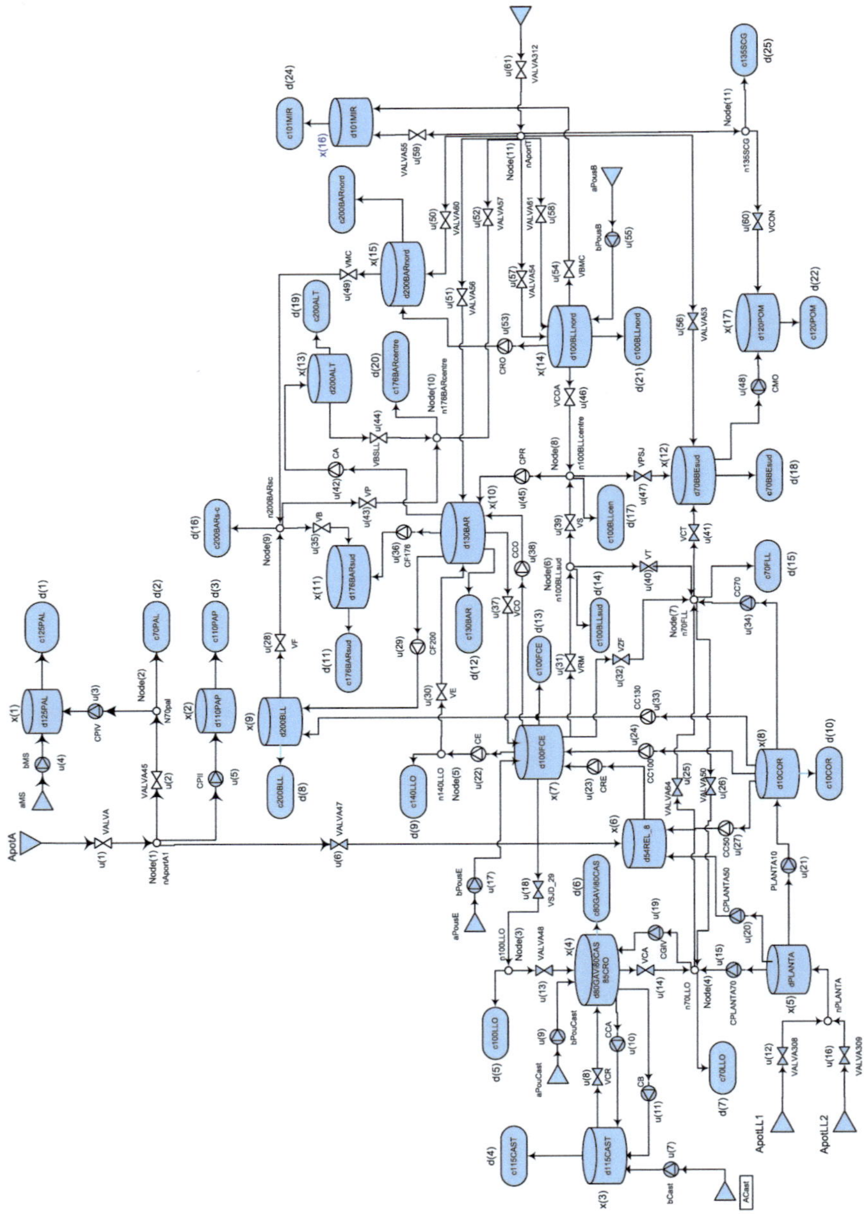

Fig. 18.6 Aggregate model of the Barcelona DWTN (BDWTN) comprised by 17 states, 61 control actions, 25 demands and 11 mass balance nodes

In the control design, the first step is the determination of cliques within the system, i.e., to make a partitioning of the system. The aforementioned partition process of the BDWTN is a problem already studied in [18]. For the BDWTN control problem, the proposed partitioning is determined based on the system mass balance constraints. Lagrange-multiplier vertices are connected to decision variables vertices from which information is needed in order to compute the fitness functions $\mathbf{F}(\mu)$. As a criterion for performing the partitioning, it is desired that all the Lagrange multipliers, and the nodes connected with them, belong to the same clique. In order to formalize this partitioning criterion, let $\mathcal{H}_j$ be the set of all the nodes that are involved in the *jth* equality constraint of the form (18.10), where $j = 1, \ldots, L$, e.g., for the BDWTN system, $\mathcal{H}_1 = \{1, 2, 5, 6\}$, and $\mathcal{H}_2 = \{2, 3\}$. Furthermore, we consider two sets of nodes for mass balance constraints $\mathcal{H}_i$, and $\mathcal{H}_j$. If $\mathcal{H}_i \cap \mathcal{H}_j \neq \emptyset$, then all the nodes $\mathcal{H}_i \cup \mathcal{H}_j$ should belong to the same clique.

Based on this idea, it is possible to determine the vertices (strategies) that should belong to the same clique (population). As an example, consider the set of nodes associated to the constraint given by mass balance node 9, i.e., $\mathcal{H}_9 = \{28, 35, 43, 49\}$, and the set of nodes corresponding to the mass balance constraint 10, i.e., $\mathcal{H}_{10} = \{43, 44, 52\}$. There is a common vertex given by $\mathcal{H}_9 \cap \mathcal{H}_{10} = \{43\}$. Now, considering the constraint corresponding to the mass balance node 11, i.e., $\mathcal{H}_{11} = \{50, 51, 52, 56, 57, 58, 59, 60, 61\}$, then it is obtained that $\mathcal{H}_{10} \cap \mathcal{H}_{11} = \{52\}$. Consequently, all the nodes $\mathcal{H}_9 \cup \mathcal{H}_{10} \cup \mathcal{H}_{11}$ should belong to the same clique.

On the other hand, there are some vertices that are not associated to any constraint, e.g., the node 4 associated to the decision variable x_4, then $4 \notin \mathcal{H}_j$ for all $j = 1, \ldots, 11$. In these cases, vertices are assigned to the closest clique. Cliques are presented in Fig. 18.7, and the nodes of each clique are shown in Table 18.3. Notice that $\{\mathcal{H}_1 \cup \mathcal{H}_2 \cup \mathcal{H}_3 \cup \mathcal{H}_5\} \in \mathcal{V}^1$, $\{\mathcal{H}_4 \cup \mathcal{H}_6 \cup \mathcal{H}_7 \cup \mathcal{H}_8\} \in \mathcal{V}^2$, and $\{\mathcal{H}_9 \cup \mathcal{H}_{10} \cup \mathcal{H}_{11}\} \in \mathcal{V}^3$.

Once the partitioning is performed, the optimization problem (18.1) is stated of the form (18.10) by adding slack variables, which may be solved by using the population and mass dynamics. In this case, the society is composed of three population (cliques). In order to analyze the performance of the data-driven controller, the obtained results are compared to a centralized MPC controller. Figure 18.8 presents the evolution of three volume tanks (i.e., x_1, x_{12}, and x_{14}), and three control inputs (i.e., u_{18}, u_{32}, and u_{40}) for both centralized MPC controller and data-driven controller based on population dynamics. In Figs. 18.8a, 18.8b and 18.8c show that, with the centralized MPC controller, the tanks are maintained with more volumes with respect to the data-driven controller based on population dynamics. This better performance of the centralized MPC controller is obtained due to the fact it disposes of the system model in comparison to the data-driven control approach. Moreover, Figs. 18.8d, 18.8e and 18.8f show the similar performance of the control inputs for both controller. This close behaviour is obtained because of the constraints, which are taken into account for both control approaches.

Table 18.4 shows the comparison of the economical costs obtained with the centralized MPC strategy and the data-driven population-games-based control approach. The results exhibit lower energy costs associated to the control inputs with the

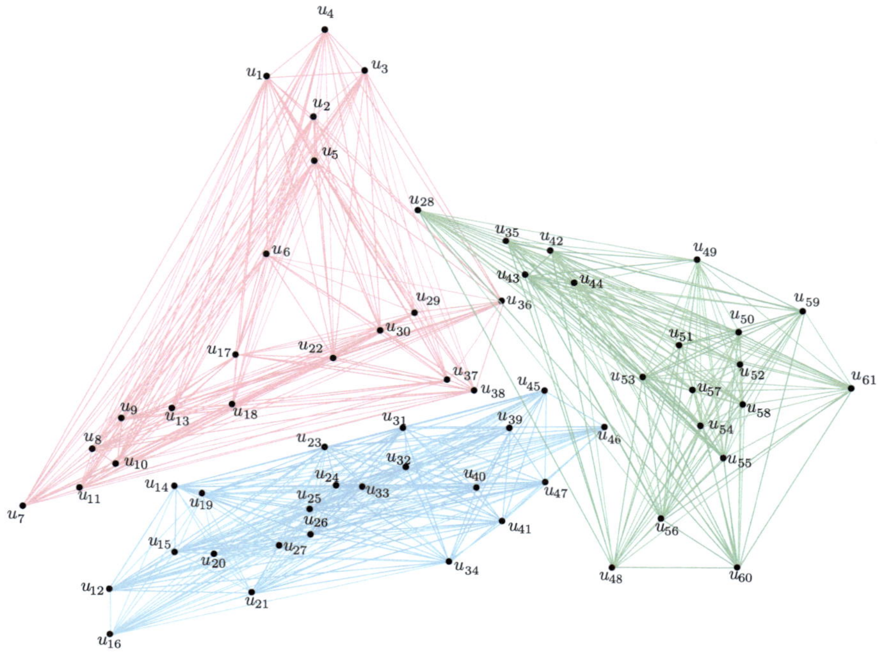

Fig. 18.7 Partitioning of the BDWTN into three cliques (see Table 18.3)

data-driven approach. However, since the MPC controller disposes of the model system to generate a prediction, the centralized MPC approach minimizes more the overall costs. In contrast, even though the minimization of costs, the data-driven control scheme is non-centralized, reducing the amount of required communication links in order to compute the final control inputs.

Table 18.3 Partitioning of the network into the three resultant cliques

Clique	Vertices u	Involved states x
1	1, 2, 3, 4, 5, 6, 7, 8, 9, 10, 11, 13, 17, 18, 22, 29, 30, 36, 37, 38	1, 2, 3, 4, 6, 7, 9, 10, 11
2	12, 14, 15, 16, 19, 20, 21, 23, 24, 25, 26, 27, 31, 32, 33, 34, 39, 40, 41, 45, 46, 47	4, 5, 6, 7, 8, 9, 10, 12, 14
3	28, 35, 42, 43, 44, 48, 49, 50, 51, 52, 53, 54, 55, 56, 57, 58, 59, 60, 61	9, 10, 11, 12, 13, 14, 15, 16, 17

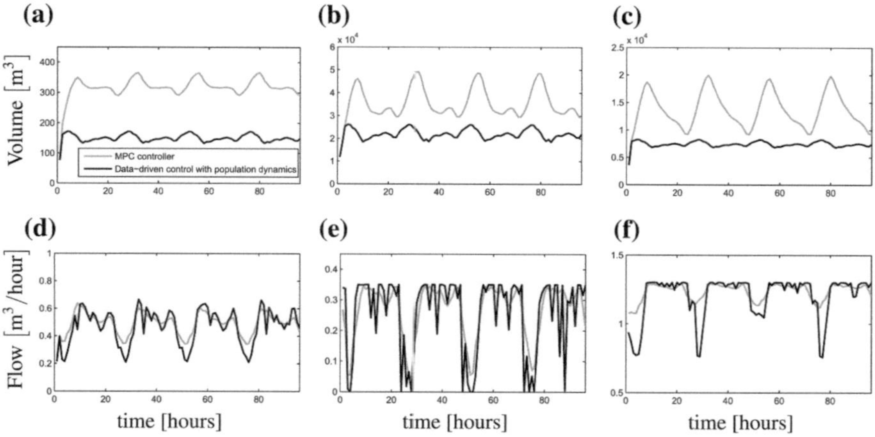

Fig. 18.8 Evolution of volumes **a** 1, **b** 12, and **c** 14. Evolution of control inputs **d** 18, **e** 32, and **f** 40

Table 18.4 Discrimination of economical costs for different control strategies

Day	Total cost in economical units (e.u.)			
	Population dynamics approach		Model predictive control	
	Data-driven controller		Model-based controller	
	Water	Energy	Water	Energy
1	45484.48	18409.34	37915.28	22096.12
2	41384.76	18131.81	28352.38	22235.15
3	40022.43	18791.73	28400.39	22288.11
4	40389.76	18387.35	28330.14	22219.59
Sum	167281.43	73720.23	122998.21	88838.97
Overall costs	241001.66		211837.17	

18.5 Conclusions

Two data-driven non-centralized control strategies to manage water flows among drinking-water networks have been presented. The proposed controllers are based on population games and have been designed using the distributed replicator dynamics and a modification of the population dynamics incorporating mass dynamics. Additionally, two partitioning approaches have been introduced in order to divide the typical centralized control problem into several subsystems. The partitioning of the system allows to reduce the computational burden required to manage the flows among the system. In the first population-games approach, the partitioning implies a decentralized control scheme since the local controllers do not communicate to each other. On the other hand, the partitioning in the second population-games implies a distributed control scheme since there is overlapping among the resulting subsystems. Both techniques have been tested using two different case studies.

References

1. Arslan G, Shamma J (2004) Distributed convergence to Nash equilibria with local utility measurements. In: Proceedings of the 43rd IEEE Conference on Decision and Control (CDC), Atlantis, Paradise Island, Bahamas, pp 1538–1543
2. Barreiro-Gomez J, Obando G, Quijano N (2016) Distributed population dynamics: optimization and control applications. Trans Syst Man Cybern 99:1–11
3. Barreiro-Gomez J, Obando G, Riaño Briceño G, Quijano N, Ocampo-Martinez C (2015) Decentralized control for urban drainage systems via population dynamics: Bogotá case study. In: Proceedings of the European Control Conference (ECC), Linz, Austria
4. Barreiro-Gomez J, Quijano N, Ocampo-Martinez C (2014) Constrained distributed optimization based on population dynamics. In: Proceedings of the 53rd IEEE Conference on Decision and Control (CDC), Los Angeles, California, USA, pp 4260–4265
5. Barreiro-Gomez J, Quijano N, Ocampo-Martinez C (2014) Distributed control of drinking water networks using population dynamics: Barcelona case study. In: Proceedings of the 53rd IEEE Conference on Decision and Control (CDC), Los Angeles, California, USA, pp 3216–3221
6. Barreiro-Gomez J, Quijano N, Ocampo-Martinez C (2016) Constrained distributed optimization: a population dynamics approach. Automatica 69:101–116
7. Bomze IM, Pelillo M, Stix V (2000) Approximating the maximum weight clique using replicator dynamics. IEEE Trans Neural Netw Learn Syst 11(6):1228–1241
8. Chong EKP, Zak SH (2011) An introduction to optimization. Series in discrete mathematics and optimization. Wiley
9. Grosso JM, Ocampo-Martinez C, Puig V, Limon D, Pereira M (2014) Economic MPC for the management of drinking water networks. In: European Control Conference (ECC), Strasbourg, France, pp 790–795
10. Grundmann Jens, Schütze Niels, Schmitz Gerd, Al-Shaqsi Saif (2012) Towards an integrated arid zone water management using simulation-based optimization. Environ Earth Sci 65(5):1381–1394
11. Hofbauer J, Sandholm WH (2009) Stable games and their dynamics. J Econ Theory 144(4):1665–1693
12. Johnston HC (1976) Cliques of a graph-variations on the Bron-Kerbosch algorithm. Int J Parallel Prog 5(3):209–238
13. Kalbus E, Kalbacher T, Kolditz O, Krüger E, Seegert J, Röstel G, Teutsch G, Borchardt D, Krebs P (2012) Integrated water resources management under different hydrological, climatic and socio-economic conditions. Environ Earth Sci 65(5):1363–1366
14. Lasaulce S, Tembine H (2011) Game theory and learning for wireless networks: fundamentals and applications. Academic Press
15. Marden J (2012) State based potential games. Automatica 48:3075–3088
16. Marden J, Shamma J (2015) Game theory and distributed control. In: Handbook of game theory with economic applications, pp 861–899. Elsevier
17. Marden JR, Shamma JS (2014) Game theory and distributed control. In: Handbook of Game Theory, vol 4, pp 861–899
18. Ocampo-Martinez C, Bovo S, Puig V (2011) Partitioning approach oriented to the decentralised predictive control of large-scale systems. J Process Control 21:775–786
19. Price RK, Vojinović Z (2011) Urban hydroinformatics: data, models, and decision support for integrated urban water management. IWA Publishing

20. Ramirez-Llanos E, Quijano N (2010) A population dynamics approach for the water distribution problem. Int J Control 83(9):1947–1964
21. Riaño-Briceño G, Barreiro-Gomez J, Ramirez-Jaime A, Quijano N, Ocampo-Martinez C (2016) MatSWMM—An open-source toolbox for designing real-time control of urban drainage systems. Environ Model Softw 83:143–154
22. Sandholm WH, Dokumaci E, Lahkar R (2008) The projection dynamic and the replicator dynamic. Games Econ Behav 64(2):666–683
23. Sandholm WH (2010) Population games and evolutionary dynamics. MIT Press, Cambridge, Mass
24. Tembine H, Altman E, El-Azouzi R, Hayel Y (2010) Evolutionary games in wireless networks. IEEE Trans Syst Man Cybern Part B Cybern 40(3):634–646
25. UNICEF (2015) Progress on sanitation and drinking water update and MDG assessment. World Health Organization, Geneva
26. Wang Y, Ocampo-Martinez C, Puig V (2015) Robust model predictive control based on Gaussian processes: application to drinking water networks. In: Proceedings of the European Control Conference (ECC), Linz, Austria
27. Weibull JW (1997) Evolutionary game theory. The MIT Press, London, England

Chapter 19
Coordinating Regional and Urban Water Networks

Congcong Sun, Gabriela Cembrano and Vicenç Puig

19.1 Introduction

Regional and urban networks normally operate to deliver water from natural sources to municipal, industrial and irrigation needs. Management of these systems from planning to operation is challenging since the problem deals with many complex modelling issues related to inflows, river delays, storage, urban, irrigation and industrial water demands as described in [1].

Regional supply and urban delivery networks are often operated separately by different authorities and/or utilities. Moreover, planning and management of these subsystems have different goals and timescales. Additionally, hydraulics involved differ considerably from one to another, in particular, between large and spatially distributed open canals and pressurized water sections for distribution to consumers. In many water systems, network operation is carried out based on the heuristic approaches and operator judgement, among other approaches, which may be quite complex for large-scale interconnected systems.

An effective management of complex water network requires a supervisory control system that takes optimal decisions regarding the operation of the whole network. Such decisions are either implemented automatically or offered as a decision support to operators and managers. Decisions of the control systems are translated into set points to individual, localized, lower distribution water systems that optimize the pressure profile to minimize losses by leakage and provide sufficient pressure [2]. The whole control system responds to changes in network topology (ruptures or changes of configuration), typical daily/weekly profiles, as well as major changes in demand.

C. Sun · G. Cembrano
Institut de Robòtica i Informàtica Industrial, CSIC-UPC, Barcelona, Spain

V. Puig (✉)
Research Center Supervision, Safety and Automatic Control (CS2AC-UPC), Terrassa, Spain
e-mail: vicenc.puig@upc.edu

V. Puig et al. (eds.), *Real-Time Monitoring and Operational Control of Drinking-Water Systems*, Advances in Industrial Control,
DOI 10.1007/978-3-319-50751-4_19

Considering different dynamics and control objectives, complex water network can be divided into two functional networks: *Regional supply network* and *Urban delivery network*:

- Regional supply network, composed of rivers, natural aquifers, and large reservoirs, which mainly conveys water from natural sources to the cities.
- Urban delivery network, which links water treatment and desalinization plants and transports water using pipes and tanks distributed along a city.

This chapter presents separately the modelling and design of MPC controllers for regional supply and urban delivery networks. Both of the regional supply and urban delivery networks must be modelled separately because of their different timescales, compositional elements and specified objectives. In order to generate coordinated control strategies including both functional networks, a temporal multi-level coordination for regional supply and urban delivery networks is presented. The Catalonia Regional Water Network is used to validate the proposed modelling and control schemes.

19.2 Problem Formulation

19.2.1 Control-Oriented Model for Regional Supply Networks

The control-oriented model for a regional supply network model can be considered as composed of a set of constitutive elements, which are shown below. Some of these elements are also relevant in Chap. 12.

Tanks and Reservoirs

Water dams/reservoirs provide the entire water system with the storage capacity of water. The mass balance expression relating the stored volume v and the manipulated inflows q_j^{in} and outflows q_h^{out} (including the demand flows as outflows) for the ith tank can be written as the discrete-time difference equation

$$v_i(k+1) = v_i(k) + \Delta t \left(\sum_j q_j^{\text{in}}(k) - \sum_h q_h^{\text{out}}(k) \right), \tag{19.1}$$

where Δt denotes the sampling time and k denotes the discrete-time instant. The physical constraint related to the range of admissible water in the ith dam/tank is expressed as follows:

$$v_i^{min} \leq v_i(k) \leq v_i^{max}, \quad \forall\ k, \tag{19.2}$$

where v_i^{min} and v_i^{max} denote the minimum and the maximum admissible storage capacities, respectively. As this constraint is physical, it is impossible to send more water to a tank than it can store, or drawing more water than the stored amount. Although v_i^{min} might correspond with an empty dam/tank, in practice this value can be set as nonzero in order to maintain an emergency stored volume enough to supply for facing extreme circumstances. Moreover, there will be restrictions in the amount of flow that can be extracted from the dam/tank depending on the volume stored according to the discharge curves.

For simplicity purposes, the dynamic behaviour of these elements is described as a function of the volume. However, in most of the cases, the measured variable is the water level (by using level sensors), which implies the computation of the water volume taking into account the element geometry.

Actuators

Several types of control actuators are considered: valves, gates and pumps (more precisely, pumping stations). It is assumed that the MPC controller provides the flow set point to a local controller that is responsible to establish the required flow through the actuator by using a closed-loop control system with a PID or a PLC. The manipulated flows through the actuators represent the manipulated variables, denoted as q_u. All considered actuators have lower and upper physical limits, which are taken into account as system constraints. As in (19.2), they are expressed as follows:

$$q_{u_i}^{min} \leq q_{u_i}(k) \leq q_{u_i}^{max}, \quad \forall\, k, \tag{19.3}$$

where $q_{u_i}^{min}$ and $q_{u_i}^{max}$ denote the minimum and the maximum flow capacities, respectively.

Nodes

These elements correspond to the network points where water flows are merged or split. Thus, the nodes represent mass balance relations, being modelled as equality constraints related to inflows (from other dams/tanks through gates/valves or pumps) and outflows, and these latter being represented not only by manipulated flows but also by demand flows. The expression of the mass conservation in these elements can be written as follows:

$$\sum_j q_j^{\text{in}}(k) = \sum_h q_h^{\text{out}}(k). \tag{19.4}$$

From now on and with some abuse of notation, node inflows and outflows are denoted by q^{in} and q^{out}, respectively, despite they can be manipulated flows and hence denoted by q_u, if correspond.

River/Canal Reaches

In order to model a river/canal into a regional supply network, a single reach can be approximated by using the modelling approach proposed by [3], which leads to the following relation between the upstream (q^{ups}) and downstream (q^{dns}) flows:

$$q^{dns}(k+1) = a_1 q^{dns}(k) + b_0 q^{ups}(k - \tau_d), \tag{19.5}$$

where $\tau_d = \lceil \tau / T_s \rceil$, τ is the downstream transport delay, T_s is the sampling time, $b_0 = 1 - a_1$, and $a_1 = e^{-\frac{T_s}{T}}$ with T the time constant of the reach canal.

Urban and Irrigation Demands

Urban and irrigation demands are considered as measured disturbance of the system at a given time instant. The demand in urban areas can be anticipated by a forecasting algorithm such as those presented in Chap. 6. On the other hand, irrigation demand is typically planned in advance with farmers. Pre-established flows for irrigation are established in the irrigation areas for determined periods of the year.

State-Space MPC Model

Given the existence of transportation delays in rivers/canals, the state-space model of a regional supply network has two kinds of states and control variables. The first kind of state variables represents reservoir/dam volumes, and the control variables correspond to actuator flows in gates/valves:

$$\mathbf{x}(k+1) = \mathbf{A}\,\mathbf{x}(k) + \mathbf{B}\,\mathbf{u}(k) + \mathbf{B}_p\,[\mathbf{d}(k) - \varepsilon(k)], \quad k \in \mathbb{Z} \tag{19.6}$$

where $\mathbf{x}(k) \in \mathbb{R}^{n_x}$ are the state variables, $\mathbf{u}(k) \in \mathbb{R}^{n_u}$ are the control variables, $\mathbf{d}(k) \in \mathbb{R}^{n_d}$ are the disturbances corresponding to the demands, and $\varepsilon(k) \in \mathbb{R}^{n_d}$ are slack variables for unsatisfied demands.

The demands of a regional supply network are expected to be satisfied by the MPC strategy with exceptional situations (e.g., drought) when some demands (especially irrigation demands) may be satisfied only partially. In (19.6), $\varepsilon(k)$ is introduced to control the amount of demand that is not satisfied.

The second kind of states and control variables represents river flows in a river reach model with delays. For simplicity and brevity of the explanation, the river reach model (19.5) is considered as a transport delay [4]:

$$q_i^{dns}(k) = q_i^{ups}(k - \tau_d), \tag{19.7}$$

For time delays associated with flows within the network, the following auxiliary state equations z are introduced:

$$z_{1,i}(k+1) = q_i^{ups}(k), \tag{19.8a}$$

$$z_{j+1,i}(k+1) = z_j, i(k), \quad j = 1, \ldots, \tau_d, \tag{19.8b}$$

where $z_j,_i\,(k) \in \mathbb{R}^{n'_x}$ are state variables representing delay flows and $q_i^{ups}(k) \in \mathbb{R}^{n'_u}$ flows, part of control variables.

More details on how this approach can be extended to the case that the river reach model (19.5) is not just considered as a delay can be found in [4].

After combining (19.8a) and (19.8b) with (19.6), a new augmented state-space representation is presented as follows:

$$\widetilde{\mathbf{x}}(k+1) = \widetilde{\mathbf{A}}\,\widetilde{\mathbf{x}}(k) + \widetilde{\mathbf{B}}\,\widetilde{\mathbf{u}}(k) + \widetilde{\mathbf{B}}_p\,[\mathbf{d}(k) - \varepsilon(k)], \quad k \in \mathbb{Z} \tag{19.9}$$

where

$$\widetilde{\mathbf{x}}(k) = \begin{bmatrix} \mathbf{x}(k) \\ \mathbf{z}_i(k) \end{bmatrix}, \quad \widetilde{\mathbf{u}}(k) = \begin{bmatrix} \mathbf{u}(k) \\ q_i^{ups}(k) \end{bmatrix}.$$

All variables are subject to the following inequality constraints:

$$\widetilde{\mathbf{x}}^{min} \leq \widetilde{\mathbf{x}}(k) \leq \widetilde{\mathbf{x}}^{max}, \tag{19.10}$$

$$\widetilde{\mathbf{u}}^{min} \leq \widetilde{\mathbf{u}}(k) \leq \widetilde{\mathbf{u}}^{max}, \tag{19.11}$$

$$\varepsilon^{min} \leq \varepsilon(k) \leq \varepsilon^{max}, \tag{19.12}$$

where $\widetilde{\mathbf{x}}^{min}$ and $\widetilde{\mathbf{x}}^{max}$ are physical limitations of the reservoirs, while $\widetilde{\mathbf{u}}^{min}$ and $\widetilde{\mathbf{u}}^{max}$ are volume limitations of the river flows. The range of ε^{min} lies between zero and the related demand.

As described in Chap. 12, the balance at every node should be satisfied, where $\mathbf{E}$, $\mathbf{E}_d$ and $\mathbf{E}_{\widetilde{x}}$ are matrices which parameters can be obtained from topology of the water network:

$$\mathbf{E}\,\widetilde{\mathbf{u}}(k) + \mathbf{E}_d\,\mathbf{d}(k) - \mathbf{E}_d(k)\,\varepsilon + \mathbf{E}_{\widetilde{x}}(k)\,\widetilde{\mathbf{x}}(k) = 0. \tag{19.13}$$

During the consumption process, water storage of reservoir should be kept above a given volume (named as water safety volume) that is used as emergency supply for drought period or emergency situations. Any situation below the emergency volume should be penalized using the soft constraint

$$\widetilde{\mathbf{x}} \geq \widetilde{\mathbf{x}}_s - \varepsilon_{\widetilde{\mathbf{x}}}, \quad \text{with} \quad \varepsilon_{\widetilde{\mathbf{x}}} \geq 0. \tag{19.14}$$

where $\widetilde{\mathbf{x}}_s$ is the water safety volume and $\varepsilon_{\widetilde{\mathbf{x}}}$ is the slack variable associated with $\widetilde{\mathbf{x}}_r$.

19.2.2 Operational Goals for Regional Supply Networks

Considering the dynamic characteristics of river water, a regional supply network operates with a 30-day horizon and a daily time interval. The main operational goals to be achieved are as follows:

- *Operational safety* (J_{safety}): To maintain appropriate water storage volumes in dams or reservoirs for emergency-handling, i.e.,

$$J_{safety}(k) = \varepsilon_{\widetilde{\mathbf{x}}}(k)^{\top}\mathbf{W}_{\widetilde{x}}\varepsilon_{\widetilde{\mathbf{x}}}(k), \tag{19.15}$$

where $\varepsilon_{\widetilde{\mathbf{x}}}(k) = \widetilde{\mathbf{x}}(k) - \widetilde{\mathbf{x}}_s$, being $\widetilde{\mathbf{x}}_s$ the safety volume for the reservoirs and ecological flow for the river (canal) reaches, and $\mathbf{W}_{\widetilde{x}}$ is the associated prioritization weight.

- *Demand management* (J_{demand}): To satisfy as much as possible irrigation demands with allowable slackness during drought seasons, i.e.,

$$J_{demand}(k) = \varepsilon(k)^{\top}\mathbf{W}_d\varepsilon(k), \tag{19.16}$$

where $\varepsilon_{\mathbf{x}}(k)$ is the slack variable associated with the relaxation of the demand satisfaction according to (19.9), and $\mathbf{W}_d$ is the associated prioritization weight.

- *Minimizing waste* (J_{waste}): To avoid unnecessary water release from reservoirs that is wasted in the receiving environment, i.e.,

$$J_{waste}(k) = (\widetilde{\mathbf{u}}(k) - \widetilde{\mathbf{u}}_s(k))^{\top}\mathbf{W}_{\widetilde{w}}(\widetilde{\mathbf{u}}(k) - \widetilde{\mathbf{u}}_s(k)), \tag{19.17}$$

where $\widetilde{\mathbf{u}}_s$ is desired flow to be released to the receiving environment and $\mathbf{W}_{\widetilde{w}}$ is the associated prioritization weight.

- *Balance management* ($J_{balance}$): To keep rivers and reservoirs exploited in a balanced way in order to overusing some of them, i.e.,

$$J_{balance}(k) = \widetilde{\mathbf{x}}^{\top}(k)\mathbf{W}_b\widetilde{\mathbf{x}}(k), \tag{19.18}$$

where $\mathbf{W}_{\widetilde{b}}$ is the associated prioritization and balancing weight.

- *Environment conservation* ($J_{ecological}$): To maintain water volume and ecological flows. Since the river flow is modelled as a part of the state vector, this control objective is included in J_{safety}.

The above-mentioned goals lead to the following cost multi-objective function for the optimal management of a regional network

$$J_{regional} = \sum_{k=0}^{H_p-1}(J_{safety}(k) + J_{demand}(k) + J_{waste}(k) + J_{balance}(k)). \tag{19.19}$$

The weights $\mathbf{W}_{\widetilde{x}}$, W_f, $\mathbf{W}_{\widetilde{w}}$, $\mathbf{W}_{\widetilde{x}}$ and $\mathbf{W}_{\widetilde{m}}$ are used to prioritize management policies (typically, established by the water network authorities) appearing as objectives in (19.19). The weight tuning method proposed in [5], based on computing the Pareto front of the multi-objective optimization problem presented in (19.20), can be used. The initial step of this tuning approach is to find what are known as the anchor points corresponding to the best possible value for each objective obtained by optimizing a single criterion at a time. Then, a normalization procedure is applied, a management point (MP) defined by establishing objective priorities is defined, and the optimal weights are determined by computing those that minimize the distance from the solutions of the Pareto front and the MP.

19.3 Coordination Scheme for Regional and Urban Networks

19.3.1 Centralized MPC

In order to generate optimal strategies for the complete system including both regional and urban networks, centralized MPC is an option. The state-space model of the complete water system includes the state-space model of the regional network modelling volumes of dams and tanks, actuator flows and also river flows in a river reach model with delays as in (19.6) and the state-space model of the urban network presented in Chap. 12.

The operational goals to be achieved in the complex water network are presented after combining control objectives using appropriate weights in both regional and urban networks as presented, respectively, $J_{regional}$ in (19.19) and J_{urban} in Chap. 12 (see (12.13)). Here, it writes

$$J = J_{regional} + J_{urban}. \tag{19.20}$$

However, considering the different dynamics and timescales of both networks, the MPC prediction horizon should be the one of the slowest system that corresponds to the regional supply network and as previously discussed is equal to 30 days. On the other hand, the sampling time should be selected as the one of the fastest systems (the urban network), which is 1 h. This implies that the MPC prediction horizon is equal to 720 h with a sampling time of 1 h. Because of this long prediction horizon, the use of a single centralized MPC could be quite difficult to apply in real time for a large-scale network. This is the main motivation for introducing the multi-layer MPC scheme presented in the next section.

19.3.2 Multi-layer MPC

A temporal multi-level MPC scheme is the alternative to optimize regional and urban networks by coordinating the separated MPC controllers controlling them. The general policy for coordinating these MPC controllers is to transfer the long-term control decisions of regional network to the urban network by *target constraints*. Besides, the short-term urban delivery network will update its daily demand information to the regional network by providing *measured disturbances*, where

- Measured disturbance (M_s) handles the daily related aggregated demands at the urban network as communication information to the regional network.
- Target constraint (T_d) expresses management policies at regional network to urban network in the form of constraints to be satisfied by its MPC controller.

Measured Disturbance

In the topology of the regional network, the whole urban network is simplified as one aggregated demand. Measured disturbance in every optimization process for a regional network should be the sum of the related demand at every prediction horizon (here is 24 h), i.e.,

$$M_s(k) = \sum_{m=1}^{24} d_t(k, m), \tag{19.21}$$

where $d_t(k, m)$ is demand vector at the urban delivery network corresponding to the k-th day.

Thus, $M_s(k)$ is considered as the demand for the regional network $d_s(k)$:

$$d_s(k) = M_s(k). \tag{19.22}$$

Target Constraints

The goal for the temporal coordination algorithm is transferring management policies from the regional network to the urban network. In order to achieve this coordination, the following constraint is added to the MPC in urban network:

$$\sum_{m=1}^{24} u_s(k, m) \leq T_d(k), \tag{19.23}$$

where u_s is the shared control vector between regional and urban networks.

This constraint is introduced in order to enforce that the amount of water decided to be transferred from the regional to the urban networks by the regional MPC is respected by the urban MPC. Without such a constraint, the urban MPC would decide the amount of water ignoring the regional MPC policy. This coordination policy is shown in Fig. 19.1.

Algorithm 19.1 shows how the constraint in (19.23) that establishes a daily limitation is generated and adapted at every time iteration of the urban MPC that operates at a hourly scale. Algorithm 19.1 takes into account the following facts when generating the constraint (19.23):

- After the application of n hourly control actions u_s corresponding to the kth day, the total remaining water for this day will be $T_d(k) - \sum_{m=1}^{n} u_s(m)$.

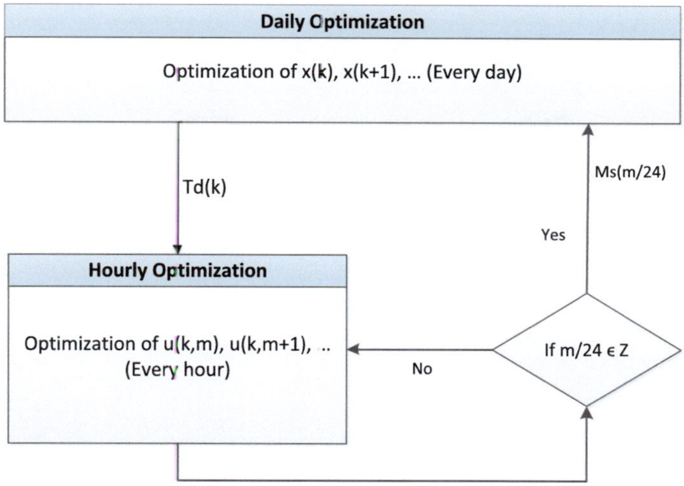

Fig. 19.1 Optimizations of multi-layer MPC

- When limiting the control actions in the prediction horizon L, there is a part of control actions $u(m)$ that corresponds to hours of the current day k that should be limited by $T_d(k)$, while the control actions correspond to hours of the next day $k+1$ that should be limited by $T_d(k) - \sum_{m=1}^{n} u(m)$.
- The generated constraints are added as additional constraints of the optimization problem associated with the urban MPC.

19.4 Results

19.4.1 Case Study: The Catalonia Regional Water Network

The Catalonia Regional Water Network, including the Catalonia Inland Basisns, is shown in Fig. 19.2. This network supplies the metropolitan area of Barcelona where most of the population of the region is concentrated. It is composed by river *Llobregat*, *Ter* and the related components. According to the functional definitions of regional and urban networks, the two rivers *Llobregat* and *Ter*, and all the connected elements compose the regional network while the urban network lies inside the centre part representing in an aggregate manner the topology of the water network of the metropolitan area.

Results are produced after applying the proposed centralized MPC and the coordination scheme to the Catalonia Regional Water Network for validation and comparison.

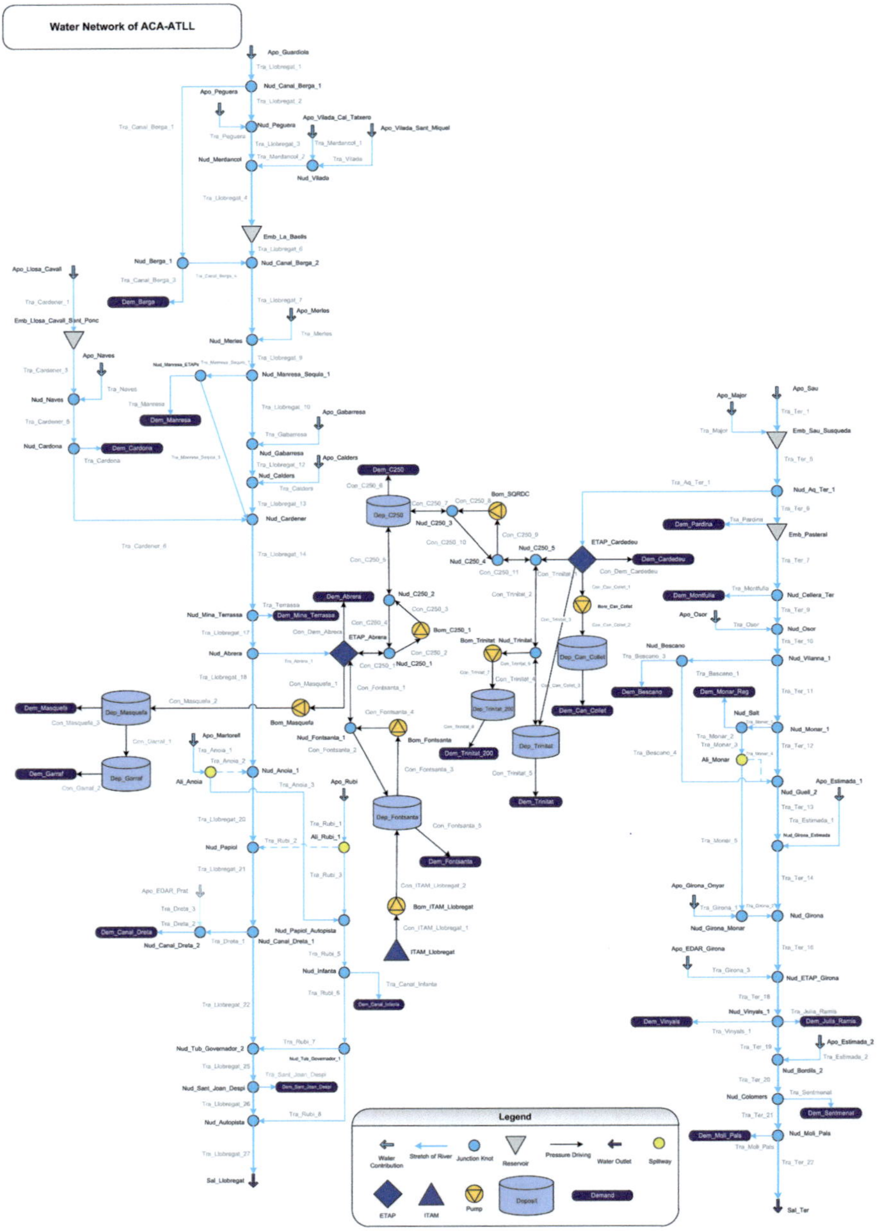

Fig. 19.2 Aggregate diagram of Catalonia Regional Water Network

Algorithm 19.1 Temporal multi-level coordinator

$L := 24$ h
$I := 24N$ h
$T_s := 1$ h
{start creating new constraints for urban delivery BOP }
for $i := 1$ to I **do**
  $d := floor(i/24)$
  $t := rem(i, 24)$
  if $t == 0$ **then**
    Update BOP by adding the following constraints:
    $u(1|k) \leq T_d(d) - \sum_{j=i-L+1}^{i-1} u_s(j|k);$
    $\sum_{j=2}^{L} u(j|k) \leq T_d(d+1);$
  end if
  if $t == 1$ **then**
    Update BOP by adding the following constraints:
    $\sum_{j=1}^{L} u(j|k) \leq T_d(d+1);$
  end if
  if $t == 2$ **then**
    Update BOP by adding the following constraints:
    $\sum_{j=1}^{L-1} u(j|k) \leq T_d(d+1);$
    $u(L|k) \leq T_d(d+2);$
  end if
  if $t \geq 3$ **then**
    Update BOP by adding the following constraints:
    $\sum_{j=1}^{L-t+1} u(j|k) \leq T_d(d+1) - \sum_{j=i-L+1}^{i-1} u_s(j|k);$
    $\sum_{j=L-t+2}^{L} u(j|k) \leq T_d(d+2);$
  end if
  Solve BOP to obtain $u(j|k), u(j+1|k), \ldots$ with the new constraints added
  $u_s(i|k) := u(1|k);$
end for
{end of $'i'$ loop}

19.4.2 MPC Results for Regional Supply Network

According to the environment conservation management for the regional supply network, ecological water volumes should be maintained in both rivers.

Figure 19.3 is one example of river reach. This plot shows that after being optimized by MPC, water flow at this reach could meet the ecological objective during the whole optimization process.

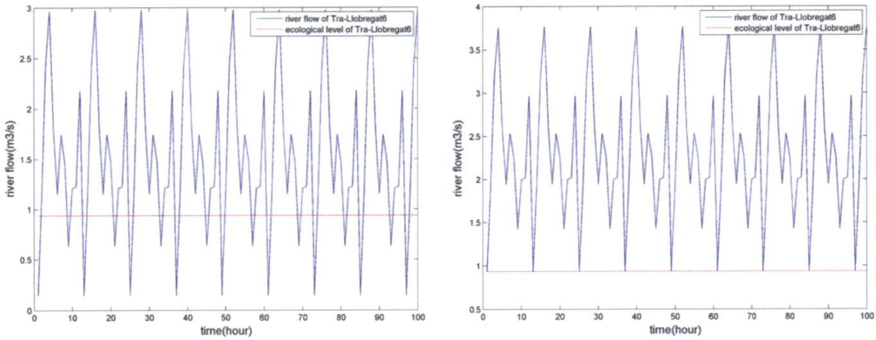

(a) before (b) after

Fig. 19.3 River flow comparing with ecological volume before and after ecological control in river Llobregat

19.4.3 MPC Results for Urban Delivery Network

In the urban delivery network, water transportation implies electricity costs when pumping water from lower elevation to a higher elevation. Figure 19.4 shows in the same plot the pump flow and the electricity tariff. It can be noticed that pump sends more water to the reservoir at the lower price period and less or no water at the higher price period.

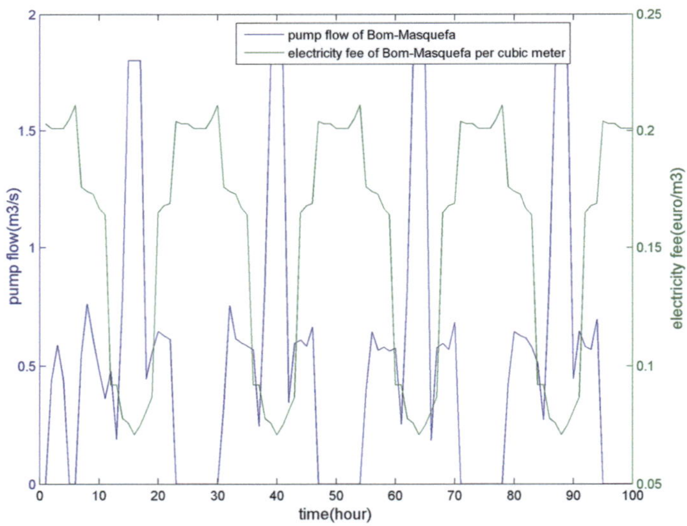

Fig. 19.4 Pump flow with electricity price

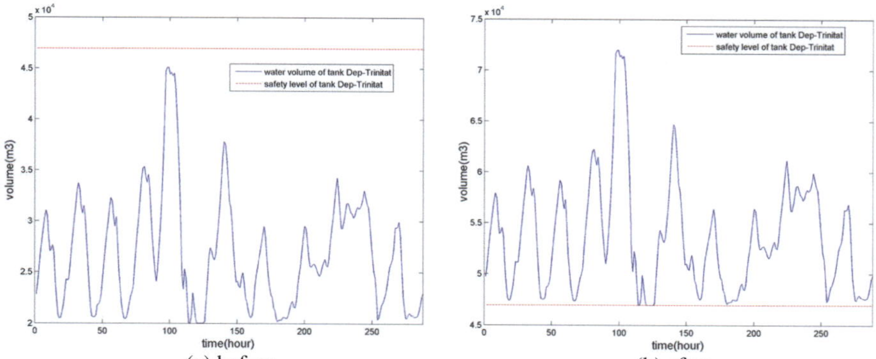

(a) before (b) after

Fig. 19.5 Water volume of tank Dep-Masquefa before and after safety control starts from the date of 01/08/2011

For the safety control objective in urban delivery network, Fig. 19.5 shows water volume of one tank $Dep\_Trinitat$ compared with its safety volume before and after the safety control, which fits the safety objective of MPC in urban delivery network.

19.4.4 Results of Centralized MPC

The centralized MPC optimizes the complex water network as a whole, which can obtain optimal strategies to fit objectives in both supply and delivery parts.

Table 19.1 provides detailed results and also the improvement of water usages in the two rivers achieved by balance management in the proposed centralized MPC scheme. In this table, *S.* means outside sources flow into rivers and *F. Demand* means fixed demands which cannot choose water source, while *V. Demand* is the demand which can receive water from more than one river. *B. Demand* is water volume that has been consumed from each of the reservoirs, and *B. Proportion* is the proportion of

Table 19.1 Balancing comparison of Catalonia Regional Water Network

Sc.	After centralized MPC control						
Es.	S.	F. demand	V. demand	B. demand	B. proportion (%)	R. proportion (%)	Supply ability (days)
L.	3008	2981	724	697	58.93	53.48	242
T.	3532	3518	1196	1182			
Sc.	*Before centralized MPC control*						
L.	3008	2981	7.6	−19.4	−1.02	53.48	177
T.	3532	3518	1914	1900			

B. Demand for the two reservoirs. *R. Proportion* is the proportion of storage capacities of the two reservoirs. The similar values for *B. Proportion* and *R. Proportion* are what the centralized MPC scheme wants to reach. And *Supply Ability* is a measure in days of how long the network might supply the demand needs, considering the case of no additional rainwater or other input. The comparisons show that after using this centralized MPC scheme, the proportion of water usage from two rivers (58.93%, which is ratio of Llobregat/Ter) is much closer to the proportion of storage capacities (53.48%). In this case, the Catalonia Regional Water Network can supply water 65 days longer than in the case without balance management, which represents an important benefit regarding the sustainable usage of water resource in the long-term perspective.

19.4.5 Results of Temporal Multi-level Coordination Scheme

Without coordination, MPC controllers in the regional supply and urban delivery networks are working separately in different timescales and control objectives. In order to manage the two controllers, the coordination scheme is applied.

In order to balance the regional supply network, water consumption in both rivers will be proportional to their supplying capacity. When coordination is used, this balance management goal will also be included in the urban delivery management problem.

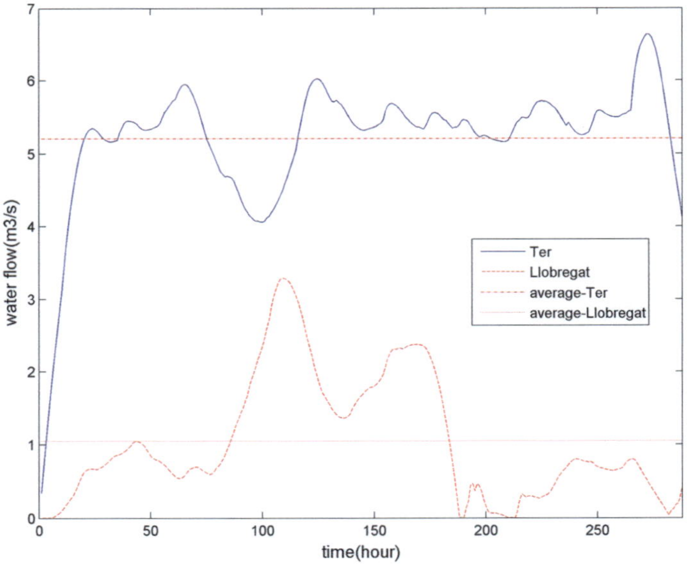

Fig. 19.6 Water consumed from two rivers by urban delivery network without coordination

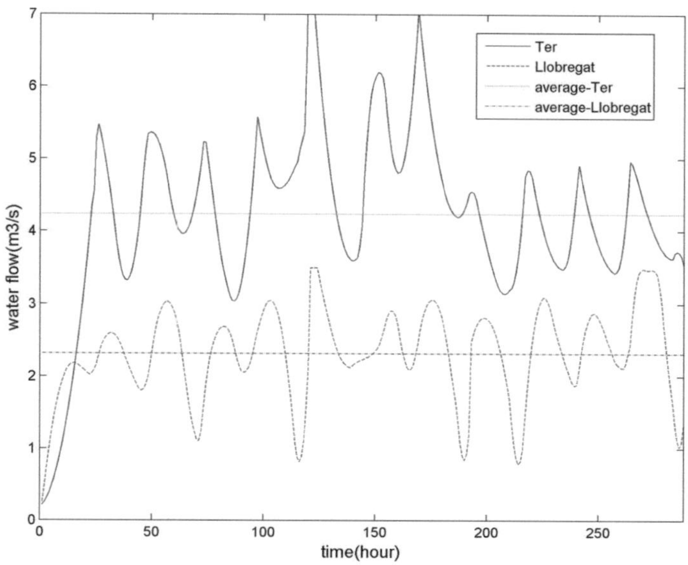

Fig. 19.7 Water consumed from two rivers by urban delivery network with coordination

Figures 19.6 and 19.7 show the amount of water consumed by the urban delivery network from different rivers with and without coordination, respectively. While both solutions meet the goals of the urban delivery network, the strategy using coordination achieves a subjective balance in the use of water of both rivers.

19.4.6 Comparisons Between Centralized MPC and Coordination Scheme

In order to operate the regional supply and urban delivery networks, which have different timescales and control objectives, centralized MPC and coordination scheme have been presented. The centralized MPC can optimize complex water network using one controller. However, this implies using the shortest time interval and the longest horizon for both problems.

The coordination scheme, in contrast, allows each system to be optimized separately with its appropriate time interval and horizon, but sharing more information.

The advantage of using coordination with MPC in these networks is apparent when considering that for a small optimality in the urban delivery management problem, a more convenient long-term strategy may be achieved.

19.5 Conclusions

In this chapter, two functional networks have been defined in a complex water network, and MPC has been presented to operate each of them. The need of partitioning definition of water network derives from the fact that different functional parts in the water network are modelled and designed separately according to different compositional dynamics and operational goals. In order to optimize water network completely, a centralized MPC scheme and a temporal multi-layer coordination scheme have been proposed to the complex water network that includes regional supply and urban delivery networks. The use of the centralized MPC techniques and coordination scheme makes possible to manage the complete water network in order to let individual operational goals affect to each other and, finally, obtain control strategies that can effectively consider objectives in both functional parts as well. Results of different operational goals show the advance of MPC for managing water network in the requirement of reality. Comparisons between centralized MPC and coordination scheme provide their usage in different situations. Improvements and limitations of MPC application have been discussed after comparing with the current control based on operators' experience.

References

1. Rani D, Moreira MM (2010) Simulation-optimizaiton modeling: a survey and potential application in reservoir systems operation. Water Resour Manage 24(1):1107–1138
2. Brdys M, Ulanicki B (1994) Operational control of water systems: structures, algorithms and applications. Prentice-Hall, Upper Saddle River
3. Litrico X, Fromion V (2004) Simplified modeling of irrigation canals for controller design. J Irrig Drain Eng 130(5):373–383
4. Evans R, Li L, Mareels I, Okello N, Pham M, Qiu W, Saleem SK (2011) Real-time optimal control of river basin networks. In: Preprints of the 18th IFAC world congress, Milano, Italy, pp 11459–11464
5. Toro R, Ocampo-Martinez C, Logist E, Van Impe J, Puig V (2011) Tuning of predictive controllers for drinking water networked systems. In: Proceedings of 18th IFAC world congress, Milano, Italy
6. Ocampo-Martinez C, Puig V, Cembrano G, Quevedo J (2013) Application of predictive control strategies to the management of complex networks in the urban water cycle. IEEE Control Syst Mag 33(1):15–45
7. Sun CC, Puig V, Cembrano G (2014) Temporal multi-level coordination techniques oriented to regional water networks: application to the Catalunya case study. J Hydroinform 16(4):952–970

Chapter 20
Big Data Analytics and Knowledge Discovery Applied to Automatic Meter Readers

Diego Garcia, Vicenç Puig, Joseba Quevedo and Miquel Angel Cugueró

20.1 Introduction

The ageing of water networks combined with an increase of the demand, caused by the growth of the population and the corresponding rise of cities urbanization, challenges the control of water distribution infrastructures. One of the main challenges is the inability to detect anomalies caused by bursts, leaks, water loss, and unaccounted activity, among other factors; additional challenge is presented when aiming to detect these anomalies in real time [15]. Water network monitoring helps not only to ensure the adequate water distribution to the final user, but also to ensure sustainability by reducing water loss at different stages in the production and pumping. Given this need, supervisory control and data acquisition (SCADA) systems have become the foundation for water utilities (WU), allowing constant data collection from crucial points in the network.

The volume of data collected by a WU is constantly growing. In this new era, data are important because guarantees the success of decisions based on the relevant values and underlying information extracted from raw data [4]. Hence, WU are adopting Big Data technologies and have started building a data strategy [8]. Even more important than obtaining precision from the data analysed is to strive for data relevance.

This data growth is motivated by different causes, e.g., the continuous evolution of the technology, with the appearance of cheaper monitoring devices, allowing the WU to monitor and observe new and more detailed information about the processes done by the company. Thus, the spread of monitoring devices along water networks or cities, such as the automatic meter reading (AMR) sensors, allows to automatically collect consumption data from water/energy metering devices (e.g., gas and electric) into a database for further analysis, billing and troubleshooting detection.

D. Garcia · V. Puig (✉) · J. Quevedo · M.A. Cugueró
Research Center Supervision, Safety and Automatic Control (CS2AC-UPC), Terrassa, Spain
e-mail: vicenc.puig@upc.edu

V. Puig et al. (eds.), *Real-Time Monitoring and Operational Control of Drinking-Water Systems*, Advances in Industrial Control,
DOI 10.1007/978-3-319-50751-4_20

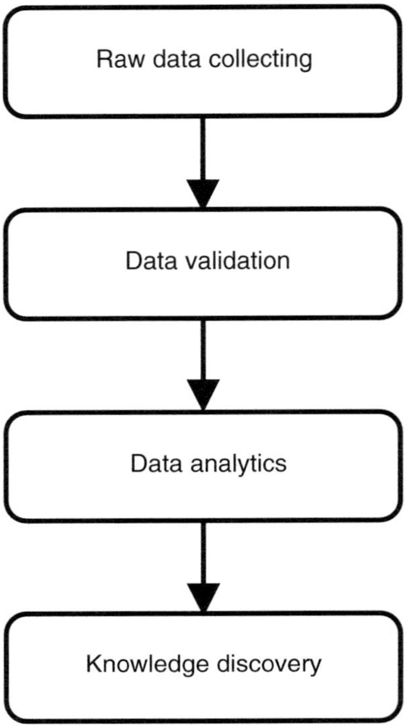

Fig. 20.1 General methodology proposed for Big Data analytics problems

Traditional tools for data collecting and processing are overwhelmed by the increasing data volume, data variety and the corresponding processing speed required by the WU systems to obtain results and a better insight. Indeed, these are the typical three degrees of freedom handled by the Big Data approach, which emerged after the new requirements of the industry. Figure 20.1 shows a general approach for the Big Data analytics problem.

This approach is divided in four stages. First, the *raw data collecting stage* handles the data gathering from multiple sources with multiple formats. Traditionally, this task is achieved by a centralized database (e.g., MySQL or Oracle), but the datasets' size and required processing speed are bottlenecks when millions of sensors are sending high-frequency data. Moreover, these databases are strong structured: data types are defined during the data model building process. In contrast, new database paradigms such a NoSQL (not only relational databases) and distributed file systems (DFS) allow to handle these limitations. For instance, NoSQL schema-free databases are capable of managing multiple formats and adapt existing data models to new types in run-time. This approach is based on a distributed architecture; hence, it is horizontally scalable and therefore capable to adapt to any growth of the dataset by adding more resources.

The second stage in the Big Data analytics problem is the *data validation stage, which guarantees* a certain grade of data reliability. Results or conclusions obtained from noisy or inconsistent data may be invalid and lead to take wrong actions. Hence, a data validation strategy is crucial to validate the data collected in order to improve next-stage performance. A monitoring system can be affected by certain problems, which may have impact in the communication system or in the sensors. On the one hand, if a problem affects the communication system module, data gaps will appear in the database. In this case, reconstruction and imputation methods are required to fill the missing data. On the other hand, there exist a wide variety of problems that impact the sensors' data accuracy and reliability. Noise and outliers are quite common problems affecting these elements. Depending on the sensor complexity, the way to face these problems may differ, e.g., an outlier can be easily detected in a temperature sensor by means of trend analysis, but a complex uncalibrated chemical water quality sensor may require different attention. Also, wrong sensor electrical connections can generate accurate but totally unreliable data.

The third stage in the Big Data analytics problem is the *data analytics, which is the process* that turns data into real values. After raw data are validated in the previous stage, here different tools allow their analysis and feature extraction. The exploratory analysis is the first step to have a deeper understanding of the data under study. At this point, it is important to have some domain expertise or a domain expert in order to understand and validate some given hypothesis, which must be chosen carefully in order to make the data analysis reliable. After the data are understood, there is a model building process in order to make the conversion from historical data to understanding. The performance of the obtained model must be evaluated and compared between suitable models in order to obtain the best model. At this point, predictive modelling will use the model provided in order to forecast future data points. An example of application of this technique is the water demand forecast in water distribution network (WDN): it is crucial for the WU to anticipate and adequate the WDN to satisfy any demand or even an eventual emergency demand which may occur, in order to achieve correct operation of the WDN.

The final stage in the Big Data analytics problem is the *knowledge discovery* [16], *where knowledge is discovered or derived* from existing information. We try to interpret the patterns and models obtained from the previous stage to describe and discover known or hidden relations between variables. The translated useful patterns must be understandable by the stakeholder in order to provide information to take better decisions.

A WU usually collects a wide variety of data types. Data collected from the exploitation of the WDN is required for the operation to supply water to the citizens, councils and industries. These data come from different sensors, such as level meters installed in tanks, flow and pressure meters installed in pipes or flow data from pumps, and valve status data. These operational data is usually handled by a SCADA system.

Also, quality data are collected to supply potable water to the citizens and accomplish the regulations, established by the government and additional regulators. These data are collected from analysers and multi-parameter sensors to measure a variety of parameters, such as chlorine, conductivity, pH, turbidity and total organic carbon (TOC). The WU also takes observations directly from the WDS and analyse them in laboratories.

The supplied water is charged to the customers based on the water used, i.e., water metering. This process to measure the use by the residential and commercial building is done using water meters. These readings are taken from the water meters installed in the supply point, usually by operators every two or three months (or even less, depending on the country). The data collected are important to bill but also to understand water demand patterns or to analyse different billing systems. Nowadays, new systems such as AMR allow to register these readings remotely and with higher frequency (e.g., hourly for residential and 15 min for industries).

Data associated with the customer and the water supply installation are important cross-sectional data to the WU. Some of the data fields collected are personal information, tariff associated, water meter model and water meter age.

Other systems generating data are the sewerage infrastructures. These systems convey the sewage to a sewage treatment plant (or a discharge point). These systems must also accomplish the regulations established by the regulators. There are sensors installed in these systems to monitor the sewage loads and avoid flooding episodes. The sensors usually used in these systems are limnimeters, rain gauges (to detect rain and flooding episodes), and actuators' state (pumps and valves).

Data collected are not only important for the WU, but also important for the customer. Giving a feedback to the customer improves the quality of the service. For instance, AMR data offer households and businesses the chance to understand and reduce their energy and water consumption much than ever before, when meter readings were taken every two months, quarterly, or even annually.

Moreover, AMR could help utility firms to improve the accuracy of billing and cut visits to properties to read meters. However, with AMR, there is an exponential growth of data: a single AMR produces 17500 readings per year, with a single reading every half hour. These data should be first processed in real-time streaming, in order to be validated before being stored and translated into a metadata model which may be usable in multiple further applications. Thus, utilities have found scaling smart meter management systems difficult to handle. This motivates the use of Big Data technologies in this application domain. On the other hand, applying data analytics and knowledge discovery tools to AMR data combined with other streams of information (data coming from the billing system, call centre service and meteorological information) could help with fraud detection, maintenance requirements prediction, water/energy user consumption patterns determination and response generation to variations in the demand.

This chapter presents novel algorithms and methodologies to carry out real-time streaming data processing, data analytics, data quality assessment and improvement, as well as prediction and visualization tasks, at extremely large scale and with diverse structured and unstructured data from multiple sources such as water, power, telecommunication and other utilities, as well as from social media. The algorithms and methodologies will be illustrated using real data coming from several WU.

20.2 Problem Statement

The AMR meters deployed by the utilities in households, industries, councils, etc., measure water consumption at regular time intervals. The sampling interval is determined by the customer's category and constrained by the AMR battery life. But not all the customers have the same needs. For instance, a household does not consume water in the same way as a paper company does, which needs intensive water use in the industrial processes involved. Hence, big consumers require more granularity and control to manage its own water consumption than households. One observation every 15 min is a common sampling interval for big consumers, while hourly is sufficient for households. It must be noted that the quantity of households handled by a utility is far higher than big consumers, and therefore, the costs, maintenance and replacement of the corresponding AMR batteries must be considered carefully.

First of all, the data from the AMR meters from a single or multiple deployments (i.e., sectors and subsectors) are collected. A concentrator gathers the data from a group of nearby AMR meters. Then, the data are sent from the concentrators to the data centre via some telecommunication system (e.g., GPRS). However, the AMR meters installed for industrial customers are different from the ones used in households: an AMR meter for an industrial customer sends the data directly to the data centre.

The data gathered can be stored in traditional relational database management system (RDMS) or files (unformatted, formatted or binary) and then fetched by Extract, Transform and Load (ETL) processes to support business applications.

Traditionally, the different business units of a company had their own custom systems storing different data formats. In fact, this is done in many companies nowadays. ETL processes follow the next procedure to allow cross-data from heterogeneous storage systems: extract the data from these systems and then transform these data in a consolidated format to load them into the end target system, usually a data warehouse.

For instance, descriptive indicators can be obtained from water meter data to summarize and analyse these data, such as the average daily consumption or the minimum night flow, or advanced data analytics such as water demand forecast or customer segmentation by their behaviour from the historical consumption records. But at this point, the first problem might be faced: the data volume.

A two-year dataset from 50 thousand households has a size around 14 gigabytes. A city such as Alicante has more than 300,000 households and Barcelona more than 1.6 millions of households; the datasets with full AMR coverage of both cities will imply volumes around 80 and 450 gigabytes, respectively. Processing a data volume over the main memory capacity is not affordable by traditional data analytic tools in a common computer. A naive solution could be increasing the main amount of memory, but this is bounded by the computer maximum capacity. Hence, this approach, known as vertical scaling, is not a general solution for arbitrarily high volumes of data.

Suppose that unlimited memory to the used computer can be added. Then, it would be possible to fit and process any dataset in the main memory. However, the processing time of a large volume of data would totally rely on the number of cores available in the computer considered. Thus, the current bottleneck is the computation resources. Here, the second problem should be faced: the velocity.

In contrast, Big Data,[1] supports horizontal scaling and provides different mechanisms to process large volumes of data in a reasonable time [17]. This allows to add computational and storage resources as the volume and time restrictions of the problem to be solved evolve along the project life. Besides, vertical scaling is more expensive than horizontal scaling, due to the cost of the high-end technology required by the first one, in contrast to the cheaper commodity hardware used by horizontal scaling.

Moreover, the raw data collected do not give valuable insights by themselves. In addition, raw data can be affected by different problems such as noise and unreliable events that may result to wrong conclusions. AMR data are affected by gaps, missing data due to the problems in the communication system, the AMR meter or a battery fault, and they can also contain outliers generated by the sensor or at any point during the transfer. Raw data can also include irregular sampling intervals that must be taken into account when applying analytical processes. These irregular sampling intervals may be caused by the noise introduced deliberately to avoid a collapse of the communication system.

After the raw data are cleaned and validated [6], the database is ready for indicators and knowledge extraction, and the conclusions obtained will drive to decisions based on reliable data. Understanding how the customers are consuming the water is crucial to the WU for WDN efficient operation, as well as to understand and foresee the evolution of the different water demand patterns to fit the assets and guarantee the service [1, 13] while reducing operational costs.

Summing up, the key challenges handled by Big Data are volume, variety and velocity, also known as the 3Vs key challenges. These are highlighted in [10] where the data management weaknesses identified in the industry, research centres and governments are collected. As commented in this section, managing the 3Vs key challenges requires two main components: a data storage system, to handle large data volumes and data variety, and a data processing framework, to query and apply operations over the large datasets within an acceptable response time. Further details of these components are discussed next.

20.2.1 Data Storage

Data storage allows to handle data volume and data variety, but also data access velocity is important when large volumes of data have to be processed.

Firstly, the different storage devices from a hardware point of view are going to be introduced, and finally, software systems such as databases and distributed storage systems are going to be also introduced.

The storage system of a computer is hierarchical. Figure 20.2 shows a top-down classification from the velocity point of view, with the fastest memory on the top and

[1] The definition of Big Data by the Oxford English Dictionary (OED) is "Computing data of a quite large size, typically to the extent that its manipulation and management present significant logistical challenges; (also) the branch of computing involving such data".

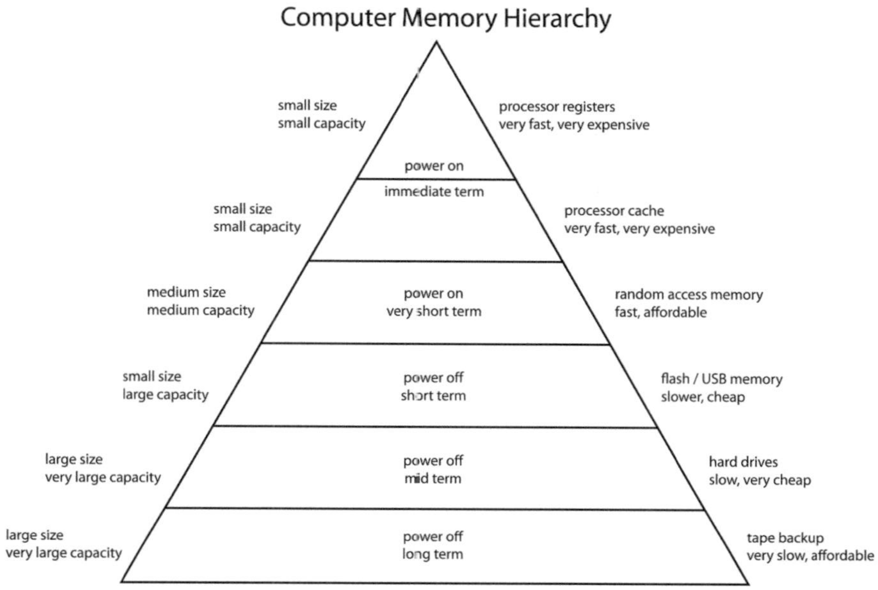

Fig. 20.2 Memory hierarchy (image source: Wikipedia at https://en.wikipedia.org/wiki/Memory_hierarchy)

the slowest in the bottom. The most popular memory type is the secondary memory. This memory is used for large size and midterm saving, since if the system power vanishes, no data are lost. The most commonly used memory devices of this kind are the hard disk drives (HDD), which are based on mechanical moving parts. These have been removed in the newer solid state disk (SDD), which have not any moving part.

Due to the mechanical components, HDD are slower than SDD. The data reading rate of the HDD is in the range of 50 to 150 MB/s, depending on the device quality and the interface. Alternatively, the SDD is able to achieve a reading data rate between 200 MB/s and 550 MB/s.

Alternatively, main memory, also called random-access memory (RAM), has higher data access speeds, with reading data rates between 2 GB/s to 30 GB/s. This type of memory has an important role in data analytics and fast data storage systems such as in-memory databases. Its main important drawback is that it is volatile; hence, data are lost when device power vanishes. RAM size limit in 32-bit systems is of 4 GB, whether is of 8 TB in 64-bit systems.

Finally, cache and registers, with shorter sizes, are closer to the central process unit (CPU). These memory types are also volatile, and their data reading rates are some orders of magnitude higher than their previous counterparts.

Conversely, Relational Database Management Systems (RDBMS) have been the main systems used to store data and, nowadays, are supporting a quite large number of applications, but when it comes to handle the exponential increase of data volume and variety, these kinds of systems are insufficient. RDBMS are centralized systems based on structured data constraints. Thus, unstructured data are not handled by these kinds of systems. Also, the data volume capacity is handled vertically, and thus, adding resources are limited and expensive.

The traditional and general approach in many companies to extract knowledge from data gathered in multiple operational systems is to establish a *data warehouse*, by means of ETL. ETL process is required to obtain a big picture of the company and to take data-driven decisions based on real information (fundamental principles of the *Business Intelligence*).

The *extraction* stage collects the relevant data subset (since putting all the data together is often unaffordable) from homogeneous or heterogeneous data sources, such as:

- Customer relationship management database (CRM).
- Sensors' raw data handled by SCADA systems.
- Financial and billing data.
- Operational incidence database.

Usually, each operational database has a different data model, type and notation. Hence, a *transformation* stage converts each format in a proper format or structure for query and analysis purposes. Finally, a *loading* stage stores the date into the final target, usually a data warehouse.

Distributed file systems and not only SQL (NoSQL) solutions emerged to overcome these shortcomings: to scale up and to handle data variety. A distributed file system allows to store files in any format and size. Files are divided in blocks in order to be accessed concurrently and replicated to be fault tolerant (e.g., in the event of some node crashing). Some popular examples of these architectures are as follows:

- HDFS.
- Google GFS.

Table 20.1 NoSQL database systems

NoSQL	Data model	Written in	Protocol	Licence
Cassandra	Column	Java	CQL3	Apache
MongoDB	Document	C++	BSON	AGPL
CouchDB	Document	Erlang	HTTP/REST	Apache
HBase	Column	Java	HTTP/REST	Apache
Redis	Key-value	C	Telnet	BSD
Accumulo	Document	Java and C++	Thrift	Apache
Hypertable	Column	C++	Thrift and C++ library	GPL 2.0
OrientDB	Graph	Java	Java and HTTP/REST	Apache
Neo4j	Graph	Java	Java API and HTTP/REST	GPL
Couchbase	Document	Erlang and C	memcached	Apache
Scalaris	Key-value	Erlang	JSON-RPC	Apache
Riak	Key-value	Erlang and C	HTTP/REST	Apache
Kyoto Tycoon	Key-value	C++	HTTP/REST	GPL

- Amazon S3.
- Tachyon (in-memory).

The main difference of NoSQL databases and traditional relational databases is that NoSQL is based on different data models besides the relational one. Some popular NoSQL examples and its underlaying data model are listed in Table 20.1.

20.2.2 Data Processing

The principal aim of data processing is to extract knowledge from the raw data to achieve the goals proposed in further stages. In the WU scope, such goals are localization of leakages, detection of malfunctions, demand prediction, fraud detection, etc. From the raw data to any of these goals achievement, there are multiple steps to be performed.

From a technical point of view, data processing requires certain computing resources working together and led by a set of instructions to achieve a final goal. These instructions are specified by means of a programming language, and there are more than 600 different programming languages. Not all of them are equally suitable for doing the same task, since in their evolution each programming language has been designed to improve certain particular aspects. Some of them, for example, are more suitable to develop data processing tasks. The most popular tools and languages to analyse and process data are MATLAB, R, Weka (a Java Machine Learning framework), Python and Scala. These tools are designed to allow the user to focus on the data and avoid further computer's technical challenges. But when the datasets are large and cannot be allocated in the main memory, these tools are not able to process them in an acceptable time. Then, several approaches, such as the Big Data techniques nowadays, allow to handle large datasets in a distributed fashion in order to allow the expansion of the computing resources used to face this problem.

Indeed, Big Data is not trying to solve a new problem: the analysis of large data volumes to extract knowledge has been carried out since long time ago. High-performance computing (HPC) was introduced in the 1960s [14]. HPC is the term to describe the use of supercomputers and parallel processing techniques to solve complex computational problems, e.g., modelling, simulation and analysis that cannot be practically addressed by conventional computers. In order to use a supercomputer by means of HPC, high programming and network skills are required. Most HPC applications are written in Fortran, C or C++, with the aid of Message Passing Interface (MPI) libraries to scale up and parallelize tasks. In addition, application authors must manage communications, synchronization, I/O, debugging, etc. Big Data tries to simplify these tasks with a new programming model called *MapReduce*. Many industries have the need and the interest to embrace and support this new paradigm to overcome the limitations of traditional tools. These industries include

- Search engines (e.g., Google and Yahoo).
- Social networks (e.g., Facebook, Twitter and Linkdln).

- Client-oriented services such as recommender systems (e.g., Amazon, Ooyala, Spotify and Netflix).
- Wearables and mobile technologies (e.g., smartphones, tablets, clocks, bracelets and clothes).
- Banks (e.g., credit approval and fraud).
- Governments (e.g., thread prediction and prevention, tax compliance, crime prediction and prevention).

This data-driven approach has also arrived to the water sector. Companies such as Takadu [2] prove the data-driven success with profits to the water utilities by means of reducing the operational costs. Furthermore, MapReduce [5] was created from Google, to process large datasets with a parallel and scalable algorithm on a cluster of commodity hardware. This new programming model can take advantage of locality of data, processing them on the storage assets in order to reduce time latencies generated by the movement of data through the network. Basically, this programming model is based on the following three stages:

1. *Map stage* splits the input data into a sequence of *key–value* pairs. The key-value pairs are created by the user-defined function passed to the map task.
2. *Shuffle stage* sorts the key-value pairs from the previous stage by key, divides them again and passes them to the next stage.
3. *Reduce stage* aggregates/combines the key-value pairs associated to the same key. The way to combine the pairs is defined in the user-defined function passed to the reduce task.

This programming model allows to process large batch-oriented data volumes (a.k.a data-at-rest). The first version of MapReduce is batch oriented because it writes each output of each stage to the disk, which has meaningful impact on disk access (input/output operations) performance. The situation is even worse regarding intensive iterative algorithms that needs to make several passes over the whole dataset (e.g., neural networks can require thousands of passes in the training stage). The highly heterogeneous data access rates affects drastically the computation time of the final result.

Furthermore, MapReduce programming model is fault tolerant. The map and reduce tasks are dispatched by a director to different computers in order to parallelize them, but if any task fails due to a node failure (e.g., communication or disk failure), the uncompleted task is derived to another node.

An important drawback of this programming model is that it also requires advanced programming skills to be implemented. Thus, new programming models have emerged to address this drawback. For instance, Hive and Pig are SQL-like languages to perform queries and advanced data pipelines in an easier fashion. In Table 20.2, some of these Big Data frameworks are listed.

In many cases, the batch-oriented approach does not allow to face applications where real-time queries are required (e.g., stock market and fraud). Hence, new data streaming processing frameworks emerge to be able to process continuous input data streams. For instance, Apache Storm is an open-source streaming framework, used

Table 20.2 Big data processing frameworks

Framework	Batch	Real time
Hadoop MapReduce	Disk-based	No
Hive	Disk-based	No
Pig	Disk-based	No
Apache storm	No	streaming
Apache spark	Disk-based and in-memory	In-memory and streaming

by some important companies such as Twitter to process massive input feeds. This approach allows to process each input data stream in real time.

An improved version of the programming model MapReduce has been performed by Algorithms, Machines and People Laboratory (AMPLab) from the University of Berkeley. The project is called Apache Spark [12]. It is an open-source cluster computing framework that aims to make intensive data analytics faster. This framework provides in-memory cluster computing, which is much quicker than disk-based systems (see Sect. 20.2.1) such as the original Hadoop MapReduce. In addition, Apache Spark provides a data streaming processing engine to process input data streams in real time.

20.3 Proposed Approach

In this section, it is proposed an approach to predict water demand and detect suspicious behaviours— e.g., a leak, a smart meter breakdown or even a fraud—by extracting water consumption patterns [7] and applying an unsupervised cluster analysis. Moreover, a software framework, based on Big Data techniques, implements this approach to tackle the barriers, mentioned in Sect. 20.2, of the traditional data storage and analysis.

Analogously, to the four-stage scheme detailed in Sect. 20.1, it is defined an approach based on the processes detailed next. At an early stage of this process, the historical AMR raw data are transferred from the WU into a distributed file system HDFS. Then, a *preprocessing* process is performed to align irregular sampling instants to sharp o'clock hours and fill some missing readings. After, a *filtering* process filters relevant information and discards useless data (e.g., empty households with none consumption), followed by a *feature extraction* process, which transforms the time series high-dimensional data into a space of fewer dimensions. Finally, an *unsupervised cluster analysis* process extracts a set of groups, minimizing the distance between the members of the same group and maximizing the distance between groups.

Although each smart meter collects hourly sampled data, in practice the interval is not exactly an hour: an observed demand is registered at 10:46 and the next one at 11:49. Also, these readings are not aligned to sharp o'clock hours (i.e., 10:00, 11:00, etc.). Moreover, some readings are missing mainly due to communication

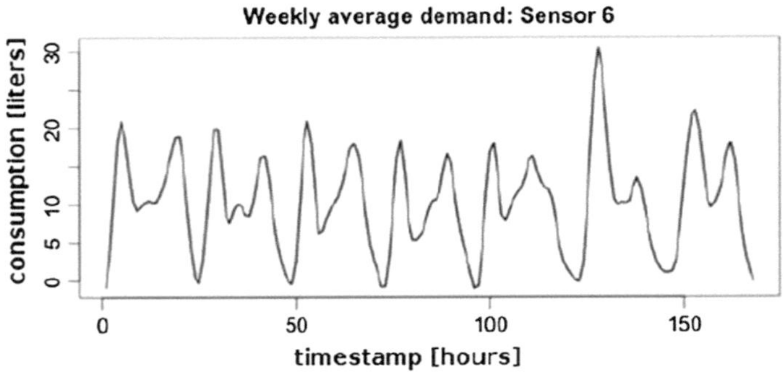

Fig. 20.3 Aggregated hourly average consumption pattern generated by the feature extraction process within a week period

problems, among others. In this work, it has been assumed the data collected is valid. The problem of data validation/reconstruction has already been addressed in [3, 6]. In this preprocessing stage, a linear interpolation is applied to homogenize the sampling interval and to align the sampling time to o'clock hours. When only less than 10 samples are missing, a simple linear interpolation is enough to provide a reliable reconstructed signal. Given a vector of observed demands of length N, $\mathbf{x} = (x_1, x_2, \ldots, x_N)$, first a linear interpolation method is applied obtaining $\hat{\mathbf{x}} = (\hat{x}_1, \hat{x}_2, \ldots, \hat{x}_p)$ with a regular sampling time.

Once applied the previous filter, the following statistical indicators are estimated over the hourly consumption vectors $\mathbf{z}$: maximum, minimum, mean and variance. These indicators are used to discard null smart meter consumption readings, probably occurring in uninhabited houses. In addition, smart meters with negative readings are also discarded, since backflow should not be occurring in the end points of district metered areas (DMAs).

The feature extraction stage aggregates a year dataset to a weekly pattern. Different techniques for representing and reducing the dimensionality of time series has been proposed in the literature [11], e.g., the Gaussian Mixture Models [13] for representing water demands. In this chapter, a feature vector $\omega_{\mathbf{i}} = (\omega_{i1}, \omega_{i2}, \ldots, \omega_{i168})$ of length 168 (i.e., the number of hours in a week period) represents the weekly pattern for a given smart meter i (see Fig. 20.3) where each component k is given by

$$\omega_{ik} = \frac{\sum_{h(j)=k} z_j}{M_k} \tag{20.1}$$

where M_k is the number of observations that satisfies $h(j) = k$, and $h(j)$ is the hour of the week with time stamp j.

The dataset considered here is composed by a set of AMR meter demand readings in an hourly sampled fashion. No additional information is available, such as the smart meter's diameter or the consumer activity.

The unsupervised clustering method applied in this chapter is the k-means [9]. This algorithm aims at partitioning n observations into k clusters in which each observation belongs to the nearest cluster. The number of clusters k is unknown, and therefore, it must be estimated previously. As will be seen in the application of this methodology (Sect. 20.4), first a cluster analysis indicator is obtained for a range of clusters, and then, the number of clusters whose indicator does not improve significantly when adding an extra cluster is selected.

We have designed and developed a framework to implement the approach presented to handle large volumes of data. In particular, it is proposed a scalable solution to adapt the resources to the problem requirements, in order to allow the application of the approach presented to a small town or even to a group of cities.

As introduced in Sect. 20.2, the volume of raw data generated by the AMR meters is too big to be analysed and stored by means of traditional technologies. Depending on the amount of data to process, ad hoc centralized traditional solutions could be satisfactorily applied in the short term. However, the increasing number of smart meters deployed each year is unsustainable to handle in the midterm and long term. Thus, it is proposed a framework based on Big Data technologies to achieve horizontal scalability without limiting the data volume. Furthermore, all the technologies involved here are open source, which substantially reduces the final implementation cost.

The framework presented here is composed by three main modules: The *storage system* module stores the raw data and is supported by Hadoop with HDFS. The *processing* module implements the methodology detailed before in this section, and it is based on a large-scale data processing engine called Spark. The preprocessing and feature extraction stages are based on built-in Spark functions, with the exception of the linear interpolator, provided by a numerical processing library called Breeze. The unsupervised cluster analysis module (i.e., the k-means algorithm) is provided by MLlib, a scalable machine learning libraryon top of Apache Spark, a general engine for large-scale data processing. Finally, the *results* module saves the output from the *processing* stage in a NoSQL distributed database (Apache Cassandra).

20.4 Results

In this section, the results obtained from Tarragona, Torremolinos and Alicante, three different locations from Spain, are presented. Figure 20.4 shows the three locations marked in a map of Spain. The data used are AMR data samples from the total population.

20.4.1 Tarragona

Two different datasets are considered here for the city of Tarragona, coming from residential and non-residential customers, respectively. The unsupervised clustering performed on each of these datasets is presented in this section. These results were helpful to determine the basic differences between these two separate groups of users.

Fig. 20.4 Case studies

The first dataset contains raw data gathered from 199 sensors of residential houses. 17 of these sensors are discarded as they do not comply with the following predefined quality requirements:

- The time series of the customer cannot have more than a 10% of missing data.
- The total consumption of the customer must be positive, to avoid empty houses.

The remaining set has 182 sensors, which will be used to obtain the corresponding demand patterns.

First, the number of groups is estimated using the within-group sum of squares measure (WSS). This is the sum of the squared differences between the centre of the cluster and its members. Hence, a set of compact groups will obtain a lower WSS than a set of groups with spread members. Finally, it is selected the number of clusters to which adding one more cluster does not provide a significant decrease of WSS.

Figure 20.5 shows the WSS plot, indicating that ten clusters are enough. And Fig. 20.6 shows the clustered patterns for the first dataset, with the cluster centre (weekly average) depicted as a solid black line. With the exception of cluster 10, whose consumption and maximum value are the highest within all the clusters, all them have more than two members. Most of the patterns show the standard morning/evening activity peaks which are the characteristic of domestic consumption and can be associated to the living habits and characteristics of the users, such as showering and cooking schedule, working week schedule, activity during the weekends and the number of inhabitants per household. At this point, no further information about the nature of the data or the sensors (e.g., activity declared by the customer,

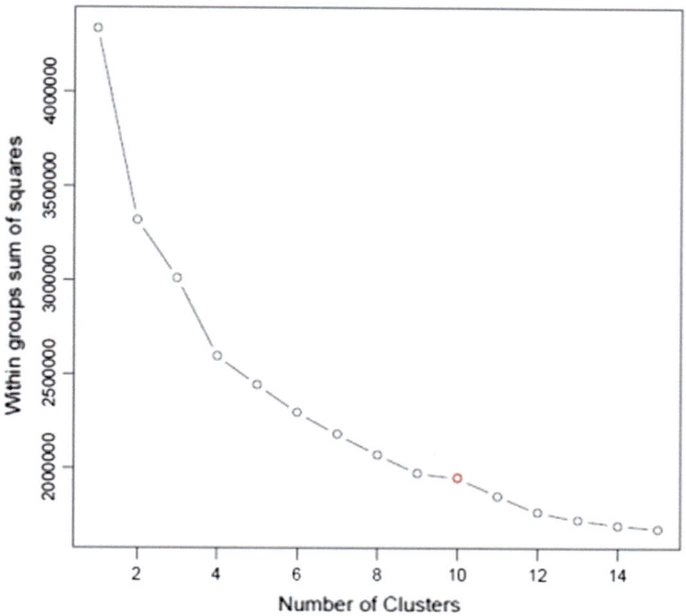

Fig. 20.5 WSS plot of the residential customers

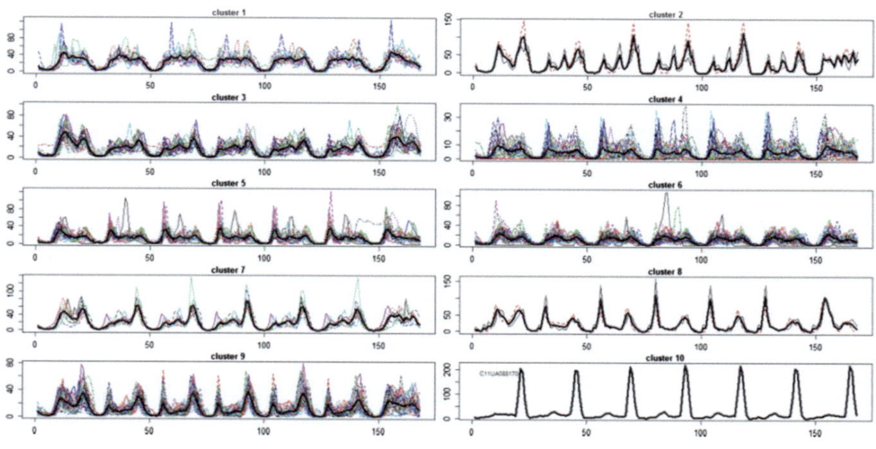

Fig. 20.6 Clustered demand patterns of Tarragona residential customers

address, sensor model, brand or calibre, among others) is included in the analysis. This will be discussed and analysed further in this chapter.

The second dataset contains raw data gathered from 110 sensors of industrial costumers. 40 of these sensors are discarded as they do not comply with the predefined quality requirements defined before. It was found that some of these had already been replaced due to malfunction in readings. This leaves a remaining set of 70 sensors,

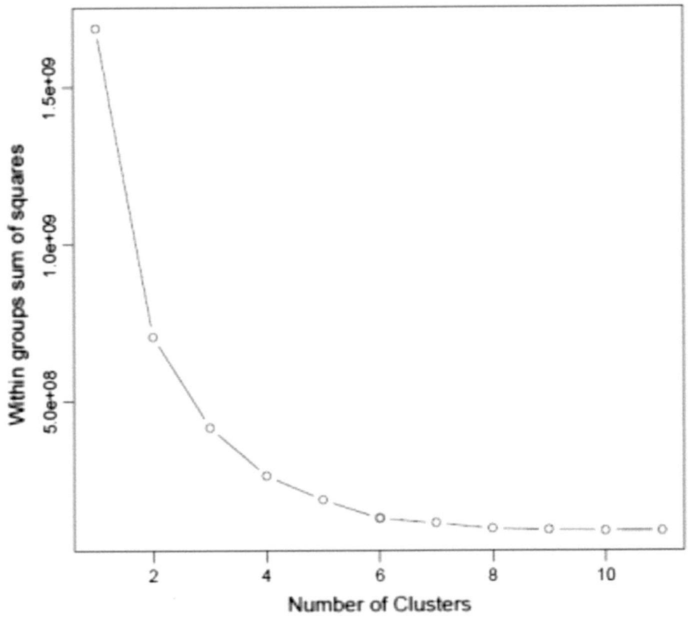

Fig. 20.7 WSS plot, non-residential customers

which can be used to obtain the corresponding demand patterns. The WSS plot for this dataset (Fig. 20.7) shows six clusters is enough in this case, because adding one more does not decrease the WSS. After the algorithm is applied (see Fig. 20.8), it is seen how almost all the sensors are classified in cluster 5, which has the lowest mean weekly demand. The remaining clusters do not contain more than six sensors.

All of these patterns show a quite high weekly consumption average, consistent with their registered use as industrial consumers, and with the exception of cluster 2, all have day-to-day activity patterns. Also, although cluster 1 shows a day-to-day basis consumption, these appear to be consistent with a part-time business, given the long inactive interval between working hours and the high average level of consumption. It also shows high consumption on Sundays. Similarly, cluster 6 also shows high consumption during the weekends, but with irregularities during the week. Cluster 2 shows uninterrupted use during the weekdays, which could correspond to a factory or business with reduced or null activity during the weekend. Cluster 4 shows a regular domestic water consumption behaviour, with morning/evening activity peaks. But this cluster shows a higher average consumption than a regular domestic consumption. This could be caused by a house with more persons than usual or an inefficient use due to the deterioration of the installation (excessive pressure, appliances inefficient appliances, etc.).

In this cluster analysis, the 81% of the sensors are classified in cluster 6 with a mean weekly demand of 54,4 l/h. The mean pattern shows regular to constant use

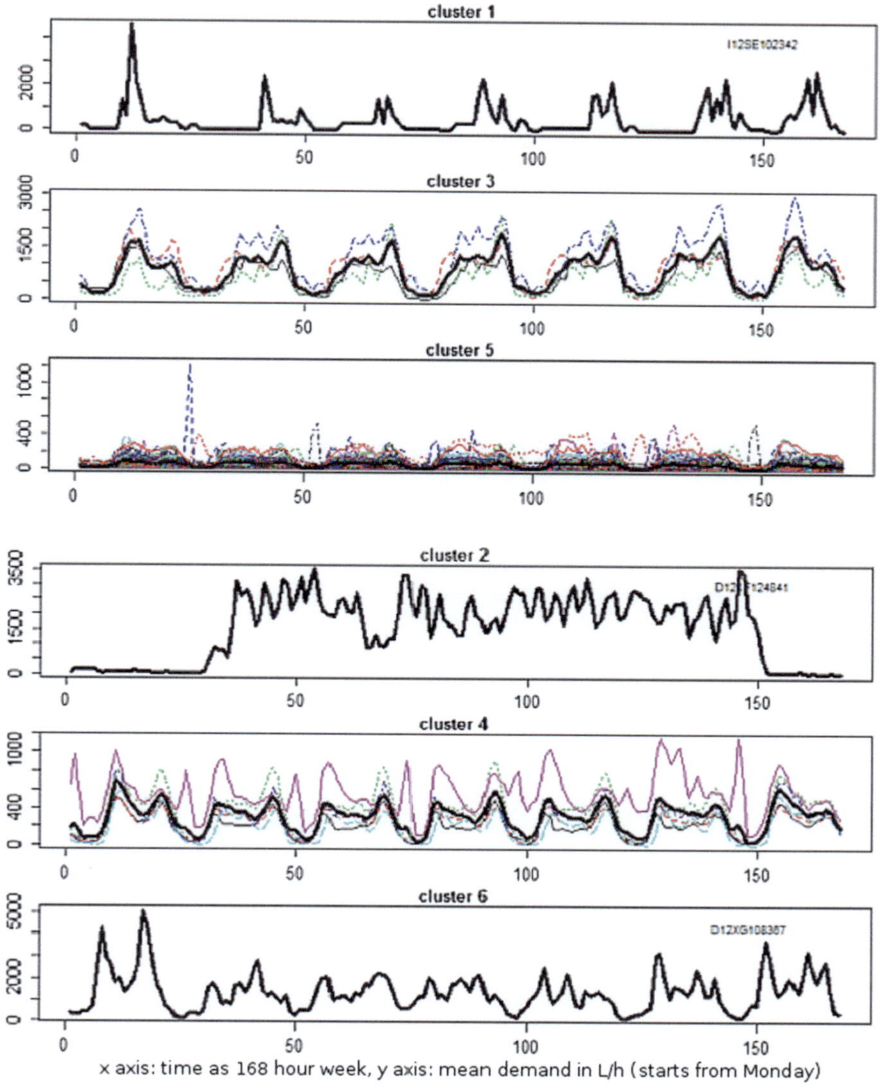

Fig. 20.8 Clustered demand patterns of Tarragona non-residential customers

during the day, in contrast to the rest of the patterns which appear noisy. These could range from local businesses to restaurants, bars and domestic businesses.

Further analysis executed using the clustering for this dataset proved that it would not be useful for seasonal, classification or regression analysis, since neither enough data nor sensors were available to find a generalized outcome; some of these data did not figure in the billing information (representing extra loss of information) so they had to be discarded, and due to the fact that 8% belong to the same class, it would prove difficult to find an underlying relationship with other variables associated to the account, or the meter.

20.4.2 Torremolinos

In this dataset, it is unknown whether the use of the sensors is domestic or industrial.

After applying the filters, detailed in the previous cases on the raw dataset, 1209 sensors are found to be useful for the analysis, while 503 show big gap rates when passed through the filtering process and thus are discarded from the analysis. This leaves 706 sensors for the unsupervised clustering stage. Figure 20.9 shows the WSS plot. Although 10 clusters could have been enough, because adding one more does not decrease the WSS significantly, it is decided to use 12 to force the extraction of irregular patterns associated with a different use of the sensor regarding the other clusters.

Figure 20.10 shows the 12 clusters after the cluster analysis. And Table 20.3 summarizes the contents of the clustered patterns. Approximately 80% of the sensors show a mean weekly consumption ranging between 1.02 and 6.2 l/h, classifying them as low consumers. These sensors belong either to clusters 7, 8 or 10, whose patterns show consistent and reduced daily use. Since Torremolinos is a coastal city, where

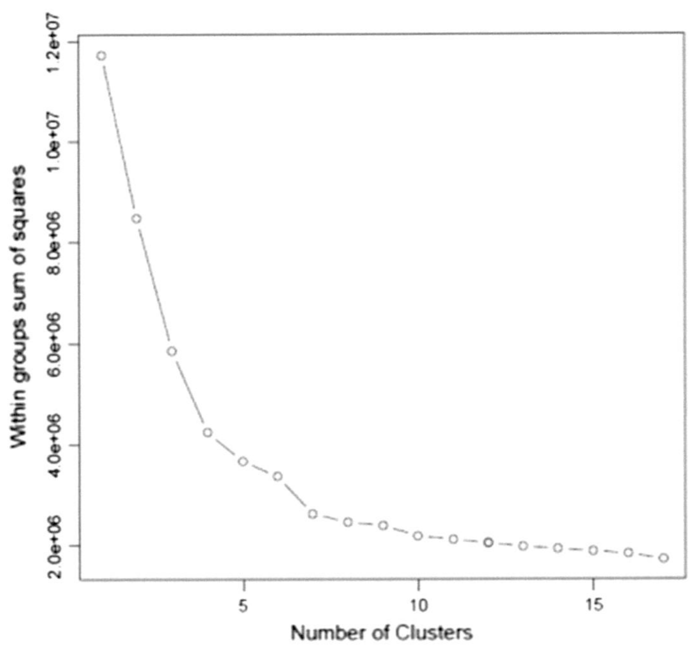

Fig. 20.9 WSS plot of Torremolinos residential and non-residential customers

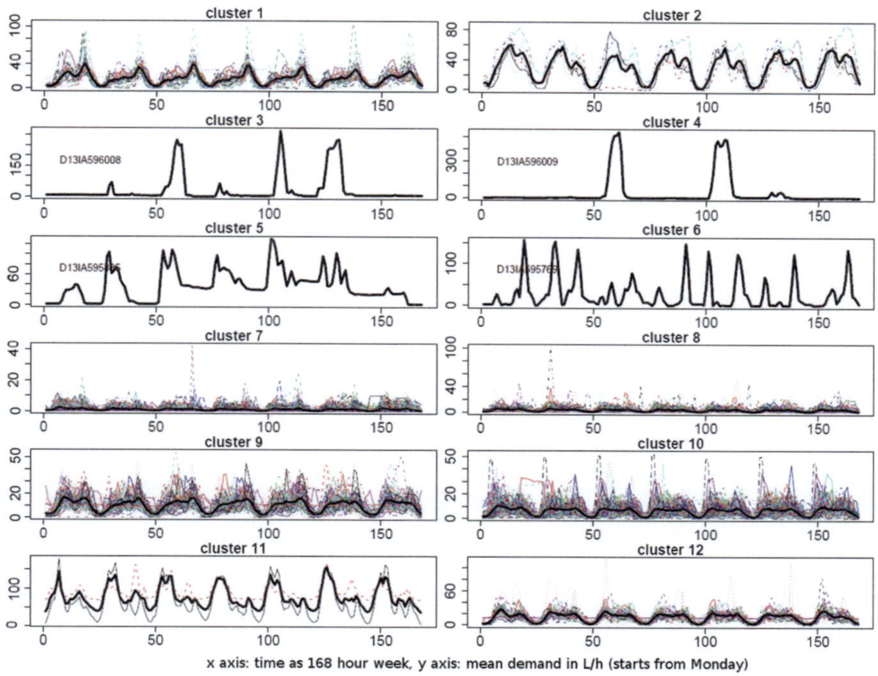

Fig. 20.10 Clustered demand patterns of Torremolinos residential and non-residential customers

some of the sensors might be installed in summer homes, more information about the user behavioural consumption pattern can be obtained from the seasonal analysis. The cluster mean morphology, which does not show the pronounced two-peak

Table 20.3 Summary statistics for unsupervised clustering of consumption patterns in Torremolinos

Cluster	Mean	# Sensors
1	16.348	29
2	30.212	5
3	32.805	1
4	41.400	1
5	37.311	1
6	27.404	1
7	1.028	252
8	3.258	174
9	10.040	65
10	6.203	137
11	77.458	2
12	14.146	38

consumption pattern associated with domestic users, is a slight indicator of atypical domestic behaviour which could be the characteristic to cities such as Torremolinos. On the other hand, clusters 3, 4, 5 and 6 contain only one sensor each, whose mean consumption is quite high compared to the rest of the clusters. Given the irregularities in the consumption pattern, these can be early classified as non-domestic users and further validated in the regression analysis and classification analysis.

20.4.3 Alicante

Finally, some results based on the Alicante city DMA (Spain) are presented. Alicante is a coastal city with a population of around 300 thousand people. The dataset used in this work is the hourly sampling observational readings from 51,117 smart meters within a one-year period (from July 2013 to July 2014). This dataset has 317, 705 and 562 observed readings that corresponds to a size of 14 GB.

Due to the limitation of space and the difficulty to visualize thousands of results, a reduced set of 100 smart meters is considered for illustrative purposes (Fig. 20.11), out of the complete initial set of 51117 smart meters.

As it has been pointed out in Sect. 20.4.1, the k-means algorithm requires the number of clusters input parameter. Hence, the WSS is obtained (see Fig. 20.12). Notice that after considering nine clusters, the sum begins to be stable.

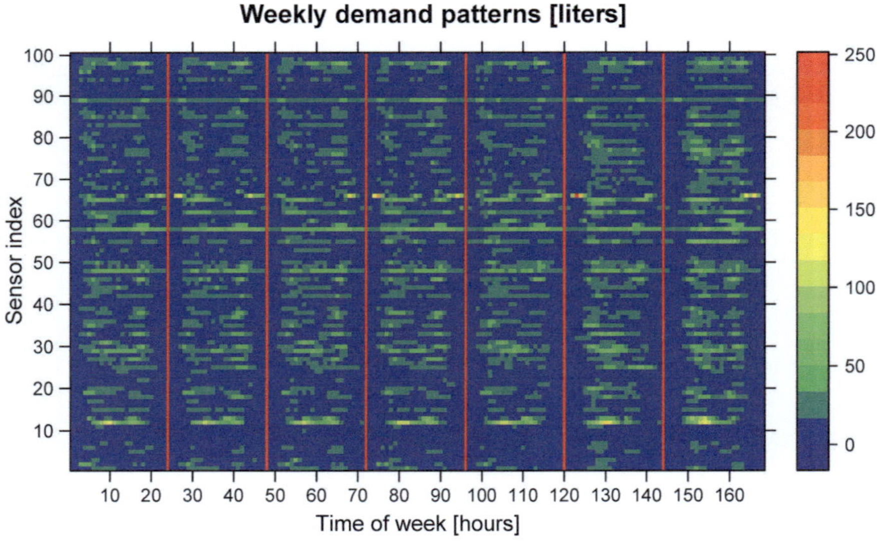

Fig. 20.11 Heatmap where each row is the weekly water demand pattern of AMR meters. The *horizontal axis* is the time index of the week. The *red vertical lines* divide each day (starts from Monday)

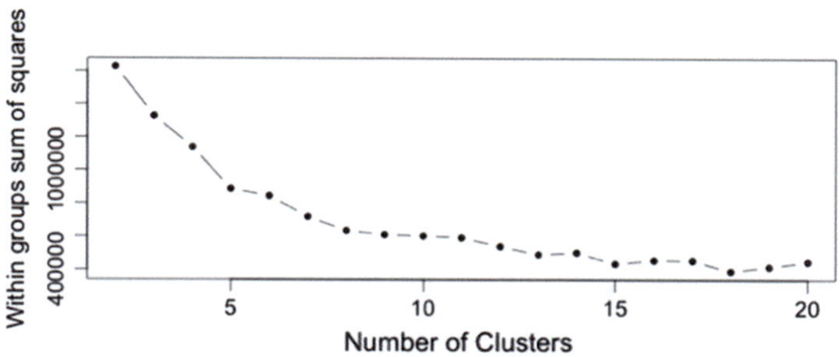

Fig. 20.12 Evolution of the WSS in the *vertical axis* against the number of clusters in the *horizontal axis*

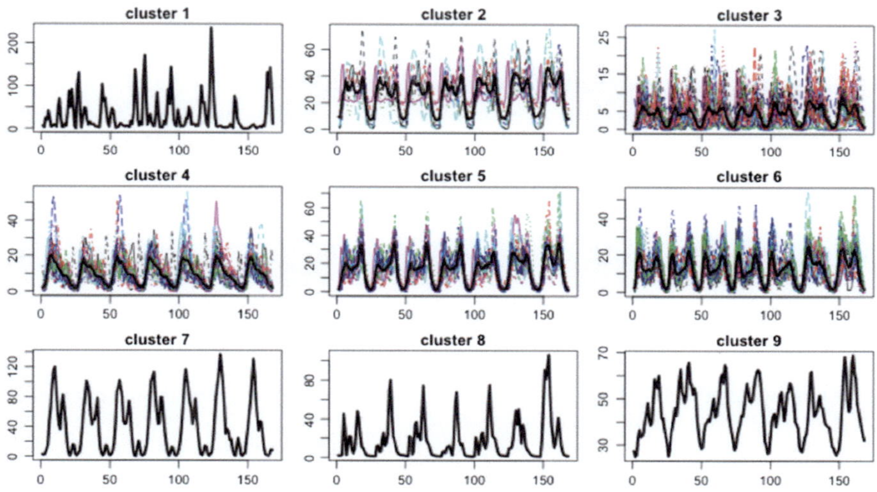

Fig. 20.13 Cluster analysis using k-means algorithm of 9 selected clusters in Alicante DMA

The clusters obtained are shown in Fig. 20.13. On the one hand, clusters 2, 3, 4, 5 and 6 have several members, allowing the WU to perform certain tasks of interest over the network, e.g., forecast the demand of a client or the DMA, improve the leak detection model or detect a pattern change different to the pattern expected to the activity declared in the contract.

On the other hand, clusters 1, 7, 8 and 9 are single-member clusters (i.e., outliers) which must be analysed in more detail. Cluster 1 is an irregular consumption pattern with several peaks between 100 to 200 l/h that could be generated by, e.g., some irrigation or washing system. Therefore, the WU could change the type of fee assigned to this consumer or could verify whether the real activity of the client is not the one declared by the client in the contract (fraud). The pattern of cluster 7 shows a regular daily shape with higher average consumption in relation to the rest of clusters; thus, the WU can change the type of fee assigned to this consumer or even give advice to this client in order to achieve a responsible water consumption. Cluster 9 shows the water demand pattern with the lowest number of consecutive low consumption hours (valleys), probably due to a leak occurring in the DMA.

20.5 Conclusions

This chapter has presented the application of Big Data analytics and knowledge discovery tools to AMR data combined with other streams of information (e.g., data coming from the billing system, call centre service and meteorological information) in order to help with fraud detection, maintenance requirements prediction, water/energy user consumption patterns determination and response generation to variations in the demand. This chapter also presents novel algorithms and methodologies to carry out real-time streaming data processing, data analytics, data quality assessment and improvement, as well as prediction and visualization tasks, at extremely large scale and with diverse structured and unstructured data from multiple sources such as water, power, telecommunication and other utilities, as well as from social media. The algorithms and methodologies presented have been illustrated using real data coming from several WU.

References

1. Aksela K, Aksela M (2011) Demand estimation with automated meter reading in a distribution network. J Water Resour Plann Manage:456–467
2. Armon A, Gutner S, Rosenberg A, Scolnicov H (2011) Algorithmic network monitoring for a modern water utility: a case study in Jerusalem. Water Sci Technol: A J Int Assoc Water Pollut Res 63(2):233–239
3. Cugueró-Escofet MÀ, García D, Quevedo J, Puig V, Espin S, Roquet J (2015) A methodology and a software tool for sensor data validation/reconstruction: application to the Catalonia regional water network. Control Eng Pract:1–14
4. Dean J (2014) Big data, data mining, and machine learning. Wiley
5. Dean J, Ghemawat S (2008) Mapreduce: simplified data processing on large clusters. Commun ACM:1–13
6. Garcia D, Quevedo J, Puig V, Saludes J, Espin S, Roquet J, Valero F (2014) Automatic validation of flowmeter data in transport water networks: application to the ATLLc water network. In: Intelligent data engineering and automated learning. IDEAL 2014 SE - 15, vol 8669. Lecture notes in computer science. Springer International Publishing, pp 118–125
7. Garcia Valverde D, González Vidal D, Quevedo Casin J, Puig Cayuela V, Saludes Closa J (2015) Water demand estimation and outlier detection from smart meter data using classification and Big Data methods. In: New developments in IT & water conference, pp 1–8
8. Han H, Wen Y, Chua T-S, Li X (2014) Toward scalable systems for big data analytics: a technology tutorial. IEEE Access 2:652–687
9. Hartigan JA, Wong MA (1979) Algorithm AS 136: a k-means clustering algorithm. Appl Stat 28(1):100–108
10. Jordan M (2013) Frontiers in massive data analysis. National Academies Press
11. Lin J, Williamson S, Borne K, Debarr D (2011) Chapter 1 Pattern recognition in time series, pp 1–28

12. Zaharia M, Chowdhury M, Franklin MJ, Shenker S, Stoica I (2010) Spark: cluster computing with working sets. In: HotCloud
13. McKenna SA, Fusco F, Eck BJ (2014) Water demand pattern classification from smart meter data. Procedia Eng 70:1121–1130
14. Rojas R (2002) The first computers: history and architectures. MIT Press
15. Scolnicov H, Horowitz G (2010) Water network monitoring: a new approach to managing and sustaining water distribution infrastructure, pp 1–6
16. Soundararajan E, Joseph JVM (2005) Knowledge discovery tools and techniques. Recent Adv Inf Technol 14:141
17. Wang C, Ming-Hui C, Elizabeth S, Wu J, Jun Y (2013) A survey of statistical methods and computing for big data, pp 1–29

Index

V. Puig et al. (eds.), *Real-Time Monitoring and Operational Control of Drinking-Water Systems*, Advances in Industrial Control,
DOI 10.1007/978-3-319-50751-4

MIX
Papier aus verantwortungsvollen Quellen
Paper from responsible sources
FSC® C105338

If you have any concerns about our products,
you can contact us on
ProductSafety@springernature.com

In case Publisher is established outside the EU,
the EU authorized representative is:
Springer Nature Customer Service Center GmbH
Europaplatz 3, 69115 Heidelberg, Germany

Printed by Libri Plureos GmbH
in Hamburg, Germany